GALOIS
THEORY AND APPLICATIONS
Solved Exercises and Problems

GALOIS

THEORY AND APPLICATIONS

Solved Exercises and Problems

Mohamed Ayad

Université du Littoral, Calais, France

World Scientific

NEW JERSEY • LONDON • SINGAPORE • BEIJING • SHANGHAI • HONG KONG • TAIPEI • CHENNAI • TOKYO

Published by

World Scientific Publishing Co. Pte. Ltd.
5 Toh Tuck Link, Singapore 596224
USA office: 27 Warren Street, Suite 401-402, Hackensack, NJ 07601
UK office: 57 Shelton Street, Covent Garden, London WC2H 9HE

Library of Congress Cataloging-in-Publication Data
Names: Ayad, Mohamed (Mathematician), author.
Title: Galois theory and applications : solved exercises and problems / by
Mohamed Ayad (Université du Littoral, Calais, France).
Description: New Jersey : World Scientific, 2018. | Includes bibliographical references.
Identifiers: LCCN 2018020885 | ISBN 9789813238305 (hardcover : alk. paper)
Subjects: LCSH: Galois theory--Problems, exercises, etc. |
Algebra, Abstract--Problems, exercises, etc.
Classification: LCC QA162 .A93 2018 | DDC 512/.32--dc23
LC record available at https://lccn.loc.gov/2018020885

British Library Cataloguing-in-Publication Data
A catalogue record for this book is available from the British Library.

Cover image: Factorizations of 90 in $\mathbb{Z}[\sqrt{-14}]$.

For any available supplementary material, please visit
http://www.worldscientific.com/worldscibooks/10.1142/10939#t=suppl

Printed in Singapore

Preface

This book arose from lectures and tutorials given by the author at Algiers University (U.S.T.H.B.) and at Université du Littoral (Calais).

To the best of my knowledge, there is no book of exercises and problems in Galois Theory except the author's books, published in French. The material contained in the present book is completely different from that which makes up the French one.

This book is intended for several types of audiences:

First, students who want to learn the material around the arithmetic of polynomials and field extensions. It can be used by beginners (3rd year students), but also more advanced students (e.g. Master or Ph. D. students working in the area).

Second, teachers who want to work on problems with their students.

Third, researchers can use it as a reference: the book contains some useful results generally known by specialists, but for which a detailed proof is not always in the literature.

Besides Galois Theory, the book contains chapters devoted to fields which make use of this theory: Finite fields, Permutation polynomials and an introduction to Algebraic Number theory. This last field occupies the longest chapter of the book.

In the solutions of the exercises or problems, when it is useful, I refer to Lang's book (Algebra) if it concerns results in Galois theory, and to Marcus's book (Number Rings) for results in Number theory. Occasionally references are made to other books cited in the bibliography.

I am indebted to several colleagues who each read some chapters and suggested corrections on a preliminary draft of the book. In alphabetical order: B. Bensebaa, D. Bitouzé, R. Bouchenna, A. Bouhamidi, H. Chapdelaine, P. Dèbes, P. Feischman, O. Kihel, F. Recher and C. Woodcock.

I express my thanks to the following students of Brock University for reading parts of the book or for their help in the use of Latex: C. Asselin, Z. Schedler, J. Larone, B. Earp-Lynch and S. Earp-Lynch.

The responsibility for any errors is solely mine and I thank in advance vigilant readers who will point out errors or will indicate alternative solutions.

I learned a lot during the preparation of this work, which was going on for seventeen years. I hope that the readers will profit from its use.

M. Ayad
ayadmohamed502@yahoo.com

Contents

Chapter 1

Polynomials, Fields, Generalities

Exercise 1.1.
Let p be a prime number and s be a positive integer. Show that for any $i \in \{0, 1, \ldots, p^s - 1\}$, $\binom{p^s-1}{i} \equiv (-1)^i \pmod{p}$.

Solution 1.1.
Let x be an indeterminate over $\mathbb{Q}$, then

$$\begin{aligned}\sum_{i=0}^{p^s-1} \binom{p^s-1}{i} x^i &= (1+x)^{p^s-1} = (1+x)^{p^s}/(1+x) = (1+x)^{p^s} \sum_{i\geq 0} (-1)^i x^i \\ &\equiv (1+x^{p^s}) \sum_{i\geq 0} (-1)^i x^i \pmod{p} \\ &\equiv \sum_{i\geq 0} (-1)^i x^i + \sum_{i\geq 0} (-1)^i x^{p^s+i} \pmod{p} \\ &\equiv \sum_{i\geq 0} (-1)^i x^i + \sum_{j\geq p^s} (-1)^{j-p^s} x^j \pmod{p} \\ &\equiv \sum_{i=0}^{p^s-1} (-1)^i x^i \pmod{p},\end{aligned}$$

hence the result.

Exercise 1.2.
Let n be a positive integer.

(1) Show that $\sum_{k=0}^{n} (-1)^k \binom{n}{k} = 0$.
(2) Show that for any $k \in \{1, \ldots, n\}$, $k\binom{n}{k} = n\binom{n-1}{k-1}$.
(3) Show that for any commutative ring R and any polynomial $f(x) \in R[x]$ of degree $d < n$, we have $\sum_{k=0}^{n} (-1)^k \binom{n}{k} f(k) = 0$.

(4) Show that the condition on d cannot be omitted.

Solution 1.2.

(1) The assertion follows immediately from the identity

$$(x-1)^n = \sum_{k=0}^{n} (-1)^k \binom{n}{k} x^{n-k}$$

after substitution of 1 for x.

(2) We have

$$k\binom{n}{k} = \frac{kn(n-1)!}{k(k-1)!(n-1-(k-1))!} = n\binom{n-1}{k-1}.$$

(3) We fix the ring R, the integer n and we prove the statement by induction on d. For $d = 0$, the result follows from **(1)**. Suppose that the statement is true for all polynomials with coefficients in R of degree less than d. To prove it for polynomials of degree d it is sufficient to prove it for $f(x) = x^d$.
Using **(2)** and the inductive hypothesis, we obtain

$$\begin{aligned}
\sum_{k=0}^{n} (-1)^k \binom{n}{k} k^d &= \sum_{k=1}^{n} (-1)^k \binom{n}{k} k^d \\
&= \sum_{k=1}^{n} (-1)^k n \binom{n-1}{k-1} k^{d-1} \\
&= \sum_{k=0}^{n-1} (-1)^{k+1} n \binom{n-1}{k} (k+1)^{d-1} \\
&= -n \sum_{k=0}^{n-1} (-1)^k \binom{n-1}{k} (k+1)^{d-1} \\
&= 0.
\end{aligned}$$

(4) For $n = 1$, the identity in **(3)** reads $f(0) - f(1) = 0$. Therefore any polynomial $f(x) \in R[x]$ such that $f(0) \neq f(1)$ does not satisfy this identity. For instance $f(x) = x^d$, with $d \geq 1$ is such a polynomial.

Exercise 1.3.
Let $0 \leq k \leq n$ be integers, $n \neq 0$ and let p be a prime number. Let $n = \sum_{i=0}^{r} n_i p^i$ and $k = \sum_{i=0}^{r} k_i p^i$ with $0 \leq k_i, n_i < p$, $n_r \neq 0$ and $k_r \leq n_r$.

(1) Show that $\binom{n}{k} \equiv \prod_{i=0}^{r} \binom{n_i}{k_i} \pmod{p}$, where (by definition) $\binom{n_i}{k_i} = 0$ if $k_i > n_i$. Compute the residue of $\binom{23}{6}$ modulo 35.

(2) Show that $\binom{n}{k} \equiv 0 \pmod p$ if and only if there exists $i \in \{1, \ldots, r\}$ such that $k_i > n_i$.

(3) Show that the number of $\binom{n}{k}$, $0 \leq k \leq n$ such that $\binom{n}{k} \not\equiv 0 \pmod p$ is equal to $\prod_{i=0}^{r}(n_i+1)$. Deduce that, reducing modulo 2, the coefficients in Pascal's triangle, then on each row the number of one's is a power of 2.

(4) Let p be a fixed prime number. Represent any non negative integer n in the form $n = n_k p^k + n_{k-1}p^{k-1} + \cdots + n_0$, where $n_i \in \{0, \ldots, p-1\}$ for $i = 0, \ldots, k$ and $n_k \neq 0$ if $n \neq 0$. Denote this representation by $n = [n_k n_{k-1} \cdots n_0]$. Define in $\mathbb{N}$ the following law of composition:

$$[0] \star [n_k n_{k-1} \cdots n_0] = [n_k n_{k-1} \cdots n_0],$$
$$[n_k n_{k-1} \cdots n_0] \star [0] = [n_k n_{k-1} \cdots n_0 0] \quad \text{and}$$
$$[n_k n_{k-1} \cdots n_0] \star [m_h m_{h-1} \cdots m_0] = [n_k n_{k-1} \cdots n_0 m_h \cdots m_0].$$

We admit that $(\mathbb{N}, \star)$ is a semigroup. Let G be the multiplicative semigroup, $G = \mathbb{Z}[x]/(x^{p-1} - 1)$. Fix a generator ξ of $(\mathbb{Z}/p\mathbb{Z})^\star$ and let $\sigma : \mathbb{N} \to G$ be the map such that $\sigma(0) = 1$ and

$$\sigma(n) = \sum_{i=0}^{p-2} r_i(n) x^i \tag{Eq 1}$$

if $n \neq 0$, where $r_i(n)$ denotes the number of integers l, $0 \leq l \leq n$ such that $\binom{n}{l} \equiv \xi^i \pmod p$. Show that σ is a morphism of semigroups.

(5) For any positive integer m let

$$R_m(x) = \sum_{i=0}^{p-2} r_i(m) x^i. \tag{Eq 2}$$

Show that

$$R_n(x) \equiv \prod_{j=1}^{p-1} R_j(x)^{t_j(n)} \pmod{x^{p-1} - 1}, \tag{Eq 3}$$

where $t_j(n)$ denotes the number of times the digit j appears in the p-ary expansion of n.

(6) Let m and n be positive integers. Suppose that any non zero digit j has the same number of apparitions in the p-ary expansions of m and of n. Show that reducing modulo p the m-th and the n-th line of Pascal's triangle, any integer $l \in \{1, \ldots, p-1\}$ has the same number of occurrences in both lines.

(7) Fix c and d in $\{1,\ldots,p-1\}$. For any $(i,j)\in\{1,\ldots,p-1\}^2$, $i<j$ let $n_{ij}=cp^i+dp^j$. Show that reducing modulo p the elements of Pascal's triangle, the number of apparitions of $l\in\{1,\ldots,p-1\}$ in the line n_{ij} does not depend on (i,j).

(8) Let $n=[n_k\cdots n_0]$ with $n_k\neq 0$. Show that $R_n(1)=\prod_{i=0}^k(n_i+1)$.

Solution 1.3.

(1) We have $(1+x)^n\equiv\prod(1+x^{p^i})^{n_i} \pmod p$. Identifying, in both sides of this identity, the coefficient of x^k, we obtain

$$\binom{n}{k}\equiv\sum_{\sum l_ip^i=k}\binom{n_0}{l_0}\binom{n_1}{l_1}\cdots\binom{n_r}{l_r} \pmod p.$$

Since k has a unique representation in base p this sum (of products) reduces to one product and the proof of the first part of **(1)** is complete. We have $23=4.5+3$ and $6=1.5+1$, hence

$$\binom{23}{6}\equiv\binom{4}{1}\binom{3}{1} \pmod 5\equiv 2 \pmod 5.$$

Similarly, we have $23=3.7+2$ and $6=0.7+6$, hence

$$\binom{23}{6}\equiv\binom{3}{0}\binom{2}{6} \pmod 7\equiv 0 \pmod 7.$$

Now Chinese remainder theorem shows that $\binom{23}{6}\equiv 7 \pmod{35}$.

(2) By **(1)**, $\binom{n}{k}\equiv 0 \pmod p$ if and only if there exists $i\in\{0,\ldots,r\}$ such that $\binom{n_i}{k_i}\equiv 0 \pmod p$. But it is clear that if $k_i\leq n_i$, then $\binom{n_i}{k_i}\not\equiv 0 \pmod p$, hence the result.

(3) We have $\binom{n}{k}\not\equiv 0 \pmod p$ if and only if for any $i\in\{0,\ldots,r\}$, $k_i\leq n_i$, hence $\binom{n}{k}\not\equiv 0 \pmod p$ if and only if $k_0=0,\ldots,n_0$, $k_1=0,\ldots,n_1$, $\ldots$ and $k_r=0,\ldots,n_r$. It follows that the number of $\binom{n}{k}$, $0\leq k\leq n$ such that $\binom{n}{k}\not\equiv 0 \pmod p$ is equal to $\prod_{i=0}^r(n_i+1)$.
We deduce that if $p=2$, then the number of coefficients congruent to 1 modulo 2 in the n-th row of the Pascal's triangle is equal to $\prod(n_i+1)$. Since $n_i=0$ or $n_i=1$, then $\prod(n_i+1)=2^e$ with $e\geq 1$.

(4) For fixed n, we extend the definition of $r_i(n)$ for any $i\geq 0$ as follows. We write i in the form $i=(p-1)q_i+u_i$ with $u_i<p-1$. We set

$$\begin{aligned} r_i(n)&=\left|\left\{l,\ 0\leq l\leq n \text{ such that } \binom{n}{l}\equiv\xi^i \pmod{p-1}\right\}\right| \\ &=\left|\left\{l,\ 0\leq l\leq n \text{ such that } \binom{n}{l}\equiv\xi^{u_i} \pmod{p-1}\right\}\right| \\ &=r_{u_i}(n). \end{aligned}$$

Let

$$m_1 = [a_k \cdots a_0],\ m_2 = [b_h \cdots b_0] \text{ and } n = m_1 \star m_2.$$

We may write any integer $g \leq n$ in one and only one way in the form $g = g_1 \star g_2$ with $g_1 = [a_k' \cdots a_0']$ and $g_2 = [b_h' \cdots b_0']$. Here some first digits a_j' of g_1 may be equal to 0. The same remark applies for g_2.
We have

$$\begin{aligned}
\sigma(m_1 \star m_2) &= \sum_{i=0}^{p-2} r_i(m_1 \star m_2) x^i \\
&= \sum_{i=0}^{p-2} \left| \left\{ g;\ \binom{m_1 \star m_2}{g} \equiv \xi^i \pmod p \right\} \right| x^i \\
&= \sum_{i=0}^{p-2} \left| \left\{ g;\ \binom{m_1}{g_1}\binom{m_2}{g_2} \equiv \xi^i \pmod p \right\} \right| x^i \\
&= \sum_{i=0}^{p-2} \left(\sum_{j=0}^{p-2} \left| \left\{ g_1;\ \binom{m_1}{g_1} \equiv \xi^j \pmod p \right\} \right| . \left| \left\{ g_2;\ \binom{m_2}{g_2} \right.\right.\right. \\
&\qquad \left.\left.\left. \equiv \xi^{i-j} \pmod p \right\} \right| \right) x^i \\
&= \sum_{i=0}^{p-2} \left(\sum_{j=0}^{p-2} r_j(m_1) r_{i-j}(m_2) \right) x^i \\
&\equiv \sum_{i=0}^{p-2} \sum_{j=0}^{p-2} (r_j(m_1)x^j).(r_{p-1+i-j}(m_2)x^{p-1+i-j}) \pmod{x^{p-1}-1} \\
&\equiv \sigma(m_1)\sigma(m_2) \pmod{x^{p-1}-1}.
\end{aligned}$$

(5) Set $n = [n_k \cdots n_0]$. Since $R_0(x) = 1$, then by **(4)** we have

$$\begin{aligned}
R_n(x) &= \sigma(n) \\
&\equiv \sigma([n_k])\sigma([n_{k-1}]) \cdots \sigma([n_0]) \pmod{x^{p-1}-1} \\
&\equiv R_{n_k}(x) R_{n_{k-1}}(x) \cdots R_{n_0}(x) \pmod{x^{p-1}-1} \\
&\equiv \prod_{j=1}^{p-1} R_j(x)^{t_j(n)} \pmod{x^{p-1}-1}.
\end{aligned}$$

(6) Our assumption on the apparition of the digit j means that $t_m(j) = t_n(j)$. Since this equality is true for any $j \in \{1, \ldots, p-1\}$, then

$$\begin{aligned} R_n(x) &\equiv \prod_{j=1}^{p-1} R_j(x)^{t_j(n)} \pmod{x^{p-1}-1} \\ &\equiv \prod_{j=1}^{p-1} R_j(x)^{t_j(m)} \pmod{x^{p-1}-1} \\ &\equiv R_m(x) \pmod{x^{p-1}-1}. \end{aligned}$$

Since the polynomials $R_n(x)$ and $R_m(x)$ have their degrees less than $p-1$, then $R_n(x) = R_m(x)$. It follows that for any $i = 0, \ldots, p-2$, $r_i(n) = r_i(m)$, that is

$$\begin{aligned} &\left|\left\{l \in \{0, \ldots, n\}; \binom{n}{l} \equiv \xi^i \pmod{p}\right\}\right| \\ &= \left|\left\{l \in \{0, \ldots, m\}; \binom{m}{l} \equiv \xi^i \pmod{p}\right\}\right|. \end{aligned}$$

Since for any digit $j \neq 0$, there exists $i \in \{0, \ldots, p-2\}$ such that $j \equiv a^i \pmod{p}$, then

$$\begin{aligned} &\left|\left\{l \in \{0, \ldots, n\}; \binom{n}{l} \equiv j \pmod{p}\right\}\right| \\ &= \left|\left\{l \in \{0, \ldots, m\}; \binom{m}{l} \equiv j \pmod{p}\right\}\right|. \end{aligned}$$

(7) Let $n_{i_1j_1} = cp^{i_1} + dp^{j_1}$ and $n_{i_2j_2} = cp^{i_2} + dp^{j_2}$, then

$$\begin{aligned} t_c(n_{i_1j_1}) &= t_c(n_{i_2j_2}) = 1 \quad \text{and} \\ t_d(n_{i_1j_1}) &= t_d(n_{i_2j_2}) = 1. \end{aligned}$$

By **(6)**, the number of apparitions of $l \in \{1, \ldots, p-1\}$ in the line $n_{i_1j_1}$ is the same as in the line $n_{i_2j_2}$.

(8) The definition of $R_n(x)$ (see **(Eq 2)**), implies that

$$R_n(1) = \sum_{i=0}^{p-2} r_i(n) = \left|\left\{l \in \{0, \ldots, p-1\} \text{ such that } \binom{n}{l} \not\equiv 0 \pmod{p}\right\}\right|.$$

By **(3)**, we get $R_n(1) = \prod_{i=0}^{k}(n_i + 1)$.

Exercise 1.4.

(1) For any non negative integers n and k define the binomial coefficient $\binom{-n}{k}$ by $\binom{-n}{0} = 1$ and $\binom{-n}{k} = \frac{(-n)(-n-1)\cdots(-n-(k-1))}{k!}$ for $n > 0$. Show that

$$\begin{aligned}&\frac{n(n+1)\cdots(n+k-1)}{k!}\\ &\quad+\frac{n(n+1)\cdots(n+k-2)}{(k-1)!}+\cdots+\frac{n}{1!}+1\\ &=\frac{(n+1)\cdots(n+k)}{k!}.\end{aligned}$$

Deduce that $(1+x)^{-n} = \sum_{k\geq 0}\binom{-n}{k}x^k$.

(2) Let p be a prime number, r, s, a, b be rational integers such that $r \geq 1$, $s \geq 0$ and $a \equiv b \pmod{p^{r+s}}$. Show that $\binom{a}{m} \equiv \binom{b}{m} \pmod{p^{s+1}}$ for any integer m such that $1 \leq m < p^r$.

Solution 1.4.

(1) We proceed by induction on k. If $k = 1$, then $\frac{n}{1!} + 1 = n + 1 = \frac{n+1}{1!}$, hence the proof in this case. Suppose that the result holds for $k - 1$, that is

$$\frac{n(n+1)\cdots(n+k-2)}{(k-1)!}+\cdots+\frac{n}{1!}+1=\frac{(n+1)\cdots(n+k-1)}{(k-1)!},$$

then

$$\begin{aligned}&\frac{n(n+1)\cdots(n+k-1)}{k!}\\ &\quad+\frac{n(n+1)\cdots(n+k-2)}{(k-1)!}+\cdots+\frac{n}{1!}+1\\ &=\frac{n(n+1)\cdots(n+k-1)}{k!}\\ &\quad+\frac{(n+1)\cdots(n+k-1)}{(k-1)!}\\ &=\frac{(n+1)\cdots(n+k-1)}{(k-1)!}(1+n/k)\\ &=\frac{(n+1)\cdots(n+k)}{k!}.\end{aligned}$$

We prove the second part of **(1)**. Notice that if $n = 1$, then $\binom{-1}{k} = (-1)^k$. We have

$$(1+x)^{-1} = 1/(1+X) = \sum_{k\geq 0}(-1)^k x^k = \sum_{k\geq 0}\binom{-1}{k}x^k,$$

so that the formula holds for $n = 1$. Suppose that it holds for the integer n, then

$$(1+x)^{-n-1} = \sum_{k\geq 0} \binom{-n}{k} x^k \sum_{k\geq 0} (-1)^k x^k$$
$$= \sum_{k\geq 0} \left(\sum_{l+t=k} (-1)^t \binom{-n}{l} \right) x^k.$$

Let $\sigma_{n,k} = \sum_{l+t=k} (-1)^t \binom{-n}{l}$, then by the first part of **(1)**,

$$\sigma_{n,k} = (-1)^k \left(\frac{n\cdots(n+k-1)}{k!} + \frac{n\cdots(n+k-2)}{(k-1)!} + \cdots + \frac{n}{1!} + 1 \right)$$
$$= (-1)^k \, \frac{(n+1)\cdots(n+k)}{k!}$$
$$= \binom{-n-1}{k},$$

hence the result.

(2) Let $c \in \mathbb{Z}$ such that $a = b + c$. Since $(1+x)^a = (1+x)^b(1+x)^c$, then from **(1)**, we have

$$\binom{a}{m} = \sum_{l\geq 0, t\geq 0, l+t=m} \binom{b}{t}\binom{c}{l},$$

for any non negative integer m, in particular for $1 \leq m < p^r$. We have $\binom{c}{l} = \frac{c}{l}\binom{c-1}{l-1}$. Since $c = a - b \equiv 0 \pmod{p^{r+s}}$ and $1 \leq m < p^r$, then $\frac{c}{l} \equiv 0 \pmod{p}^{s+1}$, thus $\binom{a}{m} \equiv \binom{b}{m} \pmod{p^{s+1}}$

Exercise 1.5.

Let $d_1, \ldots, d_n$ be rational integers and $d = \gcd(d_1, \ldots, d_n)$. Recall that there always exist integers $k_1, \ldots, k_n \in \mathbb{Z}$ such that $d = \sum_{i=1}^n k_i d_i$. Show that $\gcd(k_1, \ldots, k_n) = 1$.

Solution 1.5.

Let $x = \gcd(k_1, \ldots, k_n)$, then $xd \mid k_i d_i$ for $i = 1, \ldots, n$, hence $xd \mid d$ and therefore $x = 1$.

Exercise 1.6.

Let $f(x)$ and $g(x) \in \mathbb{Z}[x]$ be coprime. Let $u(x), v(x) \in \mathbb{Z}[x]$ and $m \in \mathbb{N}$ such that $u(x)f(x) + v(x)g(x) = m$. Let d be the greatest positive divisor of m such that there exists an integer x_0 such that $\gcd(f(x_0), g(x_0)) = d$. Let a be an integer such that $0 \leq a < d$ and $x_0 \equiv a \pmod{d}$. Set

$$f_1(x) = \frac{1}{d} f(dx+a), \quad g_1(x) = \frac{1}{d} g(dx+a),$$
$$u_1(x) = u(dx+a) \quad \text{and} \quad v_1(x) = v(dx+a).$$

Show that $u_1(x)f_1(x)+v_1(x)g_1(x)=m/d$, $f_1(x),g_1(x)\in\mathbb{Z}[x]$ and that for any integer x, $\gcd(f_1(x),g_1(x))=1$.

Solution 1.6.
Replacing x by $dx+a$ in the Bezout's Identity relating $f(x)$ and $g(x)$ and dividing by d, we obtain $u_1(x)f_1(x)+v_1(x)g_1(x)=m/d$. We show that $f_1(x)\in\mathbb{Z}[x]$ (the same argument applies for the polynomial $g_1(x)$). We have

$$\begin{aligned} f_1(x) &= \frac{f(a)+dx\frac{f'(a)}{1!}+\cdots+(dx)^n\frac{f^{(n)}(a)}{n!}}{d} \\ &= \frac{f(a)}{d}+x\frac{f'(a)}{1!}+\cdots+d^{n-1}x^n\frac{f^{(n)}(a)}{n!} \end{aligned}$$

and since $d\mid f(a)$, then $f_1(x)\in\mathbb{Z}[x]$. Suppose by contradiction that there exist an integer x_1 and an integer $\delta>1$ such that $\gcd(f_1(x_1),g_1(x_1))=\delta$. Then $\delta\mid m/d$, hence $d\delta\mid m$, which contradicts the definition of d.

Exercise 1.7.
Let K be a field, $f(x,y)$ and $g(x,y)$ be relatively prime polynomials with coefficients in K, $m=\deg_x f$ and $n=\deg_x g$. Let $x_1,\ldots,x_k$ be new variables, $A(x_1,\ldots,x_k)$ and $B(x_1,\ldots,x_k)$ be relatively prime polynomials with coefficients in K such that $A/B\notin K$. Show that $B^m f(A/B,y)$ and $B^n g(A/B,y)$ are coprime in $K[x_1,\ldots,x_k,y]$.

Solution 1.7.
By Bezout's identity applied in $k(x)[y]$, there exist $u_0,v_0\in k(x)[y]$ such that $u_0f+v_0g=1$. Multiplying this identity by the common denominator of u_0 and v_0, we obtain the following new identity $uf+vg=w$, where $u,v\in K[x,y]$ and $w\in K[x]$. Substituting A/B for x and multiplying by a power of B, say B^e, we obtain:

$$\hat{u}(x,y)B^m f(A/B,y)+\hat{v}(x,y)B^n g(A/B,y)=B^e w(A/B),$$

where $\hat{u}$ and $\hat{v}\in K[x,y]$. Any common divisor $D(x_1,\ldots,x_k,y)$ of $B^m f(A/B,y)$ and $B^n g(A/B,y)$ must divide $B^e w(A/B)$, hence it belongs to $K[x_1,\ldots,x_k]$, that is free from y. It follows that expressing $B^m f(A/B,y)$ and $B^n g(A/b,y)$ as polynomials in y, then all their coefficients are divisible by D. Set

$$f(x,y)=\sum_{j=0}^{t} f_j(x)y^j \text{ and } g(x,y)=\sum_{i=0}^{l} g_i(x)y^i,$$

then

$$B^m f(A/B,y) = \sum_{j=0}^{t} B^m f_j(A/B)y^j \text{ and } B^n g(A/B,y) = \sum_{i=0}^{l} B^n g_i(A/B)y^i.$$

Since $f(x,y)$ and $g(x,y)$ are relatively prime, then so are the polynomials $f_t(x),\ldots,f_0(x),g_l(x),\ldots,g_0(x)$. Writing a Bezout's identity relating these polynomials, substituting A/B for x and multiplying by some power of B, we obtain a new identity, which shows that D divides a power of B. Since $m = \deg_x f$, there exists $j \in \{0,\ldots,t\}$ such that $\deg f_j = m$. Set $f_j(x) = a_m x^m + \cdots + a_1 x + a_0$, then

$$B^m f_j(A/B) = a_m A^m + \cdots + a_1 AB^{m-1} + a_0 B^m.$$

Let p be a prime factor of D. Then p divides B and the left hand side, hence also A. This contradiction shows that $D = 1$.

Exercise 1.8.
Let K be a field of characteristic 0, $f(X)$ be a polynomial of degree n and $m > n$ be an integer. Show that

$$f(m) = \sum_{j=0}^{n} (-1)^{n-j} \binom{m}{j} \binom{m-j-1}{n-j} f(j).$$

Solution 1.8.
Using Lagrange's interpolation formula, [Bourbaki (1950), Application: Formule d'interpollation de Lagrange, Chap. 4.2] or [Ayad (1997), Exercice 1.14] we obtain:

$$f(X) = \sum_{j=0}^{n} f(j) \frac{\prod_{i\neq j}(x-i)}{\prod_{i\neq j}(j-i)}.$$

We deduce that

$$\begin{aligned} f(m) &= \sum_{j=0}^{n} f(j) \frac{\prod_{i\neq j}(m-i)}{\prod_{i\neq j}(j-i)} \\ &= \sum_{j=0}^{n} f(j) \frac{m(m-1)\cdots(m-(j-1))(m-(j+1))\cdots(m-n)}{j(j-1)\ldots(j-(j-1))(j-(j+1))\cdots(j-n)} \\ &= \sum_{j=0}^{n} f(j) \frac{m(m-1)\cdots(m-(j-1))}{j!} \quad \frac{(m-(j+1))\cdots(m-n)}{(-1)^{n-j}(n-j)(n-(j-1))\cdots 1} \\ &= \sum_{j=0}^{n} (-1)^{n-j} \binom{m}{j} \binom{m-j-1}{n-j} f(j) \end{aligned}$$

and the proof is complete.

Exercise 1.9.
Let K be a field, $a_0, \ldots, a_n$ be distinct elements of K, $b_0, \ldots, b_n$ be elements of K, distinct or not. Let $f(x) \in K[x]$ such that $\deg f \leq n$ and $f(a_i) = b_i$ for $i = 0, \ldots, n$. Let $g(x)$ and $h(x) \in K[x]$, both of degree at most $n-1$ and satisfying the conditions $g(a_i) = b_i$ and $h(a_j) = b_j$ for $i = 0, \ldots, n-1$ and $j = 1, \ldots, n$. Show that

$$f(x) = \frac{(x-a_0)h(x) - (x-a_n)g(x)}{a_n - a_0}.$$

Solution 1.9.
Let $F(x) = \frac{(x-x_0)h(x)-(x-x_n)g(x)}{x_n - x_0}$. One verifies easily that $F(a_i) = f(a_i)$ for $i = 0, \ldots, n$. We also have $\deg f$ and $\deg F \leq n$, hence $F(x) = f(x)$ by Lagrange's interpolation theorem [Bourbaki (1950), Application: Formule d'interpollation de Lagrange, Chap. 4.2] or [Ayad (1997), Exercice 1.14].

Exercise 1.10.
Let A be a factorial ring, K be its fraction field. Let $g(x), h(x) \in K[x]$ and $f(x) = g(x)h(x)$. Let a and b be coefficients of g and h respectively. Suppose that $f(x) \in A[x]$. Prove that $ab \in A$.

Solution 1.10.
We have $a = a' \operatorname{cont}(g)$ and $b = b' \operatorname{cont}(h)$ where a' and b' are elements of A. Then

$$ab = \operatorname{cont}(g)\operatorname{cont}(h)a'b' = \operatorname{cont}(f)a'b' \in A.$$

Exercise 1.11.
Let K be a field, n, m be positive integers, $a_0, \ldots, a_n, b_0, \ldots, b_m$ be algebraically independent variables over K. Let $f(x) = a_n x^n + \cdots + a_0$ and $g(x) = b_m x^m + \cdots + b_0$ be polynomials with coefficients in $K(a_0, \ldots, b_m)$ and $R(a_0, \ldots, a_n, b_0, \ldots, b_m) = \operatorname{Res}_x(f(x), g(x))$. Show that R is homogeneous of weight nm under either of the following assignments of weights, $w(a_i)$ and $w(b_j)$ for the a_i and the b_j.

(1) $w(a_i) = i$ for $i = 0, \ldots, n$ and $w(b_j) = j$ for $j = 0, \ldots, m$.
(2) $w(a_i) = n - i$ for $i = 0, \ldots, n$ and $w(b_j) = m - j$ for $j = 0, \ldots, m$.

Solution 1.11.

(1) Let t a new variable, we must show that

$$R(a_0, ta_1, \ldots, t^n a_n, b_0, tb_1, \ldots, t^m b_m) = t^{nm} R(a_0, \ldots, a_n, b_0, \ldots, b_m).$$

Let $\alpha_1, \ldots, \alpha_n$ (resp. $\beta_1, \ldots, \beta_m$) be the roots of $f(x)$ (rep. $g(x)$) in an algebraic closure of $K(t, a_0, \ldots, a_n, b_0, \ldots, b_m)$, then we have

$$\begin{aligned} R(a_0, ta_1, \ldots, t^n a_n, b_0, tb_1, \ldots, t^m b_m) &= \mathrm{Res}_x(f(tx), g(tx)) \\ &= (t^n a_n)^m (t^m b_m)^n \prod_{(i,j)} (\alpha_i/t - \beta_j/t) \\ &= t^{nm} a_n^m b_m^n \prod_{(i,j)} (\alpha_i - \beta_j) \\ &= t^{nm} \,\mathrm{Res}_x(f(x), g(x))) \\ &= t^{nm} R(a_0, \ldots, a_n, b_0, \ldots, b_m). \end{aligned}$$

(2) Here we must show that

$$\begin{aligned} R(t^n a_0, t^{n-1} a_1, \ldots, a_n, t^m b_0, t^{m-1} b_1, \ldots, b_m) \\ = t^{nm} R(a_0, \ldots, a_n, b_0, \ldots, b_m). \end{aligned}$$

We have

$$\begin{aligned} R(t^n a_0, t^{n-1} a_1, \ldots, a_n, t^m b_0, t^{m-1} b_1, \ldots, b_m) \\ &= \mathrm{Res}_x(t^n f(x/t), t^m g(x/t)) \\ &= (a_n)^m (b_m)^n \prod_{(i,j)} (t\alpha_i - t\beta_j) \\ &= t^{nm} a_n^m b_m^n \prod_{(i,j)} (\alpha_i - \beta_j) \\ &= t^{nm} \,\mathrm{Res}_x(f(x), g(x))) \\ &= t^{nm} R(a_0, \ldots, a_n, b_0, \ldots, b_m), \end{aligned}$$

hence the result.

Exercise 1.12.
Let K be a field, $f(x) = a_0 x^n + \cdots + a_n$ and $g(x) = b_0 x^m + \cdots + b_m$ be polynomials with coefficients in K of degree n and m respectively. Recall that the resultant of $f(x)$ and $g(x)$ is the determinant of the $(m+n) \times (m+n)$ matrix:

$$S = \begin{pmatrix} a_0 & a_1 & \ldots & a_n & 0 & \ldots & 0 \\ 0 & a_0 & \ldots & a_{n-1} & a_n & \ldots & 0 \\ \ldots & \ldots & \ldots & \ldots & \ldots & \ldots & \ldots \\ 0 & \ldots & 0 & a_0 & \ldots & \ldots & a_n \\ b_0 & b_1 & \ldots & b_m & 0 & \ldots & 0 \\ 0 & b_0 & \ldots & b_{n-1} & b_m & \ldots & 0 \\ \ldots & \ldots & \ldots & \ldots & \ldots & \ldots & \ldots \\ 0 & \ldots & 0 & b_0 & \ldots & \ldots & b_m \end{pmatrix},$$

where the first m rows are devoted to the coefficients of $f(x)$ and the remaining ones to the coefficients of $g(x)$. Let $D(x) = \gcd(f(x), g(x))$ and $d = \deg D$. Let $K_{m-1}[x]$ (resp. $K_{n-1}[x], K_{m+n-1}[x]$) be the K-vector space constituted by the polynomials with coefficients in K of degree at most $m-1$ (resp. $n-1, m+n-1$). Let $\phi : K_{m-1}[x] \times K_{n-1}[x] \to K_{m+n-1}[x]$ be the map such that $\phi(A(x), B(x)) = A(x)f(x) + B(x)g(x)$.

(1) Show that ϕ is linear. Let $f_1(x) = f(x)/D(x)$ and $g_1(x) = g(x)/D(x)$. Show that

$$\mathcal{B} = \left\{ \Big(g_1(x), -f_1(x)\Big), \Big(xg_1(x), -xf_1(x)\Big), \ldots, \Big(x^{d-1}g_1(x), -x^{d-1}f_1(x)\Big) \right\}$$

is a basis of $\operatorname{Ker}\phi$ over K.

(2) Deduce that $rank(S) = m + n - d$.

Solution 1.12.

(1) Obviously, ϕ is linear over K. Let $(A(x), B(x)) \in K_{m-1}[x] \times K_{n-1}[x]$. Suppose that $(A(x), B(x)) \in \operatorname{Ker}\phi$, then $A(x)f(x) + B(x)g(x) = 0$ hence $A(x)f_1(x) = -B(x)g_1(x)$. It follows that $g_1(x)|A(x)$. Set $A(x) = g_1(x)h(x)$ where $\deg h = \deg A - \deg g_1 \le m - 1 - (m - d) = d - 1$; then $g_1(x)h(x)f_1(x) = -B(x)g_1(x)$, hence $B(x) = -f_1(x)h(x)$. Set $h(x) = h_0 + h_1x + \cdots + h_{d-1}x^{d-1}$, where $h_i \in K$ for $i = 0, \ldots, d$. Then

$$\begin{aligned}(A(x), B(x)) &= ((h_0 + h_1x + ... + h_{d-1}x^{d-1})g_1(x) \\ &\quad - (h_0 + h_1x + ... + h_{d-1}x^{d-1}))f_1(x) \\ &= h_0(g_1(x), -f_1(x)) + h_1(xg_1(x), -xf(x)) + \cdots \\ &\quad + h_{d-1}\Big(x^{d-1}g_{d-1}(x) - x^{d-1}f_{d-1}(x)\Big)\end{aligned}$$

hence $\operatorname{Ker}\phi$ is contained in the vector space generated by $\mathcal{B}$. Obviously the elements of $\mathcal{B}$ belong to $\operatorname{Ker}\phi$. We conclude that $\operatorname{Ker}\phi$ is generated by $\mathcal{B}$. Obviously the elements of $\mathcal{B}$ are linearly independent over K. Therefore $\mathcal{B}$ is a basis of $\operatorname{Ker}\phi$ over K.

(2) Let

$$\mathcal{A} = \{(1,0), (x,0), \ldots, (x^{m-1}, 0), (0,1), (0,x), \ldots, (0, x^{n-1})\}.$$

It it easy to see that $\mathcal{A}$ is a basis of $K_{m-1}[x] \times K_{n-1}[x]$. Moreover one verifies that the matrix of ϕ relatively to $\mathcal{A}$ and the canonical basis of $K_{m+n-1}[x]$ over K is equal to the transpose of S. Therefore

$$rank(S) = rank(S^T) = m + n - \operatorname{Dim}_K \operatorname{Ker}\phi = m + n - d.$$

Exercise 1.13.
Let E and F be two fields of characteristic $\neq 2$. Suppose that there exists a map $\sigma : E \to F$ satisfying the following conditions.

(1) For any $x, y \in E$, $\sigma(x+y) = \sigma(x) + \sigma(y)$.
(2) For any $x \in E$, $\sigma(x^2) = (\sigma(x))^2$.

Show that $\sigma = 0$ or σ is a field homomorphism.

Solution 1.13.
Let x and $y \in E$, then $\sigma((x+y)^2) = (\sigma(x+y))^2$, hence

$$\sigma(x^2) + \sigma(y^2) + 2\sigma(xy) = (\sigma(x)))^2 + (\sigma(y))^2 + 2\sigma(x)\sigma(y).$$

Since the characteristic is different from 2, then the result follows.

Exercise 1.14.
Let K be a field and p be a prime number. Show that the following propositions are equivalent.

(i) p is at least equal to the characteristic of K.
(ii) For any $f(x) \in K[x]$, $f^{(p)}(x) = 0$, where $f^{(p)}(x)$ denotes the p-th derivative of $f(x)$.

Solution 1.14.

(1) $(i) \Rightarrow (ii)$. It is sufficient to show that for any integer $k \geq 0$, $(x^k)^{(p)} = 0$. If $k < p$, then $(x^k)^{(k)} = k!$, hence the result in this case. If $k \geq p$, then

$$(x^k)^{(p)} = k(k-1)\cdots(k-(p-1))x^{k-p}.$$

Clearly one of the integers $k, k-1, \ldots, k-(p-1)$ is divisible by the characteristic of K, hence the result.
(2) $(ii) \Rightarrow (i)$. Let l be the characteristic of K. Suppose that $p < l$ and let $f(x) = x^p$, then $f(x)^{(p)} = p! \neq 0$.

Exercise 1.15.
Let $f(x) = a_n x^n + a_{n-1}x^{n-1} + \cdots + a_0$ be a polynomial with integral coefficients. Suppose that there exists a prime number $p | a_0$ such that $p > \sum_{i=1}^{n} |a_i| |\frac{a_0}{p}|^{i-1}$.

(1) Show that any root α of $f(x)$ satisfies the condition $|\alpha| > |\frac{a_0}{p}|$.
(2) Deduce that if $f(x)$ is primitive then it is irreducible in $\mathbb{Z}[x]$.
(3) Show that the preceding result applies for the polynomial $f(x) = x^2 + \epsilon_1 x + 3\epsilon_2$, where ϵ_1 and $\epsilon_2 \in \{-1, 1\}$.

Solution 1.15.

(1) Suppose that $|\alpha| \le |\frac{a_0}{p}|$, then

$$\begin{aligned}|a_0| &= |a_n\alpha^n + a_{n-1}\alpha^{n-1} + \cdots + a_1\alpha| \\ &\le \sum_{i=1}^{n} |a_i| \left|\frac{a_0}{p}\right|^i \\ &= \left|\frac{a_0}{p}\right| \sum_{i=1}^{n} |a_i| \left|\frac{a_0}{p}\right|^{i-1} \\ &< a_0,\end{aligned}$$

hence a contradiction. We conclude that $|\alpha| > |\frac{a_0}{p}|$.

(2) Suppose that $f(x)$ is reducible in $\mathbb{Z}[x]$ and let

$$f(x) = (b_m x^m + \cdots + b_0)(c_k x^k + \cdots + c_0)$$

be a factorization in this ring with m and $k \ge 1$. Since $b_0c_0 = a_0$, then $p|b_0$ or $p|c_0$. We may suppose that $p|b_0$. Let $\alpha_1, \ldots, \alpha_k$ be the roots of the second factor, then $\prod_{i=1}^k \alpha_i = (-1)^k \frac{c_0}{c_k}$. Hence by **(1)**,

$$\left|\frac{a_0}{p}\right| = \left|\frac{b_0}{p}\right| |c_0| \ge \left|\frac{c_0}{c_k}\right| = \prod_{i=1}^{k} |\alpha_i| > \left|\frac{a_0}{p}\right|^k,$$

which is a contradiction, thus $f(x)$ is irreducible over $\mathbb{Z}$.

(3) Here $p = 3$ and $|a_1| + |a_2||\frac{a_0}{p}| = 2$.

Exercise 1.16.

Let $n \ge 2$ be an integer, $p \ge 5$ be a prime number and $f(x) = x^n - x^{n-1} - 2p$.

(1) Suppose that $f(x) = g(x)h(x)$ in $\mathbb{Z}[x]$, where $g(x)$ and $h(x)$ are monic of degree at least 1. Show that $\deg g = 1$ or $\deg h = 1$.
(2) Show that $f(x)$ is irreducible over $\mathbb{Z}$.
(3) Show that the condition $p \ge 5$ is necessary.

Solution 1.16.

(1) We have

$$f(x) \equiv x^{n-1}(x-1) \pmod p \equiv g(x)h(x) \pmod p.$$

We may suppose that

$$g(x) \equiv x^r(x-1) \pmod p \textit{ and } h(x) \equiv x^s \pmod p,$$

where r and s are nonnegative integers, $s \ge 1$ and $r + s = n - 1$. Suppose that $r \ge 1$, then $g(0) \equiv 0 \pmod p$ and $h(0) \equiv 0 \pmod p$, hence $f(0) \equiv 0 \pmod{p^2}$, which contradicts the relation $f(o) = -2p$. We deduce that $r = 0$, $g(x) \equiv x-1 \pmod{p}$ and $h(x) \equiv x^{n-1} \pmod{p}$. Therefore, there exists $a \in \mathbb{Z}$, $a \equiv 1 \pmod{p}$ such that $g(x) = x - a$.

(2) Suppose that $f(x)$ is reducible over $\mathbb{Z}$ and let $f(x) = g(x)h(x)$ be a nontrivial factorization, then g and h are monic and we may suppose that $g(x) = x - a$, with $a \equiv 1 \pmod{p}$ and $h(x) \equiv x^{n-1} \pmod{p}$. It follows that $f(a) = 0 = a^n - a^{n-1} - 2p$, hence $a^{n-1}(a-1) = 2p$. We deduce that $a^{n-1}\frac{a-1}{p} = 2$. This implies that $a - 1 = \epsilon p$ with $\epsilon = \pm 1$ and $|a| = 2$. These conditions are incompatible with the assumption $p \geq 5$. Therefore $f(x)$ is irreducible over $\mathbb{Z}$.

(3) Let $p = 3$, $n = 2$ and $f(x) = x^2 - x - 6$, then $f(x)$ is reducible over $\mathbb{Z}$, since $f(x) = (x+2)(x-3)$.

Exercise 1.17.

Determine the set of the couples $(a, b) \in \mathbb{Z}^2$ such that the polynomial $f(x) = x^3 - ax - b$ is reducible over $\mathbb{Q}$.

Solution 1.17.

Let (a, b) be a couple of integers such that the polynomial $f(x)$ is reducible, then it has a root $d \in \mathbb{Z}$. Moreover $d = 0$ or $d \mid b$. In the first case we conclude that $b = 0$. Conversely if $b = 0$, then $f(x)$ is reducible. In the second case, set $b = de$, then $d^3 - ad - de = 0$, hence $d^2 - a - e = 0$. It follows that $a = d^2 - e$ and $b = de$ for some $d, e \in \mathbb{Z}$. Conversely it is easy to verify that for these values of a and b, $f(x)$ is reducible over $\mathbb{Q}$.

Exercise 1.18.

Let p be a prime number, $f(x) \in \mathbb{Z}[x]$ be monic and non constant. Denote by $\overline{u(x)}$, the reduction modulo p of the polynomial $u(x)$. Suppose that $f(x)$ may be written in the form $f(x) = \left(g(x)\right)^e + ph(x)$, where $g(x)$ and $h(x) \in \mathbb{Z}[x]$, $\overline{g(x)}$ is monic, irreducible over $\mathbb{F}_p$, $\deg h < \deg f$ and $\overline{g(x)} \nmid \overline{h(x)}$. Show that $f(x)$ is irreducible over $\mathbb{Q}$. Show that the condition $\deg h < \deg f$ could not be omitted.

Solution 1.18.

Suppose that $f(x) = f_1(x)f_2(x)$ in $\mathbb{Z}[x]$, where f_1 and f_2 are monic and non constant, then $\overline{f(x)} = \overline{f_1(x)f_2(x)} = \overline{g(x)}^e$, hence $\overline{f_1(x)} = \overline{g(x)}^{e_1}$ and $\overline{f_2(x)} = \overline{g(x)}^{e_2}$. It follows that

$$\begin{aligned} f_1(x) &= g(x)^{e_1} + ph_1(x) \quad \text{and} \\ f_2(x) &= g(x)^{e_2} + ph_2(x), \end{aligned}$$

where $h_1(x)$ and $h_2(x) \in \mathbb{Z}[x]$ and e_1, e_2 are positive integers such that $e_1 + e_2 = e$. Therefore

$$f(x) = g(x)^e + p\left(h_2(x)g(x)^{e_1} + h_1(x)g(x)^{e_2} + p^2 h_1(x)h_2(x)\right),$$

and

$$h(x) = h_2(x)g(x)^{e_2} + h_1(x)g(x)^{e_1} + ph_1(x)h_2(x),$$

which shows that $\overline{g(x)} \mid \overline{h(x)}$, hence a contradiction.

Through the following example it is seen that the condition on the degree of $h(x)$ is necessary. Let $p = 2$, $f(x) = (x^2+2)(2x+1) = x^2+2(x^3+2x+1)$, $g(x) = x$, and $h(x) = x^3 + 2x + 1$, then $f(x)$ is reducible in $\mathbb{Z}[x]$ and $\overline{g(x)} \nmid \overline{h(x)}$.

Exercise 1.19.
Let $f(x) = x^4 + 3x + 7x + 4$. Factorize this polynomial into its irreducible factors in $\mathbb{Z}/2\mathbb{Z}[x]$ and in $\mathbb{Z}/11\mathbb{Z}[x]$. Show that $f(x)$ is irreducible in $\mathbb{Z}[x]$.

Solution 1.19.
We have

$$f(x) \equiv x(x^3 + x + 1) \pmod 2 \quad \text{and}$$
$$f(x) \equiv (x^2 + 5x - 1)(x^2 - 5x - 4) \pmod{11}.$$

In these factorizations, the factors over $\mathbb{F}_2$ (resp. $\mathbb{F}_{11}$) are irreducible. Suppose that $f(x) = f_1(x)f_2(x)$, where f_1 and f_2 are monic polynomials with integral coefficients and $1 \leq \deg f_1 \leq \deg f_2 \leq 3$, then $f_1(x) \equiv x \pmod 2$ and $f_2(x) \equiv x^3+x+1 \pmod 2$, so that $\deg f_1 = 1$ and $\deg f_2 = 3$. We also have $f_1(x) \equiv x^2 + 5x - 1 \pmod{11}$ and $f_2(x) \equiv x^2 - 5x - 4 \pmod{11}$ or conversely. In any case $\deg f_1 = 2$. Therefore we realized a contradiction. It follows that $f(x)$ is irreducible over $\mathbb{Z}$.

Exercise 1.20.
Let p be a prime number, $f(X,Y) \in \mathbb{Z}[X,Y]$ be a homogeneous polynomial of degree $n \geq 1$. Suppose that f is primitive and that for any $(a,b) \in \mathbb{Z}^2$, $p \mid f(a,b)$. Show that $p < n$.

Solution 1.20.

(1) **First proof.** We have $f = Y^n \tilde{f}(X/Y)$ with $\tilde{f}(X) = f(X,1)$ and where $\bar{\tilde{f}}$ induces the zero function on $\mathbb{F}_p$. It follows that $X^p - X \mid \tilde{f}(X)$, hence $Y^p(X^p/Y^- X/Y) \mid f(X,Y)$, thus $X^p - XY^{p-1} \mid f(X,Y)$. By symmetry the same holds with X, Y swapped, hence $XY(X^{p-1} - Y^{p-1})$ divides $f(X,Y)$, so $n \geq p+1$.

(2) **Second proof.** We first show that if $g(X,Y) \in \mathbb{F}_p[X,Y]$ induces the zero function over $\mathbb{F}_p$, then $XY(X^{p-1} - Y^{p-1}) \mid g(X,Y)$ in $\mathbb{F}_p[X,Y]$. Let $c \in \mathbb{F}_p$. Application of the Euclidean algorithm yields:

$$g(X,Y) = (X - cY)Q(X,Y) + d,$$

where $Q(X,Y) \in \mathbb{F}_p[X,Y]$ and $d \in \mathbb{F}_p$. Putting $Y = 1, X = c$ in this identity leads to $d = 0$, hence $X - cY \mid g(X,Y)$. Using the same argument, we obtain $Y \mid g(X,Y)$. Therefore $XY(X^{p-1} - y^{p-1}) \mid g(X,Y)$. Denote by $\bar{f}(X,Y)$ the polynomial obtained from f by reducing its coefficients modulo p. Since f is primitive, then $\bar{f} \neq 0$. According to what has been proved, we conclude that $XY(X^{p-1} - Y^{p-1})$ divides $\bar{f}(X,Y)$, hence $p+1 \leq \deg \bar{f} \leq \deg f = n$ and the proof is complete.

Exercise 1.21.
Let $N_q(n)$ be the number of monic irreducible polynomials over $\mathbb{F}_q$. Recall that $N_q(n) = \frac{1}{n}\sum_{d|n} \mu(n/d)q^d$, where μ is the Mobius function. If $q_1 < q_2$ show that $N_{q_1}(n) < N_{q_2}(n)$.

Solution 1.21.
We fix n and we consider $N_q(n)$ as a polynomial function of q. We show that this function is strictly increasing for $q \geq 2$. Let $m = \lfloor n/2 \rfloor$, then we have

$$\begin{aligned}
(N_q(n))' &= \frac{1}{n}\sum_{d|n} \mu(n/d)dq^{d-1} \\
&= q^{n-1} + \frac{1}{n}\sum_{d|n, d\neq n} \mu(n/d)dq^{d-1} \\
&\geq q^{n-1} - \frac{1}{n}\sum_{d=1}^{m} dq^{d-1} \\
&\geq q^{n-1} - \frac{m}{n}\sum_{d=1}^{m} q^{d-1} \\
&= q^{n-1} - \frac{m}{n}(q^m - 1)/(q-1) \\
&\geq q^{n-1} - \frac{m}{n}q^m \\
&\geq q^{n-1} - \frac{1}{2}q^{n-1} \\
&= \frac{1}{2}q^{n-1} > 0.
\end{aligned}$$

Exercise 1.22.
Let s and n be positive integers, K be a field, $P_0, P_1, \ldots, P_s$ be $s+1$ distinct points in K^n. Show that there exists a polynomial of degree s, $f(x_1, \ldots, x_n) \in K[x_1, \ldots, x_n]$ having the form

$$f(x_1, ..., x_n) = \prod_{i=1}^{n} (a_i x_i + b_i)^{e_i},$$

where $a_i, b_i \in K$, $a_i \neq 0$, $0 \leq e_i \leq s$ for $i = 1, ..., n$ and $e_1 + e_2 + \cdots + e_n = s$ such that $f(P_i) = 0$ for $i = 1, \ldots, s$ and $f(P_0) = 1$.

Solution 1.22.
For $i = 0, 1, \ldots, s$, set $P_i = (c_1^i, \ldots, c_n^i)$, where $c_j^i \in K$ for any (i, j). Fix $h \in \{1, \ldots, s\}$. Since $P_0 \neq P_h$, then there exists $j \in \{1, \ldots, n\}$ such that $c_j^h \neq c_j^0$. Therefore we may determine a_h and $b_h \in K$ such that $a_h \neq 0$, $a_h c_j^0 + b_h = 1$ and $a_h c_j^h + b_h = 0$, that is the polynomial $f_h(x_1, \ldots, x_n) = a_h x_j + b_h$, satisfies the condition $f_h(P_0) = 1$ and $f_h(P_h) = 0$. That a_h and b_h could be determined may be easily verified since the determinant of the system of equations is equal to $c_j^0 - c_j^h$, hence nonzero. Now the polynomial

$$f(x_1, \ldots, x_n) = \prod_{h=1}^{s} f_h(x_1, ..., x_n)$$

satisfies the stated assertions.

Exercise 1.23.
Let K be a field, E be an extension of K, I be an ideal of $K[x_1, \ldots, x_n]$ and $\hat{I}$ be the ideal of $E[x_1, \ldots, x_n]$ generated by I. Show that $\hat{I} \cap K[x_1, \ldots, x_n] = I$.

Solution 1.23.
Let $(e_j)_{j \in J}$ be a basis of E over K. We may suppose that $e_{j_0} = 1$. We have

$$E[x_1, \ldots, x_n] = \bigoplus_{j \in J} K[x_1, ..., x_n] \cdot e_j = K[x_1, \ldots, x_n] \bigoplus_{j \in J \setminus \{e_{j_0}\}} K[x_1, ..., x_n] \cdot e_j,$$

hence

$$\hat{I} = I \cdot E(x_1, \ldots, x_n) = I \bigoplus \bigoplus_{j \in J \setminus \{e_{j_0}} I e_j.$$

Thus $\hat{I} \cap K[x_1, \ldots, x_n] = I$.

Exercise 1.24.

Let K be a field, $f(x) \in K[x]$ be non constant and let $\alpha \in K$.

(1) Let $A_\alpha = \{u(x)/v(x) \in K(x), v(\alpha) \neq 0 \text{ and } \gcd(u, v) = 1\}$. Show that A_α is a local ring. Describe explicitly its maximal ideal M. Show that $A_\alpha/M \simeq K$.

(2) Let m_α be the unique nonnegative integer such that $f(x) = (x - \alpha)^{m_\alpha} g(x)$, where $g(x) \in K[x]$ and $g(\alpha) \neq 0$. Show that A_α/fA_α is a vector space over K of dimension m_α.

Solution 1.24.

(1) Let $S = \{v(x) \in k[x], v(\alpha) \neq 0\}$, then S is a multiplicative subset of $K[x]$ containing 0 and 1 and clearly $A_\alpha = S^{-1}K[x]$. We first determine the units of A_α. Let $u(x)/v(x) \in A_\alpha^*$, then there exists $a(x)/b(x) \in A_\alpha$ such that $\frac{u(x)a(x)}{u(x)b(x)} = 1$. We deduce that $\frac{a(x)}{b(x)} = \frac{v(x)}{u(x)}$, hence $u(\alpha) \neq 0$. Conversely if $u(\alpha) \neq 0$, then $u(x)/v(x) \in A_\alpha^*$. We conclude that

$$A_\alpha^* = \{u(x)/v(x) \in k(x), \gcd(u, v) = 1, v(\alpha) \neq 0 \quad \text{and} \quad u(\alpha) \neq 0\}.$$

To show that A_α is local it is sufficient to prove that the complementary set M of A_α^* in A_α is an ideal. We have

$$M = \{u(x)/v(x) \in K(x), \gcd(u, v) = 1, v(\alpha) \neq 0 \quad \text{and} \quad u(\alpha) = 0\}$$

and we omit the proof that this set is an ideal of A_α. Let $\phi : A_\alpha \to K$ be the map defined for any $u(x)/v(x) \in A_\alpha$ by $\phi(u/v) = u(\alpha)/v(\alpha)$. Clearly ϕ is a morphism of rings. For any $c \in K$, we have $c \in A_\alpha$ and $\phi(c) = c$, hence ϕ is surjective. Let $u(x)/v(x) \in A_\alpha$, then $\phi(u/v) = 0$ if and only if $u(\alpha) = 0$, that is if and only if $u(x)/v(x) \in M$. Therefore $\operatorname{Ker}\phi = M$ and then, the first isomorphism theorem implies $A_\alpha/M \simeq K$.

(2) Substituting y for $x - \alpha$ if necessary, we may suppose that $\alpha = 0$, so that we have $f(x) = x^m g(x)$ where $g(x) \in K[x]$, $g(0) \neq 0$ and $m \geq 0$. Since $g(x)$ is a unit in A_0, then

$$f(x)A_0 = x^m A_0 \text{ and } A_0/fA_0 = A_0/x^m A_0.$$

Obviously A_0 is a vector space over K and $x^m A_0$ is a subspace, hence $A_0/x^m A_0$ is a vector space over K. We show that $\{\bar{1}, \bar{x}, \ldots, \overline{x^{m-1}}\}$ is a basis of $A_0/x^m A_0$ over k. Let $u(x)/v(x) \in A_0$, then $v(0) \neq 0$. We divide $u(x)$ by $v(x)$ according to the ascending powers of x at the order

m. We obtain: $u(x) = v(x)q(x) + x^m r(x)$, where $q(x)$ and $r(x) \in K[x]$, hence

$$\frac{u(x)}{v(x)} = q(x) + x^m \frac{r(x)}{v(x)} \equiv q(x) \pmod{x^m A_0},$$

so that $\{\bar{1}, \bar{x}, ..., \overline{x^{m-1}}\}$ generates $x^m A_0$ over K. Suppose that

$$a_0\bar{1} + a_1\bar{x} + \cdots + a_{m-1}\overline{x^{m-1}} = 0,$$

with $a_0, a_1, \ldots, a_{m-1} \in K$. Then $a_0 + a_1 x + \cdots + a_{m-1}x^{m-1} \in x^m A_0$, hence

$$a_0 + a_1 x + \cdots + a_{m-1}x^{m-1} = x^m \frac{u(x)}{v(x)},$$

where $u(x), v(x) \in K[x]$, $\gcd(u, v) = 1$ and $v(0) \neq 0$. We deduce that x^m divides $a_0 + a_1 x + \cdots + a_{m-1}x^{m-1}$ in $K[x]$. Therefore $a_0 = a_1 = \ldots = a_{m-1} = 0$, which implies that $\{\bar{1}, \bar{x}, \ldots, \overline{x^{m-1}}\}$ is a basis of $A_0/x^m A_0$ over K.

Exercise 1.25.
Let n be a positive integer, K be a field, E be an extension of K and $A \in M_n(K)$. Show that the characteristic polynomials (resp. minimal polynomials) of A over K and E are equal.

Solution 1.25.
The proof of the assertion on the characteristic polynomials is obvious and will be omitted. Let $f(x)$ and $g(x)$ be the minimal polynomials of A over K and E respectively. It is clear that $g(x) \mid f(x)$ in $E[x]$. Let $\mathcal{B}$ be a basis of E over K. We may suppose that $1 \in \mathcal{B}$. Set

$$g(x) = e_0 + e_1 x + \cdots + e_{n-1}x^{m-1} + x^m,$$

where $m \leq \deg f$ and $e_i \in E$ for $i = 0, \ldots, m-1$. There is a finite set $\mathcal{C} \subset \mathcal{B}$ say $\mathcal{C} = \{1, w_1, \ldots, w_k\}$ such that each e_i, for $i = 0, \ldots, m-1$, is a linear combination of the elements of $\mathcal{C}$. Set $w_0 = 1$ and $e_i = \sum_{j=0}^{k} \lambda_{ij} w_j$ for $i = 0, \ldots, m-1$. The identity $g(A) = 0$ takes the form: $\sum_{j=0}^{k} B_j w_j$, where B_j is the $n \times n$ matrix with coefficients in K given by $B_j = \sum_{i=0}^{m-1} \lambda_{ij} A^i$ for $j = 1, \ldots, k$ and $B_0 = \sum_{i=0}^{m-1} \lambda_{i0} A^i + A^m$. We deduce that $B_j = 0$ for $j = 0, \ldots, k$. This implies that $f(x) \mid \sum_{i=0}^{m-1} \lambda_{i0} x^i + x^m$ and $f(x) \mid \sum_{i=0}^{m-1} \lambda_{ij} x^i$ for $j = 1, \ldots, k$. Since $m \leq \deg f$ and $f(x)$ is monic, then $m = \deg f$, $f(x) = \sum_{i=0}^{m-1} \lambda_{i0} x^i + x^m$ and $\lambda_{ij} = 0$ for $i = 0, \ldots, m-1$ and $j = 1, \ldots, k$. We conclude that $g(x) = f(x)$.

Exercise 1.26.
Let E be a finite dimensional vector space over K, $\sigma : E \to E$ be a linear map. Let $m_\sigma(x)$ and $\chi_\sigma(x)$ be the minimal polynomial and characteristic polynomial of σ respectively.

(1) Let $\alpha \in E$ and let $I = \{f(x) \in K[x], f(\sigma)(\alpha) = 0\}$. Show that I is a principal ideal of $K[x]$ generated by some monic polynomial. Denote this polynomial by $m_{(\sigma,\alpha)}(x)$. Show that $m_{(\sigma,\alpha)}(x)$ divide $m_\sigma(x)$ and $\chi_\sigma(x)$.
(2) Let $\{\alpha_1, \ldots, \alpha_n\}$ be a basis of E of K. Show that

$$m_\sigma(x) = \operatorname*{lcm}_{\alpha \in E} m_{(\sigma,\alpha)}(x) = \operatorname*{lcm}_{i=1}^{n} m_{(\sigma,\alpha_i)}(x).$$

Solution 1.26.

(1) Let $\phi : K[x] \to E$ be the map such that for any $f(x) \in K[x]$, $\phi(f(x)) = f(\sigma)(\alpha)$. Then obviously ϕ is a K-linear map. Since $I \neq \operatorname{Ker}(\phi)$, then I is an additive subgroup of $K[x]$. Let $f(x) \in I$ and $g(x) \in K[x]$, then

$$\phi(g(x)f(x)) = (g(\sigma)f(\sigma))(\alpha) = g(\sigma)(f(\sigma)(\alpha)) = g(\sigma)(0) = 0,$$

hence $g(x)f(x) \in I$. Therefore I is an ideal of $K[x]$. Thus, I is principal. Since $m_\sigma(x) \in I$, then $I \neq (0)$ and I is generated by a monic polynomial. Clearly $m_{(\sigma,\alpha)}(x) \mid m_\sigma(x) \mid \xi_\sigma(x)$.
(2) Since for any $\alpha \in E$, $m_{(\sigma,\alpha)}(x) \mid m_\sigma(x)$, then $\operatorname{lcm} m_{(\sigma,\alpha)}(x) \mid m_\sigma(x)$. On the other hand, let $g(x) = \operatorname*{lcm}_{\alpha \in E} m_{(\sigma,\alpha)}(x)$, then for any $\beta \in E$, we have $g(\sigma)(\beta) = 0$, hence $g(\sigma) = 0$. Therefore $m_\sigma(x) \mid g(x)$. We conclude that $m_\sigma(x) = g(x) = \operatorname*{lcm}_{\alpha \in E} m_{(\sigma,\alpha)}(x)$. In the same way, we may prove that $m_\sigma(x) = \operatorname*{lcm}_{i=1}^{n} m_{(\sigma,\alpha_i)}(x)$.

Exercise 1.27.

(1) Show that the number of monomials $x^i y^j$ with $0 \le i, j \le m$ and $i + j \le m$ is equal to $\binom{m+2}{2}$.
(2) Let K be a field, $f(t), g(t)$ be polynomials with coefficients in K of degree r and s respectively. Show that for m large enough, the family of polynomials $f(t)^i g(t)^j$ with $i + j \le m$ is dependent over K.
(3) Deduce that there exists $F(x, y) \in K[x, y] \backslash \{0\}$ irreducible such that $F(f(t), g(t)) = 0$.

Solution 1.27.

(1) We make a partition of the set of couples (i, j) with $i, j \geq 0$ and $i+j \leq m$ by putting in the same class all the couples having the same first component. The number of couples whose first component equals i is equal to $m - i + 1$. Therefore the total number of couples is equal to

$$(m+1) + m + ... + 1 = (m+1)(m+2)/2 = \binom{m+2}{2}.$$

(2) If $f(t)^i g(t)^j = f(t)^h g(t)^k$, with $i + j \leq m$ and $h + k \leq m$, then the result is true. Therefore, we may suppose that the set

$$\{f(t)^i g(t)^j \quad \text{with} \quad i + j \leq m\}$$

contains exactly $\binom{m+2}{2}$ polynomials with coefficients in K. We have: $\deg f(t)^i g(t)^j = ri + sj \leq (r+s)m$. Therefore all the elements of the set belong to the K-vector space $K_m[t]$ whose elements are the polynomials $h(t) \in K[t]$ such that $\deg h \leq (r+s)m$. The dimension over K of this space is equal to $(r+s)m+1$. Therefore if $\binom{m+1}{2} > (r+s)m+1$, that is $m > 2(r+s) - 3$, then the family is linearly dependent over K.

(3) If $m > 2(r+s) - 3$ the family $f(t)^i g(t)^j$ is linearly dependent over K, hence there exist a family $(a_{ij})_{(i,j)}$ with $a_{ij} \in K$ not all 0 such that $\sum a_{ij} f(t)^i g(t)^j = 0$. Let $G(x, y) = \sum a_{ij} x^i y^j$, then clearly $G \neq 0$ and $G(f(t), g(t)) = 0$. Let $G(x, y) = G_1(x, y) \cdots G_k(x, y)$ be the factorization into irreducible factors in $K[x, y]$ of $G(x, y)$. Then there exist $i \in \{1, \ldots, k\}$ such that $G_i(f(t), g(t)) = 0$.

Exercise 1.28.
Let K be a field, E be an extension of K, x be an indeterminate over E and $\alpha_1, \ldots \alpha_n$ be elements of E, linearly independent over K. Show that these elements are linearly independent over $K(x)$.

Solution 1.28.
Suppose that there exist $\lambda_1(x), \ldots, \lambda_n(x) \in K(x)$ such that $\sum_{i=1}^n \lambda_i(x)\alpha_i = 0$. We may suppose that $\lambda_i(x) \in K[x]$. Substituting 0 for x leads to $\lambda_1(0) = \cdots = \lambda_n(0) = 0$. This shows that $x \mid \lambda_i(x)$ for $i = 1, \ldots, n$. Suppose that $x^k \mid \lambda_i(x)$, then $\sum_{i=1}^n (\lambda_i(x)/x^k)\alpha_i = 0$. Substituting again 0 for x shows that $x^{k+1} \mid \lambda_i(x)$. It follows that all the polynomials $\lambda_i(x)$ are 0.

Exercise 1.29.
Let $f(x)$ and $g(x)$ be polynomials with integral coefficients. Show that $g(x) \mid f(x)$ in $\mathbb{Z}[x]$ if and only if $\operatorname{cont}(g) \mid \operatorname{cont}(f)$ and there exist infinitely many $n \in \mathbb{Z}$ such that $g(n) \neq 0$ and $g(n) \mid f(n)$.

Solution 1.29.
The necessity of the conditions is obvious. We prove the sufficiency of the conditions.

First proof. Let (u_n) be a sequence of integers such that

$$\lim_{n\to\infty} |u_n| = \infty,\ g(u_n) \neq 0 \quad \text{and} \quad g(u_n) \mid f(u_n).$$

Using the Euclidean algorithm, we obtain $f(x) = g(x)q(x) + r(x)$, where $q(x), r(x) \in \mathbb{Q}[x]$ and $\deg r < \deg g$. Suppose that $r(x) \neq 0$ and let m be the lcm of all the denominators of the coefficients of $g(x)$ and $r(x)$. Then

$$mf(x)/g(x) - mq(x) = mr(x)/g(x).$$

There exists a rank n_0 such that for $n > n_0$, $r(u_n) \neq 0$. For any of these n, we have $mr(u_n)/g(u_n) \in \mathbb{Z}$ and $|mr(u_n)/g(u_n)| \geq 1$. On the other hand since $\deg r < \deg g$, then $\lim_{n\to\infty} |mr(u_n)/g(u_n)| = 0$ and we have reached a contradiction. Therefore $r(x) = 0$ and $f(x) = g(x)q(x)$, hence $\operatorname{cont}(q) = \operatorname{cont}(f)/\operatorname{cont}(g) \in \mathbb{Z}$ which shows that $q(x) \in \mathbb{Z}[x]$ and $g(x) \mid f(x)$ in $\mathbb{Z}[x]$.

Second proof. Let $D(x) = \gcd(f(x), g(x))$ and let $f_1(x), g_1(x) \in \mathbb{Z}[x]$ such that

$$f(x) = D(x)f_1(x) \text{ and } g(x) = D(x)g_1(x).$$

Let $r = \operatorname{Res}_x(f_1(x), g_1(x))$, then $r \in \mathbb{Z}$ and there exist $u(x), v(x) \in \mathbb{Z}[x]$ such that $u(x)f_1(x) + v(x)g_1(x) = r$. By hypothesis the set

$$A = \{m \in \mathbb{Z}, g_1(m) \neq 0 \,\text{and}\, g_1(m) \mid f_1(m)\}$$

is infinite. Define in A the relation $\mathcal{R}$ by $n\mathcal{R}\Updownarrow$ if $g_1(n) = g_1(m)$. Clearly this relation is an equivalence relation. Let $\mathcal{C}$ be an equivalence class and $m \in \mathcal{C}$, then $g_1(m) = d$, for some nonzero integer d dividing r, hence the number of classes is finite. If $\deg g_1 \geq 1$, then $|\mathcal{C}| \leq \deg g_1$, which implies that A is finite, contradicting our assumption. It follows that there exists a divisor d of r such that $g_1(x) = d$. We have

$$\operatorname{cont}(f) = \operatorname{cont}(D)\operatorname{cont}(f_1) \text{ and } \operatorname{cont}(g) = d\operatorname{cont}(D)$$

and by hypothesis the later of these integers divides the former, hence $d \mid \operatorname{cont}(f_1)$. It follows that

$$\begin{aligned} f(x) &= f_1(x)D(x) \\ &= \big(f_1(x)/\operatorname{cont}(f_1)\big)\operatorname{cont}(f_1)D(x) \\ &= \big(f_1(x)/cont(f_1)\big)\big(\operatorname{cont}(f_1)/d\big)dD(x) \\ &= \big(f_1(x)/\operatorname{cont}(f_1)\big)\big(\operatorname{cont}(f_1)/d\big)g(x), \end{aligned}$$

hence $g(x) \mid f(x)$ in $\mathbb{Z}[x]$.

Exercise 1.30.
Let $q_1, \ldots, q_n$ be positive integers such that $\gcd(q_i, q_j) = 1$ for $i \neq j$. Let $g_1(x), \ldots, g_n(x)$ be monic polynomials with integral coefficients all of degree m. Show that there exists a monic polynomial $g(x)$ of degree m with integral coefficients such that $g(x) \equiv g_i(x) \pmod{q_i\mathbb{Z}[x]}$, for $i = 1, \ldots, n$.

Solution 1.30.
First proof. For j fixed in $\{0, \ldots, m-1\}$, let $a_j \in \mathbb{Z}$ such that $a_j \equiv c(x^j, g_i(x)) \pmod{q_i}$ for $i = 1, \ldots, n$, where $c(x^j, g_i(x))$ denotes the coefficient of x^j in the polynomial $g_i(x)$. Such elements a_j are determined by the Chinese remainder theorem. Let $g(x) = x^m + \sum_{j=0}^{m-1} a_j x^j$. Then clearly $g(x) \equiv g_i(x) \pmod{q_i\mathbb{Z}[x]}$, for $i = 1, \ldots, n$.

Second proof. We use again the Chinese remainder theorem but in a different way. For $i = 1, \ldots, n$, let $y_i \in \mathbb{Z}$ such that $y_i \equiv 1 \pmod{q_i}$ and $y_i \equiv 0 \pmod{\prod_{j \neq i} q_j}$. Let $g_0(x) = \sum_{i=1}^{n} y_i g_i(x)$, then $\deg g_0(x) = m$. Its leading coefficient c is given by $c = \sum_{i=1}^{n} y_i$ and we have $c \equiv 1 \pmod{q_i}$ for $i = 1, \ldots, n$. Now the polynomial $g(x) = g_0(x)/c$ satisfies the required conditions.

Exercise 1.31.
Let $f(x) = a_n x^n + a_{n-1}x^{n-1} + \cdots + a_0$ be a polynomial with integral coefficients, $\alpha_1, \ldots, \alpha_n$ be the roots of $f(x)$ and let $H(f) = \max(|a_n|, \ldots, |a_0|)$. Show that $H(f) \geq \frac{\max_{i=1}^{n}(\alpha_i)}{2}$.

Solution 1.31.
We show that $H(f) \geq \alpha_i/2$ for any $i \in \{1, \ldots, n\}$. Let α be any of these α_i. If $\alpha \leq 2$, then the result is true since $H(f) \geq 1$. Suppose that $\alpha > 2$, then

$$\begin{aligned}
|\alpha^n| &\leq |a_n \alpha^n| \\
&= |a_0 + \cdots + a_{n-1}\alpha^{n-1}| \\
&\leq |a_0| + |a_1||\alpha| + \cdots + |a_{n-1}||\alpha|^{n-1} \\
&\leq H(f)(1 + |\alpha| + \cdots + |\alpha|^{n-1}),
\end{aligned}$$

hence

$$\begin{aligned}
\alpha &\leq H(f)\left(\frac{1}{|\alpha|^{n-1}} + \frac{1}{|\alpha|} + 1\right) \\
&\leq H(f)\left(\frac{1}{2^{n-1}} + \cdots + \frac{1}{2} + 1\right) \\
&\leq 2H(f).
\end{aligned}$$

Chapter 2

Algebraic extension, Algebraic closure

Exercise 2.1.
Let $f(x) = x^4 - 2x^2 + 2x + 2$ and α be a root of $f(x)$ in $\mathbb{C}$. Show that $f(x)$ is irreducible over $\mathbb{Q}$ and express $\beta = 1/(\alpha^2 + \alpha + 1)$ as a polynomial in α with rational coefficients.

Solution 2.1.
The polynomial $f(x)$ is irreducible over $\mathbb{Q}$ by application of Eisenstein's irreducibility criterion [Lang (1965), Eisenstein Criterion, Chap. 5.7]. It follows that $\{1, \alpha, \alpha^2, \alpha^3\}$ is a basis of $\mathbb{Q}(\alpha) = \mathbb{Q}[\alpha]$ over $\mathbb{Q}$. Therefore β may be expressed as a polynomial in α with rational coefficients. Let $g(x) = x^2 + x + 1$, then $gcd(f(x), g(x)) = 1$, hence there exist $u(x)$ and $v(x)$ in $\mathbb{Q}[x]$ such that $u(x)f(x) + v(x)g(x) = 1$. Substituting α for x in this identity, we obtain $v(\alpha)g(\alpha) = 1$. Thus $\beta = 1/g(\alpha) = v(\alpha)$. We compute $v(x)$ as follows. The Euclidean division of $f(x)$ by $g(x)$ leads to

$$f(x) = g(x)(x^2 - x - 2) + 5x + 4.$$

Similarly, we have

$$g(x) = (5x + 4)(x/5 + 1/25) + 21/25.$$

It follows that

$$\begin{aligned} 21/25 &= g(x) - (5x + 4)(x/5 + 1/25) \\ &= g(x) - (x/5 + 1/25)\Big(f(x) - g(x)(x^2 - x - 2)\Big) \\ &= g(x)\Big(1 + (x^2 - x - 2)(x/5 + 1/25)\Big) - (x/5 + 1/25)f(x). \end{aligned}$$

Substituting α for x, we obtain

$$21 = g(\alpha)\Big(25 + (\alpha^2 - \alpha - 2)(5\alpha + 1)\Big).$$

We conclude that

$$\begin{aligned}\beta &= 1/g(\alpha)\\ &= \frac{25 + (\alpha^2 - \alpha - 2)(5\alpha + 1)}{21}\\ &= \frac{23}{21} - \frac{11}{21}\alpha - \frac{4}{21}\alpha^2 + \frac{5}{21}\alpha^3.\end{aligned}$$

Exercise 2.2.
Show that the following identities hold in the complex field.

(1) $\sqrt{11 + 6\sqrt{2}} + \sqrt{11 - 6\sqrt{2}} = 6$.
(2) $(10 + 6\sqrt{3})^{1/3} + (10 - 6\sqrt{3})^{1/3} = 2\sqrt{3}$.

Here the square roots are the positive ones and the cubic roots are the real ones.

Solution 2.2.

(1) We have

$$\begin{aligned}\sqrt{11 + 6\sqrt{2}} + \sqrt{11 - 6\sqrt{2}} &= \sqrt{(3 + \sqrt{2})^2} + \sqrt{(3 - \sqrt{2})^2}\\ &= 3 + \sqrt{2} + 3 - \sqrt{2} = 6.\end{aligned}$$

(2) Let

$$\alpha = (10 + 6\sqrt{3})^{1/3}, \quad \beta = (10 - 6\sqrt{3})^{1/3} \quad \text{and} \quad \gamma = \alpha + \beta.$$

Then

$$\gamma^3 = \alpha^3 + \beta^3 + 3\alpha\beta\gamma = 12\sqrt{3} + 6\gamma.$$

Hence $\gamma^3 - 6\gamma = 12\sqrt{3}$. Squaring both sides of this equality shows that γ is a root of $f(x) = x^6 - 12x^4 + 36x^2 - 3.12^2$. One may check that $f(2\sqrt{3}) = 0$. From the above calculations it is clear that

$$\{\gamma, -\gamma, j\alpha + j^2\beta, -j\alpha - j^2\beta, j^2\alpha + j\beta, -j^2\alpha - j\beta\}$$

is the set of the roots of $f(x)$. One and only one element of this set is real and positive, namely γ. Therefore $\gamma = 2\sqrt{3}$.

Exercise 2.3.
Let $f(x)$ and $g(x) \in \mathbb{Z}[x]$ such that $f(x)$ is monic and irreducible, α be a root of $f(x)$ in $\mathbb{C}$. Let p be a prime number and let $\overline{f(x)}$ and $\overline{g(x)}$ denote the reduced polynomials modulo p of $f(x)$ and $g(x)$ respectively. Show that $g(\alpha) \equiv 0 \pmod{p}$ in $\mathbb{Z}[\alpha]$ if and only if $\overline{f(x)} \mid \overline{g(x)}$ in $\mathbb{F}_p[x]$.

Solution 2.3.
We have

$$\begin{aligned} g(\alpha) \equiv 0 \pmod p \text{ in } \mathbb{Z}[\alpha] &\Leftrightarrow g(\alpha) = ph(\alpha) \quad \text{with} \quad h(x) \in \mathbb{Z}[x] \\ &\Leftrightarrow f(x) \mid g(x) - ph(x) \quad \text{in} \quad \mathbb{Z}[x] \\ &\Leftrightarrow g(x) - ph(x) = f(x)q(x), \text{with } q(x) \in \mathbb{Z}[x] \\ &\Leftrightarrow \overline{f(x)} \mid \overline{g(x)} \quad \text{in} \quad \mathbb{F}_p[x]. \end{aligned}$$

Exercise 2.4.
Let K be a field of characteristic not equal to 2, $a, b, d \in K$ such that $d \notin K^2$. Show that the following statements are equivalent

(i) There exist $a_1, \ldots, a_n \in K$ such that

$$\sqrt{a + b\sqrt{d}} \in K\left(\sqrt{d}, \sqrt{a_1}, \ldots, \sqrt{a_n}\right).$$

(ii) There exists $s \in K \setminus \{0\}$ such that $\sqrt{s(a + b\sqrt{d})} \in K(\sqrt{d})$.
(iii) $\sqrt{a^2 - b^2 d} \in K$.

Solution 2.4.
(i) $\Rightarrow$ *(ii)*. The proof is by induction on n, n being the number of a_i. If $n = 0$, then *(ii)* holds with $s = 1$. Suppose that the implication is true for $n - 1$ elements a_i. Without loss of generality, we assume that $\sqrt{a_n} \notin E$, where $E = K(\sqrt{d}, \sqrt{a_1}, \ldots, \sqrt{a_{n-1}})$. We may write $\sqrt{a + b\sqrt{d}}$ in the form $\sqrt{a + b\sqrt{d}} = A + B\sqrt{a_n}$, where $A, B \in E$. Then

$$a + b\sqrt{d} = A^2 + B^2 a_n + 2AB\sqrt{a_n}.$$

Since $a + b\sqrt{d}, A^2 + B^2 a_n \in E$, then $AB = 0$.

Case 1. $A = 0$.

We have $\sqrt{a + b\sqrt{d}} = B\sqrt{a_n}$, hence $\sqrt{a_n(a + b\sqrt{d})} = Ba_n \in E$. Therefore by the inductive hypothesis, there exists $s_1 \in K \setminus \{0\}$ such that $\sqrt{s_1 a_n(a + b\sqrt{d})} \in K(\sqrt{d})$. We conclude that *(ii)* holds for $s = s_1 a_n$.

Case 2. $B = 0$.

We have $\sqrt{a + b\sqrt{d}} = A \in E$, hence by the inductive hypothesis, there exists $s \in K \setminus \{0\}$ such that $\sqrt{s(a + b\sqrt{d})} \in K(\sqrt{d})$.

(ii) $\Rightarrow$ *(iii)*. Let $u, v \in K$ such that $s(a + b\sqrt{d}) = (u + v\sqrt{d})^2$, then $sa = u^2 + dv^2$ and $sb = 2uv$. If $v = 0$, then $b = 0$, $\sqrt{a^2 - b^2 d} = \pm a \in K$ and *(ii)* holds. If $v \neq 0$ then $sa = \frac{s^2 b^2}{4v2} + dv^2$. Therefore we obtain the following equation satisfied by s: $s^2 b^2 - 4av^2 s + 4dv^4 = 0$. Since $s \in K$,

it follows that the discriminant of this equation is a square in K, that is $\sqrt{a^2 - b^2 d} \in K$.

$(iii) \Rightarrow (i)$. Suppose that $\sqrt{a^2 - b^2 d} := e \in K$ and let $a_1 = 2(a + e)$. If $a_1 = 0$, then $a = -e = -\sqrt{a^2 - b^2 d}$, hence $a^2 = a^2 - db^2$, whence $b = 0$. Therefore $\sqrt{a + b\sqrt{d}} = \sqrt{a} \in K(\sqrt{d}, \sqrt{a})$. Suppose that $a_1 \neq 0$. We have

$$\begin{aligned}\left(\frac{a_1 + 2b\sqrt{d}}{2\sqrt{a_1}}\right)^2 &= \frac{a_1^2 + 4db^2 + 4a_1 b\sqrt{d}}{4a_1}\\ &= \frac{4(a+e)^2 + 4(a^2 - e^2) + 8(a+e)\sqrt{d}b}{8(a+e)}\\ &= a + b\sqrt{d},\end{aligned}$$

hence

$$\sqrt{a + b\sqrt{d}} = \frac{a_1 + 2b\sqrt{d}}{2\sqrt{a_1}} \in K(\sqrt{d}, \sqrt{a_1}).$$

Exercise 2.5.
Let K be a field, E be an algebraic extension of K of degree n and $\mathcal{B} = \{\omega_1, \ldots, \omega_n\}$ be a basis of E over K. Let $\alpha \in E$ and $\beta = u(\alpha)$, where $u(x) \in K[x]$. Let M be the matrix relative to the base $\mathcal{B}$ of the K-endomorphism m_α of E defined by $m_\alpha(\gamma) = \alpha\gamma$. Show that the matrix in the same basis of the K-endomorphism multiplication by β is equal to $u(M)$.

Solution 2.5.
Let m_β be the endomorphism of E representing the multiplication by β. To get the result it is equivalent to prove that $m_{u(\alpha)} = u(m_\alpha)$. It is sufficient to prove this identity, when $u(x)$ is a monomial, say $u(x) = ax^k$. In this case for any $\gamma \in E$, we have

$$a(m_\alpha)^k(\gamma) = a(m_\alpha)^{k-1}(\alpha\gamma) = a\alpha^k\gamma = m_{a\alpha^k}(\gamma) = m_{u(\alpha)}(\gamma),$$

hence the result.

Exercise 2.6.
Let p be a prime number. Show that the series $f(x) = \sum_{n=0}^{\infty} x^{p^n}$ is algebraic over $\mathbb{F}_p(x)$.

Solution 2.6.
Since

$$f(x) = x + f(x^p) = x + f(x)^p,$$

then $f(x)$ is a root of the polynomial $g(y) = y^p - y + x$ which is nonzero and has its coefficients in $\mathbb{F}_p(x)$, hence $f(x)$ is algebraic over $\mathbb{F}_p(x)$.

Exercise 2.7.
Let $f(x) = x^3 - x^2 - 2x - 8$, θ be a root of $f(x)$ and $K = \mathbb{Q}(\theta)$.

(1) Show that $f(x)$ is irreducible over $\mathbb{Q}$.
(2) Let $\alpha = (\theta^2 - \theta)/2$. Show that α is a primitive element of K over $\mathbb{Q}$.
(3) Set $\theta = b_0 + b_1\alpha + b_2\alpha^2$, where b_0, b_1, $b_2 \in \mathbb{Q}$. By solving a linear system of equations, compute b_0, b_1, b_2.
(4) Compute the minimal polynomial $g(x)$ of α and the dual basis $\{\omega_0, \omega_1, \omega_2\}$ of $\{1, \alpha, \alpha^2\}$. Using this basis, compute again b_0, b_1, b_2.

Solution 2.7.

(1) The polynomial $f(x)$ is irreducible in $\mathbb{F}_3[x]$, hence irreducible over $\mathbb{Q}$.
(2) Since $\mathbb{Q}(\alpha) \subset \mathbb{Q}(\theta)$, then $[\mathbb{Q}(\alpha) : \mathbb{Q}] \mid 3$, hence $[\mathbb{Q}(\alpha) : \mathbb{Q}] = 1$ or $[\mathbb{Q}(\alpha) : \mathbb{Q}] = 3$. In the first case, we conclude that $\alpha \in \mathbb{Q}$, which implies θ is a root of a monic polynomial of degree 2 with rational coefficients. This statement is false, hence $[\mathbb{Q}(\alpha) : \mathbb{Q}] = 3$ and then α generates K over $\mathbb{Q}$.
(3) Replacing α by its expression in terms of θ, leads to

$$\theta = b_0 + b_1\alpha + b_2\alpha^2 = (b_0 - 2b_2) + \left(\frac{3}{2}b_2 - \frac{1}{2}b_1\right)\theta + \left(\frac{1}{2}b_2 + \frac{1}{2}b_1\right)\theta^2,$$

hence $b_0 - 2b_2 = 0$, $3b_2 - b_1 = 2$ and $b_2 + b_1 = 0$. Therefore $b_2 = 1/2$, $b_1 = -1/2$, $b_0 = 1$ and $\theta = 1 - \alpha/2 + \alpha^2/2$.
(4) We have

$$\alpha = \frac{1}{2}\theta^2 - \frac{1}{2}\theta, \quad \alpha\theta = \theta + 4 \quad \text{and} \quad \alpha\theta^2 = \theta^2 + 4\theta.$$

We conclude that $(\theta^2, \theta, 1)$ is a nonzero solution of the following linear system of equations

$$1/2x - 1/2y - \alpha z = 0,\ (1-\alpha)y + 4z = 0,\ (1-\alpha)x + 4y = 0.$$

It follows that the determinant of this system is 0. We thus obtain

$$\begin{vmatrix} 1/2 & -1/2 & -\alpha \\ 0 & (1-\alpha) & 4 \\ (1-\alpha) & 4 & 0 \end{vmatrix} = 0.$$

Therefore the minimal polynomial of α over $\mathbb{Q}$ is given by $g(x) = x^3 - 2x^2 + 3x - 10$.
Performing the Euclidean division of $g(x)$ by $x - \alpha$, we obtain

$$g(x) = (x - \alpha))(x^2 + (\alpha - 2)x + \alpha^2 - 2\alpha + 3).$$

Therefore the dual basis $\{\omega_0, \omega_1, \omega_2\}$ of $\{1, \alpha, \alpha^2\}$ is given by

$$\omega_0 = (\alpha^2 - 2\alpha + 3)/(3\alpha^2 - 4\alpha + 3),$$
$$\omega_1 = (\alpha - 2)/(3\alpha^2 - 4\alpha + 3) \text{ and}$$
$$\omega_2 = 1/(3\alpha^2 - 4\alpha + 3),$$

see [Lang (1965), Prop. 1, Chap. 8.6]. It follows that $b_i = \mathrm{Tr}_{K/\mathbb{Q}}(\theta\omega_i)$ for $i = 0, 1, 2$. We have

$$1/(3\alpha^2 - 4\alpha + 3) = (5\alpha^2 + 32\alpha - 18)/(2.503),$$

hence

$$\begin{aligned} b_2 &= \mathrm{Tr}_{K/\mathbb{Q}}(\theta\omega_2) \\ &= \mathrm{Tr}_{K/\mathbb{Q}}\left(\frac{\theta(5\alpha^2 + 32\alpha - 18)}{2.503}\right) \\ &= \mathrm{Tr}_{K/\mathbb{Q}}\left(\frac{\theta(5(\theta^2 - \theta)^2/4 + 32(\theta^2 - \theta)/2 - 18)}{2.503}\right) \\ &= \frac{1}{8.503}\mathrm{Tr}_{K/\mathbb{Q}}(40\theta^2 + 36\theta + 592). \end{aligned}$$

Since $\mathrm{Tr}_{K/\mathbb{Q}}(\theta) = 1$ and $\mathrm{Tr}_{K/\mathbb{Q}}(\theta^2) = 5$, then

$$b_2 = \frac{5}{503}\mathrm{Tr}_{K/\mathbb{Q}}(\theta^2) + \frac{9}{2.503}\mathrm{Tr}_{K/\mathbb{Q}}(\theta) + \frac{74.3}{503} = 1/2.$$

Similarly, we have

$$\begin{aligned} b_0 &= \mathrm{Tr}_{K/\mathbb{Q}}(\theta\omega_0) \\ &= \mathrm{Tr}_{K/\mathbb{Q}}\left(\theta\frac{\alpha^2 - 2\alpha + 3}{3\alpha^2 - 4\alpha + 3}\right) \\ &= \mathrm{Tr}_{K/\mathbb{Q}}\left(\frac{\theta(\alpha^2 - 2\alpha + 3)(5\alpha^2 + 32\alpha - 18)}{2.503}\right) \\ &= \frac{1}{2.503}\mathrm{Tr}_{K/\mathbb{Q}}\left(\theta(-18\alpha^2 + 86\alpha + 266)\right) \\ &= \frac{1}{2.503}\mathrm{Tr}_{K/\mathbb{Q}}\left(\theta(-18(\theta^2 - \theta)^2)/4 + 86(\theta^2 - \theta)/2 + 266\right) \\ &= \frac{1}{2.503}\mathrm{Tr}_{K/\mathbb{Q}}(-36\theta^2 + 370\theta + 272) \\ &= \frac{-36}{2.503}\mathrm{Tr}_{K/\mathbb{Q}}(\theta^2) + \frac{370}{2.503}\mathrm{Tr}_{K/\mathbb{Q}}(\theta) + \frac{3.272}{2.503} \\ &= \frac{-5.36}{2.503} + \frac{370}{2.503} + \frac{816}{2.503} = 1. \end{aligned}$$

By the same method, we compute b_1. We have

$$\begin{aligned}
b_1 &= \mathrm{Tr}_{K/\mathbb{Q}}(\theta\omega_1)\\
&= \mathrm{Tr}_{K/\mathbb{Q}}\left(\theta\frac{\alpha-2}{3\alpha^2-4\alpha+3}\right)\\
&= \mathrm{Tr}_{K/\mathbb{Q}}\left(\frac{\theta(\alpha-2)(5\alpha^2+32\alpha-18)}{2.503}\right)\\
&= \frac{1}{2.503}\,\mathrm{Tr}_{K/\mathbb{Q}}\left(\theta(32\alpha^2-97\alpha+86)\right)\\
&= \frac{1}{2.503}\,\mathrm{Tr}_{K/\mathbb{Q}}\left(\theta(32(\theta^2-\theta)^2)/4-97(\theta^2-\theta)/2+86\right)\\
&= \frac{1}{2.503}\,\mathrm{Tr}_{K/\mathbb{Q}}\left(\frac{128\theta^2-86\theta-520}{2}\right)\\
&= \frac{128}{4.503}\,\mathrm{Tr}_{K/\mathbb{Q}}(\theta^2)-\frac{86}{4.503}\,\mathrm{Tr}_{K/\mathbb{Q}}(\theta)-\frac{3.520}{4.503}\\
&= \frac{5.64}{2.503}-\frac{43}{2.503}-\frac{3.260}{2.503}=-1/2.
\end{aligned}$$

Exercise 2.8.
Let K be a field, α be algebraic of degree n over K. Let r and s be non negative integers such that $r+s=n-1$. Show that any $\beta\in K(\alpha)$ may be written in the form $\beta=u(\alpha)/v(\alpha)$, where $u(x)$ and $v(x)$ are polynomials with coefficients in K satisfying the conditions $\deg u\le r$ and $\deg v\le s$. Moreover if we suppose that the polynomials u and v are coprime, then this representation is unique (up to a multiplication of u and v by a same nonzero constant).

Solution 2.8.
If $\beta=0$, the result is clear. Suppose next that $\beta\neq 0$. Let A and B be the linear spaces generated by $\{1,\alpha,\dots,\alpha^r\}$ and $\{\beta,\beta\alpha,\dots,\beta\alpha^s\}$ respectively. The dimensions of the spaces A and B over K are equal to $r+1$ and $s+1$ respectively. Since the family $\{1,\alpha,\dots,\alpha^r,\beta,\beta\alpha,\dots,\beta\alpha^s\}$ contains $n+1$ elements, then this family is not free. It follows that $A\cap B\neq\{0\}$. Let γ be a non zero element of this intersection, then

$$\gamma=a_0+\cdots+a_r\alpha^r=b_0\beta+\cdots+\beta\alpha^s$$

and the result follows.

Suppose that some β has two representations

$$\beta=u_1(\alpha)/v_1(\alpha)=u_2(\alpha)/v_2(\alpha),$$

where $gcd(u_1(x), v_1(x)) = gcd(u_2(x), v_2(x)) = 1$ and with the conditions on the degrees. Then $u_1(\alpha)v_2(\alpha) - u_2(\alpha)v_1(\alpha) = 0$, hence

$$u_1(x)v_2(x) - u_2(x)v_1(x) = 0.$$

We deduce that $u_1(x) \mid v_1(x)$ and $v_1(x) \mid u_1(x)$. It follows that $v_1(x) = cu_1(x)$ and $v_2(x) = cu_2(x)$ with $c \in K^\star$.

Exercise 2.9.

(1) Let K be a field of characteristic $p \geq 0$, $u(x)$ be a non constant polynomial with coefficients in K such that $u(0) = 0$, $f(x) \in K[[x]]$ satisfying the identity $f(u(x)) = f(x) - v(x)$ for some non zero polynomial $v(x)$ with coefficients in K with $\deg v < \deg u$. Show that $f(x)$ is transcendental over $K(x)$ or $p > 0$, $f(x)$ is algebraic over $K(x)$ and $p \mid [K(x, f(x)) : K(x)]$.

(2) Suppose that K is of characteristic 0 and let $d \geq 2$ be an integer. Show that $f(x) = \sum_{k \geq 0} x^{d^k}$ is transcendental over $K(x)$.

Solution 2.9.

(1) Suppose that $f(x)$ is algebraic over $K(x)$ and let

$$\phi(x, y) = y^n + a_{n-1}(x)y^{n-1} + \cdots + a_0(x) \in K(x)[y]$$

be its minimal polynomial over $K(x)$. From the following identity

$$(f(x))^n + a_{n-1}(x)(f(x))^{n-1} + \cdots + a_0(x) = 0$$

and the property of $f(x)$, we obtain

$$(f(x) - v(x))^n + a_{n-1}(u(x))(f(x) - v(x))^{n-1} + \cdots + a_0(u(x)) = 0.$$

We deduce that

$$a_{n-1}(x) = a_{n-1}(u(x)) - nv(x). \qquad \text{(Eq 1)}$$

Set $a_{n-1}(x) = a(x)/b(x)$, where $a(x)$ and $b(x) \in K[x]$. We may suppose that $b(x)$ is monic and $gcd(a(x), b(x)) = 1$. We have

$$a(x)b(u(x)) = a(u(x))b(x) - nv(x)b(x)b(u(x)),$$

hence $b(u(x)) \mid b(x)$. It follows that $b(x) = 1$ and $a(x) = a(u(x)) - nv(x)$. The assumption on the degrees of u and v implies that $a(x) \in K$ and then by **(Eq 1)**, we conclude that $n = 0$ if $p = 0$ and $p \mid n$ if $p > 0$.

(2) Here, we have $f(x^d) = f(x) - x$, so that **(1)** applies with $u(x) = x^d$ and $v(x) = x$.

Exercise 2.10.
Find the gap in the following reasoning. Let K be a field. Let α be a root of the polynomial $f(x,y) = y^3 - x^2$ in an algebraic closure of $K(x)$. $f(x,y)$ is irreducible over $K(x)$, hence $[K(x,\alpha) : K(x)] = 3$. The element x is algebraic over $K(x^2)$ since it is a root of the polynomial $g(x,y) = y^2 - x^2$ irreducible over $K(x^2)$, hence $[K(x) : K(x^2)] = 2$. Considering the chain of fields $K(x^2) \subset K(x) \subset K(x,\alpha)$, we can conclude that the degree of the minimal polynomial of α over $K(x^2)$ is equal to 6. But α is a root of $f(x,y)$, irreducible over $K(x^2)$. Therefore we have reached a contradiction.

Solution 2.10.
The gap in the proof lies in the fact that α is not a primitive element of $K(x,\alpha)$ over $K(x^2)$, that is $K(x,\alpha) \neq K(x^2,\alpha)$. To better understand the situation look at the following diagram of fields.

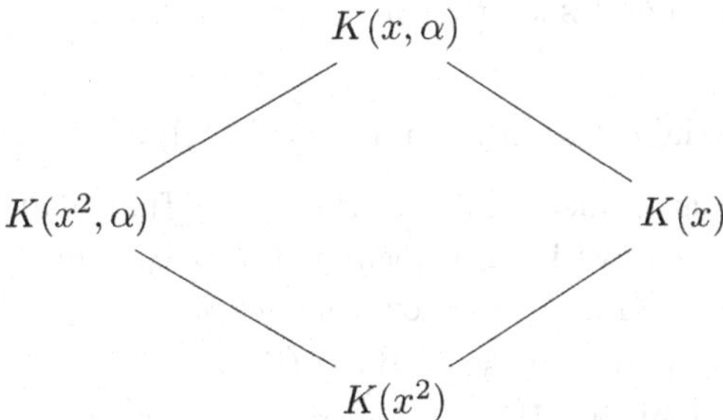

Exercise 2.11.
Let $f(x)$ be a monic irreducible polynomial of degree n with coefficients in $\mathbb{Z}$ and let p be a prime number. Let $\alpha_1, \dots, \alpha_n$ be the roots of f and $g(x) = (x - \alpha_1^p) \cdots (x - \alpha_n^p)$. Show that $p^n \mid \operatorname{Res}_x(g(x), f(x))$.

Solution 2.11.
Let $h(x) = f(x^p) - (f(x))^p$, then $h(x) = pu(x))$ with $u(x) \in \mathbb{Z}[x]$. We have

$$\begin{aligned}\operatorname{Res}_x(g(x), f(x)) &= \prod_{i=1}^{n} f(\alpha_i^p) = \prod_{i=1}^{n} (f(\alpha_i^p) - f(\alpha_i)^p) = \prod_{i=1}^{n} h(\alpha_i) \\ &= p^n \prod_{i=1}^{n} u(\alpha_i) = p^n \operatorname{Res}_x(f(x), u(x)),\end{aligned}$$

hence the result.

Exercise 2.12.
Let m be a positive integer. Show that the polynomial with integral coefficients $f_m(x) = (x^2+2)^m - 2^m x$ is irreducible over $\mathbb{Q}$.

Solution 2.12.
Let α be a root of $f_m(x)$ and let $\beta = (\alpha^2+2)/2$. Since $\alpha = \beta^m$, then $\mathbb{Q}(\alpha) = \mathbb{Q}(\beta)$. Easy computations show that $\beta^{2m} - 2\beta + 2 = 0$, hence β is a root of the polynomial $g(x) = x^{2m} - 2x + 2$ which is irreducible over $\mathbb{Q}$ by Eisenstein's criterion. We deduce that $[\mathbb{Q}(\beta) : \mathbb{Q}] = 2m$ and then $[\mathbb{Q}(\alpha) : \mathbb{Q}] = 2m$. We conclude that $f_m(x)$ is irreducible over $\mathbb{Q}$.

Exercise 2.13.
Let $f(x)$ and $g(x)$ be monic polynomials with integral coefficients such that $\gcd(f(x), g(x)) = 1$.

(1) Show that there exist $c \in \mathbb{Z} \setminus \{0\}$, $u(x)$ and $v(x) \in \mathbb{Z}[x]$ with $\deg u < \deg g$, $\deg v < \deg f$ such that $u(x)f(x) + v(x)g(x) = c$.
(2) Let
$$I = \{c \in \mathbb{Z}, \text{ there exist } u(x) \text{ and } v(x) \in \mathbb{Z}[x];\ u(x)f(x)+v(x)g(x) = c\}.$$
Show that I is an ideal of $\mathbb{Z}$ and that $\mathrm{Res}_x(f(x), g(x)) \in I$.
(3) Let c_0 be a positive integer such that $I = c_0\mathbb{Z}$ and let p be a prime factor of $\mathrm{Res}_x(f(x), g(x))$. Show that $p \mid c_0$.
(4) Show that we may have $c_0 < |\mathrm{Res}_x(f(x), g(x))|$.
(5) Recall that if we set $f(x) = \sum_{i=0}^{n} a_i x^{n-i}$ and $g(x) = \sum_{j=0}^{m} b_j x^{m-j}$, then $\mathrm{Res}_x(fx), g(x)$ is the following $(m+n) \times (m+n)$ determinant, where the first m-th rows are devoted to the coefficients of $f(x)$ and the remaining one's to the coefficients of $g(x)$,
$$R = \begin{vmatrix} a_0 & a_1 & \dots & a_n & 0 & \dots & 0 \\ 0 & a_0 & \dots & a_{n-1} & a_n & \dots & 0 \\ \dots & \dots & \dots & \dots & \dots & \dots & \dots \\ 0 & 0 & \dots & 0 & a_0 & \dots & a_n \\ b_0 & b_1 & \dots & b_m & 0 & \dots & 0 \\ 0 & b_0 & \dots & b_{m-1} & b_m & \dots & 0 \\ \dots & \dots & \dots & \dots & \dots & \dots & \dots \\ 0 & 0 & \dots & 0 & b_0 & \dots & b_m \end{vmatrix}$$
and that
$$\begin{aligned} \mathrm{Res}_x(f(x), g(x)) = {} & (A_0 + A_1x + \dots + A_{m-1}x^{m-1})f(x) \\ & + (B_0 + B_1x + \dots + B_{n-1}x^{n-1})g(x), \end{aligned}$$

where A_i (resp. B_j) is the cofactor of the coefficient appearing in the first column at the $i+1$ (resp. $m+j+1$)-th row. Show that $c_0 = |\operatorname{Res}_x(f(x), g(x))|/d$, where $d = \gcd(A_0, \dots, A_{m-1}, B_0, \dots, B_{n-1})$.

Solution 2.13.

(1) Since $f(x)$ and $g(x)$ are relatively prime in $\mathbb{Q}[x]$, then by Bezout's Identity, there exist $u_1(x)$ and $v_1(x) \in \mathbb{Q}[x]$ with $\deg u_1 < \deg g$, $\deg v_1 < \deg f$ such that $u_1(x)f(x) + v_1(x)g(x) = 1$. If we multiply the two sides of this identity by the common denominator, say c, of all the coefficients of $u_1(x)$ and $v_1(x)$, then we obtain the required identity.

(2) Easy.

(3) For any polynomial $h(x)$, with integral coefficients, denote by $\overline{h(x)}$ the reduction modulo p of $h(x)$. Since $p \mid \operatorname{Res}_x(f(x), g(x))$, then $\overline{f(x)}$ and $\overline{g(x)}$ have a common root in an algebraic closure of $\mathbb{F}_p$. Let $u_0(x)$ and $v_0(x) \in \mathbb{Z}[x]$ such that $u_0(x)f(x) + v_0(x)g(x) = c_0$. Reducing modulo p the two sides of this identity shows that $\overline{c_0} = 0$, hence $p \mid c_0$.

(4) Let $f(x) = x^2 + 1$ and $g(x) = x^2 + 3$, then

$$\operatorname{Res}_x(f(x), g(x)) = \left(\sqrt{-1}^2 + 3\right)^2 = 4.$$

On the other hand, we have $2 = (x^2+3) - (x^2+1)$ so that $c_0 \mid 2$. Since the polynomials $\overline{f(x)}$ and $\overline{g(x)}$ have a common root, then $c_0 \neq 1$, hence $c_0 = 2$.

(5) It is clear that $\operatorname{Res}_x(f(x), g(x))/d \in I$, thus $c_0 \mid \operatorname{Res}_x(f(x), g(x))/d$. Let $\lambda \in \mathbb{Z}$ such that $\operatorname{Res}_x(f(x), g(x))/d = \lambda c_0$. Write c_0 in the form

$$c_0 = (A'_0 + A'_1 x + \cdots + A'_{m-1}x^{m-1})f(x) + (B'_0 + B'_1 x + \cdots + B'_{n-1}x^{n-1})g(x),$$

where the coefficients A'_i, B'_j are integers. Multiply this identity by $d\lambda$ and subtract it from the following identity

$$\begin{aligned}\operatorname{Res}_x(f(x), g(x)) = {} & (A_0 + A_1 x + \cdots + A_{m-1}x^{m-1})f(x) \\ & + (B_0 + B_1 x + \cdots + B_{n-1}x^{n-1})g(x).\end{aligned}$$

We obtain

$$0 = \left(\sum_{i=0}^{m-1}(A_i - d\lambda A'_i)x^i\right) f(x) + \left(\sum_{j=0}^{n-1}(B_j - d\lambda B'_j)x^j\right) g(x).$$

Since $f(x)$ and $g(x)$ are coprime, then

$$f(x) \mid \sum_{j=0}^{n-1}(B_j - d\lambda B'_j)x^j \text{ and } g(x) \mid \sum_{i=0}^{m-1}(A_i - d\lambda A'_i)x^i.$$

It follows that $A_i = d\lambda A'_i$ for $i = 1, \ldots, m-1$ and $B_j = d\lambda B'_j$ for $j = 1, \ldots, n-1$. According to the definition of d, we conclude that $\lambda = \pm 1$, so that $|\operatorname{Res}_x(f(x), g(x))|/d = c_0$.

Exercise 2.14.
Let K be a field of characteristic 0, $f(x) = x^n + \cdots + a_1x + a_0$ and $g(x) = x^m + \cdots + b_1x + b_0 \in K[x]$. Let y a new variable and $R(y) = \operatorname{Res}_x(x^n + \cdots + a_1x + y, g(x))$. Let e be the exact number of roots β of $g(x)$, each of them counted with its multiplicity, such that $f(\beta) = 0$. Show that e is the multiplicity of a_0 as a root of $R(y)$.

Solution 2.14.
Let $\beta_1, \ldots, \beta_m$ be the roots (distinct or not) of $g(x)$ in an algebraic closure of K and let

$$F(x, y) = f(x) - a_0 + y = x^n + \cdots + a_1x + y.$$

Let $P(x, y) = \prod_{i=1}^m (x - F(\beta_i, y))$, then $P(x, y)$ has the form:

$$P(x, y) = x^m - A_{m-1}(y)x^{m-1} + \cdots + (-1)^{m-1}A_1(y)x + (-1)^m A_0(y),$$

where $A_i(y) = \sum_{|J|=m-i} \prod_{j\in J} F(\beta_j, y)$, for $i = 0, \ldots, m-1$. We have

$$A_0(y) = \prod_{j=1}^m F(\beta_j, y) = Res_x(g(x), F(x, y)) = R(y).$$

Since $\frac{d}{dy}F(\beta_j, y) = 1$, then

$$R'(y) = \sum_{i=1}^m \prod_{j\neq i} F(\beta_j, y) = A_1(y).$$

By induction we may prove that $R^{(k)}(y) = A_k(y)$ for $k = 0, \ldots, m-1$. This preparation being done, let e' be the multiplicity of a_0 as a root of $R(y)$. We have $R(a_0) = R'(a_0) = \cdots = R^{(e'-1)}(a_0) = 0$ and $R^{(e')}(a_0) \neq 0$, hence $A_0(a_0) = A_1(a_0) = \cdots = A_{e'-1}(a_0) = 0$ and $A_{e'}(a_0) \neq 0$. It follows that 0 is a root of $P(x, a_0)$ of order e'. But the roots of $P(x, a_0)$ are $F(\beta_1, a_0), \ldots, F(\beta_m, a_0)$. Since $F(\beta_i, a_0) = f(\beta_i)$ for $i = 1, \ldots, m$, then there are exactly e' elements among $\beta_1, \ldots, \beta_m$ such that $f(\beta_i) = 0$, hence $e = e'$.

Exercise 2.15.
Let K be a field, x_1, x_2, x_3 be the roots of $P(X) = X^3 - s_1X^2 + s_2X - s_3 \in K[X]$ in an algebraic closure of K. Let $D(s_1, s_2, s_3)$ be the discriminant of $P(X)$. Recall that

$$D(s_1, s_2, s_3) = -4s_1^3s_3 + s_1^2s_2^2 + 18s_1s_2s_3 - 27s_3^2 - 4s_2^3.$$

Let

$$A = x_2x_1^2 + x_3x_2^2 + x_1x_3^2 \text{ and } B = x_1x_2^2 + x_2x_3^2 + x_3x_1^2.$$

(1) Show that $A + B = s_1s_2 - 3s_3$ and $AB = 9s_3^2 + s_2^3 - 6s_1s_2s_3 + s_1^3s_3$.
(2) Show that $A = B$ if and only if $D(s_1, s_2, s_3) = 0$.

Solution 2.15.

(1) We first compute $\sigma_j = \sum_{i=1}^{3} x_i^j$ for $j = 2$ and $j = 3$. We have $\sigma_2 = s_1^2 - 2s_2$,

$$\begin{aligned}\sigma_3 &= \sum_{i=1}^{3}(s_1x_i^2 - s_2x_i + s_3)\\ &= s_1\sigma_2 - s_1s_2 + 3s_3\\ &= s_1^3 - 3s_1s_2 + 3s_3.\end{aligned}$$

We have

$$\begin{aligned}A + B &= s_1\sigma_2 - \sigma_3 = s_1s_2 - 3s_3 \text{ and}\\ AB &= \sum_{i,j,k \text{ distinct}} x_i^4x_jx_k + \sum_{i<j} x_i^3x_j^3 + 3s_3.\end{aligned}$$

It is clear that

$$\sum_{i,j,k \text{ distinct}} x_i^4x_jx_k = s_3\sigma_3 = s_1^3s_3 - 3s_1s_2s_3 + 3s_3^2.$$

We have

$$\sum_{i<j} x_i^3x_j^3 = \sum_{i<j}(s_1x_i^2 - s_2x_i + s_3)(s_1x_j^2 - s_2x_j + s_3).$$

(2) We have $A = B$ if and only if $(A + B)^2 - 4AB = 0$, if and only if $-4s_1^3s_3 + s_1^2s_2^2 + 18s_1s_2s_3 - 27s_3^2 - 4s_2^3 = 0$, if and only if $D(s_1, s_2, s_3) = 0$.

Exercise 2.16.
Let K be a field, α and β be algebraic elements over K of degree m and n respectively. Suppose that the elements $\alpha^i\beta^j$ for $i = 0\ldots, m-1$ and $j = 0, \ldots, n-1$ are ordered (for some given order). Denote these elements by $\omega_1, \ldots, \omega_{mn}$. Let $\mu = \alpha + \beta$. Show that for any $h \in \{1, \ldots, mn\}$, $\mu\omega_h$ may be written in the form $\mu\omega_h = \sum_{i=1}^{mn} a_h^i\omega_i$, where $a_h^i \in K$. Deduce that it is possible to use a determinant in order to compute a monic polynomial $f(x) \in K[x]$ of degree mn such that $f(\mu) = 0$. Compute this polynomial when $K = \mathbb{Q}$, $\alpha = \sqrt{2}$ and $\beta = \sqrt{3}$.

Solution 2.16.

We have

$$\mu\omega_h = (\alpha+\beta)\omega_h = (\alpha+\beta)\alpha^{i(h)}\beta^{j(h)} = \alpha^{i(h)+1}\beta^{j(h)} + \alpha^{i(h)}\beta^{j(h)+1},$$

where $i(h) \in \{0,\dots,m-1\}$ and $j(h) \in \{0,\dots,n-1\}$. If $i(h) < m-1$, then $\alpha^{i(h)+1}\beta^{j(h)}$ is equal to some $\omega_{h'}$. If $i(h) = m-1$, then

$$\alpha^{i(h)+1}\beta^{j(h)} = \left(-\sum_{i=0}^{m-1} a_i\alpha^i\right)\beta^{j(h)}$$

which is a linear combination of the ω_j. The same argument applies for the second term of $\mu\omega_h$. Consider the set of all the expressions of $\mu\omega_h$ for $h = 1,\dots,mn$ as linear combinations of the ω_i,

$$\mu\omega_h = a_1^h\omega_1 + \cdots + a_{mn}^h\omega_{mn},\ h = 1,\dots,mn,$$

where $a_j^h \in K$ for $j = 1,\dots,mn$

We may write these equations in the form:

$$a_1^h\omega_1 + \cdots + (a_h^h - \mu)\omega_h + \cdots + a_{mn}^h\omega_{mn} = 0, \qquad h = 1,\dots,mn.$$

We may view this system as an homogeneous linear system of equations with a non trivial solution $(\omega_1,\dots,\omega_n)$. Hence its determinant vanishes. We thus obtain

$$\begin{vmatrix} (a_1^1-\mu) & a_2^1 & \dots & a_{mn}^1 \\ a_1^2 & (a_2^2-\mu) & \dots & a_{mn}^1 \\ \dots & \dots & \dots & \dots \\ a_1^{mn} & a_2^{mn} & \dots & (a_{mn}^{mn}-\mu) \end{vmatrix} = 0.$$

This shows that μ is a root of a monic polynomial of degree mn with coefficients in K. When $K = \mathbb{Q}$, $\alpha = \sqrt{2}$, $\beta = \sqrt{3}$ and $\mu = \alpha+\beta$, we have $\omega_1 = 1$, $\omega_2 = \sqrt{3}$, $\omega_3 = \sqrt{2}$ and $\omega_4 = \sqrt{6}$. We obtain the following relations:

$$\mu\omega_1 = \omega_2 + \omega_3, \quad \mu\omega_2 = 3\omega_1 + \omega_4,$$
$$\mu\omega_3 = 2\omega_1 + \omega_4 \quad \text{and} \quad \mu\omega_4 = 2\omega_2 + 3\omega_3.$$

It follows that $1, \omega_1, \omega_2, \omega_3, \omega_4$ is a non trivial solution of the following homogeneous, linear system of equations:

$$-\mu\omega_1 + \omega_2 + \omega_3 = 0$$
$$3\omega_1 - \mu\omega_2 + \omega_4 = 0$$
$$2\omega_1 - \mu\omega_3 + \omega_4 = 0$$
$$2\omega_2 + 3\omega_3 - \mu\omega_4 = 0.$$

Equating to 0, the determinant of this system, we get the equation $\mu^4 - 10\mu^2 + 1 = 0$.

Exercise 2.17.
Let K be a field of characteristic 0, $f(x)$ be a non constant polynomial with coefficients in K. Suppose that $f(0) = 0$ and that $f'(x) \mid (f(x))^2$. Show that there exist $c \in K$ and a positive integer d such that $f(x) = cx^d$.

Solution 2.17.
Let $f(x) = c\prod_{i=1}^{k}(x-\alpha_i)^{d_i}$ be the factorization of $f(x)$ into irreducible factors over an algebraic closure of K. We have

$$f'(x) = cg(x)\prod_{i=1}^{k}(x-\alpha_i)^{d_i-1},$$

where $g(x) = \sum_{i=1}^{k} d_i \prod_{j\neq i}(x-\alpha_j)$. Clearly, for any $i = 1,\ldots,k$, $g(\alpha_i) \neq 0$. Since $g(x) \mid (f(x))^2$, then $g(x)$ is a constant. Since $\deg g(x) = k-1$, then $k = 1$. The hypothesis $f(0) = 0$ implies the result.

Exercise 2.18.
Let (a_i, b_i) for $i = 1,\ldots,k$ be distinct couples of non negative integers such that $0 \leq a_i < b_i$. We say that this set of couples is a disjoint covering system if for any non negative integer n, there exists a unique $i \in \{1,\ldots,k\}$ such that $n \equiv a_i \pmod{b_i}$.

(1) Let $b = \operatorname{lcm}_{i=1}^{k}(b_i)$. Show that (a_i, b_i), $i = 1,\ldots,k$ is a disjoint covering system if and only if for any integer n, $0 \leq n \leq b-1$, there exists a unique i such that $n \equiv a_i \pmod{b_i}$.

(2) Show that (a_i, b_i), $i = 1,\ldots,k$ is a disjoint covering system if and only if

$$\sum_{i=1}^{k}\frac{x^{a_i}}{1-x^{b_i}} = \frac{1}{1-x}.$$

(3) Let (a_i, b_i) for $i = 1,\ldots,k$ be distinct couples of non negative integers such that $0 \leq a_i < b_i$. Let $m \geq 2$ be an integer. Define the couples (a'_i, b'_j) by

  (i) $b'_i = b_i$ and $a'_i = a_i$ for $i = 1,\ldots,k-1$.
  (ii) $b'_{k+j} = mb_k$ and $a'_{k+j} = a_k + jb_k$ for $j = 0,\ldots,m-1$.

  Show that (a_i, b_i), $i = 1,\ldots,k$ is a disjoint covering system if and only if (a'_i, b'_i), $i = 1,\ldots,k+m-1$ is a disjoint covering system.

(4) If (a_i, b_i), $i = 1,\ldots,k$ is a disjoint covering system, show that there exists one and only one i such that $a_i = 0$.

(5) If (a_i, b_i), $i = 1,\ldots,k$ is a disjoint covering system, show that for any i, j such that $i \neq j$, we have $\gcd(b_i, b_j) \neq 1$.

(6) If (a_i, b_i), $i = 1, \ldots, k$ is a disjoint covering system, show that there exists one and only one i such that $a_i = b_i - 1$.

(7) Show that (a_i, b_i), $i = 1, \ldots, k$ is a disjoint covering system if and only if for any positive divisor d of some b_i and for any primitive d-th root of unity ω, we have

$$\sum_{j;d|b_j} \frac{\omega^{a_j}}{b_j} = \begin{cases} 1 & \text{if } d = 1 \\ 0 & \text{if } d \neq 1 \end{cases}.$$

(8) For any divisor $d \geq 2$ of some b_i, we denote by $\phi_d(x)$, the d-th cyclotomic polynomial and we set $\psi_d(x) = \sum_{j;d|b_j} \frac{x^{a_j}}{b_j}$. Show that (a_i, b_i), $i = 1, \ldots, k$ is a disjoint covering system if and only if $\sum_{i=1}^{k} b_i^{-1} = 1$ and $\phi_d(x) \mid \psi_d(x)$ for any divisor $d > 1$ of some b_i. By using this characterization of disjoint covering system, show that the couples $(0, 4)$, $(2, 4)$, $(1, 6)$, $(3, 6)$, $(5, 12)$, $(11, 12)$ form such a system.

(9) If (a_i, b_i), $i = 1, \ldots, k$ is a disjoint covering system, show that $\sum_{i=1}^{k}(a_i/b_i) = (k-1)/2$.

Solution 2.18.

(1) The necessity of the condition is trivial. We prove the sufficiency. Suppose that the integers in the range $[0, b-1]$ are covered and let n be a non negative integer. Let q and r be non negative integers such that $n = bq + r$ with $r \in \{0, \ldots, b-1\}$. By assumption, there exists a unique i such that $r \equiv a_i \pmod{b_i}$. We deduce that $n \equiv r \pmod{b_i} \equiv a_i \pmod{b_i}$. The uniqueness of i may be proved in a similar way.

(2) We have the following equivalences.

$$\begin{aligned}(a_i, b_i) \quad \text{is a disjoint covering system} &\iff \sum_{n \geq 0} x^n = \sum_{i=1}^{k} \sum_{t \geq 0} x^{a_i + tb_i} \\ &\iff \frac{1}{1-x} = \sum_{i=1}^{k} x^{a_i} \sum_{t \geq 0} \left(x^{b_i}\right)^t \\ &\iff \frac{1}{1-x} = \sum_{i=1}^{k} \frac{x^{a_i}}{1 - x^{b_i}}.\end{aligned}$$

(3) It is sufficient to prove that for any $x \in \mathbb{N}$, $x \equiv a_k \pmod{b_k}$ if and only if there exists $j \in \{0, \ldots, m-1\}$ such that $x \equiv a_{k+j} \pmod{mb_k}$. This equivalence is true since $a_k + b_k\mathbb{Z} = \cup_{j=0}^{m-1}(a_{k+j} + mb_k\mathbb{Z})$.
An alternative way to prove this result is to use the identity proved in **(2)**. Actually, it is easy to prove that $\frac{x^{a_k}}{1-x^{b_k}} = \sum_{j=0}^{m-1} \frac{x^{a_{k+j}}}{1-x^{mb_k}}$.

(4) Suppose that (a_i, b_i), $i = 1, \dots, k$, is a disjoint covering system, then setting $x = 0$ in the identity proved in **(2)**, we get $\sum_{a_i=0} 1 = 1$, hence the conclusion of the proof.

(5) By contradiction, suppose that there exist i and j with $i \neq j$ such that $gcd(b_i, b_j) = 1$. By the Chinese remainder theorem there exists a non negative n such that $n \equiv a_i \pmod{b_i}$ and $n \equiv a_j \pmod{b_j}$, which contradicts the uniqueness in the definition of a disjoint covering system.

(6) Replacing x by x^{-1}, in the identity proved in (2), we obtain $\sum_{i=1}^{k} \frac{x^{b_i - a_i - 1}}{x^{b_i} - 1} = \frac{1}{x-1}$. Putting $x = 0$ leads to $\sum_{i; b_i - a_i - 1 = 0}(-1) = -1$, hence the result.

(7) Decomposing into partial fractions, we have $\sum_{i=1}^{k} \frac{x^{a_i}}{1 - x^{b_i}} = \sum_{\omega \in \Omega} \frac{A(\omega)}{x - \omega}$, where

$$\Omega = \{\omega \in \mathbb{C}, \quad \text{there exists} \quad i \quad \text{such that} \quad \omega^{b_i} = 1\}$$

and

$$\begin{aligned} A(\omega) &= \lim_{x \to \omega} \sum_{i=1}^{k} \frac{(x - \omega)x^{a_i}}{1 - x^{b_i}} \\ &= -\lim_{x \to \omega} \sum_{i, d^{b_i} = 1} \frac{x^{a_i}}{\prod_{\substack{\alpha^{b_i} = 1 \\ \alpha \neq \omega}} (x - \alpha)} \\ &= -\sum_{i, d^{b_i} = 1} \frac{\omega^{a_i}}{\prod_{\substack{\alpha^{b_i} = 1 \\ \alpha \neq \omega}} (\omega - \alpha)}. \end{aligned}$$

Let i and d such that $d \mid b_i$ and let ω be a root of

$$g_i(x) = x^{b_i} - 1 = \prod_{\alpha, \alpha^{b_i} = 1} (x - \alpha).$$

We have

$$g_i'(\omega) = \prod_{\alpha, \alpha^{b_i} = 1, \alpha \neq \omega} (\omega - \alpha) = b_i \omega^{b_i - 1} = b_i \omega^{-1}.$$

Therefore $A(\omega) = -\sum_{i, d^{b_i} = 1} \frac{\omega^{a_i + 1}}{b_i}$. These computations being made, we conclude, using (1), that (a_i, b_i), $i = 1, \dots, k$ is a disjoint covering system if and only if $A(1) = -1$ and $A(\omega) = 0$ if $\omega \neq 1$, thus by the preceding calculations, if and only if for any positive divisor d of some b_j, we have

$$\sum_{j; d \mid b_j} \frac{\omega^{a_j}}{b_j} = \begin{cases} 1 & \text{if } d = 1 \\ 0 & \text{if } d \neq 1 \end{cases}.$$

(8) The proof is a simple consequence of **(7)** and the observation that $\phi_d(x)$ is the minimal polynomial of ω over $\mathbb{Q}$. We verify that the given set of couples is a disjoint covering system. We have

$$\sum b_i^{-1} = \frac{1}{4} + \frac{1}{4} + \frac{1}{6} + \frac{1}{6} + \frac{1}{12} + \frac{1}{12} = 1.$$

The set of all the divisors greater than 1 of some b_i is $\{2, 3, 4, 6, 12\}$. The corresponding cyclotomic polynomials are given by:

$$\phi_2(x) = x + 1,\ \phi_3(x) = x^2 + x + 1,\ \phi_4(x) = x^2 + 1,$$
$$\phi_6(x) = x^2 - x + 1,\ \phi_{12}(x) = x^4 - x^2 + 1.$$

We compute the polynomials $\psi_d(x)$. We have

$$\begin{aligned}
\psi_2(x) &= x^{11}/12 + x^5/12 + x^3/6 + x^2/4 + x/6 + 1/4,\\
\psi_3(x) &= x^{11}/12 + x^5/12 + x^3/6 + x/6,\\
\psi_4(x) &= x^{11}/12 + x^5/12 + x^2/4 + 1/4,\\
\psi_6(x) &= x^{11}/12 + x^5/12 + x^3/6 + x/6,\\
\psi_{12}(x) &= x^{11}/12 + x^5/12.
\end{aligned}$$

Let i and j be the roots of $\phi_4(x)$ and $\phi_3(x)$ respectively. Then $-j$ is a primitive 6-th root of unity and ij is a primitive 12-th root of unity. We have

$$\begin{aligned}
\phi_2 \mid \psi_2 &\Rightarrow \psi_2(-1) = 0\\
\phi_3 \mid \psi_3 &\Rightarrow \psi_3(j) = 0\\
\phi_4 \mid \psi_4 &\Rightarrow \psi_4(i) = 0\\
\phi_6 \mid \psi_6 &\Rightarrow \psi_6(-j) = 0\\
\phi_{12} \mid \psi_{12} &\Rightarrow \psi_{12}(ij) = 0.
\end{aligned}$$

We omit the verifications that $\phi_d(x) \mid \psi_d(x)$.

(9) Let

$$B(x) = \frac{1}{1-x} - \sum_{i=1}^{k} \frac{1}{1-x^{b_i}},$$

then $B(x) = \frac{\theta(x)}{1-x}$ with

$$\theta(x) = 1 - \sum_{i=1}^{k} \frac{1}{1 + x + \cdots + x^{b_i - 1}}.$$

Since, by **(7)**, $\lim_{x\to 1}(\theta(x)) = 0$, then

$$\begin{aligned}\lim_{x\to 1}(B(x)) &= \theta^{'}(1)\\ &= -\sum_{i=1}^{k}\frac{1+2+\cdots+(b_i-1)}{b_i^2}\\ &= -\frac{1}{2}\sum_{i=1}^{k}\left(1-\frac{1}{b_i}\right)\\ &= -\frac{1}{2}(k-1).\end{aligned}$$

On the other hand, using **(2)**, we get $B(x) = \sum \frac{-1+x^{a_i}}{1-x^{b_i}}$. Dividing numerator and denominator of this fraction by $1-x$ and letting x go to 1, we obtain $\lim_{x\to 1}(B(x)) = -\sum_{i=1}^{k}\frac{a_i}{b_i}$. Equating the limits obtained by the two ways, we get $\sum_{i=1}^{k}\frac{a_i}{b_i} = \frac{1}{2}(k-1)$.

Exercise 2.19.

Let $n \geq 2$ be an integer, $\xi \in \mathbb{C}$ be a primitive n-root of unity and $\Phi_n(x) = \prod_{\substack{1\leq i\leq n-1\\ gcd(i,n)=1}}(x-\xi^i)$.

(1) Show that $\prod_{j=0}^{n-1}(x-y^j) \equiv x^n - 1 \pmod{\Phi_n(y)}$.
(2) Deduce that $\prod_{j=1}^{n-1}(x-y^j) \equiv x^{n-1}+\cdots+x+1 \pmod{\Phi_n(y)}$.
(3) Suppose that $n \geq 3$. Show that n is prime if and only if

$$\prod_{j=1}^{n-1}(2^j+1) \equiv 1 \pmod{2^n-1}.$$

Solution 2.19.

(1) Let

$$F(x,y) = \prod_{j=0}^{n-1}(x-y^j) - (x^n-1),$$

then $F(x,\xi) = 0$. Since $\Phi_n(y)$ is the minimal polynomial of ξ over $\mathbb{Q}$, then $F(x,y) \equiv 0 \pmod{\Phi_n(y)}$, thus

$$\prod_{j=0}^{n-1}(x-y^j) \equiv x^n - 1 \pmod{\Phi_n(y)}. \qquad \text{(Eq 1)}$$

(2) Easy.

(3) Notice that the congruence obtained in **(2)** is valid in $\mathbb{Z}[x,y]$, hence we may substitute integral values for x and y and obtain a congruence in $\mathbb{Z}$. Suppose that n is prime and substituting -1 for x and 2 for y, we obtain

$$\prod_{j=1}^{n-1}(-2^j-1) \equiv 1-1+\cdots+(-1)^{n-1} \pmod{\Phi_n(2)}.$$

Since n is an odd prime and $\Phi_n(2) = 1+2+\cdots+2^{n-1}$, then

$$\prod_{j=1}^{n-1}(2^j+1) \equiv 1 \pmod{2^n-1}.$$

To prove the converse we suppose first that n is even, say $n = 2m$. Then clearly $2^m+1 \mid 1$, which is impossible. So we may suppose that n is odd. By contradiction, suppose that n is composite and let p be the smallest prime divisor of n and $d = n/p$. We write any $j \in \{0,\dots,n-1\}$ in the form $j = dq+r$, where $0 \le r < d$ and $0 \le q \le p-1$. Our hypothesis is equivalent to

$$\prod_{j=0}^{n-1}(2^j+1) \equiv 2 \pmod{2^n-1},$$

that is

$$\prod_{q=0}^{p-1}\prod_{r=0}^{d-1}(2^{dq+r}+1) \equiv 2 \pmod{2^n-1}. \tag{Eq 2}$$

Since $\Phi_d(x) \mid x^n-1$, then $\Phi_d(2) \mid 2^n-1$. Therefore

$$\prod_{q=0}^{p-1}\prod_{r=0}^{d-1}(2^{dq+r}+1) \equiv 2 \pmod{\Phi_d(2)}.$$

We also have $\Phi_d(2) \mid 2^d-1$, hence $2^d \equiv 1 \pmod{\Phi_d(2)}$. Notice that this implies that the order, say t of 2 in $(\mathbb{Z}/\phi_d(2))^*$ divides d. Since d is odd, then using **(1)**, we deduce that for any $q \in \{0,\dots,p-1\}$,

$$\prod_{r=0}^{d-1}(2^{dq+r}+1) \equiv \prod_{r=0}^{d-1}(2^r+1) \pmod{\Phi_d(2)}.$$

From **(Eq 2)**, we get

$$\prod_{r=0}^{d-1}(2^r+1) \equiv 2 \pmod{\Phi_d(2)}.$$

It follows that $2^p \equiv 2 \pmod{\Phi_d(2)}$, hence $2^{p-1} \equiv 1 \pmod{\Phi_d(2)}$. This implies that t divides $p-1$, thus $t < p$. Since $t \mid d$ and p is the smallest prime divisor of d, then $t = 1$. Thus $2^1 = 2 \equiv 1 \pmod{\phi_d(2)}$. This is impossible unless $\phi_d(2) = \pm 1$. Suppose that $\phi_d(2) = \pm 1$, then $\prod(2-\omega) = \pm 1$, where the product runs over the primitive d-th roots of unity. Let

$$Z = \{\omega \in \mathbb{C}, \quad \omega \quad \text{primitive} \quad d-\text{th root of unity}, \quad \operatorname{Im}(\omega > 0)\},$$

then

$$\prod_{\omega \in Z} (2-\omega)(2-\bar{\omega}) = \pm 1,$$

hence $\prod_{\omega \in Z} |2-\omega|^2 = 1$. Set $\omega = a + ib$. Since $|\omega| = 1$, then

$$|2-\omega|^2 = (2-a)^2 + b^2 = 5 - 4a.$$

Since $|a| < 1$, then $5 - 4a > 1$, thus $\prod_{\omega \in Z} |2-\omega|^2 > 1$ which is a contradiction.

Remark. We may exclude the possibility $\phi_d(2) = \pm 1$, by using Bang's Theorem [Roitman (1997), Th. 3] which asserts that for any integer $d > 1$, $d \neq 6$, there is a prime l (called primitive prime divisor) dividing $2^d - 1$ but not $2^k - 1$ for any $k < d$. Since in our situation d is odd then this theorem applies and we conclude that the prime l divides $\phi_d(2)$. It follows that $\phi_d(2) \neq \pm 1$.

Exercise 2.20.
Let K be a field, $a(x)$ and $f(x,y)$ be polynomials with coefficients in K. Suppose that $f(x,y) \in K(a(x))[y]$. Show that $f(x,y) \in K[a(x), y]$.

Solution 2.20.
We may write $f(x,y)$ in the form

$$f(x,y) = \left(\sum_{i=0}^{n} c_i(a(x))y^i\right) / d(a(x)),$$

where $d(x)$ and $c_i(x) \in K[x]$ for $i = 0, \ldots, n$ and with $\gcd\big(c_0(x), \ldots, c_n(x), d(x)\big) = 1$. We may suppose also that $d(x)$ is monic.

First proof. We make use of the contents of polynomials. We have $\operatorname{cont}(f) \in K[x]$, hence $d(a(x)) \mid c_i(a(x))$ for $i = 0, \ldots, n$. Since $gcd\big(c_0(x), \ldots, c_n(x), d(x)\big) = 1$, then there exist $u_0(x), \ldots, u_n(x), v(x) \in K[x]$ such that $\sum_{i=0}^{n} u_i(x)c_i(x) + v(x)d(x) = 1$. It follows that

$$\sum_{i=0}^{n} u_i(a(x))c_i(a(x)) + v(a(x))d(a(x)) = 1.$$

Therefore $d(a(x)) = 1$ and then

$$f(x,y) = \sum_{i=0}^{n} c_i(a(x))y^i \in K[a(x), y].$$

Second proof. Suppose that $d(x)$ is not constant. We have

$$d(a(x))f(x,y) = \sum_{i=0}^{n} c_i(a(x))y^i.$$

Let α and $\beta \in \bar{K}$ such that $a(\alpha) = \beta$ and $d(\beta) = 0$, then

$$\sum_{i=0}^{n} c_i(a(\alpha))y^i = \sum_{i=0}^{n} c_i(\beta)y^i = 0.$$

It follows that $c_0(\beta) =, \dots, = c_n(\beta) = d(\beta) = 0$ which is a contradiction to $\gcd\big(c_0(x), \dots, c_n(x), d(x)\big) = 1$.

Exercise 2.21.
Let K be a field of characteristic $p \geq 0$, $P(x)$ be a polynomial with coefficients in K of degree $n \geq 2$. If $p > 0$, suppose that $p \nmid n$. For any $m \geq 1$, define $P_m(x)$ by $P_1(x) = P(x)$ and for $m \geq 2$, $P_m(x) = P_{m-1}(P(x))$. Let $Q(x) \in K[x]$. For any positive integer m such that $\deg Q < n^m$ and any $z \in K$, we set $f_p(Q) = n^{-m} \sum_{P_m(\xi)=z} Q(\xi)$, where the sum runs over all the roots (distinct or not) of the polynomial $p_m(x) - z$ in an algebraic closure of K.

(1) Show that $f_P(Q)$ does not depend on z nor on m (satisfying $n^m > \deg Q$).
(2) Show that the map $f_P : K[x] \to K$ which maps $Q(x)$ onto $f_p(Q)$ is K-linear and surjective.
(3) Show that $K[x] = K \bigoplus \operatorname{Ker} f_P$. Deduce that, given $Q(x) \in K[x]$, there exists one and only one $\lambda \in K$ such that $f_P(Q - \lambda) = 0$.

Solution 2.21.

(1) Set $Q(x) = a_k x^k + a_{k-1}x^{k-1} + \cdots + a_0$ where $a_0, a_1, \dots, a_k \in K$, $a_k \neq 0$ and $k < n^m$. Let $\xi_1, \dots, \xi_{n^m}$ be the roots of $P_m(x) - z$ in an algebraic

closure of K, then

$$
\begin{aligned}
f_P(Q) &= n^{-m} \sum_{j=1}^{n^m} Q(\xi_j) \\
&= n^{-m} \sum_{j=1}^{n^m} \sum_{i=0}^{k} a_i \xi_j^i \\
&= n^{-m} \sum_{i=0}^{k} a_i \left(\sum_{j=1}^{n^m} \xi_j^i \right) \\
&= n^{-m} \sum_{i=0}^{k} a_i S_i,
\end{aligned}
$$

where $S_i = \sum_{j=1}^{n^m} \xi_j^i$. For any $h \geq 1$, let σ_h be the elementary symmetric function of the ξ_j. Using the formulas of Newton [Small (1991), Prop. 2.3, Chap. 2]:

$$s_1 - \sigma_1 = 0,$$

$$S_d - \sigma_1 S_{d-1} + \sigma_2 S_{d-2} - \cdots + (-1)^{d-1} \sigma_{d-1} S_1 + (-1)^d d \sigma_d = 0,$$

for $2 \leq d \leq n$, we see that S_d is a homogeneous polynomial in $\sigma_1, \ldots, \sigma_d$ of degree d if we attribute the weight h for σ_h. Since $k = \deg Q < n^m$, then $f_P(Q)$ is independent of σ_{n^m}. Therefore $f_P(Q)$ is independent of z. We next show that $f_P(Q)$ is independent of m. For any $l \in \mathbb{N}$, we have:

$$
\begin{aligned}
\frac{1}{n^{l+m}} \sum_{P_{l+m}(\xi)=z} Q(\xi) &= \frac{1}{n^l} \frac{1}{n^m} \sum_{P_l \circ P_m(\xi)=z} Q(\xi) \\
&= \frac{1}{n^l} \sum_{P_l(\xi')=z} \frac{1}{n^m} \sum_{P_m(\xi)=\xi'} Q(\xi) = \frac{1}{n^l} \sum_{P_l(\xi')=z} f_P(Q) \\
&= f_P(Q).
\end{aligned}
$$

(2) Let $Q_1(x), Q_2(x) \in K[x]$, m be a positive integer such that $n^m > Max(\deg Q_1, \deg Q_2)$ and let $\xi_1, \ldots,_x i_{n^m}$ be the roots of $P_m(x)$, then

$$
\begin{aligned}
f_P(Q_1 + Q_2) &= n^{-m} \sum_{j=1}^{n^m} (Q_1(\xi_j) + Q_2(\xi_j)) \\
&= n^{-m} \sum_{j=1}^{n^m} \left(Q_1(\xi_j) + n^{-m} \sum_{j=1}^{n^m} Q_2(\xi_j) \right) \\
&= f_p(Q_1) + f_p(Q_2).
\end{aligned}
$$

We also have for any $\lambda \in K$

$$f_P(\lambda Q_1) = n^{-m}\sum_{j=1}^{n^m} \lambda Q_1(\xi_j) = \lambda n^{-m}\sum_{j=1}^{n^m} Q_1(\xi_j) = \lambda f_P(Q).$$

We conclude that f_P is a linear map. Obviously $f_P(\lambda) = \lambda$ for any $\lambda \in K$, hence f_P is surjective.

(3) Let $Q(x) \in K[x]$, then $f_P(Q - f_P(0)) = 0$, hence

$$Q_1(x) := Q(x) - f_P(0) \in \operatorname{Ker} f_P.$$

Therefore

$$Q(x) = f_P(Q) + Q_1(x) \in K + \operatorname{Ker} f_P.$$

Since $f_P(\lambda) = \lambda$ for any $\lambda \in K$, then $K \cap \operatorname{Ker} f_P = \{0\}$. It follows that $K[x]$ is the direct sum of K and $\operatorname{Ker}\phi$. The last statement is therefore trivial.

Exercise 2.22.
Let K be a field, $\bar{K}$ be an algebraic closure of K, $f(x)$ be a non constant polynomial with coefficients in K. Suppose that there exist $g(x) \in K[x]$, $1 \le \deg g < \deg f$ such that $g(x) - cg(y)$ divides $f(x) - f(y)$ in $\bar{K}[x, y]$.

(1) Show that c is a root of unity and there exists $h(x) \in K[x]$ such that $f(x) = h(g(x))$.
(2) Let d be the order of c in $\bar{K}^\star$. Show that $h(x) \in K[x^d]$.

Solution 2.22.

(1) Let $n = \deg f$, $m = \deg g$ and $k = \lfloor \frac{n}{m} \rfloor$. Let $f(x) = \sum_{i=0}^{k} a_i(x)g(x)^i$ be the g-adic expansion of $f(x)$ in $K[x]$ [Lang (1965), Th. 9, Chap. 5.5]. We have

$$\begin{aligned} 0 &\equiv f(x) - f(y) \pmod{g(x) - cg(y)} \\ &\equiv \sum_{i=0}^{k} a_i(x)g(x)^i - \sum_{i=0}^{k} a_i(y)g(y)^i \pmod{g(x) - cg(y)} \\ &\equiv \sum_{i=0}^{k} g(y)^i(c^i a_i(x) - a_i(y)) \pmod{g(x) - cg(y)}, \end{aligned}$$

hence

$$\sum_{i=0}^{k}(cg(y))^i a_i(x) - \sum_{i=0}^{k} g(y)^i a_i(y) = (g(x) - cg(y))F(x, y),$$

where $F(x,y) \in K[x,y]$. Since the degree in x of the polynomial appearing in the left side of this identity is at most equal to $m-1$, then $F(x,y)=0$. We deduce that

$$\sum_{i=0}^{k}(cg(y))^i a_i(x) = \sum_{i=0}^{k} g(y)^i a_i(y).$$

Equating the degrees in y of the two sides of this equality, we obtain $mk = mk + \deg a_k$, hence deg $a_k = 0$, thus $a_k \in K$. It follows that $c^k = 1$.

Let l be an integer such that $0 \leq l < k$. We prove by induction that $a_l = 0$ or $a_l \in K^\star$ and $c^l = 1$. This claim is proved for $l = k$. Fix an integer l and suppose that the claim holds for all t such that $l < t \leq k$. Then the above equation reads:

$$(cg(y))^l a_l(x) + \cdots + cg(y)a_1(x) = g(y)^l a_l(y) + \cdots + cg(y)a_1(y).$$

It follows that $a_l(y) = 0$ or $lm = lm + \deg a_l(y)$. In this last case, we have $a_l(y) \in K^*$ and $c^l = 1$. We conclude that $f(x) = h(g(x))$, where $h(x) = \sum_{i=0}^{k} a_i x^i$.

(2) We have seen in the preceding proof that if $a_j \neq 0$, then $c^j = 1$, hence $d \mid j$. It follows that $h(x) \in K[x^d]$.

Exercise 2.23.

Let K be a field not algebraically closed, n be a positive integer. For any subset S of $K[x_1,\ldots,x_n]$ denote by $V(S)$, the set

$$V(S) = \{(a_1,\ldots,a_n) \in K^n, g(a_1,\ldots,a_n) = 0 \quad \text{for any} \quad g \in S\}.$$

(1) Show that there exists $f \in K[x_1,\ldots,x_n]$ such that $V(f) = \{(0,...,0)\}$. Deduce that there exists $f \in K[x_1,\ldots,x_n]$ irreducible such that $V(f) = \{(0,\ldots,0)\}$.
(2) Let m be a positive integer and let $f_1(x_1,\ldots,x_n),\ldots,f_m(x_1,\ldots,x_n) \in K[x_1,\ldots,x_n]$. Show that there exists $f(x_1,\ldots,x_n) \in K[x_1,\ldots,x_n]$ such that $V(f_1,\ldots,f_m) = V(f)$.

Solution 2.23.

(1) We proceed by induction on n. If $n=1$, the result is clear by taking $f(x_1) = x_1$. Suppose the result is true for $n-1$ indeterminates. Let $\phi(t) \in K[t]$, $\deg\phi \geq 1$ such that ϕ has no root in K. Set $\phi(t) = a_k t^k + \cdots + a_1 t + a_0$ with $a_0,\ldots,a_k \in K$ and $a_k \neq 0$. Let

$$\begin{aligned}\psi(x,y) &= y^k \phi\left(\frac{x}{y}\right)\\ &= a_k x^k + a_{k-1}x^{k-1}y + \cdots + a_1 x y^{k-1} + a_0 y^k.\end{aligned}$$

It is clear that $V(\psi) = \{(0,0)\}$ (in K^2). Let $g(x_1,\ldots,x_{n-1}) \in K[x_1,\ldots,x_{n-1}]$ such that $V(g) = \{0\}$ (in K^{n-1}). We show that the polynomial

$$f(x_1,\ldots,x_n) = \psi(g(x_1,\ldots,x_{n-1}),x_n)$$

satisfies the required property. For any $(a_1,\ldots,a_n) \in K^n$, we have

$$\begin{aligned} f(a_1,\ldots,a_n) = 0 &\Leftrightarrow g(a_1,\ldots,a_{n-1}) = 0 \quad \text{and} \quad a_n = 0 \\ &\Leftrightarrow a_1 = \cdots = a_{n-1} = a_n = 0, \end{aligned}$$

hence $V(f) = \{(0,\ldots,0)\}$ (in K^n). We prove the existence of $F \in k[x_1,\ldots,x_n]$ irreducible such that $V(f) = \{(0,\ldots,0)\}$. Let $f \in K[x_1,\ldots,x_n]$ such that $V(f) = \{0\}$ and let

$$f(x_1,\ldots,x_n) = F_1(x_1,\ldots,x_n)\cdots F_k(x_1,\ldots,x_n)$$

be the factorization of f into a product of irreducible factors over K. For any $i = 1,\ldots,k$, $V(F_i) = \emptyset$ or $V(F_i) = \{0,\ldots,0\}$ and at least for one index i_0, $V(F_i) = \{0,\ldots,0\}$. The conclusion follows by taking $F = F_{i_0}$.

(2) By **(1)**, let $g(x_1,\ldots,x_m) \in K[x_1,\ldots,x_m]$ such that $V(g) = (0,\ldots,0)$ (in K^m). Let

$$f(x_1,\ldots,x_n) = g(f_1(x_1,\ldots,x_n),\ldots,f_m(x_1,\ldots,x_n)),$$

then for any $(a_1,\ldots,a_n) \in K^n$ we have

$$\begin{aligned} f(a_1,\ldots,x_n) = 0 &\Leftrightarrow f_1(x_1,\ldots,x_n) = \cdots = f_m(x_1,\ldots,x_n) = 0 \\ &\Leftrightarrow (x_1,\ldots,x_n) \in V(f_1,\ldots,f_m). \end{aligned}$$

Question. Is the polynomial f in **(2)** unique?

Exercise 2.24.
Let K be a field, A be a subring of $K[x_1,\ldots,x_n]$ and F be its fraction field. We say that A satisfies the property $(\mathcal{P})$ if for any $f(x_1,\ldots,x_n)$ and $g(x_1,\ldots,x_n) \in K[x_1,\ldots,x_n]$, $f \cdot g \in A \Rightarrow f \in A$ and $g \in A$. If A satisfies $(\mathcal{P})$, show that A is integrally closed in $K[x_1,\ldots,x_n]$ and F is algebraically closed in $K(x_1,\ldots,x_n)$.

Solution 2.24.
Let $f \in K[x_1,\ldots,x_n]$ such that f is integral over A, then there exist a positive integer m and $a_0,\ldots,a_{m-1} \in A$ such that

$$f^m + a_{m-1}f^{m-1} + \cdots + a_0 = 0.$$

This implies $f|a_0$. Let $h \in K[x_1, \ldots, x_n]$ such that $a_0 = f \cdot h$, then $f \cdot h \in A$. Since A satisfies $(\mathcal{P})$, then $f \in A$ and $h \in A$. Therefore A is integrally closed in $K[x_1, \ldots, x_n]$. Let $\frac{u}{v} \in K(x_1, \ldots, x_n)$, where $u, v \in K[x_1, \ldots, x_n]$, $gcd(u, v) = 1$. Suppose that $\frac{u}{v}$ is algebraic over F, then there exist a positive integer m and $(a_0, b_0), (a_1, b_1), \ldots, (a_{m-1}, b_{m-1}) \in A^2$ such that $\gcd(a_i, b_i) = 1$ for $i = 1, \ldots, m-1$ and

$$\left(\frac{u}{v}\right)^m + \frac{a_{m-1}}{b_{m-1}}\left(\frac{u}{v}\right)^{m-1} + \cdots + \frac{a_0}{b_0} = 0.$$

We deduce that there exist $d_0, \ldots, d_{m-1}, d_m \in A$ such that

$$d_m u^m + d_{m-1}u^{m-1}v + \cdots + d_0 v^m = 0.$$

This implies $v|d_m$. Let $v' \in K[x_1, \ldots, x_n]$ such that $d_m = vv'$. Since $d_m \in A$ and A satisfies $(\mathcal{P})$, we conclude that $v \in A$ and $v' \in A$. Similarly, we prove that $u \in A$. Therefore $\frac{u}{v} \in F$ and F is algebraically closed in $K(x_1, \ldots, x_n)$.

Exercise 2.25.
Let K be a field, $P(x)$ and $A(x)$ be polynomials with coefficients in K such that $P(x)$ is monic and non constant. Let $\phi_{(A,P)}$ be the K-endomorphism of $K[x]/P(x)K[x]$ such that $\phi_{(A,P)}(\overline{f(x)}) = \overline{A(x)f(x)}$ for any $f(x) \in K[x]$.

(1) Show that $\operatorname{Det} \phi_{(A,P)} = \operatorname{Res}_x(P(x), A(x))$.
(2) Let $Q(x) \in K[x]$ be monic. Suppose that $P(x)$ and $Q(x)$ are irreducible over K. Let α and β be roots of $P(x)$ and $Q(x)$ respectively in an algebraic closure of K. Suppose that $K(\beta) \subset K(\alpha)$ and let $u(x) \in K[x]$ such that $\beta = u(\alpha)$.
   (a) Show that for any $A(x) \in K[x]$,
   $$\operatorname{Res}_x(P(x), A(u(x))) = \operatorname{Res}_x(Q(x), A(x)^m),$$
   where $m = [K(\alpha) : K(\beta)]$.
   (b) Show that for any $A(x) \in K[x]$, there exists $B(x) \in K[x]$ such that
   $$\operatorname{Res}_x(P(x), A(x)) = \operatorname{Res}_x(Q(x), B(x)).$$

Solution 2.25.

(1) Suppose that the result is true for an algebraically closed field and let Ω be an algebraic closure of K. Clearly the resultant has the same value when we consider $P(x)$ and $A(x)$ as polynomials with coefficients in K or in Ω. On the other hand let $\psi_{(A,P)}$ be the Ω-endomorphism of $\Omega[x]/P(x)\Omega(x)$ such that $\psi_{(A,P)}(\overline{f}(x)) =$

$\overline{A(x)f(x)}$, then clearly $\phi_{(A,P)}$ and $\psi(A,P)$ have the same matrix in the basis $\{\bar{1}, \bar{x}, \ldots, \overline{x^{d-1}}\}$ of $K[x]/P(x)k[x]$ over K and in the basis $\{\bar{\bar{1}}, \bar{\bar{x}}, \ldots, \overline{\overline{x^{d-1}}}\}$ of $\Omega[x]/P(x)\Omega[x]$ over Ω, where $d = \deg P$ hence $\operatorname{Det}\phi_{(A,P)} = \operatorname{Det}\psi_{(A,P)}$. Therefore, we may suppose that K is algebraically closed. For any $A_1(x), A_2(x) \in K[x]$, we have:

$$\operatorname{Res}_x(P(x), A_1(x)A_2(x)) = \operatorname{Res}_x(P(x), A_1(x)) \cdot \operatorname{Res}_x(P(x), A_2(x))$$

and

$$\operatorname{Det}\phi_{(A_1A_2,P)} = \operatorname{Det}\phi_{(A_1,P)} \cdot \operatorname{Det}\phi_{(A_2,P)}.$$

Hence, it is sufficient to prove the formula when $A(x) = a \in K$ or $A(x) = x - b$ with $b \in K$.

- If $A(x) = a$, then $\operatorname{Res}_x(P(x), a) = a^{\deg P}$. The matrix of $\phi_{(a,P)}$ in the basis $\{\bar{1}, \bar{x}, \ldots, \overline{x^{d-1}}\}$, where $d = \deg P$ is the diagonal matrix aI_d, hence $\operatorname{Det}\phi_{(a,P)} = a^d = a^{\deg P}$. The formula is proved in this case.
- If $A(x) = x - b$, with $b \in K$, then

$$\operatorname{Res}_x(P(x), x-b) = (-1)^{\deg P}\operatorname{Res}_x(x-b, P(x)) = (-1)^{\deg P}P(b).$$

On the other hand, let $y = x - b$ and consider the following diagram

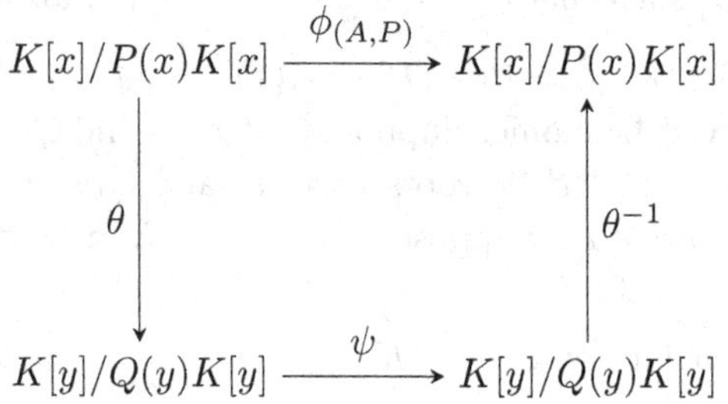

$$\begin{array}{ccc} K[x]/P(x)K[x] & \xrightarrow{\phi_{(A,P)}} & K[x]/P(x)K[x] \\ \downarrow \theta & & \uparrow \theta^{-1} \\ K[y]/Q(y)K[y] & \xrightarrow{\psi} & K[y]/Q(y)K[y] \end{array}$$

where

$$Q(y) = P(y+b), \quad \psi(\overline{g(y)}) = \overline{yg(y)} \quad \text{and} \quad \theta(\overline{f(x)}) = \overline{f(y+b)},$$

for any $g(y) \in K[y]$ and any $f(x) \in K[x]$. Then obviously it is a commutative diagram of K-algebras. Moreover θ is an isomorphism and θ^{-1} is its inverse. Since $\phi_{(A,P)} = \theta^{-1} \circ \psi \circ \theta$, then $\operatorname{Det}\phi_{(A,P)} = \operatorname{Det}\psi$. Set $P(x) = x^d + a_{d-1}x^{d-1} + \ldots + a_0$, then

$$\begin{aligned} Q(y) &= P(y+b) \\ &= (y+b)^d + a_{d-1}(y+b)^{d-1} + \cdots + a_0 \\ &= y^d + c_{d-1}y^{d-1} + \cdots + c_0. \end{aligned}$$

The matrix of ψ in the basis $\{\bar{1}, \bar{y}, \ldots, \overline{y^{d-1}}\}$ of $K[y]/Q(y)K[y]$ is given by

$$\begin{pmatrix} 0 & 0 & \ldots & -c_0 \\ 1 & 0 & \ldots & -c_1 \\ 0 & 1 & \ldots & -c_2 \\ \ldots & \ldots & \ldots & \ldots \\ 0 & 0 & \ldots & -c_{d-1} \end{pmatrix}.$$

Therefore

$$\begin{aligned} \operatorname{Det} \phi_{(A,P)} &= \operatorname{Det} \psi \\ &= -c_0(-1)^{1+d} \\ &= (-1)^d c_0 \\ &= (-1)^{\deg P} Q(0) \\ &= (-1)^{\deg P} P(b). \end{aligned}$$

We conclude that the formula holds in this case also.

(2)(a) Let $k = [K(\beta) : K]$ and $m = [K(\alpha) : K(\beta)]$, then

$$\mathcal{B} = \{1, \beta, \ldots, \beta^{k-1}, \alpha, \alpha\beta, \ldots, \alpha\beta^{k-1}, \ldots, \alpha^{m-1}, \ldots, \alpha^{m-1}\beta^{k-1}\}$$

is a basis of $K(\alpha)$ over K. Let M be the matrix of the endomorphism, multiplication by $A(\beta)$ in $K(\beta)$ relatively to the basis $\{1, \beta, \ldots, \beta^{k-1}\}$ and let N be the matrix of the endomorphism, multiplication by $A(\beta)$ in $K(\alpha)$ relatively to the basis $\mathcal{B}$, then N is a block matrix, where the blocks are filled by zeros except on the principal diagonal, where M is repeated. Now we have

$$\begin{aligned} &\operatorname{Res}_x(Q(x), A(x)^m) \\ &= \operatorname{Res}_x(Q(x), A(x))^m \\ &= (\operatorname{Det} M)^m \\ &= \operatorname{Det} N \\ &= \text{Det. of the endo. multiplication by } A(\beta) \text{ in } K(\alpha) \\ &= \text{Det. of the endo. multiplication by } A(u(\alpha)) \text{ in } K(\alpha) \\ &= \operatorname{Res}_x(P(x), A(u(x))). \end{aligned}$$

(b) For any fields E, F such that $F \subset E$ and any $\gamma \in E$, we denote by $m_\gamma^{E/F}$, the F-endomorphism of E such that $m_\gamma^{E/F}(a) = \gamma a$ for any $a \in E$. Let $B(x) \in K[x]$ such that

$$\mathrm{N}_{K(\alpha)/K(\beta)}\big(-A(\alpha)\big) = B(\beta),$$

then

$$\begin{aligned}
\operatorname{Res}_x(P(x), A(x)) &= \operatorname{Det} \phi_{(A,P)} \\
&= \operatorname{Det} m_{A(\alpha)}^{K(\alpha)/K} \\
&= (-1)^{mk} \operatorname{N}_{K(\alpha)/K}(A(\alpha)) \\
&= \operatorname{N}_{K(\beta)/K}\left((-1)^m \operatorname{N}_{K(\alpha)/K(\beta)}(A(\alpha))\right) \\
&= \operatorname{N}_{K(\beta)/K}\left(\operatorname{N}_{K(\alpha)/K(\beta)}(-A(\alpha))\right) \\
&= \operatorname{N}_{K(\beta)/K}\left(B(\beta)\right) \\
&= \operatorname{Det} m_{B(\beta)}^{K(\beta)/K} \\
&= \operatorname{Res}_x(Q(x), B(x)).
\end{aligned}$$

Exercise 2.26.
Let K be a field, Ω be an algebraic closure of K, d be a positive integer and $f(x)$ be a non constant polynomial with coefficients in K.

(1) Show that the following conditions are equivalent.
   (i) $y^d - f(x)$ is absolutely irreducible.
   (ii) For any $c \in K^*, y^d - cf(x)$ is absolutely irreducible.
   (iii) Let $f(x) = a(x-\alpha_1)^{d_1} \cdots (x-\alpha_s)^{d_s}$ be the factorization of $f(x)$ in $\Omega[x]$ into linear factors, then $\gcd(d, d_1, \ldots, d_s) = 1$.

(2) Show that the above equivalent conditions hold if $\gcd(d, \deg f) = 1$.

Solution 2.26.

(1) • *(i)* $\Rightarrow$ *(ii)*. Suppose that *(ii)* does not hold and let $c \in K^*$ such that $y^d - cf(x)$ is reducible over Ω, say $y^d - cf(x) = A(x)B(x)$ where $A(x), B(x) \in \Omega[x]$ and $\deg A$, $\deg B \geq 1$. Then

$$\left(\frac{y}{c^{1/d}}\right)^d - f(x) = \frac{A(x)}{c} B(x),$$

hence $z^d - f(x)$ is reducible over Ω, contradicting *(i)*.

• *(ii)* $\Rightarrow$ *(iii)*. Suppose that *(iii)* does not hold and let $e = \gcd(d, d_1, \ldots, d_s), d' = d/e$ and $d_i' = \frac{d_i}{e}$ for $i = 1, \ldots, s$. Then $e \geq 2$ and

$$\begin{aligned}
y^d - f(x) &= y^{d'e} - a(x-\alpha_1)^{ed_1'} \cdots (x-\alpha_s)^{ed_s'} \\
&= (y^{d'})^e - [a^{1/e}(x-\alpha_1)^{d_1'} \ldots (x-\alpha_s)^{d_s'}]^e.
\end{aligned}$$

Set $g(x) = a^{1/e}(x-\alpha_1)^{d_1'} \cdots (x-\alpha_s)^{d_s'}$. Then

$$y^d - f(x) = (y^{d'} - g(x))((y^{d'})^{e-1} + (y^{d'})^{e-2}g(x) + \ldots + g(x)^{e-2}).$$

Therefore $y^d - f(x)$ is reducible over Ω contradicting *(ii)*.

- *(iii)* $\Rightarrow$ *(i)*. Suppose that $y^d - f(x)$ is reducible over Ω and let η be a primitive d-th root of unity in Ω, and α be a root of $y^d - f(x)$ in an algebraic closure of $K(x)$, then
$$y^d - f(x) = (y-\alpha)(y-\eta\alpha)\cdots(y-\eta^{d-1}\alpha)$$
and there exist non negative integers $i_1, \ldots, i_r$ with $1 \leq r < d$ such that $\prod_{j=1}^{r}(y - \alpha\eta^{i_j}) \in \Omega[x,y]$. It follows that $\alpha^r \prod_{j=1}^{r} \eta^{i_j} \in \Omega[x]$, hence $\alpha^r \in \Omega[x]$. Let l be the smallest positive integer such that $\alpha^l \in \Omega[x]$. We show that for any positive integer m such that $\alpha^m \in \Omega[x]$, we have $l|m$. Let m be a such integer and let $q, t \in \mathbb{Z}$ such that $m = lq + t$ with $0 \leq t < l$. We have $\alpha^m = \alpha^{lq} \cdot \alpha^t$, and since $\alpha^m, \alpha^l \in \Omega[x]$, then $\alpha^t \in \Omega[x]$, hence $t = 0$. We conclude, in particular, that $l|d$ and $l|r$. Therefore $l < d$. Let $g(x) \in \Omega[x]$ such that $\alpha^l = g(x)$. Then $f(x) = \alpha^d = (\alpha^l)^{d/l} = g(x)^{d/l}$. Therefore d/l divides d_i for $i = 1, ..., s$ and then $\gcd(d, d_1, ..., d_s) > 1$, contradicting *(iii)*.

(2) Since $\deg f = d_1 + d_2 + \cdots + d_s$, then if $\gcd(\deg f, d) = 1$, we have $\gcd(d, d_1, \ldots, d_s) = 1$. Therefore **(iii)** holds.

Exercise 2.27.

Let K be an algebraically closed field of characteristic 0. A k-tuple $(x_0, x_1, \ldots, x_{k-1}) \in K^k$, with $x_i \neq x_j$ for $i \neq j$ is called a cycle of $f(x) \in K[x]$ if $f(x_i) = x_{i+1}$ for $i = 0, \ldots, k-2$ and $f(x_{k-1}) = x_0$. The integer k is called the length of the cycle.

(1) Let $(x_0, x_1, \ldots, x_{k-1})$ be a cycle of $f(x)$ of length k. Show that k is smallest positive integer such that $f_k(x_0) = x_0$, where $f_k(x)$ is the k-th iterate of $f(x)$. Show that for any positive integer n, $f_n(x_0) = x_0$ if and only if $k|n$.
(2) Let $g(x) = ax + b$ with $(a, b) \in K^2$ and $a \neq 0$. Determine the possible lengths of the cycles of $g(x)$.
(3) Let $h(x) = x^2 - x$. Show that $h(x)$ has no cycle of length 2.
(4) Let $f(x), g(x) \in K[x]$ such that $g(ax+b) = af(x) + b$ with $(a, b) \in K^2$, $a \neq 0$ and let $k \geq 2$ be an integer. Show that $f(x)$ has a cycle of length k if and only if $g(x)$ has a cycle of the same length.
(5) Let $d = \deg f$ and suppose that $d \geq 2$.
    (a) Show that $f(x)$ has cycles of length 1.
    (b) Fix two integers $n > k > 1$. Suppose that $f(x)$ has no cycle of length k nor n and let $A(x)/B(x)$ be the irreducible form of the rational function $F(x) = (f_n(x) - x)/(f_{n-k}(x) - x)$.

(i) Show that the number of solutions $x \in K$, counting multiplicities, satisfying one of the equations $F(x) = 0$ or $F(x) = 1$ is equal to $2 \deg A$.

(ii) Show that the number of distinct zeros of $A(x)/B(x)$ is at most equal to $\sum_{\substack{l|n \\ l<n}} d^l$.

(iii) Show that the number of distinct zeros of the equation $A(x)/B(x) = 1$ does not exceed $\sum_{\substack{t|k \\ t<k}} d^{n-k+t}$.

(iv) Conclude from **(ii)**, **(iii)** that the number of solutions $x \in K$ (counting multiplicities) satisfying one of the equation $A(x)/B(x) = 0$ or $A(x)/B(x) = 1$ is bounded by

$$\sum_{\substack{l|n \\ l<n}} d^l + \sum_{\substack{t|k \\ t<k}} d^{n-k+t} + \deg A + \deg B - 1.$$

(v) Show that **(i)** and **(iv)** lead to a contradiction. Conclude that a polynomial of degree at least 2 has cycles of any length with possibly one exception.

Solution 2.27.

(1) Let k_1 be the smallest positive integer such that $f_{k_1}(x_0) = x_0$ and let n be a positive integer. If $k_1 \mid n$, then obviously $f_n(x_0) = x_0$. Conversely, suppose that $f_n(x_0) = x_0$. Let q and r be the quotient and the remainder respectively of the Euclidean division of n by k_1. Then

$$0 = f_n(x_0) = f_{k_1 q + r}(x_0) = f_r(f_{qk_1} x_0) = f_r(x_0),$$

hence $r = 0$, thus $k_1 \mid n$. In particular $k_1 \mid k$, which implies $k_1 \leq k$. If $k_1 < k$, then $x_{k_1} = f_{k_1}(x_0) = x_0$, contradicting the fact that the x_i, $i = 0, 1, \ldots, k-1$, are distinct. Therefore $k_1 = k$.

(2) We have $g_n(x) = a^n x + b(a^{n-1} + \cdots + a + 1)$ for every integer $n \geq 1$. Suppose first that a is not a root of unity and let $x_0 \in K$ such that $g_n(x_0) = x_0$, then $x_0(a^n - 1) + b(\frac{a^n-1}{a-1}) = 0$ hence $x_0 = \frac{b}{(1-a)}$. Since $g(b/(1-a)) = b/(1-a)$, then $g(x)$ has no cycle of length $k \geq 2$. If $a = 1$ then $g_n(x) = x + nb$. If moreover $b \neq 0$ then $g(x)$ has no cycle of length k for any $k \geq 1$. If $a = 1, b = 0$ then $g(x) = x$ and $g_n(x) = x$. Hence $g(x)$ has one and only one cycle namely $\{0\}$. Suppose now that a is a root of unity of order $m \geq 2$. Obviously $\{b/(1-a)\}$ is a cycle of length 1. If moreover $b \neq 0$ then $\{0, g(0), \ldots, g_{m-1}(0)\}$ is a cycle of

length m. If $b = 0$ and a is a root of unity of order $m \geq 2$, then $g(x)$ has one and only one cycle namely $\{0\}$.

(3) We have $h_2(x) = x^4 - 2x^3 + x$, hence $h_2(x_0) = x_0$ if and only if $x_0^3(x_0 - 2) = 0$, thus if and only if $x_0 = 0$ or $x_0 = 2$. Since $h(0) = 0$ and $h(2) = 2$, then $\{0\}$ and $\{2\}$ are cycles of length 1. Therefore there does not exist cycles of length 2 for $h(x)$.

(4) Let $L(x) = (x - b)/a$ and $\{x_0, g(x_0), \dots, g_{k-1}(x_0)\}$ be a cycle of length k of $g(x)$, then $\{l(x_0), l(g(x_0)), \dots, l(g_{k-1}(x_0))\}$ is a cycle of $f(x)$ of the same length. Using $L^{-1}(x)$, we obtain that if $f(x)$ has a cycle of length k, then so does $g(x)$. Therefore $f(x)$ and $g(x)$ have the same set of length of cycles.

(5)(a) Since $f(x)$ is not constant and K is algebraically closed, then the equation $f(x_0) = x_0$ has at least one solution in K, hence $f(x)$ has at least one cycle of length 1.

(b) (i) Obviously the number of solutions of the equation $A(x)/B(x) = 0$, counting multiplicities is equal to $\deg A$. The equation $A(x)/B(x) = 1$ is equivalent to $A(x) - B(x) = 0$. Since $\deg B < \deg A$, the number of solutions, here also, is equal to $\deg A$. Since there is no common solution to the two equations, the number of $x \in K$ satisfying one of them is equal to $2 \deg A$.

(ii) Let $x_0 \in K$, then x_0 is a zero of $A(x)/B(x) = 0$ if and only if $f_n(x_0) = x_0$. Since there does not exists a cycle of length n for $f(x)$, then this is equivalent to the existence of a positive integer $l < n$ such that $l|n$ and $f_l(x_0) = x_0$. Therefore the number of such x_0 is at most equal to $\sum\limits_{\substack{l|n \\ l<n}} d^l$.

(iii) Let $x_0 \in K$, then $A(x_0)/B(x_0) - 1 = 0$ is equivalent to $f_n(x_0) = f_{n-k}(x_0)$, that is $f_k(f_{n-k}(x_0)) = f_{n-k}(x_0)$. Since $f(x)$ has no cycle of length k, this is equivalent to the existence of a positive integer $t < k$ such that $t \mid k$ and $f_t(f_{n-k}(x_0)) = f_{n-k}(x_0)$. Hence the number of such x_0 does not exceed $\sum\limits_{\substack{t|k \\ t<k}} d^{n-k+t}$.

(iv) Let F (resp. G) be the set of zeros of $A(x)/B(x)$ (resp. $\frac{A(x)}{B(x)} - 1$), then $F \cap G = \phi$. Let $H \subset F \cup G$ be the set of simple zeros of $A(x)/B(x)$ or $\frac{A(x)}{B(x)} - 1$. For any $x_0 \in (F \cup G) \backslash H$, let $e(x_0)$ be its multiplicity as a zero of one (and only one) of the above rational function, then x_0 is a zero of $(A(x)/B(x))'$ with multiplicity equal to $e(x_0) - 1$. Let q be the number of zeros of $A(x)/B(x)$

or $\frac{A(x)}{B(x)} - 1$ counted with their multiplicities, then

$$\begin{aligned} q &= |H| + \sum_{x_0 \in (F \cup G) \setminus H} e(x_0) \\ &= |H| + |(F \cup G) \setminus H| + \sum_{x_0 \in (F \cup G) \setminus H} (e(x_0) - 1) \\ &= |F| + |G| + \sum_{x_0 \in (F \cup G) \setminus H} (e(x_0) - 1). \end{aligned}$$

Using **(ii)** and **(iii)**, we conclude that

$$q \leq \sum_{\substack{l|n \\ l<n}} d^l + \sum_{\substack{t|k \\ t<k}} d^{n-k+t} + \deg A + \deg B - 1.$$

(v) Let $m = \deg B$, then $\deg A = d^n - d^{n-k} + m$. From (i) and (iv), we conclude that

$$2 \deg A \leq \sum_{\substack{l|n \\ l<n}} d^l + \sum_{\substack{t|n \\ t<n}} d^{n-k+t} + \deg A + \deg B - 1,$$

hence

$$d^n - d^{n-k} \leq \sum_{\substack{l|n \\ l<n}} d^l + \sum_{\substack{t||k \\ t<k}} d^{n-k+t} - 1.$$

We have

$$\begin{aligned} \sum_{\substack{l|n \\ l<n}} d^l &\leq d + d^2 + \cdots + d^{n-3} \\ &= d(1 + d + \cdots + d^{n-4}) \\ &= d(d^{n-3} - 1)/(d-1) \\ &\leq d^{n-2} \end{aligned}$$

and

$$\begin{aligned} \sum_{\substack{t|k \\ t<k}} d^{n-k+t} &= d^{n-k+1} + d^{n-k+2} + \cdots + d^{n-k+k-2} \\ &= d^{n-k+1}(1 + d + \cdots + d^{k-3}) \\ &= d^{n-k+1}(d^{k-2} - 1)/(d-1) \\ &= (d^{n-1} - d^{n-k+1})/(d-1) \\ &\leq d^{n-1}, \end{aligned}$$

hence $d^n - d^{n-k} \le d^{n-2} + d^{n-1}$. Therefore $d^k - 1 \le d^{k-2} + d^{k-1}$. Thus $d^{k-2}(d^2 - d - 1) \le 1$, which is a contradiction except in the case $k = 2$ and $d = 2$. In this last case, we have

$$2^n - 2^{n-2} \le \sum_{\substack{l|n\\ l<n}} 2^l + 2^{n-1},$$

hence

$$2^n + 2^{n-2} \le 2 + 2^2 + \cdots + 2^{n-3} + 2^{n-1}.$$

We deduce that

$$2^n \le 2 + 2^2 + \cdots + 2^{n-1} = 2^n - 1$$

and we reach a contradiction in this case also. We have shown that for a polynomial of degree $d \ge 2$, it is impossible that two positive and distinct integers may lie outside the set of lengths of the polynomial. Therefore this set is equal to $\mathbb{N}$ or $\mathbb{N}\backslash\{k_0\}$ for some positive integer k_0, where $\mathbb{N}$ is the set of positive integers.

Exercise 2.28.
Let K be a field of characteristic $p \ge 0$. A recurrence sequence (u_n) of elements of K is called linear and homogeneous if there exist a positive integer k and $a_1, \dots, a_k \in K$ such that $a_k \ne 0$ and

$$u_n = a_1 u_{n-1} + a_2 u_{n-2} + \cdots + a_k u_{n-k} \quad \text{for} \quad n \ge k. \qquad \text{(Eq 1)}$$

The polynomial $P(x) = x^k - a_1 x^{k-1} - \dots - a_k$ and the series $F(x) = \sum_{n=0}^{\infty} u_n x^n$ are called the characteristic polynomial and the generating function of the sequence (u_n). Let $P(x) = (x - \alpha_1)^{e_1} \cdots (x - \alpha_r)^{e_r}$ be the complete factorization of $P(x)$ in an algebraic closure of K.

(1) Show that $F(x)$ may be written in the form $F(x) = g(x)/x^k P(1/x)$, where $g(x)$ is a polynomial with coefficients in K of degree at most $k - 1$.
(2) Deduce that if $p = 0$ or $p \ge \max e_i$, then, for any $n \ge 0$, $u_n = \sigma_1(n)\alpha_1^n + \cdots + \sigma_r(n)\alpha_r^n$, where $\sigma_i(x)$ is a polynomial with coefficients in K of degree at most $e_i - 1$.
(3) Let S be the K-vector space consisting in sequences of elements of K and E be subspace whose elements are the sequences satisfying **(Eq 1)**. Show that if K is algebraically closed and the assumptions on p in **(2)** are satisfied, then E is generated by the sequences $(u_{i,j_i}(n))$ for $i = 1, \dots, r$ and $j_i = 0, \dots, e_i - 1$, where $u_{i,j_i}(n) = n^{j_i}\alpha_i^n$.

(4) Let $t \geq 2$ be an integer and

$$Q(x) = (x - \alpha_1^t)^{e_1} \cdots (x - \alpha_r^t)^{e_r} = x^k - b_1 x^{k-1} - \cdots - b_k.$$

Let (v_n) be the sequence defined by $v_n = u_{tn}$ for $n \geq 0$. If the conditions on p in (2) are satisfied, show that (v_n) is linear homogeneous that satisfies the following relation $v_n - b_1 v_{n-1} - \cdots - b_k v_{n-k} = 0$ for $n \geq k$.

Solution 2.28.

(1) We have $x^k P(1/x) = 1 - a_1 x - \cdots - a_k x^k$, hence

$$x^k P(1/x) F(x) = (1 - a_1 x - \cdots - a_k x^k) \sum_{n=0}^{\infty} u_n x^n.$$

For $n \geq k$, we compute the coefficient c_n of x^n in the right side of this identity and we show that it is 0. Since $c_n = u_n - a_1 u_{n-1} - \cdots - a_k u_{n-k}$, then $c_n = 0$. Set $g(x) = x^k P(1/x) F(x)$, then $g(x)$ is a polynomial with coefficients in K, of degree at most $k - 1$ and $F(x) = g(x)/x^k P(1/x)$.

(2) Since $P(x) = (x - \alpha_1)^{e_1} \cdots (x - \alpha_r)^{e_r}$, then

$$x^k P(1/x) = (1 - \alpha_1 x)^{e_1} \cdots (1 - \alpha_r x)^{e_r},$$

hence

$$F(x) = g(x)/(1 - \alpha_1 x)^{e_1} \cdots (1 - \alpha_r x)^{e_r}.$$

Since $\deg g \leq k - 1$, then the partial fractions decomposition of $F(x)$ takes the form:

$$F(x) = \sum_{i=1}^{r} \left(\frac{A_{i1}}{(1 - \alpha_i x)} + \cdots + \frac{A_{ie_i}}{(1 - \alpha_i x)^{e_i}} \right),$$

where $A_{ij} \in K$ and $A_{ie_i} \neq 0$. The partial fraction $\frac{A_{ij}}{(1-\alpha_i x)^j}$ is the $(j-1)$-th derivative of the fraction $\frac{A_{ij}\alpha_i^{j-1}}{(j-1)!(1-\alpha_i x)}$, which may be written in the form $\frac{A_{ij}\alpha_i^{j-1}\sum_{n=0}^{\infty}\alpha_i^n x^n}{(j-1)!}$. Thus the coefficient of x^n in $\frac{A_{ij}}{(1-\alpha_i x)^j}$ is equal to $\frac{A_{ij}n(n-1)\cdots(n-j+2)\alpha_i^n}{(j-1)!}$. Since $j - 1 \leq e_i - 1 \leq p - 1$, then this coefficient has the form $q_{ij}(n)\alpha_i^n$, where $q_{ij}(x)$ is a polynomial with coefficients in K such that $q_{ij}(x) = 0$ or $\deg q_{ij}(x) = j - 1$. Since $A_{ie_i} \neq 0$, then $\deg q_{ie_i}(x) = e_i - 1$. It follows that the coefficient of x^n in $\frac{A_{i1}}{1-\alpha_i x} + \cdots + \frac{A_{ie_i}}{(1-\alpha_i x)^{e_i}}$ is equal to $\sum_j q_{ij}(n)\alpha_i^n := q_i(n)\alpha_i^n$, where $q_i(x)$ is a polynomial of degree $e_i - 1$ with coefficients in K. Therefore, $u_n = \sum_{i=1}^{r} q_i(n)\alpha_i^n$.

(3) Thanks to **(2)**, we only need to show that the sequence given by $u_{ij}(n) = n^j\alpha_i^n$ satisfies **(Eq 1)**. Here $1 \le i \le r$ and $0 \le j \le e_i - 1$. Fix $n \ge k$ and consider the sequence of polynomials

$$P_0(x) = x^{n-k}P(x) = x^n - a_1x^{n-1} - \cdots - a_kx^{n-k},$$

$P_1(x) = xP_0'(x)$ and in general $P_i(x) = xP_{i-1}'(x)$. Since $\alpha_i \neq 0$ and α_i is a root of $P(x)$, with multiplicity equal to e_i, then it is a root of $P_0(x)$ with multiplicity equal to e_i. It follows that α_i is a root of $P_j(x)$ for $j = 1, \ldots, e_i - 1$. Thus

$$n^j\alpha_i^n - a_1(n-1)^j\alpha_i^{n-1} - \cdots - a_k(n-k)^j\alpha_i^{n-k} = 0.$$

This implies that the sequence $(n^j\alpha_i^n)$ satisfies **(Eq 1)**.

(4) By **(2)**, we have

$$u_n = (b_{e_1-1}^1 n^{e_1-1} + \cdots + b_0^1)\alpha_1^n + \cdots + (b_{e_r-1}^r n^{e_r-1} + \cdots + b_0^r)\alpha_r^n,$$

hence

$$u_{nt} = (b_{e_1-1}^1 (nt)^{e_1-1} + \cdots + b_0^1)\alpha_1^{nt} + \cdots + (b_{e_r-1}^r (nt)^{e_r-1} + \cdots + b_0^r)\alpha_r^{nt}.$$

From **(3)**, we conclude that the sequence $v_n = u_{nt}$ satisfies the relation $v_n - b_1v_{n-1} - \cdots - b_kv_{n-k} = 0$, where $1, -b_1, \ldots, -b_k$ represent the coefficients of $Q(x)$.

Exercise 2.29.
Let K be a field of characteristic $p \ge 0$, m and n be positive integers. Consider the polynomial with coefficients in K, $f(x) = \sum_{i=0}^{mn-1} a_ix^i$. Suppose that $p \nmid mn$ in the case $p \neq 0$. Show that the following assertions are equivalent.

(i) $a_s = a_r$ for any r, s such that $r \equiv s \pmod m$.
(ii) The polynomial $g(x) = \sum_{i=0}^{n-1} x^{im}$ divides $f(x)$.
(iii) Let ξ be a primitive mn-th root of unity in an algebraic closure of K. Then ξ^s is a root of $f(x)$ for any integer s such that $0 \le s \le mn - 1$ and $s \not\equiv 0 \pmod n$.

Solution 2.29.

- *(i)* $\Rightarrow$ *(ii)*. By **(i)**, we may write $f(x)$ in the form

$$f(x) = a_0 \sum_{i\equiv 0 \pmod m} x^i + a_1 \sum_{i\equiv 1 \pmod m} x^i + \cdots + a_{m-1} \sum_{i\equiv m-1 \pmod m} x^i,$$

hence

$$f(x) = a_0 \sum_{i\equiv 0 \pmod m} x^i + a_1 x \sum_{i\equiv 1 \pmod m} x^{i-1} + \cdots + a_{m-1}x^{m-1} \sum_{i\equiv m-1 \pmod m} x^{i-(m-1)}$$
$$= \left(\sum_{i=0}^{n-1} x^{im}\right)(a_0 + a_1 x + \cdots + a_{m-1}x^{m-1})$$

and **(ii)** is proved.

- $(ii) \Rightarrow (i)$. Easy.
- $(ii) \Leftrightarrow (iii)$. Observe that

$$g(x) = \sum_{i=0}^{n-1} x^{im} = (x^{mn-1} - 1)/(x^m - 1)$$

and that, from the hypothesis on the characteristic of K, we may conclude that the ξ^s, for s satisfying the conditions $0 \leq s \leq mn - 1$ and $s \not\equiv 0 \pmod n$, are precisely the roots of $g(x)$. The equivalence follows easily from these remarks.

Exercise 2.30.
Let $f(t)$, $g(t)$ and $h(t)$ be non constant polynomials with coefficients in K. Let $q(t)$ and $r(t)$ be the quotient and the remainder respectively in the Euclidean division of $f(t)$ by $g(t)$ in $K[t]$. If $f(t) \in K[h(t)]$ and $g(t) \in K[h(t)]$, show that $q(t) \in K[h(t)]$ and $r(t) \in K[h(t)]$. Show that the converse does not hold.

Solution 2.30.
Let $f_1(t)$ and $g_1(t) \in K[t]$ such that $f(t) = f_1(h(t))$ and $g(t) = g_1(h(t))$ and let $q_1(t)$ and $r_1(t)$ be the quotient and the remainder respectively in the Euclidean division of $f_1(t)$ by $g_1(t)$ in $K[t]$, then $f_1(t) = g_1(t)q_1(t) + r_1(t)$ with $\deg r_1(t) < \deg g_1(t)$, hence by substitution of $h(t)$ for t, we have $f(t) = g(t)q_1(h(t)) + r_1(h(t))$. Since

$$\begin{aligned}\deg r_1(h(t)) &= \deg r_1 \deg h_1\\ &< \deg g_1 \deg h(t)\\ &= \deg g,\end{aligned}$$

then

$$q(t) = q_1(h(t)) \in K[h(t)] \text{ and } r(t) = r_1(h(t)) \in K[h(t)].$$

Second proof. The following proof is not really different in nature from the preceding one. Let $\theta : K[t] \to K[t]$ be the unique K-morphism of rings such that $\theta(t) = h(t)$ and let $u(t) \in K[t] \setminus K$. Suppose that $\theta(u(t)) = 0$, then $u(h(t)) = 0$. Let $\alpha_1, \dots \alpha_n$ be the roots (distinct or not) of $u(t)$ in an algebraic closure of K. Then $(h(t)-\alpha_1)\cdots(h(t)-\alpha_n) = 0$, hence $h(t) = \alpha_i$ for some $i \in \{1, \dots, n\}$, which is a contradiction.

Consider the polynomials $f(t) = t^5$ and $g(t) = t^3$. Then they are not polynomials in t^2, while the quotient and the remainder of the Euclidean division of $f(t)$ by $g(t)$ are.

Exercise 2.31.
Let K be a field of characteristic 0.

(1) Let k be a positive integer. Show that the series $\alpha(x) = \sum_{j\geq 0} \binom{1/k}{j} x^j$ is a root of the polynomial $\phi(y) = y^k - (1+x)$ in an algebraic closure of $K(x)$.

(2) Let $f(x)$ be a polynomial with coefficients in K of degree $n = km$, with k and $m \geq 2$. Let a be the leading coefficient of $f(x)$ and let ρ be a k-th root of unity in $\bar{K}$. Show that the equation $y^k = f(x)$ has a solution y in $K(\rho)((1/x))$ of the form $y = P_m(x) + \sum_{i\geq 0} a_i(1/x)^i$, where $P_m(x)$ is a polynomial with coefficients in $K(\rho)$, whose constant and leading coefficients are equal to 0 and ρ respectively. Call this polynomial a polynomial part of the Laurent expansion at infinity of $f(x)^{1/k}$.

(3) Let $f_1(x)$ be a monic polynomial with coefficients in K of degree $n = km$ with k and $m \geq 2$. Suppose that $f_1(0) = 0$. Let $Q_m(x)$ be the polynomial part of the unique Laurent expansion at ∞ of $f(x)^{1/k}$ which is monic. Show that $Q_m(x)$ is the unique monic polynomial $R(x) \in K[x]$ such that $R(0) = 0$ and $f_1(x) = R(x)^k + x^{n-m}u(1/x)$, where $u(1/x) \in K[[1/x]]$.

(4) Suppose that $f_1(x) = g_1(h_1(x))$, where $g_1(x)$ and $h_1(x) \in K[x]$ of degree k and m respectively. Suppose that $h(x)$ is monic and that $h_1(0) = 0$. Show that $h_1(x) = Q_m(x)$.

(5) Let $f(x)$ as in **(2)** and let λ and $\mu \in K$ such that if we set $L(x) = \lambda f(x) + \mu$, then the polynomial $f_1(x) = L(f(x))$ is monic and has zero constant term. Suppose that $f(x) = g(h(x))$ with $\deg g = k$ and $\deg h = m$. Let $P_m(x)$ be the polynomial part of a Laurent expansion at ∞ of $f_(x)^{1/k}$. Show that there exist γ and $\delta \in \bar{K}$ such that $h(x) = \gamma P_m(x) + \delta$.

(6) Show that $f(x) = 3x^4 + 6x^3 + 5x^2 + 2x$ may be expressed in the form $f(x) = g(h(x))$, with $g(x), h(x) \in \mathbb{Q}[x]$ and $\deg g = \deg h = 2$. Determine $h(x)$ and $g(x)$.

Solution 2.31.

(1) Let $\beta(x)$ be a root of $\phi(y)$ in an algebraic closure of $\mathbb{Q}(x)$ such that $\beta(0) = 0$. We show that for any positive integer m, $\beta^{(m)}(0) = \alpha^{(m)}(0)$. This will imply that $\alpha(x) = \beta(x)$. Since $\beta(x) = (1+x)^{1/k}$, then

$$\beta^{(m)}(0) = (1/k)(1/k-1)\cdots(1/k-(m-1)) = \alpha^{(m)}(0).$$

Now $\alpha(x) \in \overline{\mathbb{Q}(x)} \subset \overline{F(x)}$ is a root of $\phi(y)$, hence $\alpha(x)$ is a root of $\phi(y)$ in an algebraic closure of $F(x)$.
Obviously the roots of $\phi(y)$ are given by $\alpha_i(x) = \rho_i\alpha(x)$, where ρ_i is any k-th root of unity in $\bar{F}$.

(2) Set $f(x) = a_nx^n + a_{n-1}x^{n-1} + \cdots + a_0$, with $a_n = a$. We have

$$\begin{aligned} y &= f(x)^{1/k} \\ &= (a_nx^n + a_{n-1}x^{n-1} + \cdots + a_0)^{1/k} \\ &= a^{1/k}x^m\left(1 + \frac{a_{n-1}}{a}\frac{1}{x} + \cdots + \frac{a_0}{a}\frac{1}{x^n}\right)^{1/k}. \end{aligned}$$

Putting $z(x) = \frac{a_{n-1}}{a}\frac{1}{x} + \cdots + \frac{a_0}{a}\frac{1}{x^n}$ and using **(1)**, we obtain

$$\begin{aligned} y &= a^{1/k}x^m(1+z(x))^{1/k} \\ &= \rho x^m\left(1 + \binom{1/k}{1}z(x) + \binom{1/k}{2}z(x)^2 + \cdots + \binom{1/k}{q}z(x)^q + \cdots\right) \\ &= \rho x^m\left(1 + \sum_{j=1}^{m-1}\binom{1/k}{j}z(x)^j + \sum_{j\geq m}\binom{1/k}{j}z(x)^j\right). \end{aligned}$$

Since $z(x)$ has the form $z(x) = \frac{1}{x}u(1/x)$ with $u(1/x) \in K[1/x]$, then for any positive integer j, $z(x)^j = \frac{1}{x^j}u_j(1/x)$, where $u_j(1/x) \in K[1/x]$. It follows that $x^mz(x)^j \in K[1/x]$ for $j \geq m$. For $j \leq m-1$, $x^mz(x)^j$ is the sum of polynomial in x of degree at most $m-j$ and of a polynomial in $1/x$, thus y has the form $y = P_m(x) + \sum_{i\geq 0} a_i(1/x)^i$, where the polynomial $P_m(x)$ satisfies the conditions: its degree is equal to m, with leading coefficient equal to ρ and $P_m(0) = 0$.

(3) We apply **(2)**. Since $f_1(x)$ is monic, then we may take $\rho = 1$ and then the equation $y^k = f_1(x)$ has a solution y in $K((1/x))$ of the form $y = Q_m(x) + \sum_{i\geq 0} b_i(1/x)^i$, where $Q_m(x)$ is a monic polynomial of

degree m with coefficients in K and such that $Q_m(0) = 0$. It follows that

$$f_1(x) = y^k = \left(Q_m(x) + \sum_{i\geq 0} b_i(1/x)^i\right)^k$$

$$= (Q_m(x))^k + \sum_{j=0}^{k-1}\binom{k}{j}(Q_m(x))^k \sum_{i\geq 0} b_i(1/x)^i$$

$$= (Q_m(x))^k + \sum_{j=0}^{k-1}\binom{k}{j}(x^m + q_{m-1}x^{m-1} + \cdots + q_0)^j \sum_{i\geq 0} b_i(1/x)^i$$

$$= (Q_m(x))^k + x^{m(k-1)}\sum_{j=0}^{k-1}\binom{k}{j}\frac{(x^m + q_{m-1}x^{m-1} + \cdots + q_0)^j}{x^{m(k-1)}}B(x)$$

$$= (Q_m(x))^k + x^{n-m}\sum_{j=0}^{k-1}\binom{k}{j}\frac{1}{x^{n-m-mj}}\left(1 + q_{m-1}\frac{1}{x} + \cdots + q_0\frac{1}{x^m}\right)^j B(x)$$

$$= (Q_m(x))^k + x^{n-m}u(1/x),$$

where $B(x) = \sum_{i\geq 0} b_i(\frac{1}{x})^i$ and $u(1/x) \in K[[1/x]]$. It follows that one solution of the equation $f_1(x) = R(x)^k + x^{n-m}u(1/x)$ is satisfied by $R(x) = Q_m(x)$. We now show that this equation has at most one solution $R(x)$, where $R(x)$ is a monic polynomial of degree m with coefficients in K such that $R(0) = 0$. Set

$$R(x) = r_m x^m + r_{m-1}x^{m-1} + \cdots + r_1 x$$

and

$$f_1(x) = a_n x^n + a_{n-1}x^{n-1} + \cdots + a_1 x$$

with $r_m = a_n = 1$. We have

$$\sum_{i=1}^{n} a_i x^i = \left(\sum_{j=1}^{m} r_j x^j\right)^k + x^{n-m}u(1/x)$$

$$= \sum \frac{k!}{e_1!\cdots e_m!}(r_1 x)^{e_1}\cdots(r_m x^m)^{e_m} + x^{n-m}u(1/x), \quad \text{(Eq 1)}$$

where the sum runs over all the m-tuples $(e_1, \ldots, e_m)$ of non negative integers satisfying the condition

$$e_1 + \cdots + e_m = k. \quad \text{(Eq 2)}$$

Identifying the coefficients of x^{km-m+t}, for $t = 1, \ldots, m-1$, we obtain

$$a_{km-m+t} = \sum \frac{k!}{e_1! \cdots e_m!} r_1^{e_1} \cdots r_m^{e_m}, \qquad \text{(Eq 3)}$$

where the sum is over all $(e_1, \ldots, e_m)$ with $e_i \geq 0$,

$$\sum_{i=1}^{m} e_i = k \qquad \text{(Eq 4)}$$

and

$$\sum_{i=1}^{m} ie_i = km - m + t. \qquad \text{(Eq 5)}$$

Using **(Eq 4)** and **(Eq 5)**, we get the following equation $\sum_{i=1}^{m} ie_i = \sum_{i=1}^{m} me_i - m + t$, that is

$$\sum_{i=1}^{m-1} (m-i)e_i = m - t. \qquad \text{(Eq 6)}$$

(Eq 2) may be written in the form

$$e_m = k - (e_1 + \cdots + e_{m-1}). \qquad \text{(Eq 7)}$$

(Eq 6) implies that $m - t \geq (m-i)e_i$ for $i = 1, \ldots, m$. It follows that if $i < t$, then $e_i = 0$, thus $e_1 = \cdots = e_{t-1} = 0$. Using the preceding inequality for $i = t$, we get $e_t = 0$ or $e_t = 1$. In the later case, by **(Eq 6)**, we have $e_{t+1} = \cdots = e_{m-1} = 0$. We deduce that

$$a_{km-m+t} = kr_t r_m^{k-1} + \sum \frac{k!}{e_{t+1}1! \cdots e_m!} r_{t+1}^{e_{t+1}} \cdots r_m^{e_m}, \qquad \text{(Eq 8)}$$

where $r_m = 1$ and the sum runs over all tuples of non negative integers $(e_{t+1}, \ldots, e_m)$ such that

$$e_{t+1} + \cdots + e_m = k \,\text{and}\, (t+1)e_{t+1} + \cdots + me_m = km - m + t.$$

In the particular case $t = m - 1$ the sum in **(Eq 8)** is empty, thus

$$a_{km-1} = kr_{m-1}. \qquad \text{(Eq 9)}$$

In any case **(Eq 8)** has the form

$$a_{km-m+t} = kr_t + F(r_{t+1}, \ldots, r_m), \qquad \text{(Eq 10)}$$

where F_t is a polynomial with integral coefficients depending on k, t and m. **(Eq 9)** determines r_{m-1}. Using **(Eq 10)**, it is seen that we may compute r_j by induction for $j = 1, \ldots, m-1$.

(4) Since f_1 and h_1 are monic then so is g_1. Moreover $g(0) = 0$. Set $g_1(x) = x^k + b_{k-1}x^{k-1} + \cdots + b_1 x$, then

$$\begin{aligned} f_1(x) &= h(x)^k + b_{k-1}h(x)^{k-1} + \cdots + b_1 h(x) \\ &= h(x)^k + x^{(k-1)m}\frac{b_{k-1}h(x)^{k-1} + \cdots + b_1 h(x)}{x^{(k-1)m}} \\ &= h(x)^k + x^{n-m}u(1/X), \end{aligned}$$

where $u(1/x) \in K[1/x]$. Since $h-1(x)$ is monic and has a its constant coefficient equal to 0, then by **(3)**, we conclude that $h_1(x) = Q_m(x)$.

(5) Let α, β, λ, and μ, be elements of K such that if we set $L_1(x) = \alpha x + \beta$, $L_2(x) = \lambda x + \mu$, then the polynomials $h_1(x) = L_1(h(x))$ and $f_1(x) = L_2(f(x))$ are monic and they satisfy the conditions $h_1 0 = 0$, and $f_1(x) = 0$. We have

$$f_1 = L_2 \circ f = L_2 \circ g \circ h = (L_2 \circ g \circ L_1^{-1}) \circ L_1 \circ h = (L_2 \circ g \circ L_1^{-1}) \circ h_1.$$

Let $g_1 = L_2 \circ g \circ L_1^{-1}$, then g_1 is monic, $g_1(0) = 0$ and $f_1(x) = g_1(h_1(x))$. Therefore we may apply **(4)**; thus $h_1(x) = Q_m(x)$, where $Q_m(x)$ is the polynomial part of the unique Laurent expansion at ∞ of $f_1(x)^{1/k}$ which is monic. We deduce that $h(x) = L_1^{-1}(h(x)) = L_1^{-1}(Q_m(x))$.

(6) Let $f_1(x) = f(x)/3 = x^4 + 2x^3 + 5/3x^2 + 2/3x$, then

$$\begin{aligned} f_1(x)^{1/2} &= x^2\left(1 + 2/x + 5/(3x^2) + 2/(3x^3)\right)^{1/2} \\ &= x^2\left(1 + \binom{1/2}{1}(2/x + 5/(3x^2) + 2/(3x^3)) \right. \\ &\quad + \binom{1/2}{2}(2/x + 5/(3x^2) + 2/(3x^3))^2 + \cdots \\ &\quad \left. + \binom{1/2}{q}(2/x + 5/(3x^2) + 2/(3x^3))^q + \cdots \right) \\ &= x^2 + x + u(1/x), \end{aligned}$$

hence $Q_m(x) = x^2 + x$. Therefore, if $f(x)$ has a right composition factor of degree 2, then $Q_m(x)$ is one such right composition factor. The answer to this question is positive if and only if there exist d and $e \in K$ such that $f(x) = d(x^2+x)^2 + e(x^2+x)$. Obviously this equation has one and only one solution, namely $(d, e) = (3, 2)$. It follows that $f(x) = g(Q_m(x))$ with $g(x) = 3x^2 + 2x$.

Chapter 3

Separability, Inseparability

Exercise 3.1.
Let K be a field of characteristic $p \geq 0$ and $f(x)$ be a polynomial with coefficients in K of degree $n \geq 1$. In the case $p > 0$, we suppose that $|K| \geq n$ and $f(x) \notin K[x^p]$. Show that there exists $a \in K$ such that $f(x) - a$ is separable over K.

Solution 3.1.
Since $f(x) \notin K[x^p]$, then $f^{'}(x) \neq 0$. Let $m = \deg f^{'}(x)$ and $\alpha_1, \ldots, \alpha_m$ be the distinct roots of $f^{'}(x)$ in an algebraic closure of K, then $m \leq n - 1 < |K|$. Therefore we may choose $a \in K$ such that $f(\alpha_i) \neq a$ for any $i = 1, \ldots, m$. For this element, we have $\gcd(f(x) - a, f'(x)) = 1$, hence $f(x) - a$ is separable over K.

Exercise 3.2.
Let K be a field of characteristic $p \geq 0$, and α be a root of $f(x) = (x-1)^4 - 2$ in an algebraic closure of K. Show that $K(\alpha)$ is a separable extension of K.

Solution 3.2.
Since $f^{'}(x) = 4(x-1)^3$, then $f(x)$ is separable over K if and only if $p \neq 2$. If $p = 2$, then $f(x) = (x-1)^4$, $\alpha = 1$ and $K(\alpha) = K$, hence the answer is positive in this case. Suppose that $p \neq 2$. Since the minimal polynomial, say $g(x)$, of α over K divides $f(x)$ and this last polynomial is separable, then $g(x)$ is separable over K. It follows that α and then $K(\alpha)$ is separable over K.

Exercise 3.3.

(1) Let K be a field, E be an extension of K and $f, g, h \in E[x_1, \ldots, x_n]$.

Suppose that $fg = h$, $h \neq 0$ and $g, h \in K[x_1, \ldots, x_n]$. Show that $f \in K[x_1, \ldots, x_n]$.

(2) Let L be a field, E be an extension of L. Let $f, g, h \in E[x_1, \ldots, x_n]$ such that $fg = h$. Suppose that the coefficients of g and h are algebraic and separable over L. Show the coefficients of f are algebraic and separable over L.

Solution 3.3.

(1) • **First method.** For any $i = 1, \ldots, n$, performing the Euclidean division in $K(x_1, \ldots, \hat{x}_i, \ldots, x_n)[x_i]$ of h by g, we obtain the identity $h = gq + r$, where $q, r \in K(x_1, \ldots, \hat{x}_i, \ldots, x_n)[x_i]$ and $\deg_{x_i} r < \deg_{x_i} g$ and where the hated variable inside the brackets is missed. We may consider this identity as a Euclidean division in the ring $E(x_1, \ldots, \hat{x}_i, \ldots, x_n)[x_i]$. Since the identity $h = gf$ is also an Euclidean division in the same ring, we conclude that $r = 0$. We now have $f = h/g = q$, hence

$$f \in \cap_{i=1}^{n} K(x_1, \ldots, \hat{x}_i, \ldots, x_n)[x_i] = K[x_1, \ldots, x_n].$$

• **Second method.** Let $d = \deg f$ and let $F = \sum_{i_1+\cdots+i_n \leq d} \lambda_{i_1\cdots i_n} x_1^{i_1} \cdots x_n^{i_n}$, where the $\lambda_{i_1\cdots i_n}$ are unknowns. We consider the equation $Fg = h$, which is equivalent to a system of linear equations with coefficients in K. This system has one and only one solution, namely $F = f$. Let r be the number of unknowns, then the rank of the system is equal to r and all the characteristic determinants vanish. Choose r principal equations and solve the resulting Cramer system. We obtain $\lambda_{i_1\cdots i_n} = \frac{D_{i_1\cdots i_n}}{D}$, where $D_{i_1\cdots i_n}$ and D are $r \times r$ determinants with coefficients in K. Therefore $\lambda_{i_1\cdots i_n} \in K$ for any $i_1 \cdots i_n$ and then $f \in K[x_1, \ldots, x_n]$.

(2) Let A be the set of coefficients of g and of h and let $K = L(A)$, then $g, h \in K[x_1, \ldots, x_n]$ and $fg = h$, hence $f \in K[x_1, \ldots, x_n]$ by **(1)**. Since K/L is algebraic and separable, the coefficients of f are algebraic and separable over L.

Exercise 3.4.

Let K be a field, $f(x)$ and $g(x)$ be polynomials with coefficients in K such that $g(x)$ is separable and $\deg g \leq \deg f$. Let a be the leading coefficient of $g(x)$ and $\alpha_1, \ldots, \alpha_n$ be the roots of $g(x)$ in an algebraic closure of K. Let $r(x)$ be the remainder in the Euclidean division of $f(x)$ by $g(x)$. Show that $r(x) = a \sum_{i=1}^{n} \frac{f(\alpha_i)}{g'(\alpha_i)} \prod_{j \neq i}(x - \alpha_j)$.

Solution 3.4.

Since $r(\alpha_i) = f(\alpha_i)$ and $\deg r < n$, then we may apply Lagrange's interpolation formula [Bourbaki (1950), Application: Formule d'interpollation de Lagrange, Chap. 4.2] or [Ayad (1997), Exercice 1.14] and obtain

$$r(x) = \sum_{i=1}^{n} f(\alpha_i) \prod_{j \neq i}(x - \alpha_j) / \prod_{j \neq i}(\alpha_i - \alpha_j).$$

Since

$$g^{'}(x) = a \sum_{i=1}^{n} \prod_{j \neq i}(x - \alpha_j),$$

then $g^{'}(\alpha_i) = a \prod_{j \neq i}(\alpha_i - \alpha_j)$. Thus

$$r(x) = a \sum_{i=1}^{n} \frac{f(\alpha_i)}{g^{'}(\alpha_i)} \prod_{j \neq i}(x - \alpha_j).$$

Exercise 3.5.

Let K be a field of characteristic $p > 0$ and

$$f(x) = a_n x^n + a_{n-1}x^{n-1} + \cdots + a_0$$

be a polynomial with coefficients in K. Suppose that $f(x)$ is irreducible over K.

(1) If K is perfect, show that for any positive integer λ, $f(x^{p^\lambda})$ is reducible over K.

(2) Suppose that K is not perfect.
  (a) If $a_n = 1$ and $a_i \notin K^p$ for some $i \in \{0, 1, \ldots, n-1\}$, show that for any non negative integer λ, $f(x^{p^\lambda})$ is irreducible over K.
  (b) Give an example, where the conclusion of **(a)** does not hold if we omit the condition $a_n = 1$.

Solution 3.5.

(1) Since K is perfect, then for any $a \in K$, there exists $b \in K$ such that $a = b^p$. By induction on λ, it is seen that for any $\lambda \geq 1$, there exists $b_\lambda \in K$ such that $a = b_\lambda^{p^\lambda}$. It follows that $a_i = b_i^{p^\lambda}$, with $b_i \in K$, for $i = 0, \ldots, n$. Therefore

$$\begin{aligned} f(x^{p^\lambda}) &= b_n^{p^\lambda} x^{np^\lambda} + b_{n-1}^{p^\lambda} x^{(n-1)p^\lambda} + \cdots + b_0^{p^\lambda} \\ &= \left(b_n x^n + b_{n-1}x^{n-1} + \cdots + b_0\right)^{p^\lambda}, \end{aligned}$$

thus $f(x^{p^\lambda})$ is reducible over K.

(2)(a) The proof proceeds by induction on λ. The result for $\lambda = 0$ is true by assumption. Suppose that $\lambda \geq 1$, $f(x^{p^{\lambda-1}})$ is irreducible while $f(x^{p^\lambda})$ is reducible over K. Then $f(x^{p^\lambda})$ has a factorization over K of the form

$$f(x^{p^\lambda}) = g(x)^k h(x), \qquad \text{(Eq 1)}$$

where $g(x)$ is irreducible, $k \geq 1$ $h(x)$ is monic and $\gcd(g(x), h(x)) = 1$. Moreover, we may suppose that $k \geq 2$ or $h(x) \notin K$. Differentiating **(Eq 1)**, we obtain

$$kh(x)g'(x) = -g(x)h'(x).$$

It follows that $h(x) \mid h'(x)$, which implies that $h(x) = h_1(x)^p$, where $h_1(y) \in K[y]$ and either $k = pq$ or $g'(x) = 0$. In any case, we have $g(x)^k = g_1(x^p)^l$ with $g_1(y) \in K[y]$. Now setting $y = x^p$, **(Eq 1)** may be written in the form $f(y^{p^{\lambda-1}}) = g_1(y)^l h_1(y)$. The inductive hypothesis shows that $h_1(y) = 1$ and $l = 1$. Thus $f(x^{p^\lambda}) = (g(x)^l)^p$, which implies that each a_i is a p-th power in K, contradicting the assumptions.

(b) Let $K = \mathbb{F}_2(t)$, where t is a variable algebraically independent from x. Let $f(x) = t(x - t^2)$, then clearly $f(x)$ is irreducible over K and its coefficients, t and t^3 do not belong to K^2. On the other hand, we have

$$f(x^2) = t(x^2 - t^2) = t(x - t)^2,$$

so that $f(x^2)$ is reducible over K.

Exercise 3.6.
Let K be a field, $f(x)$ be a monic, irreducible polynomial of degree n with coefficients in K and θ be a root of $f(x)$ in an algebraic closure of K. Let $u(x) = A(x)/B(x)$ be a rational function with coefficients in K, such that $\gcd(A(x), B(x)) = 1$ and $\gcd(B(x), f(x)) = 1$. Let S be the set whose elements are the coefficients of $A(x)$ or $B(x)$ and $\alpha = u(\theta)$. Let $R(S, y) = \mathrm{Res}_x(f(x), yB(x) - A(x))$.

(1) Show that $R(S, y) = \mathrm{N}_{K(\theta)/K}(B(\theta)) \,\mathrm{Char}(\alpha, K, y)$.
(2) Suppose that $u(x) = b_0 + \cdots + b_{n-1}x^{n-1}$ (i.e. $u(x) = A(x)$ and $B(x) = 1$). Consider the subset V of K^n given by:

$$V = \{(b_0, \ldots, b_{n-1}) \in K^n,\ u(\theta) \text{ does not generate } K(\theta) \text{ over } K\}.$$

Show that V defines a hypersurface in K^n (for the Zariski topology).

Solution 3.6.

(1) Let $\theta_1 = \theta, \theta_2, \ldots, \theta_n$ be the roots of $f(x)$, distinct or not, in an algebraic closure of K, then

$$\begin{aligned} R(S,y) &= \mathrm{Res}_x(f(x), yB(x) - A(x)) \\ &= \prod_{i=1}^{n}(yB(\theta_i) - A(\theta_i)) \\ &= \prod_{i=1}^{n} B(\theta_i) \prod_{i=1}^{n}(y - A(\theta_i)/B(\theta_i)) \\ &= \mathrm{N}_{K(\theta)/K}(B(\theta))\,\mathrm{Char}(\alpha, K, y). \end{aligned}$$

(2) Let $D(b_0, \ldots, b_{n-1})$ be the discriminant of $R(b_0 \ldots, b_{n-1}, y)$. Let $(b_0, \ldots, b_{n-1}) \in K^n$ and $u(x) = b_0 + \cdots + b_{n-1}x^{n-1}$ then

$$\begin{aligned} (b_0, \ldots, b_{n-1}) \in V &\Leftrightarrow u(\theta) \text{ is not a primitive element of } K(\theta) \text{ over } K. \\ &\Leftrightarrow \mathrm{Char}(u(\theta), K, y) \text{ is a strict power of } \mathrm{Irr}(u(\theta), K, y). \\ &\Leftrightarrow S(y) \quad \text{is inseparable over} \quad K. \\ &\Leftrightarrow R(b_0 \ldots, b_{n-1}, y) \quad \text{is inseparable over} \quad K. \\ &\Leftrightarrow D(b_0, \ldots, b_{n-1}) = 0. \end{aligned}$$

This last equation relating the b_i defines a hypersurface in K^n.

Exercise 3.7.
Consider the polynomials with coefficients in $\mathbb{F}_2$, $f(x,y) = x^2 + y + 1$ and $g(x,y) = x^2+xy+y^2$. Let $A = \mathbb{F}_2[x,y]/f\mathbb{F}_2[x,y]$ and $B = \mathbb{F}_2[x,y]/g\mathbb{F}_2[x,y]$.

(1) Show that A and B are integral domains.
(2) Let $E = \mathrm{Frac}(A)$ and $F = \mathrm{Frac}(B)$. Show that E and F are isomorphic as vector spaces over $\mathbb{F}_2$.
(3) Show that there exists no morphism of fields $\Phi : E \to F$.

Solution 3.7.

(1) Since $f(x,y)$ is irreducible in $\mathbb{F}_2[x,y]$, then $f(x,y)\mathbb{F}[x,y]$ is a prime ideal of A hence A/fA is an integral domain. The same arguments work for B, thus B is an integral domain.
(2) We first show that any $\alpha \in E$ may be represented in a unique way in the form $\alpha = \frac{\overline{h(x,y)}}{\overline{D(y)}}$, where $D(y) \in \mathbb{F}_2[y]$ and $h(x,y) \in \mathbb{F}_2[x,y]$, $\deg_x h \leq 1$. Let $\frac{\bar{u}}{\bar{v}} \in E$. Since $\bar{v} \neq 0$, then there exist $a(x,y), b(x,y) \in \mathbb{F}_2[x,y]$ and $D(y) \in \mathbb{F}_2[y]$ such that $au+bv = D$, hence $\bar{b}\bar{v} = \bar{D}$. Since $\bar{D} \in A^*$, then

$\bar{v} \in A^*$ and $\frac{1}{\bar{v}} = \frac{\bar{b}}{D}$. It follows that $\frac{\bar{u}}{\bar{v}} = \frac{\bar{u}\bar{b}}{D}$. Since $f(x,y)$ is monic in x, then we may proceed to a Euclidean division of ub by f in $\mathbb{F}_2[y][x]$ and obtain $ub = fq + r$ where $q, r \in \mathbb{F}_2[x,y]$ with $0 \neq \deg_x r < \deg_x f = 2$. It follows that $\overline{ub} = \bar{r}$, hence $\frac{\bar{u}}{\bar{v}} = \frac{\bar{r}}{D}$.
Consider the map $\Psi : E \to \mathbb{F}_2(y)[x]/(f)$, such that for any $\alpha = \frac{\overline{h(x,y)}}{\overline{D(y)}} \in E$ with $\deg_x(h) \leq 1$, we have $\Psi(\alpha) = \frac{h(x,y)}{D(y)} + (f)$. Then Ψ is an isomorphism of fields, fixing point wise the elements of $\mathbb{F}_2(y)$. In the same way we prove that $\mathbb{F}$ and $\mathbb{F}_2(y)[x]/(g(x,y))$ are isomorphic over $\mathbb{F}_2(y)$. Since $\{1+(f), x+(f)\}$ and $\{1+(g), x+(g)\}$ are basis of the $\mathbb{F}_2(y)$-vector spaces then $\mathbb{F}_2(y)[x]/(f)$ and $\mathbb{F}_2(y)[x]/(g)$ are isomorphic as vector spaces over $\mathbb{F}_2(y)$. It follows that E and F are isomorphic as vector spaces over $\mathbb{F}_2(y)$, hence isomorphic also as vector spaces over $\mathbb{F}_2$.

(3) Suppose that there exists a morphism of fields $\Phi : F \to E$, then obviously $\Phi(a) = a$ for any $a \in \mathbb{F}_2$. Denote by $\bar{x}, \bar{y}$ (resp. $\dot{x}, \dot{y}$) the images of x and y in E (resp. in F). The field F contains the cube roots of unity $1, \frac{\dot{x}}{\dot{y}}, \frac{\dot{x}}{\dot{y}} + 1$, hence E contains $\Phi(\dot{1}) = \bar{1}, \Phi(\frac{\dot{x}}{\dot{y}})$ and $\Phi(\frac{\dot{x}}{\dot{y}} + \dot{1})$ which are necessarily cube roots of unity. Suppose that $\Phi(\frac{\dot{x}}{\dot{y}}) = \Phi(\dot{1})$, then $\Phi(\frac{\dot{x}}{\dot{y}} + \dot{1}) = 2\Phi(\dot{1}) = 0$ which is a contradiction. Similarly, we have $\Phi(\frac{\dot{x}}{\dot{y}} + 1) \neq \Phi(\dot{1})$. Suppose that $\Phi(\frac{\dot{x}}{\dot{y}}) = \Phi(\frac{\dot{x}}{\dot{y}} + 1)$, then $\Phi(\dot{1}) = 0$ which is a contradiction. It follows that E contains a cube root of unity $\alpha \neq 1$. Set $\alpha = \frac{\overline{h_0(y) + h_1(y)x}}{\overline{D(y)}}$. Using the relations $\bar{x}^2 = \bar{y} + \bar{1}$ and $\bar{x}^3 = \bar{x}(\bar{y} + \bar{1})$, we conclude that $\alpha^3 = 1$ is equivalent to

$$(\bar{y} + \bar{1})\overline{h_1(y)}^2(\overline{xh_1(y)} + \overline{h_0(y)}) + \overline{h_0(y)}^3 + \overline{h_0(y)}^2\overline{h_1(y)x} + \overline{D(y)}^3 = 0.$$

We deduce the following two equations

$$\overline{h_1(y)}\big(\overline{h_0(y)^2} + (\bar{y} + \bar{1})\overline{h_1(y)^2}\big) = 0$$
$$(\bar{y} + \bar{1})\overline{h_1(y)^2 h_0} + \overline{h_0(y)}^3 + \overline{D(y)}^3 = 0.$$

Obviously $\overline{h_1(y)} \neq 0$, hence by the first equation $y + 1$ is a square in $\mathbb{F}_2[y]$, which is a contradiction.

Exercise 3.8.
Let K be a field of characteristic $p > 0$, E be an algebraic and separable extension of K. Let $\alpha_1, \ldots, \alpha_n \in E$.

(1) Show that $K(\alpha_1, \ldots, \alpha_n) = K(\alpha_1^p, \ldots, \alpha_n^p)$.

(2) Suppose that $\{\alpha_1, \ldots, \alpha_n\}$ is a basis of E over K. Show that $\{\alpha_1^p, \ldots, \alpha_n^p\}$ is also a basis of E over K.

Solution 3.8.

(1) Obviously $K(\alpha_1^p, \ldots, \alpha_n^p) \subset K(\alpha_1, \ldots, \alpha_n)$. Let Ω be an algebraic closure of K. Consider the map Φ from the set of the K-embeddings $\sigma : K(\alpha_1, \ldots, \alpha_n) \to \Omega$ into the set of the K- embeddings $\tau : K(\alpha_1^p, \ldots, \alpha_n^p) \to \Omega$ such that $\Phi(\sigma)$ is the restriction of σ to $K(\alpha_1^p, \ldots, \alpha_n^p)$. It is easy to see that Φ is injective. Moreover since $K(\alpha_1, \ldots, \alpha_n)/K(\alpha_1^p, \ldots, \alpha_n^p)$ is algebraic, then Φ is surjective. Therefore, since E is separable, we have

$$\begin{aligned}[K(\alpha_1, \ldots, \alpha_n) : K] &= [K(\alpha_1, \ldots, \alpha_n) : K]_s \\ &= [K(\alpha_1^p, \ldots, \alpha_n^p) : K]_s \\ &= [K(\alpha_1^p, \ldots, \alpha_n^p) : K].\end{aligned}$$

We conclude that $K(\alpha_1^p, \ldots, \alpha_n^p) = K(\alpha_1, \ldots, \alpha_n)$.

(2) Since $n = [K(\alpha_1, \ldots, \alpha_n) : K] = [K(\alpha_1^p, \ldots, \alpha_n^p) : K]$, then to prove that $\{\alpha_1^p, \ldots, \alpha_n^p\}$ is a basis of $K(\alpha_1^p, \ldots, \alpha_n^p)$, it is sufficient to show that these elements generate the vector space $K(\alpha_1^p, \ldots, \alpha_n^p)$. Let $\gamma \in K(\alpha_1^p, \ldots, \alpha_n^p) = K[\alpha_1^p, \ldots, \alpha_n^p]$, then γ may be written in the form

$$\gamma = \sum a_{i_1 \ldots i_n} (\alpha_1^p)^{i_1} \cdots (\alpha_n^p)^{i_n} = \sum a_{i_1 \ldots i_n} (\alpha_1^{i_1} \cdots \alpha_n^{i_n})^p.$$

Express $\alpha_1^{i_1} \cdots \alpha_n^{i_n}$ as a linear combination of $\alpha_1, \ldots, \alpha_n$ with coefficients in K, say: $\alpha_1^{i_1} \cdots \alpha_n^{i_n} = \sum_{j=1}^n b_j^{(i_1, \ldots, i_n)} \alpha_j$. Then

$$\gamma = \sum a_{i_1 \ldots i_n} \left(\sum_{j=1}^n b_j^{(i_1, \ldots, i_n)} \alpha_j \right)^p = \sum a_{i_1 \ldots i_n} \sum_{j=1}^n \left(b_j^{(i_1, \ldots, i_n)} \right)^p (\alpha_j)^p.$$

Therefore $\gamma \in K\alpha_1^p + \cdots + K\alpha_n^p$.

Exercise 3.9.

Let K be a field, $f(x) \in K[x]$ be monic irreducible of degree n and let α be a root of $f(x)$ in an algebraic closure of K.

(1) Let $r_K(f)$ be the number of roots of $f(x)$ lying in $K(\alpha)$. Show that $r_K(f) = |\operatorname{Aut}_K(K(\alpha))|$. Deduce that $r_K(f)$ does not depend on the choice of α.

(2) Let $s_K(f)$ be the number of distinct conjugates of the field $K(\alpha)$ over K. Show that $r_K(f) s_K(f) = [K(\alpha) : K]_s$. Deduce that $r_K(f) \mid n$.

(3) Let E be an extension of K. Show that the number of roots of $f(x)$ lying in E (resp. in the complement of E in its algebraic closure) is a multiple of $r_K(f)$.

(4) Let p be a prime number, $f(x) \in \mathbb{Z}[x]$ be an irreducible polynomial of the form

$$f(x) = x^{2p} + a_{2p-2}x^{2p-2} + \cdots + a_4x^4 + a_0.$$

Show that $r_{\mathbb{Q}}(f) = 2$.

Solution 3.9.

(1) Let $A = \{\beta \in K(\alpha), f(\beta) = 0\}$ and let $\phi : A \to \operatorname{Aut}_K(K(\alpha))$ be the map such that $\phi(\beta)$ is the unique K-automorphism σ of $K(\alpha)$ satisfying the condition $\sigma(\alpha) = \beta$. Then clearly, ϕ is one to one, hence $r_K(f) = |\operatorname{Aut}_K(K(\alpha))|$. Let β be a root of $f(x)$ and let $\sigma : K(\alpha) \to K(\beta)$ be the unique K-isomorphism such that $\sigma(\alpha) = \beta$. Let $\psi : \operatorname{Aut}_K(K(\alpha)) \to \operatorname{Aut}_K(K(\beta))$ be the map defined by $\psi(\rho) = \sigma\rho\sigma^{-1}$ for any $\rho \in \operatorname{Aut}_K(K(\alpha))$, then ψ is an isomorphism of groups. Therefore $|\operatorname{Aut}_K(K(\alpha))| = |\operatorname{Aut}_K(K(\beta))|$, which implies that $r_K(f)$ does not depend on the choice of α.

(2) Let Ω be the set of roots of $f(x)$ in an algebraic closure of K. Define the following relation in Ω. Two elements $\alpha, \beta \in \Omega$ are related if $K(\alpha) = K(\beta)$. Clearly this defines an equivalence relation. All the classes have the same cardinality, namely $r_K(f)$. Moreover $s_K(f)$ represents the number of classes for this relation. It follows that

$$r_K(f)s_K(f) = |\Omega| = [K(\alpha) : K]_s.$$

Since this last integer divides $n = [K(\alpha) : K]$, then $r_K(f) \mid n$.

(3) If E contains no root of $f(x)$, then the result is obvious. Suppose that E contains some root α of $f(x)$, then E contains the class $\bar{\alpha}$ of α. Let k be the number of classes contained in E, then E contains exactly $kr_K(f)$ roots of $f(x)$. The number of roots of $f(x)$ not in E is equal to the difference between: the total number of distinct roots of $f(x)$ (which is equal to $[K(\alpha) : K]_s$) and the number of roots of the polynomial lying in E. Since $r_K(f)$ divides both of these integers, it divides their difference.

(4) Obviously, if α is a root of $f(x)$, then $-\alpha$ is also a root of $f(x)$. Therefore $r_{\mathbb{Q}}(f)$ is even and $r_{\mathbb{Q}}(f) \geq 2$. Since $r_{\mathbb{Q}}(f) \mid 2p$, then $r_Q(f) = 2$ or $2p$. By **(3)**, $r_{\mathbb{Q}}(f)$ divides the number of real roots of $f(x)$. Looking at the derivative of $f(x)$, it is seen that the number of real roots of $f(x)$ is at most equal to $2p - 2$. Therefore $r_{\mathbb{Q}}(f) = 2$.

Exercise 3.10.
Let K be a field, Ω be an algebraic closure of K, $f(x,y) \in \Omega[x,y]$, E be the field generated by the coefficients of f over K. Suppose that E/K is separable and let $[E:K]=n$ and $\sigma_1,\dots\sigma_n$ be the distinct K-isomorphisms of E into Ω. Let α be a primitive element of E/K and let $g_i(x,y) = \sum_{j=1}^{n}\sigma_j(\alpha^i)f^{\sigma_j}(x,y)$ for $i=0,\dots,n-1$. Show that $g_i(x,y) \in K[x,y]$. Suppose that $f^{\sigma_1},\dots,f^{\sigma_n}$ are linearly independent over Ω. Show that $g_0,\dots,g_{n-1}$ are linearly independent over K.

Solution 3.10.
Let σ be a K-isomorphism of E into Ω. Then

$$g_i^{\sigma}(x,y) = \sum_{j=1}^{n}\sigma(\sigma_j(\alpha^i))f^{\sigma\circ\sigma_j}(x,y) = g_i(x,y),$$

hence $g_i(x,y) \in K[x,y]$. We have $\begin{pmatrix} g_0 \\ g_1 \\ \cdot \\ \cdot \\ \cdot \\ g_{n-1}\end{pmatrix} = A\begin{pmatrix} f^{\sigma_1} \\ f^{\sigma_2} \\ \cdot \\ \cdot \\ \cdot \\ f^{\sigma_n}\end{pmatrix}$, where A is the matrix

$$A = \begin{pmatrix} 1 & 1 & \cdots & 1 \\ \sigma_1(\alpha) & \sigma_2(\alpha) & \cdots & \sigma_n(\alpha) \\ \cdots & \cdots & \cdots & \cdots \\ \sigma_1(\alpha)^{n-1} & \sigma_2(\alpha)^{n-1}\cdots & \cdots & \sigma_n(\alpha)^{n-1}\end{pmatrix}.$$

Clearly this matrix represents a non singular endomorphism of Ω^n and also of $(\Omega[x,y])^n$. Therefore if the g_j, where linearly dependent over K, then the f^{σ_i} would be linearly dependent over Ω.

Exercise 3.11.
Let $n \geq 4$. We say that the number field K of degree n over $\mathbb{Q}$ satisfies the property $\mathcal{P}$ if it is generated by some $\theta \in K$ such that its minimal polynomial over $\mathbb{Q}$ has the form $f(x) = x^n + ax + b$. Let r be the number of real embeddings of K into $\mathbb{C}$. Show that if this property holds for K, then $r < 4$.

Solution 3.11.
Suppose that $\mathcal{P}$ holds for K. By looking at the sign of $f'(x)$, it is easy to conclude that if n is even then $f(x)$ has 2 real roots or none and if n is is odd $f(x)$ has 1 or 3 real roots. Hence in any case, the number of real roots of f is at most 3. Therefore $r \leq 3$.

Exercise 3.12.
Let K be an algebraically closed field of characteristic $p > 0$, $E = K(x, y)$ and $F = K(x^{1/p}, y^{1/p})$. For any $c \in K$, let $F_c = E\big((x + cy)^{1/p}\big)$.

(1) Show that if $c \neq c'$, then $F_c \neq F_{c'}$.
(2) Deduce that although F/E is algebraic of finite degree, there may be infinitely many fields Ω such that $E \subset \Omega \subset F$.

Solution 3.12.

(1) Suppose that $c \neq c'$ and $F_c = F_{c'}$. Let
$$\alpha_c = (x + cy)^{1/p} = x^{1/p} + c^{1/p}y^{1/p},$$
then we have
$$y^{1/p} = (\alpha_c - \alpha_{c'})/(c^{1/p} - c'^{1/p}) \quad \text{and}$$
$$x^{1/p} = (c^{1/p}\alpha_c' - c'^{1/p}\alpha_c)/(c^{1/p} - c'^{1/p}),$$
hence $F_c = F$. But we have $E \subset F_c \subset F$, $[F_c : E] = p$ and $[F : E] = p^2$ and this contradicts the equality of the fields F_c and F.
(2) Clearly F/E is algebraic and $[F : E] = p^2$. The family $(F_c)_{c\in K}$ is an infinite family of distinct fields intermediate between E and F. Therefore the assumption of separability is necessary for the conclusion of finiteness of the intermediate fields.

Exercise 3.13.
Let K be a field, E be a separable extension of K of degree n, $\{\alpha_1, \ldots, \alpha_n\}$ be a basis of E over K and $\{\beta_1, \ldots, \beta_n\}$ be the dual basis (for the trace map). Let $\sigma_1, \ldots, \sigma_n$ be the distinct K-embeddings of E into an algebraic closure of K. Show that for any $(i, j) \in \{1, \ldots, n\}^2$,
$$\sigma_i(\beta_1)\sigma_j(\alpha_1) + \cdots + \sigma_i(\beta_n)\sigma_j(\alpha_n) = \delta_{ij},$$
in particular $\alpha_1\beta_1 + \cdots + \alpha_n\beta_n = 1$.

Solution 3.13.
Let $A = \Big(\sigma_j(\alpha_i)\Big)_{(i,j)\in\{1,\ldots,n\}^2}$, $B = \Big(\sigma_j(\beta_i)\Big)_{(i,j)\in\{1,\ldots,n\}^2}$ and B^t be the transpose of B, then
$$\begin{aligned} AB^t &= \Big(\mathrm{Tr}_{E/K}(\alpha_i\beta_j)\Big)_{(i,j)\in\{1,\ldots,n\}^2} \\ &= \Big(\delta_{ij}\Big)_{(i,j)\in\{1,\ldots,n\}^2} \\ &= I_n, \end{aligned}$$

where I_n is the identity matrix. This implies that B^t is a right inverse of A, hence fully an inverse of A. It follows that $B^tA = I_n$, which gives the stated relations.

Exercise 3.14.
Let K be a field, E be a finite extension of K, $L(E,K)$ be the set of linear forms $\rho : E \to K$. Let H be the set of the maps $\rho : E \to K$ for which there exists $\theta \in E$ such that $\rho(\gamma) = \mathrm{Tr}_{E/K}(\theta\gamma)$, for any $\gamma \in E$.

(1) Show that H is a subgroup of $L(E,K)$.
(2) Show that E/K is separable if and only if $H = L(E,K)$.

Solution 3.14.

(1) For any $\alpha \in E$, denote by T_α the map such that $T_\alpha(\gamma) = \mathrm{Tr}_{E/K}(\alpha\gamma)$ for any $\gamma \in E$. Obviously $H \neq \emptyset$. Let T_α and T_β be elements of H and $\lambda \in K$, then

$$\begin{aligned}(T_\alpha + T_\beta)(\gamma) &= T_\alpha(\gamma) + T_\beta(\gamma)\\ &= \mathrm{Tr}_{E/K}(\alpha\gamma) + \mathrm{Tr}_{E/K}(\beta\gamma)\\ &= \mathrm{Tr}_{E/K}((\alpha+\beta)\gamma)\\ &= T_{\alpha+\beta}(\gamma)\end{aligned}$$

and

$$\lambda T_\alpha(\gamma) = \lambda\,\mathrm{Tr}_{E/K}(\alpha\gamma) = T_{\lambda\alpha}(\gamma).$$

Hence H is a subgroup of $L(E,K)$.

(2) Suppose that E/K is separable and let $\rho : E \to K$ be a linear form. Let $\{\alpha_1,\dots,\alpha_n\}$ be a basis of E over K. We show that there exists $\theta \in A$ such that $\rho(\alpha_i) = \mathrm{Tr}_{E/K}(\theta\alpha_i)$ for any $i \in \{1,\dots,n\}$. Set $\theta = \sum_{i=1}^n a_j\alpha_j$. We must show that the following system of linear equations has a solution $(a_1,\dots,a_n)$ in K^n,

$$\rho(\alpha_i) = \sum_{j=1}^n a_j\,\mathrm{Tr}_{E/K}(\alpha_i\alpha_j) \quad \text{for} \quad i = 1,\dots,n.$$

The determinant D of this system is given by $D = \left|\mathrm{Tr}_{E/K}(\alpha_i\alpha_j)\right|$. Let $\sigma_1,\dots,\sigma_n$ be the distinct K-embeddings of E into an algebraic closure of K and let let $\Delta = \left|\sigma_i(\alpha_j)\right|$, then it is seen that $D = \Delta^2$. Since $\alpha_1,\dots,\alpha_n$ are linearly independent over K, then so are the column vectors of Δ, hence $\Delta \neq 0$ and then $D \neq 0$. It follows that the above system has a unique solution $(a_1,\dots,a_n) \in K^n$.

Suppose that E/K is not separable, then the characteristic of K is a prime number, say p. Since

$$[E : K] = [E : K]_i[E : K]_s = p^e[E : K]_s,$$

where e is a positive integer [Lang (1965), Cor. 1, Chap. 7.4], then the trace map from E into K is the zero map. Hence $H = \{0\}$. Obviously $L(E, K) \neq \{0\}$. We conclude that $H \neq L(E, K)$.

Chapter 4

Normal extensions

Exercise 4.1.
Let $\alpha \in \mathbb{C}$ be a root of $f(x) = x^4 - 2x^3 + 2x^2 - 2x + 1$. Is $\mathbb{Q}(\alpha)$, a normal extension of $\mathbb{Q}$?

Solution 4.1.
We have $f(x) = (x-1)^2(x^2+1)$, hence $\alpha = 1$ or $\alpha = \pm i$. In the first case $\mathbb{Q}(\alpha) = \mathbb{Q}$, hence normal over $\mathbb{Q}$. In the second case $\mathbb{Q}(\alpha) = \mathbb{Q}(i)$, so that it is the splitting field of the polynomial $x^2 + 1$, hence normal over $\mathbb{Q}$.

Exercise 4.2.
Let K be a field, E be a normal extension of K, $g(x) \in E[x]$ be monic and irreducible. Let F be the field generated over K by the coefficients of $g(x)$ and let $f(x) = \mathrm{N}_{F/K}(g(x))$. Suppose that F is separable over K.

(1) Show that $f(x) \in K[x]$ and $f(x)$ is irreducible over K.
(2) Show that neither of the conditions E/K is normal, F/K is separable can be omitted.

Solution 4.2.

(1) Let $n = [F : K] = [F : K]_s$ and let $\sigma_1, \ldots, \sigma_n$ be the distinct embeddings of F into E and σ_1 is the canonical injection. Then $f(x) = \prod_{i=1}^n g^{\sigma_i}(x)$. Since for any $i = 1, \ldots, n$, $g^{\sigma_i}(x) \in E[x]$, then $f(x) \in E[x]$. Therefore to prove that $f(x) \in K[x]$, it is equivalent to show that for any K-automorphism τ of E, we have $f^\tau(x) = f(x)$. Let τ be one such automorphism, then $f^\tau(x) = \prod_{i=1}^n g^{\tau\circ\sigma_i}(x)$. It is clear that $\tau \circ \sigma_i$ is a K-embedding of F into E and $\tau \circ \sigma_i \neq \tau \circ \sigma_j$ if $i \neq j$. Thus $f^\tau(x) = f(x)$ and $f(x) \in K[x]$.

We now prove the irreducibility of $f(x)$ over K. Since $g(x)$ is irreducible over E, then for any $i = 1, \ldots, n$, $g^{\sigma_i}(x)$ is irreducible over E, thus the factorization $f(x) = \prod_{i=1}^{n} g^{\sigma_i}(x)$ is a factorization of $f(x)$ into a product of irreducible factors in $E[x]$. Let $h(x)$ be an irreducible factor of $f(x)$ in $K[x]$, then there exists $i \in \{1, \ldots, n\}$ such that $g^{\sigma_i}(x) \mid h(x)$. We may suppose that $i = 1$, that is $g(x) \mid h(x)$. It follows that $\prod_{i=1}^{n} g^{\sigma_i}(x) \mid h(x)$, thus $f(x) \mid h(x)$ and then $f(x) = h(x)$ is irreducible over K.

(2) We discuss the normality and separability conditions.

- **Normality condition.** Let $E = \mathbb{Q}\left(j, 2^{1/3}\right)$, where j is a primitive cube of unity and

$$g(x) = x^2 + x2^{1/3} + 4^{1/3} = \left(x - j2^{1/3}\right)\left(x - j^2 2^{1/3}\right).$$

Then $F = \mathbb{Q}\left(2^{1/3}\right)$ and

$$\begin{aligned} f(x) &= \left(x^2 + x2^{1/3} + 4^{1/3}\right)\left(x^2 + xj2^{1/3} + j^2 4^{1/3}\right)\left(x^2 + xj^2 2^{1/3} + j4^{1/3}\right) \\ &= \left(x^3 - 2\right)^2. \end{aligned}$$

This shows that $f(x)$ is not irreducible over K.

- **Separability condition.** Let $K = \mathbb{F}_2(s, t)$, where s and t are algebraically independent variables over $\mathbb{F}_2$. Let $E = K(\sqrt{s}, \sqrt{t})$ and $g(x) = x + \sqrt{s} + \sqrt{t}$, then $F = E$ and since F/K is purely inseparable, then

$$f(x) = (x + \sqrt{s} + \sqrt{t})^4 = (x^2 + s + t)^2,$$

hence $f(x)$ is reducible over K.

Exercise 4.3.

Let K be a field, Ω be an algebraic closure of K, $\overrightarrow{\alpha} = (\alpha_1, \ldots, \alpha_n)$ and $\overrightarrow{\beta} = (\beta_1, \ldots, \beta_n)$ be elements of Ω^n. We say that these n-tuples are conjugate over K if there exists a K-automorphism σ of Ω such that $\sigma(\alpha_i) = \beta_i$ for $i = 1, \ldots, n$.

(1) Show that $\overrightarrow{\alpha}$ and $\overrightarrow{\beta}$ are conjugate over K if and only if

$$\text{for any } P(x_1, \ldots, x_n) \in K[x_1, \ldots, x_n],\ P(\overrightarrow{\alpha}) = 0 \Rightarrow P(\overrightarrow{\beta}) = 0.$$

(2) Determine the conjugates of $\overrightarrow{\alpha} = (\sqrt{2}, 2^{1/4})$ over $\mathbb{Q}$.

Solution 4.3.

(1) • **Necessity of the condition.** Let $P(x_1,\dots,x_n) \in K[x_1,\dots,x_n]$, such that $P(\overrightarrow{\alpha}) = 0$ and let σ be a K-automorphism of Ω such that $\sigma(\alpha_i) = \beta_i$ for $i = 1,\dots,n$. We denote by $\sigma(\overrightarrow{\alpha})$ the element $(\sigma(\alpha_1),\dots,\sigma(\alpha_n))$. We have $\sigma(P(\overrightarrow{\alpha})) = 0$, hence $P^\sigma(\sigma(\overrightarrow{\alpha})) = 0$, where P^σ denotes the polynomial obtained from P by applying σ to its coefficients. Therefore $P(\overrightarrow{\beta}) = 0$.

• **Sufficiency of the condition.** Let $P_1(x)$ be the minimal polynomial of α_1 over $\mathbb{Q}$, then $P_1(\beta_1) = 0$. Therefore there exist a K-isomorphism $\sigma : K(\alpha_1) \to \Omega$ such that $\sigma_1(\alpha_1) = \beta_1$. Suppose that we have constructed a K-isomorphism $\sigma_k : K(\alpha_1,\dots\alpha_k) \to \Omega$ such that $\sigma_k(\alpha_i) = \beta_i$ for $i = 1,\dots,k$. Since α_{k+1} is algebraic over $K(\alpha_1,\dots\alpha_k) = K[\alpha_1,\dots\alpha_k]$, there exists $P_{k+1} \in K[x_1,\dots,x_k][x_{k+1}]$, irreducible such that

$$P_{k+1}(\alpha_1,\dots\alpha_k)(\alpha_{k+1}) = 0,$$

hence $P_{k+1}(\beta_1,\dots\beta_k)(\beta_{k+1}) = 0$. Therefore we may extend σ_k to a K-isomorphism $\sigma_{k+1} : K(\alpha_1,\dots\alpha_{k+1}) \to \Omega$ such that $\sigma_{k+1}(\alpha_i) = \beta_i$ for $i = 1\dots,k+1$. We apply this conclusion for $k = n$ and extend σ_k to a K-isomorphism $\sigma : \Omega \to \Omega$. Since Ω/K is normal, then σ is a K-automorphism of Ω.

(2) Let $\sigma : \bar{\mathbb{Q}} \to \bar{\mathbb{Q}}$ be an automorphism. Since the minimal polynomial of $2^{1/4}$ over $\mathbb{Q}$ is $x^4 - 2$, then its conjugates are $\pm 2^{1/4}$, $\pm i2^{1/4}$. Therefore $\sigma(2^{1/4}) = i^j 2^{1/4}$, where $j = 0,1,2,3$. Since $\sqrt{2} = (2^{1/4})^2$, then

$$\sigma(\sqrt{2}) = (i^j 2^{1/4})^2 = (-1)^j\sqrt{2}.$$

Thus the list of conjugates of $(\sqrt{2}, 2^{1/4})$ is given by

$$(\sqrt{2}, 2^{1/4}), (-\sqrt{2}, i2^{1/4}), (\sqrt{2}, -2^{1/4}) \text{ and } (-\sqrt{2}, -i2^{1/4}).$$

Exercise 4.4.
Let K be a field, E be a finite normal extension of K. If E is separable over K, show that E is the splitting field of some monic irreducible polynomial $f(x)$ over K. Moreover $f(x)$ may be chosen such that $\deg f = [E : K]$. Show that the condition of separability cannot be omitted.

Solution 4.4.
Let $\alpha \in E$ be a primitive element over K, $f(x)$ be its minimal polynomial and $\alpha_1 = \alpha,\dots,\alpha_n$ be its conjugates. Then $E = K(\alpha) = K(\alpha_1,\dots,\alpha_n)$, hence E is the splitting field of $f(x)$ over K. Here $\deg f = [E : K]$.

For the condition of separability, let α (resp. β) be a root of $x^2 - T$ (resp. $x^2 - Z$) in an algebraic closure of $K = \mathbb{F}_2(T, Z)$. Let $E = K(\alpha, \beta)$, then E is normal over K. Suppose that E is the splitting field of some monic irreducible polynomial of degree d over K. Then obviously $d = 4$. On the other hand, let

$$\gamma = A(T, Z) + B(T, Z)\alpha + C(T, Z)\beta + D(T, Z)\alpha\beta$$

be an element of E, then

$$\gamma^2 = A^2 + B^2T + C^2Z + D^2TZ,$$

thus γ is a root of a monic polynomial of degree 2 over K.

Exercise 4.5.
Let K be a field of characteristic $p > 0$, and E be an algebraic extension of K. Let

$$E_s = \{\alpha \in E, \alpha \quad \text{separable over} \quad K\} \quad \text{and}$$
$$E_i = \{\alpha \in E, \alpha \quad \text{purely inseparable over} \quad K\}.$$

We say that E splits over K if $E = E_i \cdot E_s$. For any $\alpha \in E$ let $s_k(\alpha)$ denote the elementary symmetric function of degree k of the distinct conjugates of α.

(1) Show that the following propositions are equivalent:
   (i) E splits over K.
   (ii) E is separable over E_i.
   (iii) For any $\alpha \in E, K(s_1(\alpha), \dots, s_m(\alpha)) \subset E$, where $m = [K(\alpha) : K]_s$.
   (iv) For any $\alpha \in E$, $\mathrm{Irr}(\alpha, K, x)$ is a p^e-th power in $E(x)$ where e is a non negative integer defined by the relation $p^e = [K(\alpha) : K]_i$.
(2) Show that if E is normal over K, then E splits over K.
(3) Let $\alpha \in E, m = [K(\alpha) : K]_s$ and $p^e = [K(\alpha) : K]_i$. Show that $K(\alpha)$ splits over K if and only if there exists $l \in \{1, \dots, m\}$ such that $K(s_1(\alpha), \dots, s_m(\alpha)) = K(s_l(\alpha))$.

Solution 4.5.

(1)
- *(i)* $\Rightarrow$ *(ii)*. Since $E = E_i(E_s)$ and since any element of E_s is separable over K, thus separable over E_i, then E is separable over E_i.
- *(ii)* $\Rightarrow$ *(iii)*. Let $\alpha \in E$, $m = [K(\alpha) : K]_s$, $\alpha_1, \dots, \alpha_m$ be the distinct conjugates of α over K and $s_k(\alpha)$ be the elementary symmetric function of $\alpha_1, \dots, \alpha_m$ of degree k. Consider the following diagram of fields.

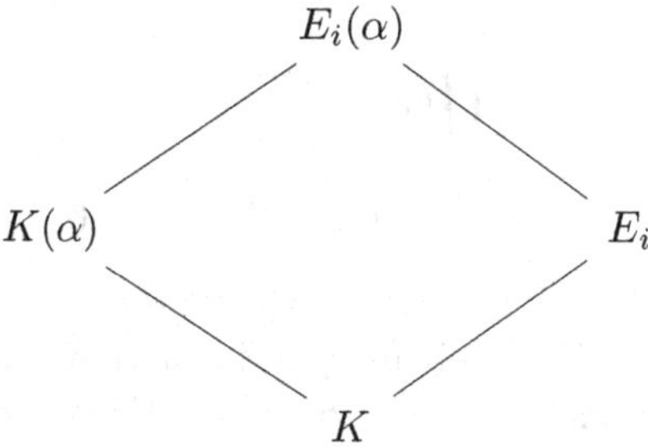

Since $E_i(\alpha) = K(\alpha)(E_i)$, then $E_i(\alpha)$ is purely inseparable over $K(\alpha)$. It follows that

$$[E_i(\alpha) : E_i]_s = [K(\alpha) : K]_s = m.$$

Since α is separable over E_i, then $[E_i(\alpha) : E_i] = m$. We deduce that $\mathrm{Irr}(\alpha, E_i, x) = \prod_{i=1}^{m}(x - \alpha_i)$ and then $s_k(\alpha) \in E_i \subset E$ for $k = 1, \ldots, m$.

- $(iii) \Rightarrow (iv)$. Let $\alpha \in E$, $p^e = [K(\alpha) : K]_i$ and $m = [K(\alpha) : K]_s$. Let $f(x) = \mathrm{Irr}(\alpha, K, x)$, then

$$\begin{aligned} f(x) &= \prod_{i=1}^{m}(x - \alpha_i)^{p^e} \\ &= \left(\prod_{i=1}^{m}(x - \alpha_i)\right)^{p^e} \\ &= (x^m - s_1(\alpha)x^{m-1} + \cdots + (-1)^m s_m(\alpha))^{p^e}. \end{aligned}$$

Since $s_k(\alpha) \in E$ for $k = 1, \ldots, m$ then $f(x)$ is a p^e-th power in $E(x)$.

- $(iv) \Rightarrow (i)$. Let $\alpha \in E$, $m = [K(\alpha) : K]_s$, $p^e = [K(\alpha) : K]_i$,

$$\begin{aligned} f(x) &= \mathrm{Irr}(\alpha, K, x) \\ &= \prod_{i=1}^{m}(x - \alpha_i)^{p^e} \\ &= \left(\prod_{i=1}^{m}(x - \alpha_i)\right)^{p^e} \\ &= (g(x))^{p^e}, \end{aligned}$$

where $g(x) = x^m - s_1(\alpha)x^{m-1} + \cdots + (-1)^m s_m(\alpha)$.

Let

$$h(x) = \prod_{i=1}^{m}(x - \alpha_i^{p^e})$$
$$= x^m - c_1x^{m-1} + \cdots + (-1)^m c_m,$$

then $c_k = (s_k)^{p^e}$. Since $h(x^{p^e}) = f(x)$, then $c_k \in K$ for $k = 1, ..., m$. It follows that s_k is purely inseparable over K. By **(iv)**, we conclude that $s_k \in E_i$. Since α is a root of $g(x) \in E_i[x]$, then $[E_i(\alpha) : E_i] \leq m$. On the other hand, we use the following diagram:

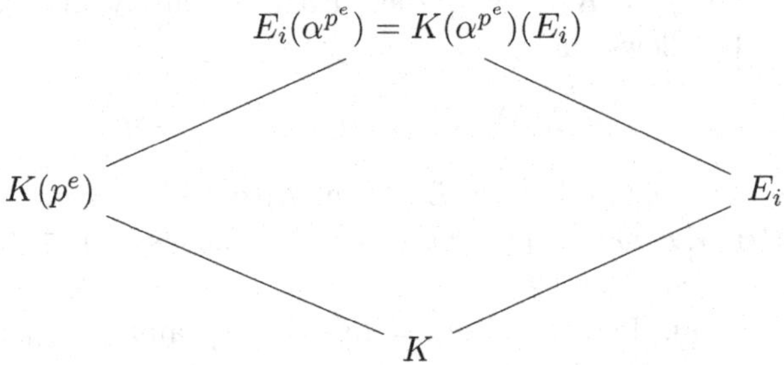

Since E_i/K and $E_i(\alpha^{p^e})/K(\alpha^{p^e})$ are purely inseparable, then

$$m = [K(\alpha^{p^e}) : K] = [K(\alpha^{p^e}) : K]_s = [E_i(\alpha^{p^e}) : E_i]_s.$$

Now since $E_i \subset E_i(\alpha^{p^e}) \subset E_i(\alpha)$, then $E_i(\alpha^{p^e}) = E_i(\alpha)$, thus $\alpha \in E_i(\alpha^{p^e}) \subset E_i \cdot E_s$

(2) We prove that **(iii)** is satisfied. Let $\alpha \in E, m = [K(\alpha) : K]_s$ and $\alpha_1, \ldots, \alpha_m$ the distinct conjugates of α over K, then $\alpha_i \in E$ for $i = 1, \ldots, m$, hence $s_k(\alpha) \in E$ for $k = 1, \ldots, m$.

(3) Suppose first that $e = 0$, that is α is separable over K. In this case $s_k(\alpha) \in K$ for $k = 1, \ldots, m$ and $K(s_1(\alpha), \ldots, s_m(\alpha)) = K$. We also have $K(\alpha)_i = K$. We conclude that the two conditions $K(\alpha)$ is separable over $K(\alpha)_i$ and

$$K(s_1(\alpha), \ldots, s_m(\alpha)) = K = K(s_1(\alpha))$$

hold. Suppose now that $e \geq 1$. We begin with the necessity of the condition. Let $\alpha_1, \ldots, \alpha_m$ be the the distinct conjugates of α,

$$g(x) = \prod_{i=1}^{m}(x - \alpha_i) = x^m - s_1(\alpha)x^{m-1} + \cdots + (-1)^m s_m(\alpha)$$

and

$$h(x) = \prod_{i=1}^{m}(x - \alpha_i^{p^e}) = x^m - c_1 x^{m-1} + \cdots + (-1)^m c_m.$$

Since $f(x) = h(x^{p^e}) = (g(x))^{p^e}$, then $c_k \in K$ and $c_k = (s_k(\alpha))^{p^e}$ for $k = 1, \ldots, m$.
Since α is not separable over K, there exists $k_0 \in \{1, \ldots, m\}$ such that $s_{k_0}(\alpha) \notin K$. Capelli's theorem [Lang (1965), Th. 16, Chap. 8.9] implies that the polynomial $x^{p^e} - c_{k_0}$ is irreducible over K. Therefore $[K(s_{k_0}(\alpha)) : K] = p^e$. On the other hand since $K(\alpha)$ splits over K, then $K(s_1(\alpha), \ldots, s_m(\alpha)) \subset K(\alpha)$. We have

$$\begin{aligned}[K(s_1(\alpha), \ldots, s_m(\alpha)) : K] &= [K(s_1(\alpha), \ldots, s_m(\alpha)) : K]_i \\ &\leq [K(\alpha) : K]_i.\end{aligned}$$

It follows that $K(s_1(\alpha), \ldots, s_m(\alpha)) = K(s_{k_0}(\alpha))$. We now prove the sufficiency of the condition. Since $K(s_1(\alpha), \ldots, s_m(\alpha))$ is purely inseparable over K, then

$$[K(s_1(\alpha), \ldots, s_m(\alpha), \alpha) : K(s_1(\alpha), \ldots, s_m(\alpha))]_s = m.$$

Since α is a root of

$$g(x) = x^m - s_1(\alpha)x^{m-1} + \cdots + (-1)^m s_m(\alpha),$$

then

$$[K(s_1(\alpha), \ldots, s_m(\alpha), \alpha) : K(s_1(\alpha), \ldots, s_m(\alpha))] \leq m.$$

It follows that

$$\begin{aligned}&[K(s_1(\alpha), \ldots, s_m(\alpha), \alpha) : K(s_1(\alpha), \ldots, s_m(\alpha))] \\ &\quad = [K(s_1(\alpha), \ldots, s_m(\alpha), \alpha) : K(s_1(\alpha), \ldots, s_m(\alpha))]_s \\ &\quad = m.\end{aligned}$$

Therefore,

$$\begin{aligned}&[K(s_1(\alpha), \ldots, s_m(\alpha), \alpha) : K] \\ &\quad = [K(s_1(\alpha), \ldots, s_m(\alpha), \alpha) : K(s_1(\alpha), \ldots, s_m(\alpha))] \\ &\qquad .[K(s_1(\alpha), ..., s_m(\alpha), \alpha) : K] \\ &\quad = m[K(s_{k_0}) : K] \\ &\quad = mp^e = [K(\alpha) : K].\end{aligned}$$

We deduce that

$$K(s_1(\alpha), \ldots, s_m(\alpha), \alpha) = K(\alpha)$$

that is $K(s_1(\alpha), \ldots, s_m(\alpha)) \subset K(\alpha)$, which means that $K(\alpha)$ splits over K.

Exercise 4.6.
Let K be a field, $f(t)$ be a non constant polynomial with coefficients in K. Suppose that $f(x) - f(y) = f_1(x, y) \cdots f_r(x, y)$, where $f_i(x, y) \in K[x, y]$ and $\deg f_i = 1$ for $i = 1, \dots, r$. Show that $K(t)/k(f(t))$ is normal.

Solution 4.6.
Clearly $f_i(x, y)$ has the form $f_i(x, y) = a_i x + b_i y + c_i$, where a_i, b_i, $c_i \in K$ and $a_i b_i \neq 0$. Let $F(x) = f(x) - f(t)$, then $F(x)$ is the minimal polynomial of t over $K(f(t))$. From the factorization of $f(x) - f(y)$, it is seen that the roots of $F(x)$ in an algebraic closure of $K(f(t))$ are given by $t_i = (-b_i t - c_i)/a_i$, hence

$$K(t) = K(t_1, \dots, t_r) = K(f(t))(t_1, \dots, t_r)$$

is the decomposition field of $F(x)$ over $K(f(t))$. Thus $K(t)/K(f(t))$ is normal.

Exercise 4.7.

(1) Let $\overline{\mathbb{Q}}$ be the algebraic closure of $\mathbb{Q}$ contained in $\mathbb{C}$. Show that $\mathbb{R} \cap \overline{\mathbb{Q}}$ is not normal over $\mathbb{Q}$.
(2) Let K be the subset of $\mathbb{R} \cap \overline{\mathbb{Q}}$ whose elements satisfy the condition: all its conjugates are real numbers. Show that K is a field and $K/\mathbb{Q}$ is normal.

Solution 4.7.

(1) The irreducible polynomial over $\mathbb{Q}$, $f(x) = x^3 - 2$ has a root, namely $\sqrt[3]{2}$, in $\mathbb{R} \cap \overline{\mathbb{Q}}$ but not all its roots in this field. Hence $(\mathbb{R} \cap \overline{\mathbb{Q}})/\mathbb{Q}$ is not normal.
(2) Let α and β be elements of K and let $\sigma : K \to \overline{\mathbb{Q}}$ be an embedding, then since $\sigma(\alpha)$ and $\sigma(\beta)$ are real, so are $\sigma(\alpha + \beta)$ and $\sigma(\alpha\beta)$, hence $\alpha + \beta$ and $\alpha\beta \in K$. Similarly we may prove that $1/\alpha \in K$ if $\alpha \neq 0$. Thus K is a subfield of $\overline{\mathbb{Q}}$. Let $f(x)$ be a monic, irreducible polynomial with rational coefficients. Suppose that it has a root θ in K. Then the conjugates of θ are real, hence $f(x)$ splits into linear factors over K.

Chapter 5

Galois extensions, Galois groups

Exercise 5.1.
A Pythagorean triple is a triple (a, b, c) of positive integers such that $a^2+b^2 = c^2$. Let a, b, c be positive integers. By using Hilbert's Theorem 90, show that (a, b, c) is a Pythagorean triple if and only if, permuting if necessary a and b, there exist positive integers m, n, λ such that $\gcd(m, n) = 1$,

$$a = \lambda(m^2 - n^2), \quad b = 2\lambda mn \quad \text{and} \quad c = \lambda(m^2 + n^2).$$

Solution 5.1.

- **Sufficiency of the condition.** Obvious.
- **Necessity of the condition.** Let $i = \sqrt{-1}$ and $K = \mathbb{Q}(i)$. The identity $a^2 + b^2 = c^2$ may be written in the form $N_{K/\mathbb{Q}}(a/c+ib/c) = 1$. Since $K/\mathbb{Q}$ is a cyclic extension, then by Hilbert's Theorem 90 [Lang (1965), Hilbert's Theorem 90, Chap. 8.6], there exist rational integers m_1, n_1 such that

  $$a/c + ib/c = (m_1 + in_1)/(m_1 - in_1) = (m_1 + in_1)^2/(m_1^2 + n_1^2).$$

  It follows that

  $$a/c = (m_1^2 - n_1^2)/(m_1^2 + n_1^2) \quad \text{and}$$
  $$b/c = 2m_1n_1/(m_1^2 + n_1^2).$$

  Let $d = \gcd(m_1, n_1)$and let m and n such that $m_1 = dm$ and $n_1 = dn$, then

  $$a/c = (m^2 - n^2)/(m^2 + n^2) \quad \text{and}$$
  $$b/c = 2mn/(m^2 + n^2).$$

  It follows that

  $$a = \lambda(m^2 - n^2) \quad \text{and} \quad b = \lambda(2mn)$$

with $\lambda = c/(m^2+n^2) \in \mathbb{Q}^+$. It is easy to see that a or b is even. Transposing them if necessary, we may suppose that b is even. This being noticed, let p be a prime number dividing the denominator of λ. Then $p \mid m^2 - n^2$ and $p \mid 2mn$. If $p \neq 2$, then $p \mid m$ and $p \mid n$, which is a contradiction. If $p = 2$, then since $2 \nmid m$ and $2 \nmid n$, then b is odd, which is a contradiction. Thus λ is a positive integer and we get the announced formulas.

Exercise 5.2.
A field K is said to be formally real if -1 is not a sum of squares in K. If moreover K has no proper algebraic extension which is formally real, it is said to be real closed. A field K is said to be an ordered field if K is equipped with a total order $<$ satisfying the following conditions: For any $a, b, c \in K$, if $a < b$ then $a + c < b + c$ and if $a < b$ and $0 < c$ then $ac < bc$.

(1) Let K be an ordered field and $a \in K^\star$. Show that $a^2 > 0$. Show that the characteristic of K is 0.
(2) Let K be a formally real field, $f(x)$ be a monic irreducible polynomial with coefficients in K and α be a root of $f(x)$ in an algebraic closure of K. show that if $\deg f$ is odd, then $K(\alpha)$ is formally real.
(3) Suppose that K is formally real and that for any $a \in K$, a or $-a$ is a square in K. Show that any sum of squares of K is a square in K.
(4) Let K be a formally real field. Show that there exists an algebraic extension of K which is real closed.
(5) Let K be a real closed field. Show that any sum of squares in K is a square in K.
(6) Let K be a real closed field, $f(x)$ be a monic irreducible polynomial of degree n with coefficients in K. Let $g(x) \in K[x]$ be an irreducible factor of $f(x)$ of degree m. If m is odd, show that $m = 1$. Deduce that if n is odd, then $f(x)$ has a root in K.
(7) Let K be a field and $S_K = \{a \in K, a \text{ is a sum of squares in } K\}$. Show that the following conditions are equivalent.
    (i) K is formally real.
    (ii) $S_K \neq K$ and the characteristic of K is not equal to 2.
    (iii) If $\sum a_i^2 = 0$, where the sum is finite, then $a_i = 0$ for all i.
    (iv) K may be equipped with a total order such that it will be an ordered field.
(8) Show that $(\mathbb{R} \cap \overline{\mathbb{Q}})(i) = \overline{\mathbb{Q}}$ and that $\mathbb{R} \cap \overline{\mathbb{Q}}$ is real closed.
(9) Let K be a formally real field, m be a positive integer. Let $a_1, \ldots, a_m$

be positive elements of K such that for any $1 \leq i \leq m-1$, $\sqrt{a_{i+1}} \notin K(\sqrt{a_1}, \ldots, \sqrt{a_i})$. Let $E = K(\sqrt{a_1}, \ldots, \sqrt{a_m})$. Show that E is formally real.

(10) Let K be a real closed field.

  (a) Show that the polynomial x^2+1 is irreducible over K.

  (b) Let i be a root of x^2+1 in an algebraic closure of K. Show that any polynomial $f(x) = x^2 + ux + v$, with coefficients in $K(i)$, has its roots in $K(i)$.

  (c) Show that $K(i)$ is algebraically closed.

Solution 5.2.

(1) Notice that, since for any $a \in K^\star$, $a+(-a) = 0$, then $a > 0$ if and only if $-a < 0$. Now to prove that $a^2 > 0$, we may suppose that $a > 0$. Then multiplying by a, we get the desired result. Suppose that the characteristic of K is a prime number p. Then $1+1+\cdots+1 = 0$. We have $0 < 1 < 1+1 < \cdots < 1+1+\cdots+1 = 0$, which is a contradiction. Thus the characteristic of K is 0.

(2) Suppose that the implication does not hold and let $f(x) \in K(x)$ be monic irreducible such that the implication fails for this polynomial. We may suppose that $f(x)$ is chosen so that $\deg f$ is minimal among all counterexamples. We will get a contradiction. Set $\deg f = 2n+1$. Then $-1 = \sum_{i=1}^r u_i(\alpha)^2$, where r is a positive integer, $u_i(x) \in K[x]$ and $\deg u_i \leq 2n$. It follows that

$$1 + \sum_{i=1}^{r} u_i(x)^2 = f(x)g(x), \qquad \text{(Eq 1)}$$

where $g(x) \in K[x]$. Let $d = \max_{i=1}^r \deg u_i$ and $i_1, \ldots, i_s$ such that $\deg u_{i_j} = d$. For $j = 1, \ldots, s$, let a_{i_j} be the leading coefficient of u_{i_j}. Then the leading coefficient of the left side of **(Eq 1)** is equal to $\sum a_{i_j}^2$ and since K is formally real, this coefficient does not vanish. Hence the degree of the left side of **(Eq 1)** is equal to $2d$ thus at most equal to $4n$. We deduce that $\deg g$ is odd and $\deg f + \deg g \leq 4n$, hence $\deg g \leq 2n-1$. Let $h(x)$ be a monic irreducible factor of $g(x)$ in $K[x]$ and β be any of its roots. From **(Eq 1)**, we have $-1 = \sum_{i=1}^r u_i(\beta)^2$, showing that -1 is a sum of squares in $K(\beta)$ contradicting the minimality property of the degree of $f(x)$.

(3) Let $a_1^2 + \cdots + a_m^2$ be a sum of squares in K, where we may suppose that $m \geq 2$ and $a_i \neq 0$ for $i = 1, \ldots, m$. Suppose that this sum is not a square in K, then by assumption the opposite is a square, that is

$-(a_1^2+\cdots+a_m^2) = a^2$ with $a \in K$. It follows that $a_1^2+\cdots+a_m^2+a^2 = 0$, contradicting the fact that K is formally real.

(4) Let Ω be an algebraic closure of K. Let $\mathcal{E}$ be the set of subfields of Ω which are formally real. This set is ordered by the inclusion and any chain of its elements has an upper bound, namely their union. By Zorn's Lemma, let E be a maximal element of $\mathcal{E}$. We show that E is real closed. Obviously, E is formally real. Let F be an algebraic extension of E which is formally real, then F is algebraic over K. Thus we may suppose that $F \subset \Omega$ and then $F = E$, which implies that E is real closed.

(5) Let $a \in K$. Suppose that a is a sum of squares in K but not a square in K. Then the polynomial $x^2 - a$ is irreducible over K. It follows that $K(\sqrt{a})$ is not formally real. Thus
$$-1 = (a_1 + b_1\sqrt{a})^2 + \cdots + (a_m + b_m\sqrt{a})^2,$$
where m is a positive integer and $a_i, b_i \in K$ for $i = 1, \ldots, m$. We deduce that
$$-1 = \sum_{i=1}^{m} a_i^2 + a\sum_{i=1}^{m} b_i^2.$$
Setting $a = \sum_{j=1}^{k} c_j^2$, we obtain
$$-1 = \sum_{i=1}^{m} a_i^2 + \sum_{(i,j)} (c_j b_i)^2,$$
contradicting the fact that K is formally real.

(6) Let $g(x)$ be an irreducible factor of $f(x)$ in $K[x]$ of degree m. Suppose that m is odd and let α be a root of $g(x)$ in an algebraic closure of K. Then, by **(2)**, $K(\alpha)$ is formally real. Since K is real closed then $K(\alpha) = K$, hence $m = 1$.
If n is odd, then $f(x)$ has an irreducible factor $g(x)$ in $K[x]$ of odd degree. Apply what was just proved and conclude that $\deg g = 1$, that is $f(x)$ has a root in K.

(7) • **(ii) ⇒ (i)**. We prove the transpose of this implication. Suppose that -1 is a sum of squares in K and the characteristic of K is not equal to 2. Set $-1 = \sum a_i^2$. Let $a \in K$, then
$$2a = (a+1)^2 - (a^2+1) = (a+1)^2 + \left(\sum a_i^2\right)(a^2+1),$$
hence
$$a = 2\left(\left(\frac{a+1}{2}\right)^2 + \sum\left(\frac{aa_i}{2}\right)^2 + \sum a_i^2\right).$$
Since $2 = 1^2 + 1^2$, then a is a sum of squares, thus $S_K = K$.

- **(i) ⇒ (ii)**. Obvious.
- **(i) ⇒ (iii)**. Suppose that $\sum a_i^2 = 0$, where the sum is finite, and one of the a_i, say a_1, is non zero. We then get $-1 = \sum_{i \neq 1}(a_i/a_1)^2$, contradicting **(i)**.
- **(iii) ⇒ (i)**. Obvious.
- **(iv) ⇒ (i)**. Denote by $\leq$ be the order in K and suppose that $-1 = \sum a_i^2$, where the sum is finite and $a_i \neq 0$ for any i. By **(1)**, $a_i^2 > 0$ for any i, hence $\sum a_i^2 > 0$, a contradiction.
- **(i) ⇒ (iv)**. By **(4)**, let E be an algebraic extension of K which is real closed. If we can define an order in E, which makes this field as an ordered field, then K itself will be an ordered field. Thus we deal with E. Define in E the relation $a \leq b$ if $b-a$ is a square in E. Obviously this relation is reflexive, antisymmetric. We prove that the relation is transitive. Suppose that $a \leq b$ and $b \leq c$, then $b-a$ and $c-b$ are squares in E. It follows that their sum is a sum of squares in E, hence a square in E by **(5)**. If $a \leq b$, then obviously $a+c \leq b+c$. Suppose that $a \leq b$ and $c > 0$, then $b-a$ and c are squares, hence $cb - ca$ is a square; thus $ca \leq cb$.

(8) The inclusion $(\mathbb{R} \cap \overline{\mathbb{Q}})(i) \subset \overline{\mathbb{Q}}$ is obvious. For the reverse inclusion, let $z = a + bi \in \overline{\mathbb{Q}}$ with a and $b \in \mathbb{R}$. Then
$$a = (z+\bar{z})/2 \in (\mathbb{R} \cap \overline{\mathbb{Q}}) \text{ and } b = (z-\bar{z})/2i \in (\mathbb{R} \cap \overline{\mathbb{Q}}).$$
It follows that $z \in (\mathbb{R} \cap \overline{\mathbb{Q}})(i)$. Since the field $(\mathbb{R} \cap \overline{\mathbb{Q}})$ is a subfield of $\mathbb{R}$, then it is formally real. According to **(4)**, we may consider an algebraic extension of it which is real closed. Denote this real closed field by E. Then
$$(\mathbb{R} \cap \overline{\mathbb{Q}}) \subset E \subset \overline{\mathbb{Q}} = (\mathbb{R} \cap \overline{\mathbb{Q}})(i),$$
hence $\mathbb{R} \cap \overline{\mathbb{Q}} = E$ and the conclusion follows.

(9) Notice that the set of elements of E of the form $\prod_{i=1}^m \sqrt{a_i}^{e_i}$ with $e_i = 0$ or $e_i = 1$ is a basis of E over K. Thus any element α of E has the form $\alpha = \sum b_{e_1,\ldots,e_m} \prod_{i=1}^m \sqrt{a_i}^{e_i}$, where $b_{e_1,\ldots,e_m} \in K$. The square of this element has the form
$$\alpha^2 = \sum b_{e_1,\ldots,e_m}^2 \prod_{i=1}^m a_i^{e_i} + \sum c_{u_1,\ldots,u_m} \prod_{i=1}^m \sqrt{a_i}^{u_i},$$
where the u_i have similar meaning as the e_i. Suppose that -1 is a sum of squares in E, then expressing each square in the basis of E over K which was explicitly given above, we conclude that -1 is a sum of the form $-1 = \sum b_{e_1,\ldots,e_m}^2 \prod_{i=1}^m a_i^{e_i}$. Since a_i is positive for $i = 1, \ldots, m$, then the right side of this identity is positive, hence a contradiction.

(10)(a) Since -1 is not a sum of squares in K, then it is not a square, hence x^2+1 is irreducible over K.

(b) The polynomial $f(x)$ has its roots in K if and only if u^2-4v is a square in $K(i)$. Thus to get the result, it is sufficient to prove that any $a+bi \in K(i)$ is a square in $K(i)$. Let $a+bi \in K(i)$, then the equation $a+bi=(x+iy)^2$ is equivalent to $a=x^2-y^2$ and $b=2xy$. If $b=0$, $a>0$ then $a=x^2$ and $(x,0)$ is a solution of this system of equations. If $b=0$, $a<0$, then $a=-y^2$ and $(0,y)$ is a solution. Suppose now that $b \neq 0$. Then $y \neq 0$ and multiplying $x+iy$ by -1 if necessary, we may suppose that $x>0$. From the second equation appearing in the system, we get $x=b/2y$. Replacing in the first equation, we obtain that x is a solution of the following: $4x^4-4ax^2-b^2=0$. Set $x^2=s$, then $4s^2-4as-b^2=0$. The discriminant of this equation is equal to $16(a^2+b^2)$. By **(5)** any sum of squares in K is a square in K. Thus $a^2+b^2=c^2$ for some $c \in K$, which we may suppose to be positive. It follows that $s=(a \pm c)/2$. Choose the sign $+$, then since $c^2>a^2$, we have $c>|a|$, which implies that $c+a>0$ and the equation $x^2=s=(a+c)/2$ is solvable in K. We now have

$$x=\sqrt{(a+c)/2} \text{ and } y=\frac{1}{2}\frac{b}{\sqrt{(a+c)/2}}.$$

(c) Suppose that there exists a strict algebraic extension E of $K(i)$ and let F be the Galois closure of E over K. Let $n=[F:K]$ and $G=\mathrm{Gal}(F,K)$. Set $n=2^k m$, where k is a non negative integer and m is a positive, odd integer. Let H be a 2-Syllow's subgroup of G, $L=K(\theta)$ be its fixed field and $f(x)=\mathrm{Irr}(\theta,K)$. Since $\deg f=[L:K]=m$ and m is odd, then by **(6)**, $f(x)$ has a root in K. Thus $m=1$ and then $n=2^k$. We deduce that $|\mathrm{Gal}(F,K(i))|=2^{k-1}$ and since F strictly contains $K(i)$, then $k \geq 2$. Let H_1 be a subgroup of $\mathrm{Gal}(F,K(i))$ of order 2^{k-2} and L_1 be its fixed field. Then $K(i) \subset L_1 \subset F$ and $[L_1:K(i)]=2$. This implies that there exists a monic irreducible polynomials of degree 2 with coefficients in $K(i)$, contradicting **(10)**, **(b)**.

Exercise 5.3.

Let K be a field and G be a group of ring-automorphisms of $K[t]$ such that, for any $\sigma \in G$, $\sigma(K)=K$. Extend in a natural way the action of G on

$K(t)$ and let

$$K^G = \{a \in K, \sigma(a) = a\},$$
$$K[t]^G = \{f(t) \in K[t], \sigma(f(t)) = f(t)\} \quad \text{and}$$
$$K(t)^G = \{f(t) \in K(t), \sigma(f(t)) = f(t)\}.$$

(1) Show that $K(t)^G = \mathrm{Frac}(K[t]^G)$.
(2) Show that there exists $\beta(t) \in K[t]^G$ such that $K(t)^G = K^G(\beta(t))$.

Solution 5.3.

(1) We first prove that if $\sigma \in G$, then $\sigma(t) = at+b$ for some $a, b \in K$, $a \neq 0$. Since $\sigma^{-1}(\sigma(t)) = t$, then $\deg \sigma(t) = \deg \sigma^{-1}(t) = 1$, hence the result. It is obvious that $Fr(K[t]^G) \subset K(t)^G$. Let $u(t)$ and $v(t) \in K[t]$ such that $\gcd(u(t), v(t)) = 1$. Suppose that $u(t)/v(t) \in K(t)^G$. Then, we may suppose that $v(t)$ is not constant and $v(t)\sigma(u(t)) = u(t)\sigma(v(t))$, hence $u(t) \mid \sigma(u(t))$. If we set $\sigma(u(t)) = u(t)q(t)$ with $q(t) \in K[t]$, we obtain $\sigma(v(t)) = v(t)q(t)$. Writing down a Bezout's identity

$$a(t)u(t) + b(t)v(t) = 1,$$

yields the similar one

$$\sigma(a(t))\sigma(u(t)) + \sigma(b(t))\sigma((v(t)) = 1,$$

which implies $q(t) \in K^\star$. We now have

$$\sigma(u(t)) = cu(t) \quad \text{and} \quad \sigma(v(t)) = cv(t)$$

for some $c \in K^\star$. Suppose that we have proved that there exists $e \in K^\star$ such that $c = e/\sigma(e)$, then we have

$$u/v = (eu)/(ev),$$
$$\sigma(eu) = \sigma(e)\sigma(u) = eu \quad \text{and}$$
$$\sigma(ev) = \sigma(e)\sigma(v) = ev$$

and the proof is finished. We prove the existence of such an e. By performing euclidean divisions, we obtain:

$$u(t) = v(t)q_1(t) + r_1(t),$$
$$v(t) = r_1(t)q_2(t) + r_2(t) \quad \text{and}$$
$$r_i(t) = r_{i+1}(t)q_{i+2}(t) + r_{i+2}(t), \quad \text{for} \quad i = 1, \dots, k,$$

with $r_{k+2}(t) \in K^\star$ and $\deg r_{i+1}(t) < \deg r_i(t)$ for all i. Apply σ to the first equation and obtain $cu = cv\sigma(q_1) + \sigma(r_1)$. We subtract from this equation the first one multiplied by c and we obtain:

$$-cv\big(\sigma(q_1) - q_1\big) = \big(\sigma(r_1) - cr_1\big).$$

Since v is non constant, $\deg \sigma(q_1) = \deg q_1$ and $\deg \sigma(r_1) = \deg r_1$, then $\sigma(q_1) = q_1$ and $\sigma(r_1) = cr_1$. By induction we may prove that $\sigma(q_i) = q_i$ and $\sigma(r_i) = cr_i$. Now let $e = 1/r_{k+2}$, then it satisfies the required properties.

(2) By **(1)**, it is sufficient to prove that there exists $\beta(t) \in K[t]$ such that $K[t]^G = K^G[\beta(t)]$. To prove this we distinguish two cases. If $K[t]^G$ contains no non constant polynomial, then we may take $\beta(t) = 1$. Suppose next that $K[t]^G$ contains non constant polynomials. Let $\beta(t)$ be such a polynomial whose degree is minimal. Clearly $K^G[\beta(t)] \subset K[t]^G$. We prove the other inclusion. Let $f(t) \in K[t]^G$. Expand this polynomial in the $\beta(t)$-basis [Lang (1965), Th. 9, Chap. 5.5]. We obtain $f(t) = \sum_{i=0}^{k} a_i(t)\beta^i(t)$, where k is a non negative integer and $a_i(t) \in K[t]$, with $\deg a_i < \deg \beta(t)$ for $i = 0, \ldots, k$. For any $\sigma \in G$, we have $f(t) = \sigma(f(t)) = \sum_{i=0}^{k} \sigma(a_i(t))\beta^i(t)$, hence $\sigma(a_i(t)) = a_i(t)$ for any $\sigma \in G$ and for $i = 0, \ldots, k$. Since the degree of $\beta(t)$ is minimal among non constant polynomials in $K[t]^G$, then $a_i(t) \in K$. Therefore $a_i(t) \in K^G$ and then $f(t) \in K^G[\beta(t)]$.

Exercise 5.4.
Let $f(x) = x^4 - 2x^2 + 2$ and $\alpha \in \mathbb{C}$ be a root of $f(x)$.

(1) Show that $f(x)$ is irreducible over $\mathbb{Q}$. Without the knowledge of $\text{Gal}(f(x), \mathbb{Q})$, determine the fields F (if any) such that

$$\mathbb{Q} \subsetneqq F \subsetneqq \mathbb{Q}(\alpha). \qquad \text{(Eq 1)}$$

(2) Find $\text{Gal}(f(x), \mathbb{Q})$ and compute again the fields F satisfying the conditions **(Eq 1)**.

Solution 5.4.

(1) By Eisenstein's irreducibility theorem, applied with the prime 2, we conclude that $f(x)$ is irreducible over $\mathbb{Q}$. Let F be a field satisfying the conditions **(Eq 1)** and let $g_F(x)$ be the minimal polynomial of α over F. Since $[F : \mathbb{Q}] = 2$, then $g_F(x) = (x-\alpha)(x-\beta) = x^2 + ax + b$, where $a, b \in F$ and β is a root of $f(x)$ distinct from α. Moreover $F = \mathbb{Q}(a, b)$. Since the roots of f are given by

$$\alpha_1 = \alpha = \sqrt{\frac{\sqrt{2}+1}{2}} + i\sqrt{\frac{\sqrt{2}-1}{2}}, \; \alpha_2 = -\alpha, \; \alpha_3 = \bar{\alpha} \text{ and } \alpha_4 = -\bar{\alpha},$$

where $\bar{\alpha}$ is the complex conjugate of α. We must consider the following three possibilities for $g_F(x)$.

- $g_F(x) = (x-\alpha)(x+\alpha) = x^2 - \alpha^2 = x^2 - (1+i)$.
- $g_F(x) = (x-\alpha)(x-\bar{\alpha}) = x^2+(\alpha+\bar{\alpha})x+|\alpha|^2 = x^2-2\sqrt{\frac{\sqrt{2}+1}{2}}x+\sqrt{2}$.
- $g_F(x) = (x-\alpha)(x+\bar{\alpha}) = x^2+(\bar{\alpha}-\alpha)x-|\alpha|^2 = x^2-2i\sqrt{\frac{\sqrt{2}-1}{2}}x-\sqrt{2}$.

In the first case, $F = \mathbb{Q}(\alpha^2) = \mathbb{Q}(i)$ and this field satisfies the conditions **(Eq 1)**.

In the second case $F = \mathbb{Q}(\sqrt{\frac{\sqrt{2}+1}{2}}, \sqrt{2})$, which is not a quadratic field.

In the third case $F = \mathbb{Q}(i\sqrt{\frac{\sqrt{2}-1}{2}}, \sqrt{2})$ and this field is not a quadratic field.

We conclude that there exists one and only one field F satisfying the conditions **(Eq 1)**, namely $F = \mathbb{Q}(i)$.

(2) Since the roots of $f(x)$ are α, $-\alpha$, $\bar{\alpha}$, $-\bar{\alpha}$ and since $\bar{\alpha} = \sqrt{2}/\alpha$, then $\mathbb{Q}(\alpha, \sqrt{2})$ is the splitting field of $f(x)$ over $\mathbb{Q}$.

We show that $\sqrt{2} \notin \mathbb{Q}(\alpha)$. Suppose that $\sqrt{2} \in \mathbb{Q}(\alpha)$, then $\mathbb{Q}(\alpha) = \mathbb{Q}(\sqrt{2}, i)$. we deduce that there exist $a, b, c, d \in \mathbb{Q}$ such that

$$\alpha = a + b\sqrt{2} + ci + di\sqrt{2}. \qquad \text{(Eq 2)}$$

Using the complex conjugation, we obtain $\bar{\alpha} = a + b\sqrt{2} - ci - di\sqrt{2}$. Adding these two identities, we obtain $a + b\sqrt{2} = \sqrt{\frac{\sqrt{2}+1}{2}}$. It follows that

$$a^2 + 2b^2 + 2ab\sqrt{2} = 1/2 + \sqrt{2}/2,$$

thus

$$a^2 + 2b^2 = 1/2 \quad \text{and} \quad ab = 1/4.$$

These equations may be written in the form $2a^2 + 4b^2 = 1$ and $(2a^2)(4b^2) = 1/8$. This implies that $2a^2$ and $4b^2$ are solutions of the equation $y^2 - y + 1/8 = 0$. Since this equation has no rational solution, then we get a contradiction, thus $\sqrt{2} \notin \mathbb{Q}(\alpha)$ and $\mathbb{Q}(\alpha, \sqrt{2})$ is the splitting field of $f(x)$ over $\mathbb{Q}$.

Any field automorphism σ of $\mathbb{Q}(\alpha, \sqrt{2})$ is completely determined by its action on α and $\sqrt{2}$. We have $\sigma(\sqrt{2}) = \pm\sqrt{2}$ and $\sigma(\alpha) = \pm\alpha$ or $\pm\,\bar{\alpha}$. Since

$$|Gal(f(x), \mathbb{Q})| = [\mathbb{Q}(\alpha, \sqrt{2}) : \mathbb{Q}] = 8,$$

then all the possibilities for σ given above give rise to automorphisms. Let σ_1 be the identity. Label the other automorphisms as follows

$$\begin{cases} \sigma_2(\alpha) = & -\alpha \\ \sigma_2(\sqrt{2}) = & \sqrt{2} \end{cases}, \quad \begin{cases} \sigma_3(\alpha) = & \bar{\alpha} \\ \sigma_3(\sqrt{2}) = & \sqrt{2} \end{cases},$$

$$\begin{cases}\sigma_4(\alpha) = & -\bar{\alpha} \\ \sigma_4(\sqrt{2}) = & \sqrt{2}\end{cases}, \quad \begin{cases}\sigma_5(\alpha) = & \alpha \\ \sigma_5(\sqrt{2}) = & -\sqrt{2}\end{cases},$$

$$\begin{cases}\sigma_6(\alpha) = & -\alpha \\ \sigma_6(\sqrt{2}) = & -\sqrt{2}\end{cases}, \quad \begin{cases}\sigma_7(\alpha) = & \bar{\alpha} \\ \sigma_7(\sqrt{2}) = & -\sqrt{2}\end{cases},$$

and

$$\begin{cases}\sigma_8(\alpha) = & -\bar{\alpha} \\ \sigma_8(\sqrt{2}) = & -\sqrt{2}\end{cases}.$$

It is easy to verify that $\mathrm{Gal}(f(x), \mathbb{Q})$ is generated by σ_3 and σ_7. The first generator is of order 2, while the second is of order 4. Moreover, we have

$$\sigma_3\sigma_7\sigma_3 = \sigma_7^{-1} = \sigma_7^3 = \sigma_8.$$

It follows that $\mathrm{Gal}(f(x), \mathbb{Q})$ is isomorphic to the dihedral group of order 8. Let F be a field satisfying **(Eq 1)** and let $H = \mathrm{Gal}(\mathbb{Q}(\alpha, \sqrt{2}), F)$, then $|H| = 4$ and

$$\mathrm{Gal}(\mathbb{Q}(\alpha, \sqrt{2}), \mathbb{Q}(\alpha)) \subseteq \mathrm{Gal}(\mathbb{Q}(\alpha, \sqrt{2}), F).$$

Therefore H contains σ_1 and σ_5. It is seen that the only subgroup of $\mathrm{Gal}(f(x), \mathbb{Q})$ of order 4 containing σ_5 is $H = \{\sigma_1, \sigma_5, \sigma_2, \sigma_6\}$ and $\mathrm{Inv}(H) = \mathbb{Q}(\alpha^2) = \mathbb{Q}(i)$. We conclude that $\mathbb{Q}(\alpha)$ contains one and only one field F satisfying **(Eq 1)**, namely $\mathbb{Q}(i)$.

Remark. We may prove that $\sqrt{2} \notin \mathbb{Q}(\alpha)$ by using number theory. Let A and B be the rings of integers of $\mathbb{Q}(\sqrt{2})$ and $\mathbb{Q}(\alpha)$ respectively. Since $x^2 - 2$ is irreducible over $\mathbb{F}_5$, then $5A$ is prime. If $\mathbb{Q}(\sqrt{2})$ was contained in $\mathbb{Q}(\alpha)$, then the splitting of $5B$ in B would be $5B = \mathcal{P}_2\mathcal{P}_2'$ or $5A = \mathcal{Q}_4$, where the indices of the ideals $\mathcal{P}_2$, $\mathcal{P}_2'$ and $\mathcal{Q}_4$ denote their respective inertial degree. On the other hand, the factorization of $f(x)$ over $\mathbb{F}_5$ is given by $f(x) = (x-2)(x+2)(x^2+2)$. By **Exercise 10.30**, we have $5B = \mathcal{Q}_1\mathcal{Q}_1'\mathcal{Q}_2$ and then we reach a contradiction. It follows that $\sqrt{2} \notin \mathbb{Q}(\alpha)$.

Exercise 5.5.
Let $f(x) = x^4 - 6x^2 + 6$, α be a root of $f(x)$ in $\mathbb{C}$, E be a splitting field of $f(x)$ and $G = \mathrm{Gal}(f(x), \mathbb{Q})$.

(1) Show that $E = \mathbb{Q}(\alpha, \sqrt{2})$ and G is isomorphic to the Dihedral group of order 8, $D_4 =< \sigma, \tau, \sigma^4 = \tau^2 = 1, \tau\sigma = \sigma^3\tau >$.

(2) Determine a monic irreducible polynomial $g(x) \in \mathbb{Q}[x]$ satisfying the following conditions.
   (a) $\deg g = 4$.
   (b) The splitting field over $\mathbb{Q}$ of $g(x)$ is contained in E.
   (c) The Galois group over $\mathbb{Q}$ of $g(x)$ is not isomorphic to G.

Solution 5.5.

(1) We compute the solutions of the equation $f(x) = 0$ by setting $x^2 = y$. We obtain the quadratic equation $y^2 - 6y + 6 = 0$, whose roots are $3 \pm \sqrt{3}$. We conclude that the roots of $f(x)$ are given by $\epsilon_1\sqrt{3 + \epsilon_2\sqrt{3}}$, where ϵ_1 and $\epsilon_2 = \pm 1$. We may suppose that $\alpha = \sqrt{3 + \sqrt{3}}$. Let $\beta = \sqrt{3 - \sqrt{3}}$. We have $\alpha\beta = \sqrt{3}\sqrt{2}$. Since $\sqrt{3} = \alpha^2 - 3$, then $\sqrt{2} = \alpha\beta/(\alpha^2 - 3)$. We deduce that $E = \mathbb{Q}(\alpha, \beta) = \mathbb{Q}(\alpha, \sqrt{2})$. We show that $\sqrt{2} \notin \mathbb{Q}(\alpha)$. Suppose that the contrary holds, then since $\sqrt{3} \in \mathbb{Q}(\alpha)$, we have
$$\mathbb{Q}(\sqrt{3}, \sqrt{2}) = \mathbb{Q}(\alpha) = \mathbb{Q}(\sqrt{3}, \sqrt{3 + \sqrt{3}}).$$
We use the following result.
Claim. Let F be a field of characteristic not equal to 2 and let a and b be elements of F, not squares. Suppose that $F(\sqrt{a}) = F(\sqrt{b})$. Then a/b is a square in F.
Proof. We have $\sqrt{a} = \lambda + \mu\sqrt{b}$, with λ and $\mu \in F$, then $a = \lambda^2 + b\mu^2 + 2\lambda\mu\sqrt{b}$, hence $\lambda = 0$. It follows that $(\sqrt{a})/(\sqrt{b}) = \mu \in F$, thus a/b is a square in F.
Applying this result for $F = \mathbb{Q}(\sqrt{3})$, we get $(3 + \sqrt{3})/2$ is a square in $\mathbb{Q}(\sqrt{3})$, which implies that $\mathrm{N}_{\mathbb{Q}(\sqrt{3})/\mathbb{Q}}((3+\sqrt{3})/2)$ is a square in $\mathbb{Q}$, that is $3/2$ is a rational square, hence a contradiction. Therefore $\sqrt{2} \notin \mathbb{Q}(\alpha)$.
We now have
$$\begin{aligned}[E : \mathbb{Q}] &= [\mathbb{Q}(\alpha, \sqrt{2}) : \mathbb{Q}] \\ &= [\mathbb{Q}(\alpha, \sqrt{2}) : \mathbb{Q}(\alpha)][\mathbb{Q}(\alpha) : \mathbb{Q}] \\ &= 8.\end{aligned}$$
We deduce that $(G : 1) = 8$. To define any element $\rho \in G$, it is sufficient to determine explicitly $\rho(\alpha)$ and $\rho(\sqrt{2})$. Obviously $\rho(\alpha) = \pm\alpha$ or $\rho(\alpha) = \pm\beta$ and $\rho(\sqrt{2}) = \pm\sqrt{2}$. By taking all possible combinations of signs, we obtain the eight elements of G. Among them consider the automorphisms σ and τ of E such that
$$\begin{cases} \sigma(\alpha) &= \beta \\ \sigma(\sqrt{2}) &= \sqrt{2} \end{cases} \quad \text{and} \quad \begin{cases} \tau(\alpha) &= \alpha \\ \tau(\sqrt{2}) &= -\sqrt{2} \end{cases}.$$

We compute the values of $\sigma(\beta)$ and $\tau(\beta)$. We have

$$\sigma(\beta) = \sigma\left(\frac{\sqrt{2}(\alpha^2-3)}{\alpha}\right) = \frac{\sqrt{2}(\beta^2-3)}{\beta} = -\sqrt{2}\sqrt{3}/\beta = -\alpha \quad \text{and}$$
$$\tau(\beta) = \tau\left(\frac{\sqrt{2}(\alpha^2-3)}{\alpha}\right) = -\frac{\sqrt{2}(\alpha^2-3)}{\alpha} = -\beta.$$

We deduce that

$$\begin{aligned}
\sigma^2(\alpha) &= \sigma(\beta) = -\alpha,\\
\sigma^3(\alpha) &= -\sigma(\alpha) = -\beta,\\
\sigma^4(\alpha) &= -\sigma(\beta) = \alpha,\\
\sigma^2(\sqrt{2}) &= \sigma^3(\sqrt{2}) = \sigma^4(\sqrt{2}) = \sqrt{2} \quad \text{and}\\
\tau^2(\alpha) &= \alpha,\\
\tau^2(\sqrt{2}) &= \sqrt{2}.
\end{aligned}$$

Thus $\sigma^4 = \tau^2 = 1$. We also have

$$\begin{aligned}
\sigma^3\tau(\alpha) &= \sigma^3(\alpha) = -\beta,\\
\tau\sigma(\alpha) &= \tau(\beta) = -\beta,\\
\sigma^3\tau(\sqrt{2}) &= \sigma^3(-\sqrt{2}) = -\sqrt{2} \quad \text{and}\\
\tau\sigma(\sqrt{2}) &= \tau(\sqrt{2}) = -\sqrt{2},
\end{aligned}$$

hence $\sigma^3\tau = \tau\sigma$. We conclude that $G \simeq D_4$.

(2) Since

$$\sqrt{3} = \alpha^2 - 3 \text{ and } \sqrt{2} = \alpha\beta/(\alpha^2-3),$$

then $\sqrt{3} \in E$ and $\sqrt{2} \in E$. Let $\gamma = \sqrt{3} + \sqrt{2}$ and $g(x) = \mathrm{Irr}(\gamma, \mathbb{Q}, x)$. It is seen that $g(x) = x^4 + 2x^2 - 11$ and the splitting field of $g(x)$ is equal to $\mathbb{Q}(\sqrt{3}, \sqrt{2})$. Moreover $\mathrm{Gal}(\gamma, \mathbb{Q}) \simeq \mathbb{Z}/2\mathbb{Z} \times \mathbb{Z}/2\mathbb{Z}$, so that the conditions **(a), (b) and (c)** hold.

Exercise 5.6.
Let K be a field, $f(X) = X^n - a_1X^{n-1} + \cdots + (-1)^n a_n \in K[X]$ be separable and let $\alpha_1, \ldots, \alpha_n$ be its roots in an algebraic closure of K.

(1) Let

$$\mathcal{M} = \{P(x_1, \ldots, x_n) \in K[x_1, \ldots, x_n]; P(\alpha_1, \ldots, \alpha_n) = 0\}.$$

Show that $\mathcal{M}$ is a maximal ideal of $K[x_1, \ldots, x_n]$.

(2) Let $G = \text{Gal}(f(x), K)$ and $\tau \in \mathcal{S}_n$. Show that the following assertions are equivalent.

(i) $\tau \in \text{Gal}(f(x), K)$.

(ii) For any polynomial $P(x_1, \ldots, x_n) \in K[x_1, \ldots, x_n]$, we have $P(\alpha_1, \ldots, \alpha_n) = 0 \Rightarrow P(\tau(\alpha_1), \ldots, \tau(\alpha_n)) = 0$.

(3) For any $\tau \in \mathcal{S}_n \setminus G$, let

$$S_\tau = \{P(x_1, \ldots, x_n) \in \mathcal{M}; P(\tau(\alpha_1), \ldots, \tau(\alpha_n)) \neq 0\}$$

(a) Show that S_τ is a multiplicative semigroup of $K[x_1, \ldots, x_n]$ contained in $\mathcal{M}$.

(b) If $|K| \geq n! - 1$, show that for any $\tau_1, \ldots, \tau_m \in \mathcal{S}_n \setminus G$, we have $S_{\tau_1} \cap \cdots \cap S_{\tau_m} \neq \emptyset$.

(4) Let $g(x) = x^4 - 6x^2 + 6$. By using **(2)**, show that the Galois group of $g(x)$ over $\mathbb{Q}$ is isomorphic to the dihedral group of order 8.

Solution 5.6.

(1) Let $\Phi : K[x_1, \ldots, x_n] \to K[\alpha_1, \ldots, \alpha_n] = K(\alpha_1, \ldots, \alpha_n)$ be the map such that $\Phi(P(x_1, \ldots, x_n)) = P(\alpha_1, \ldots, \alpha_n)$, then Φ is a morphism of rings. We have $\text{Ker}\,\Phi = \mathcal{M}$ and then

$$K[x_1, \ldots, x_n]/\mathcal{M} \simeq \text{Im}\,\Phi = K(\alpha_1, \ldots, \alpha_n).$$

It follows that $\mathcal{M}$ is maximal.

(2) • *(i)* $\Rightarrow$ *(ii)*. Let $P(x_1, \ldots, x_n) \in K[x_1, \ldots, x_n]$ such that $P(\alpha_1, \ldots, \alpha_n) = 0$. Then since $\tau \in \text{Gal}(f(x), K)$, we have

$$0 = \tau(P(\alpha_1, \ldots, \alpha_n)) = P(\tau(\alpha_1), \ldots, \tau(\alpha_n)),$$

thus *(ii)*.

• *(ii)* $\Rightarrow$ *(i)*. Let $E = K(\alpha_1, \ldots, \alpha_n)$ be the splitting field of $f(x)$ over K and let τ be a permutation of $\alpha_1, \ldots, \alpha_n$ satisfying the condition *(ii)*. We extend the definition of τ to the whole E and obtain an application from E into E. Let

$$x \in E = K(\alpha_1, \ldots, \alpha_n) = K[\alpha_1, \ldots, \alpha_n].$$

Set $x = P(\alpha_1, \ldots, \alpha_n)$, where P is a polynomial with coefficients in K. Define $\tau(x)$ by $\tau(x) = P(\tau(\alpha_1), \ldots, \tau(\alpha_n))$. We show that τ is well defined. Suppose that

$$P(\alpha_1, \ldots, \alpha_n) = Q(\alpha_1, \ldots, \alpha_n),$$

then

$$(P - Q)(\alpha_1, \ldots, \alpha_n) = 0$$

and by **(ii)**,

$$(P - Q)(\tau(\alpha_1), \ldots, \tau(\alpha_n)) = 0,$$

thus

$$P(\tau(\alpha_1), \ldots, \tau(\alpha_n)) = Q(\tau(\alpha_1), \ldots, \tau(\alpha_n)).$$

Obviously τ is a morphism of fields. Moreover it is a K-morphism of E into E and since E/K is algebraic, it is an automorphim [Lang (1965), Chap. 7.2 Lemma 1], hence $\tau \in \mathrm{Gal}(E, K)$

(3)(a) From **(2)** it is seen that, for any $\tau \in \mathcal{S}_n \setminus G$, $S_\tau \neq \emptyset$. Obviously S_τ is a semigroup of $K[x_1, \ldots, x_n]$ contained in $\mathcal{M}$.

(b) The proof is done by induction on m. For $m = 1$ the result is proved in **(a)**. Suppose that $S_{\tau_1} \cap \cdots \cap S_{\tau_{m-1}} \neq \emptyset$ and let P be an element of this intersection. The proof is finished if $P \in S_{\tau_m}$. If not let $Q \in S_{\tau_m}$. We may suppose that $Q \notin S_{\tau_1} \cap \cdots \cap S_{\tau_{m-1}}$. If for all $i \in \{1, \ldots, m-1\}$, $Q \notin S_{\tau_i}$, the result is established since $P + Q \in S_{\tau_1} \cap \cdots \cap S_{\tau_m}$. It remains to consider the case where for some non empty subset J of $\{1, \ldots, m-1\}$, $|J| \leq m-2$, we have $Q \in S_{\tau_j}$ for any $j \in J$. For any $j \in \{1, m-1\}$, consider the equation (in x):

$$Q(\tau(\alpha_1), \ldots, \tau(\alpha_n)) + xP(\tau(\alpha_1), \ldots, \tau(\alpha_n)) = 0.$$

Any of these equations has at most one solution in K. Since $m-1 < n! - 1 \leq |K|$, we may choose $\lambda \in K$ distinct from all the roots of the preceding equations. Let $R = Q + \lambda P$, then one verifies that $R \in S_{\tau_1} \cap \cdots \cap S_{\tau_m}$.

(4) Eisenstein's theorem, shows that $g(x)$ is irreducible over $\mathbb{Q}$. The roots of $g(x)$ are given by

$$\alpha_1 = \sqrt{3 + \sqrt{3}},\ \alpha_2 = -\alpha_1,\ \alpha_3 = \sqrt{3 - \sqrt{3}} \text{ and } \alpha_4 = -\alpha_3.$$

We will use the following non trivial relations between the roots of $g(x)$: $\alpha_1 + \alpha_2 = 0$ and $\alpha_3 + \alpha_4 = 0$. Let $\tau \in S_4$, then since $\alpha_1 + \alpha_2 + \alpha_3 + \alpha_4 = 0$, we have $\tau(\alpha_1) + \tau(\alpha_2) = 0$ if and only if $\tau(\alpha_3) + \tau(\alpha_4) = 0$. Thus, for the determination of the Galois group of $g(x)$, we use the relation $\alpha_1 + \alpha_2 = 0$ and forget the second relation. We compute $\tau(\alpha_1) + \tau(\alpha_2)$

for the 24 permutations of $\alpha_1, \ldots, \alpha_4$. We find that these expressions vanish for the following permutations:

$$\begin{aligned}
&\tau_0 = \mathrm{Id}, \tau_1 = (\alpha_1\, \alpha_3\, \alpha_2\, \alpha_4), \tau_2 = (\alpha_3\, \alpha_4), \tau_3 = \tau_1^2 = (\alpha_1\, \alpha_2)(\beta_1\, \beta_2),\\
&\tau_4 = \tau_1^3 = (\alpha_1\, \alpha_4\, \alpha_2\, \alpha_3), \tau_5 = \tau_1\tau_2 = (\alpha_1\, \alpha_3)(\alpha_2\, \alpha_4),\\
&\tau_6 = \tau_1^2\tau_2 = (\alpha_1\, \alpha_2), \tau_7 = \tau_1^3\tau_2 = (\alpha_1\, \alpha_4)(\alpha_2\, \alpha_3).
\end{aligned}$$

For all the other permutations, these expressions do not vanish since the value of each of them is one of the followings:

$$\alpha_1 + \alpha_3,\ \alpha_1 + \alpha_4,\ \alpha_2 + \alpha_4,\ \alpha_2 + \alpha_3.$$

By **(2)**, it follows that the Galois group of $g(x)$ is contained in the set formed by $\tau_0, \ldots, \tau_7$. Since the splitting field of $g(x)$ contains $\mathbb{Q}(\alpha_1)$, then the order of its Galois group is a multiple of 4 no greater than 8. The identity $\alpha_1\alpha_3 = \sqrt{6}$, and the observation that $\sqrt{6}$ does not belong to $\mathbb{Q}(\alpha_1)$, show that the degree of the splitting field is at least equal to 8. It follows the Galois group is the set whose elements are the permutations $\tau_0, \ldots, \tau_7$. Now, τ_1 is of order 4, τ_2 is of order 2, and we have $\tau_1^3\tau_2 = \tau_2\tau_1$, hence $\mathrm{Gal}(g(x), \mathbb{Q})$ is isomorphic to the dihedral group of order 8.

Exercise 5.7.

(1) Show that $\mathbb{Q}(\sqrt{2})/\mathbb{Q}$ and $\mathbb{Q}(\sqrt[4]{2})/\mathbb{Q}(\sqrt{2})$ are Galois but $\mathbb{Q}(\sqrt[4]{2})/\mathbb{Q}$ is not.

(2) Let K, E, F be fields such that $K \subset E \subset F$, E/K and F/E are finite Galois extensions with Galois groups M and N respectively. Suppose that the following conditions hold.

   (a) For any $\sigma \in M$, there exist automorphisms of F extending σ. Denote by z_σ one fixed such extension and let $\hat{M} = \{z_\sigma, \sigma \in M\}$.
   (b) For any $\sigma, \tau \in M$, we have $z_\sigma z_\tau = z_{\sigma\tau} h$ with $h \in N$.
   (c) For each $\sigma \in M$ and each $h \in N$, we have $z_\sigma h z_\sigma^{-1} \in N$.

   Let G be the subgroup of $\mathrm{Aut}_K(F)$ generated by N and $\hat{M}$. Show that F/K is Galois and $\mathrm{Gal}(F, K) = G$.

Solution 5.7.

(1) Since the extensions $\mathbb{Q}(\sqrt{2})/\mathbb{Q}$ and $\mathbb{Q}(\sqrt[4]{2})/\mathbb{Q}(\sqrt{2})$ are quadratic, then they are Galois. We show that $\mathbb{Q}(\sqrt[4]{2})/\mathbb{Q}$ is not normal. Let $\rho : \mathbb{Q}(\sqrt[4]{2}) \to \mathbb{C}$ be the unique embedding such that $\rho(\sqrt[4]{2}) = i\sqrt[4]{2}$. Then

$\rho(\mathbb{Q}(\sqrt[4]{2})) = \mathbb{Q}(i\sqrt[4]{2}) \neq \mathbb{Q}(\sqrt[4]{2})$. Hence $\mathbb{Q}(\sqrt[4]{2})/\mathbb{Q}$ is not normal. Alternatively, we could argue that the irreducible polynomial $x^4 - 2$ over $\mathbb{Q}$ has a root in the extension, but not all its roots.

(2) We first compute $(G : 1)$. The conditions **(a)**, **(b)**, **(c)** show that any $\rho \in G$ may be written in the form $\rho = z_\sigma h$ with $h \in N$ and $z_h \in \hat{M}$. Suppose that $z_{\sigma_1} h_1 = z_{\sigma_2} h_2$, then $z_{\sigma_1} h_1 h_2^{-1} = z_{\sigma_2}$. For any $e \in E$, we have

$$\sigma_2(e) = z_{\sigma_2}(e) = z_{\sigma_2} h_2(e) = z_{\sigma_1} h_1(e) = z_{\sigma_1}(e) = \sigma_1(e),$$

hence $\sigma_1 = \sigma_2$ and then $z_{\sigma_1} = z_{\sigma_2}$ which in turn implies $h_1 = h_2$. It follows that the representation of the elements of G in the form $z_\sigma h$ is unique, hence $(G : 1) = |\hat{M}||N| = |M||N|$. By [Lang (1965), Th. 9, Chap. 7.4], F/K is separable. We show that this extension is normal. Let Ω be an algebraic closure of K, then it is clear that the number of K-embeddings $\rho : F \to \Omega$ is equal to $[F : E][E : K] = |N||M|$, hence equal to $|G|$. Let $\mathcal{A}$ be the set of all these K-embeddings and let $\phi : G \to \mathcal{A}$ be the map such that $\phi(\rho)$ is the K-embedding $F \xrightarrow{\rho} F \xrightarrow{i} \Omega$, where i is the canonical injection. Then ϕ is one to one, hence $\phi(G) = \mathcal{A}$. Therefore, for any K-embedding $\rho : F \to \Omega$, we have $\rho(F) = F$, that is F/K is normal. Therefore

$$|\operatorname{Gal}(F, K)| = [F : K] = |N||M| = |G|.$$

Since $G \subset \operatorname{Gal}(F, K)$, then $G = \operatorname{Gal}(F, K)$.

Exercise 5.8.

Let K be a field, E be a Galois extension of K and $G = \operatorname{Gal}(E, K)$. Let $\sigma \in G$.

(1) Let n be the smallest positive integer such that $\sigma^n = Id_E$. Show that $n = [E : F]$, where F is the invariant field of σ.

(2) Let $\beta \in E$ and m be the smallest positive integer such that $\sigma^m(\beta) = \beta$. Show that $m \mid n$ and m is the smallest positive integer such that $\sigma^m \operatorname{Gal}(E, K(\beta)) = \operatorname{Gal}(E, K(\beta))$.

(3) Let $f(x) \in K[x]$ be irreducible and separable, β be a root of f and let F be the normal closure of $K(\beta)$ over K.
Show that $F = K(\beta)$ if and only if for any $\sigma \in \operatorname{Gal}(F, K)$ we have $n = m$, where n and m are the integers defined in **(1)** and **(2)** respectively.

Solution 5.8.

(1) By the fundamental theorem of Galois theory, [Lang (1965), Th. 1, Chap. 8.1], E is Galois over F and its Galois group is cyclic generated by σ. Therefore $n = |<\sigma>| = [E : F]$.

(2) Obviously $m \mid n$. We have

$$\sigma^m \operatorname{Gal}(E, K(\beta)) = \operatorname{Gal}(E, K(\beta)) \Leftrightarrow \sigma^m \in \operatorname{Gal}(E, K(\beta)) \Leftrightarrow \sigma^m(\beta) = \beta,$$

hence the result.

(3) Suppose that $F = K(\beta)$ and let $\sigma \in \operatorname{Gal}(F, K)$. We already know that $m \leq n$. Since $\sigma^m(\beta) = \beta$, then $\sigma^m = Id_F$, hence $m \geq n$, thus $m = n$. Suppose now that for any $\sigma \in \operatorname{Gal}(F, K)$, the related integers m and n are equal. Let $\tau \in \operatorname{Gal}(F, K(\beta))$, then $\tau(\beta) = \beta$, hence $1 = m = n = [F : K(\beta)]$. Therefore $F = K(\beta)$.

Exercise 5.9.
Let K be a field, $f(x)$ be an irreducible and separable polynomial of degree n with coefficients in K, $\gamma_1, \ldots, \gamma_n$ be the roots of $f(x)$ in an algebraic closure of K, $E = K(\gamma_1, \ldots, \gamma_n)$ and $G = \operatorname{Gal}(E, K)$.

(1) For any $i = 1, \ldots, n$, let $G_i = \{\sigma \in G, \sigma(\gamma_1) = \gamma_i\}$. Show that G_1 is a subgroup of G and explain what is G_i.

(2) Let $\theta \in E$ be an element whose conjugates over K constitute a normal basis of E/K. For any $i = 1, \ldots, n$, choose and fix $\sigma_i \in G_i$ and let $\beta_i = \sum_{\sigma \in G_1} \sigma_i(\sigma(\theta))$. Show that $\beta_i \neq \beta_j$ if $i \neq j$.

(3) Let $F(x_1, \ldots, x_n)$ be a non zero polynomial with coefficients in E and suppose that K is infinite. Show that there exist distinct elements $\alpha_1, \ldots, \alpha_n \in E$ such that
   (a) any $\sigma \in G$ permutes $\alpha_1, \ldots, \alpha_n$,
   (b) $F(\alpha_1, \ldots, \alpha_n) \neq 0$.

Solution 5.9.

(1) Since $G_1 = \operatorname{Gal}(E, K(\gamma_1))$, then G_1 is a subgroup of G. Let $i \in \{1, \ldots, n\}$ and let σ_i be a fixed element of G_i, then we have
   - **Claim.** $G_i = \sigma_i G_1$.
   - **Proof.** Let $\sigma \in G_i$, then $\sigma = \sigma_i(\sigma_i^{-1}\sigma)$ and clearly $\sigma_i^{-1}\sigma \in G_1$. We prove the reverse inclusion. Let $\sigma \in \sigma_i G_1$, $\sigma = \sigma_i h$ with $h \in G_1$, then $\sigma(\gamma_1) = \sigma_i(\gamma_1) = \gamma_i$, thus $\sigma \in G_i$.

   The claim implies that G_i is the left coset of G_1 in G with respect to γ_i.

(2) We may write β_i in the form $\beta_i = \sum_{\tau \in \sigma_i G_1} \tau(\theta)$. Since the conjugate of θ form a basis of E/K and since $\cup_{i=1,\ldots,n} \sigma_i G_1$ is a partition of G, then the β_i are all distinct. Moreover, since G operates on G/G_1 by translation, then $\beta_1, \ldots, \beta_n$ are conjugate over K.

(3) For $i = 1, \ldots, n$, let

$$y_i = \sum_{j=1}^{n} \beta_i^{j-1} x_j := L_i(x_1, \ldots, x_n)$$

and let

$$G(x_1, \ldots, x_n) = F(L_1(x_1, \ldots, x_n), \ldots, L_n(x_1, \ldots, x_n)) = F(y_1, \ldots, y_n).$$

The determinant of the transition matrix from $\{x_1, \ldots, x_n\}$ to $\{y_1, \ldots, y_n\}$ is equal to $\prod_{i<j}(\beta_i - \beta_j)$, hence non zero, thus $y_1, \ldots, y_n$ are algebraically independent over K. It follows that $G(x_1, \ldots, x_n) \neq 0$. The element $L_1(x_1, \ldots, x_n)$ belongs to the field $K(x_1, \ldots, x_n)(\beta_1)$ which is an algebraic extension of $K(x_1, \ldots, x_n)$. Let $\phi(x_1, \ldots, x_n)(x)$ be its characteristic polynomial and let $D(x_1, \ldots, x_n)$ be the discriminant of this last polynomial. We have $D(x_1, \ldots, x_n) \neq 0$. Otherwise

$$x_1 + x_2\beta_i + \cdots + x_n\beta_i^{n-1} = x_1 + x_2\beta_j + \cdots + x_n\beta_j^{n-1}$$

for $i \neq j$, thus

$$x_2(\beta_i - \beta_j) + \cdots + x_n(\beta_i^{n-1} - \beta_j^{n-1}) = 0.$$

But then $x_1, \ldots x_n$ would be algebraically dependent over E, which is a contradiction.

Let

$$H(x_1, \ldots, x_n) = G(x_1, \ldots, x_n)D(x_1, \ldots, x_n).$$

Since $H(x_1, \ldots, x_n) \neq 0$ and K is infinite, there exists $(a_1, \ldots, a_n) \in K^n$ such that $H(a_1, \ldots, a_n) \neq 0$, thus $G(a_1, \ldots, a_n) \neq 0$ and $D(a_1, \ldots, a_n) \neq 0$. Let $\alpha_i = L_i(a_1, \ldots, a_n)$ for $i = 1, \ldots, n$, then $F(\alpha_1, \ldots, \alpha_n) = G(a_1, \ldots, a_n) \neq 0$ and the α_i are distinct since $D(a_1, \ldots, a_n) \neq 0$. Since $\beta_1, \ldots, \beta_n$ are conjugate over K, then so are $\alpha_1, \ldots, \alpha_n$.

Exercise 5.10.

Let K be a field, E be a Galois extension of K, $G = \text{Gal}(E, K)$, H be a subgroup of G and $k = (G : H)$. Let $G = \cup_{i=1}^{k} Hg_i$ be the right coset decomposition of G modulo H. Let $\alpha \in E$ such that $\{\sigma(\alpha), \sigma \in G\}$ is a basis of E over K.

(1) Let $\xi = \sum_{\sigma \in G} a_\sigma \sigma(\alpha)$ with $a_\sigma \in K$ for any $\sigma \in G$. Show that the following conditions are equivalent.

 (i) $\xi \in \text{Inv}(H)$.
 (ii) For any $(\sigma, \tau) \in G^2$, if $H\sigma = H\tau$ then $a_\sigma = a_\tau$.

(2) For $i = 1, \ldots, k$, let $\xi_i = \sum_{h \in H} h(g_i(\alpha))$. Show that $\{\xi_1, \ldots, \xi_k\}$ is a basis of $\text{Inv}(H)$ over K.
(3) If H is normal in G, show that $\xi_1, \ldots, \xi_k$ are conjugates over K. In this case deduce that $\{\xi_1, \ldots, \xi_k\}$ is a normal basis of $\text{Inv}(H)$ over K.
(4) Let p be a prime number and β be a primitive p-th root of unity in $\mathbb{C}$.

 (a) Show that $\mathbb{Q}(\beta)/\mathbb{Q}$ is Galois and its Galois group is cyclic of order $p-1$ generated by the unique automorphism ρ of $\mathbb{Q}(\beta)$ such that $\rho(\beta) = \beta^a$, where $a \in \{2, \ldots, p-2\}$ and $a + p\mathbb{Z}$ generates the group $(\mathbb{Z}/p\mathbb{Z})^\star$. Show that $\{\beta, \beta^2, \ldots, \beta^{p-1}\}$ is a normal basis of $\mathbb{Q}(\beta)$ over $\mathbb{Q}$.
 (b) Let k be a positive divisor of $p-1$. Show that $\mathbb{Q}(\beta)$ contains a unique subfield F of degree k over $\mathbb{Q}$. Determine a normal basis of F over $\mathbb{Q}$.

Solution 5.10.

(1) • *(i)* $\Rightarrow$ *(ii)*. Let $(\sigma_1, \sigma_2) \in G^2$ such that $H\sigma_2 = H\sigma_1$. Let $h_0 \in H$ such that $\sigma_2 = h_0\sigma_1$. Since $h_0(\xi) = \xi$, then

$$\sum_{\sigma \in G} a_\sigma h_0 \circ \sigma(\alpha) = \sum_{\sigma \in G} a_\sigma \sigma(\alpha).$$

Since $\sigma_2(\alpha) = h_0 \circ \sigma_1(\alpha)$, then $a_{\sigma_1} = a_{\sigma_2}$.

• *(ii)* $\Rightarrow$ *(i)*. Let $h_0 \in H$ then

$$\begin{aligned} h_0(\xi) &= \sum_{\sigma \in G} a_\sigma h_0 \circ \sigma(\alpha) \\ &= \sum_{\tau \in G} a_{h_0^{-1}\tau} \tau(\alpha) \\ &= \sum_{\tau \in G} a_\tau \tau(\alpha) = \xi, \end{aligned}$$

hence $\xi \in \text{Inv}(H)$.

(2) By **(1)**, for any $i \in \{1, \ldots, k\}$, $\xi_i \in \text{Inv}(H)$. Since $(\text{Inv}(H) : K) = k$, then in order to prove that $\xi_1, \ldots, \xi_k$ is a basis of $\text{Inv}(H)$ over K, it is sufficient to prove that this family is free over K. Suppose that there

exist $a_1, \ldots, a_k \in K$ such that $\sum_{i=1}^{k} a_i\xi_i = 0$, then

$$\sum_{i=1}^{k} \sum_{h\in H} h \circ g_i(\alpha) = 0.$$

Since α generates a normal basis of E over K and

$$\{h \circ g_i,\ h \in H \text{ and } i \in \{1, \ldots, k\}\} = G,$$

then $a_i = 0$ for $i = 1, \ldots, k$.

(3) We may suppose that $g_1 = Id_E$ and then $\xi_1 = \sum_{h\in H} h(\alpha)$. Since H is normal in G, then

$$\xi_i = \sum_{h\in H} h \circ g_i(\alpha) = \sum_{h'\in H} g_i \circ h'(\alpha) = g_i\left(\sum_{h'\in H} h'(\alpha)\right) = g_i(\xi_1),$$

hence ξ_i is conjugate to ξ_1 over K. By **(2)**, $\{\xi_1, \ldots, \xi_k\}$ is a normal basis of $\mathrm{Inv}(H)$ over K.

(4)(a) The statements about the Galois group follow from [Lang (1965), Chap. 8.3]. Since $[\mathbb{Q}(\beta) : \mathbb{Q}] = p - 1$, it is sufficient to prove that $\beta, \beta^2, \ldots, \beta^{p-1}$ are linearly independent over $\mathbb{Q}$. Suppose that

$$a_1\beta + a_2\beta^2 + \cdots + a_{p-1}\beta^{p-1} = 0,$$

where $a_0, \ldots, a_{p-1} \in \mathbb{Q}$, then $a_1 + a_2\beta + \cdots + a_{p-1}\beta^{p-2} = 0$. Therefore

$$a_1 + a_2x + \cdots + a_{p-1}x^{p-2} \equiv 0 \pmod{(1 + x + \cdots + x^{p-1})},$$

which implies $a_1 = a_2 = \cdots = a_{p-1} = 0$ and then $\{\beta, \beta^2, \ldots, \beta^{p-1}\}$ is a normal basis of $\mathbb{Q}(\beta)$ over $\mathbb{Q}$.

(b) Since the Galois group of $\mathbb{Q}(\beta)$ over $\mathbb{Q}$ is cyclic of order $p - 1$, it contains a unique subgroup H of order k. Therefore $\mathbb{Q}(\beta)$ contains a unique subfield F of degree k over $\mathbb{Q}$, namely $F = \mathrm{Inv}(H)$. We have

$$H = \{1, \rho^k, \rho^{2k}, \ldots, \rho^{(n-1)k}\}.$$

We may take $g_j = \rho^j$ for $j = 1, \ldots, k$. By **(3)**, $\{\xi_1, \ldots, \xi_k\}$ is a normal basis of F over K, where

$$\begin{aligned}\xi_i &= \sum_{h\in h} h(g_i(\beta)) = \sum_{j=0}^{n-1} \rho^{kj}\rho^i(\beta)\\ &= \sum_{j=0}^{n-1} \rho^{kj+i}(\beta) = \sum_{j=0}^{n-1} \beta^{a^{kj+i}}.\end{aligned}$$

Exercise 5.11.
Let K be a field, E be a finite Galois extension of K of degree n. Show that there exists a normal basis $\{\omega_1, \ldots, \omega_n\}$ over K and $i \in \{1, \ldots, n\}$ such that $\mathrm{Tr}_{E/K}(\omega_i) = 1$.

Solution 5.11.
By the normal basis theorem [Lang (1965), Th. 20, Chap. 8.12], E has a normal basis $\{\alpha_1, \ldots, \alpha_n\}$ over K. Since the trace map from E into K is not the zero map [Lang (1965), Th. 9, Chap. 8.5], then there exists $i \in \{1, \ldots, n\}$ such that $\mathrm{Tr}_{E/K}(\alpha_i) \neq 0$. For $j = 1, \ldots, n$, let $\omega_j = \frac{\alpha_j}{\mathrm{Tr}_{E/K}(\alpha_i)}$, then $\{\omega_1, \ldots, \omega_n\}$ is also a normal basis and we have $\mathrm{Tr}_{E/K}(\omega_i) = 1$.

Exercise 5.12.
Let K be a field, E a finite Galois extension of K and $\theta \in E$ be an element generating a normal basis of E. Let F be a field such that $K \subset F \subset E$ and let $\alpha = \mathrm{Tr}_{E/F}(\theta)$.

(1) Show that α is a primitive element of F over K.
(2) Show that the condition, θ generates a normal basis of E over K, cannot be replaced by the condition θ is a primitive element of E over K.

Solution 5.12.

(1) Let

$$G = \mathrm{Gal}(E, K), \quad H = \mathrm{Gal}(E, F),$$
$$F_1 = K(\alpha) \subset F \quad \text{and} \quad H_1 = \mathrm{Gal}(E, F_1) \supset H.$$

Let $\sigma_1 \in H_1$, then $\sigma_1(\alpha) = \alpha$, hence

$$\sum_{\sigma \in H} \sigma(\theta) = \sum_{\sigma \in H} \sigma_1 \circ \sigma(\theta).$$

Since the conjugates of θ form a basis of E/K, then for any $\sigma \in H$, there exists $\tau \in H$ such that $\sigma(\theta) = \sigma_1 \circ \tau(\theta)$. Since θ generates E/K, then for any $\sigma \in H$, there exists $\tau \in H$ such that $\sigma = \sigma_1 \circ \tau$. Hence $H \subset \sigma_1 H$ and then $H = \sigma_1 H$. Therefore $\sigma_1 \in H$. This proves that $H_1 = H$, which in turn implies $K(\alpha) = F_1 = F$.

(2) Let $K = \mathbb{Q}$, $E = \mathbb{Q}(2^{1/3}, j)$, $F = \mathbb{Q}(j)$ and $\theta = 2^{1/3} - j2^{1/3}$, where j is a primitive cube root of unity. The conjugates of θ over $\mathbb{Q}$ are given by

$$\theta_1 = \theta,\ \theta_2 = j2^{1/3} - j^2 2^{1/3},\ \theta_3 = j^2 2^{1/3} - 2^{1/3},$$
$$\theta_4 = 2^{1/3} - j^2 2^{1/3},\ \theta_5 = j2^{1/3} - 2^{1/3},\ \theta_6 = j^2 2^{1/3} - j2^{1/3},$$

thus θ is a primitive element of K over $\mathbb{Q}$. But we have

$$\mathrm{Tr}_{E/F}(\theta) = (\theta_1 + \theta_2 + \theta_3) = 0,$$

so that this trace does not generate $F/\mathbb{Q}$.

Exercise 5.13.
Let $f(x)$ be a monic irreducible polynomial over $\mathbb{F}_p$ of degree n and let θ be a root of f. Suppose that there exist $i, j \in \{0, \ldots, p-1\}$, $i \neq j$ such that $\theta + i$ and $\theta + j$ are conjugate over $\mathbb{F}_p$.

(1) Show that $f(x+1) = f(x)$.
(2) Deduce that $p \mid n$ and $f(x) = u(x^p - x + 1)$, where $u(x)$ is a monic polynomial of degree n/p with coefficients in $\mathbb{F}_p$.

Solution 5.13.

(1) We may suppose that $i > j$. By assumption, there exists a positive integer e such that

$$\theta + i = (\theta + j)^{p^e} = \theta^{p^e} + j,$$

hence $\theta + (i - j) = \theta^{p^e}$. Let $h = i - j$ and let $k \in \{1, \ldots, p-1\}$ such that $kh \equiv 1 \pmod p$ then θ and $\theta + h$ are conjugate over $\mathbb{F}_p$, hence there exists a unique automorphism σ of $\mathbb{F}_p(\theta)$ such that $\sigma(\theta) = \theta + h$. We deduce that

$$\sigma^k(\theta) = \theta + kh = \theta + 1.$$

It follows that θ and $\theta + 1$ are conjugate over $\mathbb{F}_p$. This implies that $f(\theta + 1) = 0$, hence $f(x+1) = f(x)$.

(2) Consider the group G of the $\mathbb{F}_p$-automorphisms of $\mathbb{F}_p(x)$ generated by the automorphism τ such that $\tau(x) = x + 1$. Then from **(1)** we conclude that $f(x)$ belongs to the invariant field say E of G. Clearly $x^p - x + 1 \in E$, hence

$$\mathbb{F}_p(x^p - x + 1) \subset E \subset \mathbb{F}_p(x).$$

Obviously

$$[\mathbb{F}_p(x) : \mathbb{F}_p(x^p - x + 1)] = p$$

and by Artin's theorem [Lang (1965), Th. 2, Chap. 8.1], $[\mathbb{F}_p(x) : E] = p$, hence $E = \mathbb{F}_p(x^p - x + 1)$ and $f(x) = u(x^p - x + 1)$, where $u(x)$ is a polynomial having the properties stated in the exercise.

Exercise 5.14.

Let K be a field, $f(x) \in K[x]$ be monic of degree n and separable over K. Let $\alpha_1, \ldots \alpha_n$ be the roots of f in an algebraic closure of K, $E = K(\alpha_1, \ldots \alpha_n)$ and $G = \mathrm{Gal}(f(x), K)$. Fix an integer $k \in \{1, \ldots, n-1\}$. For $j = 1, \ldots, n$, let

$$F_j(x_1, \ldots, x_k) = \sum_{\sigma \in G} \sigma(\alpha_j) \prod_{i=1}^{k} \frac{f(x_i)}{x_i - \sigma(\alpha_i)}.$$

(1) Show that for $j = 1, \ldots, n$, $F_j(x_1, \ldots, x_k) \in K[x_1, \ldots, x_k]$.

(2) Suppose that $E = K(\alpha_1, \ldots \alpha_k)$ and let $i_1, \ldots, i_k \in \{1, \ldots, n\}$ be distinct integers. Show that the following propositions are equivalent.

   (i) There exists $\tau \in G$ such that, for $t = 1, \ldots, k$, $\tau(\alpha_t) = \alpha_{i_t}$.

   (ii) For $j = 1, \ldots, n$, $F_j(\alpha_{i_1}, \ldots, \alpha_{i_k}) \neq 0$.

   (iii) There exists $j \in \{1, \ldots, n\}$ such that $F_j(\alpha_{i_1}, \ldots, \alpha_{i_k}) \neq 0$.

(3) Suppose that the equivalent conditions in **(2)** hold.

   (a) Show that for all $j \in \{1, \ldots, n\}$

$$\tau(\alpha_j) = \frac{F_j(\alpha_{i_1}, \ldots, \alpha_{i_k})}{\prod_{t=1}^{k} f'(\alpha_{i_t})}.$$

   (b) Deduce that

$$\alpha_j = \frac{F_j(\alpha_1, \ldots, \alpha_k)}{\prod_{t=1}^{k} f'(\alpha_t)}.$$

(4) Show that $\left(f(x)\right)^{k-1} \mid F_j(x, \ldots, x)$.

(5) Suppose that $K = \mathbb{Q}$, $f(x) = x^3 - 3x + 1$. Show that $G \simeq \mathbb{Z}/3\mathbb{Z}$ and $E = K(\alpha_1)$. Compute $F_j(\alpha_1)$ for $j = 1, 2, 3$ and express α_2 and α_3 as rational functions of α_1.

Solution 5.14.

(1) Let $\tau \in G$ then since the map $\sigma \to \tau \circ \sigma$ is a bijection of G onto G, then we have

$$F_j^{\tau}(x_1, \ldots, x_k) = \sum_{\tau \circ \sigma \in G} \tau \circ \sigma(\alpha_j) \prod_{i=1}^{k} \frac{f(x_i)}{x_i - \tau \circ \sigma(\alpha_i)} = F_j(x_1, \ldots, x_k),$$

hence $F_j(x_1, \ldots, x_k) \in K[x_1, \ldots, x_k]$.

(2) • *(i)* $\Rightarrow$ *(ii)*. $F_j(\alpha_{i_1}, \ldots, \alpha_{i_k})$ is a sum of terms of the form

$$T_{j,\sigma}(\alpha_{i_1}, \ldots \alpha_{i_k}) = \sigma(\alpha_j) \frac{f(\alpha_{i_1})}{\alpha_{i_1} - \sigma(\alpha_1)} \cdots \frac{f(\alpha_{i_k})}{\alpha_{i_k} - \sigma(\alpha_k)}.$$

Clearly this term is non zero if and only if $\sigma = \tau$. It follows that $F_j(\alpha_{i_1}, \ldots \alpha_{i_k}) = T_{j,\tau}(\alpha_{i_1}, \ldots \alpha_{i_k}) \neq 0$.

- *(ii)* $\Rightarrow$ *(iii)*. Obvious.
- *(iii)* $\Rightarrow$ *(i)*. We have

$$F_j(\alpha_{i_1},\dots,\alpha_{i_k}) = \sum_{\sigma\in G} T_{j,\sigma}(\alpha_{i_1},\dots\alpha_{i_k}) \neq 0,$$

hence there exists $\tau \in G$ such that $T_{j,\tau}(\alpha_{i_1},\dots\alpha_{i_k}) \neq 0$. This automorphism τ satisfies $\tau(\alpha_t) = \alpha_{i_t}$ for $t = 1,\dots,k$.

(3)(a) We have

$$\begin{aligned} F_j(\alpha_{i_1},\dots,\alpha_{i_k}) &= T_{j,\tau}(\alpha_{i_1},\dots,\alpha_{i_k}) \\ &= \tau(\alpha_j)\prod_{u\neq 1}(\alpha_{i_1} - \tau(\alpha_u))\cdots\prod_{u\neq k}(\alpha_{i_k} - \tau(\alpha_u)) \\ &= \tau(\alpha_j)f'(\alpha_{i_1})\cdots f'(\alpha_{i_k}) \end{aligned}$$

and this allows to express $\tau(\alpha_j)$ as a rational function of $\alpha_{i_1},\dots,\alpha_{i_k}$.

(b) Apply **(a)** when $\tau = Id_E$.

(4) We continue to use the polynomials $T_{j,\sigma}$ defined in **(2)**. We have

$$\begin{aligned} T_{j,\sigma}(x,\dots,x) &= \sigma(\alpha_j)\prod_{u\neq 1}(x - \sigma(\alpha_u))\cdots\prod_{u\neq k}(x - \sigma(\alpha_u)) \\ &\equiv 0 \pmod{(f(x))^{k-1}} \end{aligned}$$

in $\bar{K}[x]$. Therefore $F_j(x,\dots,x) \equiv 0 \pmod{(f(x))^{k-1}}$.

(5) The discriminant of the polynomial $f(x) = x^3 - 3x + 1$ is equal to 9^2 hence its Galois group is cyclic of order 3 [Lang (1965), Example 2, Chap. 8.2] and its field of decomposition over $\mathbb{Q}$ is $E = \mathbb{Q}(\alpha_1)$, where α_1 is any of its roots. We have

$$F_1(x_1) = \alpha_1 f(x_1)/(x_1 - \alpha_1) + \alpha_2 f(x_1)/(x_1 - \alpha_2) + \alpha_3 f(x_1)/(x_1 - \alpha_3).$$

Denote by s_1, s_2, s_3 the elementary symmetric functions of the roots of f of degree $1, 2, 3$ respectively. Then

$$F_1(x_1) = s_1 x_1^2 - 2s_2 x_1 + 3s_3 = -6x_1 - 3.$$

Similarly we obtain

$$\begin{aligned} F_2(x_1) &= \alpha_2 f(x_1)/(x_1 - \alpha_1) \\ &\quad + \alpha_3 f(x_1)/(x_1 - \alpha_2) \\ &\quad + \alpha_1 f(x_1)/(x_1 - \alpha_3) \\ &= s_1 x_1^2 - (\alpha_1^2 + \alpha_2^2 + \alpha_3^2 + \alpha_1\alpha_2 + \alpha_1\alpha_3 + \alpha_2\alpha_3)x_1 \\ &\quad + \alpha_1^2\alpha_2 + \alpha_2^2\alpha_3 + \alpha_3^2\alpha_1 \\ &= s_2 x_1 + \alpha_1^2\alpha_2 + \alpha_2^2\alpha_3 + \alpha_3^2\alpha_1 \\ &= -3x_1 + \alpha_1^2\alpha_2 + \alpha_2^2\alpha_3 + \alpha_3^2\alpha_1 \end{aligned}$$

and

$$\begin{aligned}F_3(x_1) &= \alpha_3(x_1)/(x_1-\alpha_1)\\ &\quad +\alpha_1 f(x_1)/(x_1-\alpha_2)\\ &\quad +\alpha_2 f(x_1)/(x_1-\alpha_3)\\ &= s_1x_1^2 - (\alpha_1^2+\alpha_2^2+\alpha_3^2+\alpha_1\alpha_2+\alpha_1\alpha_3+\alpha_2\alpha_3)x_1\\ &\quad +\alpha_1^2\alpha_3+\alpha_2^2\alpha_1+\alpha_3^2\alpha_2\\ &= s_2x_1+\alpha_1^2\alpha_3+\alpha_2^2\alpha_1+\alpha_3^2\alpha_2\\ &= -3x_1+\alpha_1^2\alpha_3+\alpha_2^2\alpha_1+\alpha_3^2\alpha_2.\end{aligned}$$

We must compute the non symmetric functions

$$A = \alpha_1^2\alpha_2+\alpha_2^2\alpha_3+\alpha_3^2\alpha_1 = \mathrm{Tr}_{E/\mathbb{Q}}(\alpha_1^2\alpha_2)$$

and

$$B = \alpha_1^2\alpha_3+\alpha_2^2\alpha_1+\alpha_3^2\alpha_2 = \mathrm{Tr}_{E/\mathbb{Q}}(\alpha_1^2\alpha_3).$$

We have $A+B=\sum_{i\neq j}\alpha_i^2\alpha_j$. Since

$$0=\sum\alpha_i^2\sum\alpha_j = A+B+\sum\alpha_i^3,$$

then $A+B=-\sum\alpha_i^3=-\sum(3\alpha_i-1)=3$. We also have

$$AB = \sum_{i,j,l\ \text{distinct}}\alpha_i^4\alpha_j\alpha_l + 3s_3^2 + \sum_{i\neq j}\alpha_i^3\alpha_j^3.$$

We have

$$\sum_{i,j,l\ \text{distinct}}\alpha_i^4\alpha_j\alpha_l = s_3\sum\alpha_i^3 = s_3\sum(3\alpha_i-1)=3$$

and

$$\sum_{i\neq j}\alpha_i^3\alpha_j^3=\sum_{i\neq j}(3\alpha_i-1)(3\alpha_j-1)=9s_2+3=-24,$$

hence $AB=-18$. It follows that A and B are roots of the equation $y^2-3y-18=0$, hence we may suppose (i.e. there is a labeling of the roots of f) that $A=6$ and $B=-3$. We have

$$F_1(x_1)=6x_1-3,\ F_2(x_1)=-3x_1+6 \text{ and } F_3(x_1)=-3x_1-3,$$

hence

$$\alpha_2=F_2(\alpha_1)/f'(\alpha_1)=(-3\alpha_1+6)/(3\alpha_1^2-3)=(-\alpha_1+2)/(\alpha_1^2-1)$$

and

$$\begin{aligned}\alpha_3 &= F_3(\alpha_1)/f'(\alpha_1)=(-3\alpha_1-3)/(3\alpha_1^2-3)\\ &=(-\alpha_1-1)/(\alpha_1^2-1)=1/(1-\alpha).\end{aligned}$$

Exercise 5.15.
Let K be a field, $f(x,y) \in K[x,y]$ be irreducible and separable over $K(x)$ and Ω be an algebraic closure of $K(x)$. Let $E \subset \Omega$ be the splitting field of f over $K(x)$, K_0 be the algebraic closure of K in E, $G = \mathrm{Gal}(f, K(x))$ and $H = \mathrm{Gal}(f, K_0(x))$. Show that K_0/K is Galois, H is a normal subgroup of G and $G/H \simeq \mathrm{Gal}(K_0, K)$.

Solution 5.15.
Since E is separable over $K(x)$, then $K_0(x)/K(x)$ is separable. Let $a \in K_0$, then $\mathrm{Irr}(a, K) = \mathrm{Irr}(a, K(x)$ and since a is separable over $K(x)$, it is separable over K. We show that K_0/K is normal. Let $\sigma : K_0 \to \Omega$ be any K-morphism of fields. We must prove that $\sigma(K_0) = K_0$. Extend σ to $K_0(x)$ and obtain σ_1, by setting $\sigma_1(x) = x$. Since E is algebraic over K_0, then we may extend σ_1 to E and obtain a $K(x)$-morphism σ_2 of fields. Moreover since E is normal over $K(x)$, then $\sigma_2(E) = E$. Let $\mu \in K_0$ and let $g(y) = y^m + b_{m-1}y^{m-1} + \cdots + b_0$ be the minimal polynomial of μ over K. Since $g(\mu) = 0$, then

$$\sigma_2(\mu)^m + b_{m-1}\sigma_2(\mu)^{m-1} + \cdots + b_0 = 0.$$

We conclude that $\sigma_2(\mu) \in E$. Thus $\sigma(\mu)$ is algebraic over K and belongs to E. It follows that $\sigma(\mu) \in K_0$. This implies that σ is a K-morphism of fields from K_0 into K_0. Since K_0 is algebraic over K, then σ is onto; that is $\sigma(K_0) = K_0$.

Let $\phi : G \to \mathrm{Gal}(K_0(x), K(x))$ be the map such that for any $\sigma \in G$, $\phi(\sigma) = \sigma_{|K_0(x)}$. Clearly ϕ is a morphism of groups. Since $E/K_0(x)$ is algebraic, then any $K(x)$-automorphism of $K_0(x)$ may be extended to E, hence ϕ is surjective. It is clear that $H \subset \mathrm{Ker}\,\phi$. Let $\sigma \in \mathrm{Ker}\,\phi$, then σ is an automorphism of E which fixes every element of $K_0(x)$, hence $\sigma \in H$ and then $H = \mathrm{Ker}\,\phi$. The first isomorphism theorem implies that $G/H \simeq \mathrm{Gal}(K_0(x), K(x))$. Since $\mathrm{Gal}(K_0(x), K(x)) \simeq \mathrm{Gal}(K_0, K)$, then $G/H \simeq \mathrm{Gal}(K_0, K)$.

Exercise 5.16.
Give examples of two polynomials $f(x)$ and $g(x)$ irreducible and separable over K such that $\deg f < \deg g$ and $\mathrm{Gal}(f(x), K) = \mathrm{Gal}(g(x), K)$.

Solution 5.16.
Let $f(x) \in [K[x]$ be irreducible and separable over K of degree n and let G be its Galois group. Suppose that $|G| > n$ and let θ be a primitive element of the splitting field of f over K, then the polynomial $g(x) = \mathrm{Irr}(\theta, K)$ satisfies the required conditions.

Exercise 5.17.
Let K be an field, Ω be an algebraic closure of K, $f(x,y) \in K[x,y]$ be separable over $K(x)$ and satisfying the conditions $\deg_x f \geq 1$ and $\deg_y f \geq 1$. Let $G = \mathrm{Gal}(f(x,y), K(x))$ and $G^* = \mathrm{Gal}(f(x,y), \Omega(x))$.

(1) Show that G^* is a subgroup of G.
(2) Considering the case $K = \mathbb{Q}$, and $f(x,y) = y^4 - 2x^2$, show that we may have the strict inclusion $G^* \subsetneq G$.

Solution 5.17.

(1) Let $\alpha_1, \ldots, \alpha_n$ be the roots of $f(x,y)$ in an algebraic closure of $K(x)$ and $E = K(x, \alpha_1, \ldots, \alpha_n)$. Then $\Omega(x, \alpha_1, \ldots, \alpha_n) = \Omega.E$ is a splitting field of $f(x,y)$ over $\Omega(x)$. Let $\sigma \in G^*$ and
$$\bar{\sigma} : K(x, \alpha_1, \ldots, \alpha_n) \to \Omega(x, \alpha_1, \ldots, \alpha_n)$$
be its restriction to $K(x, \alpha_1, \ldots, \alpha_n)$. Then $\bar{\sigma}$ is a $K(x)$-isomorphism of E. Clearly the map $\sigma \to \bar{\sigma}$ is an isomorphism of groups from G^* into G.

(2) Let α be a root of $f(x,y)$ in an algebraic closure of $K(x)$, then the roots of f are $\pm\alpha, \pm i\alpha$, hence $E = \mathbb{Q}(x, \alpha, i)$. Since $2x^2$ is not a square in $\mathbb{Q}(x)$ then $f(x,y)$ is irreducible over $\mathbb{Q}(x)$, hence $[\mathbb{Q}(x,\alpha) : \mathbb{Q}(x)] = 4$. We show that $i \notin \mathbb{Q}(x,\alpha)$, and this will imply that $[E : \mathbb{Q}(x)] = 8$. Suppose that $i \in \mathbb{Q}(x,\alpha)$. Since $\alpha^4 = 2x^2$, then $x\sqrt{2} = \pm\alpha^2$, hence $\sqrt{2} \in \mathbb{Q}(x,\alpha)$. We have
$$\mathbb{Q}(x,\sqrt{2})(\alpha) = \mathbb{Q}(x,\alpha) = \mathbb{Q}(x,\sqrt{2},i) = \mathbb{Q}(x,\alpha)(i).$$
Since the minimal polynomials of α and i over $\mathbb{Q}(x,\sqrt{2})$ are $y^2 \mp x\sqrt{2}$ and $y^2 + 1$ respectively, then $\frac{\alpha}{i} \in \mathbb{Q}(x,\sqrt{2})$. Set $\frac{\alpha}{i} = \lambda$ where $\lambda \in \mathbb{Q}(x,\sqrt{2})$ then $-\alpha^2 = \lambda^2 = \mp x\sqrt{2}$. This implies that $\mp x\sqrt{2}$ is a square in $\mathbb{Q}(\sqrt{2})(x)$, which is a contradiction. We conclude that $i \notin \mathbb{Q}(x,\alpha)$. Therefore $[E : \mathbb{Q}(x)] = 8$. The elements of G are determined by there action on α and on i. Indeed any $\sigma \in G$ satisfies the following conditions:
$$\sigma(\alpha) = \pm\alpha,\ \pm i\alpha \text{ ,and } \sigma(i) = \pm i,$$
where all the possible combinations of signs are acceptable.
Let $\overline{\mathbb{Q}}$ be an algebraic closure of $\mathbb{Q}$. Over this field, we have
$$f(x,y) = (y^2 - x\sqrt{2})(y^2 + x\sqrt{2}),$$
hence
$$G^* = \mathrm{Gal}(f(x,y), \bar{\mathbb{Q}}(x)) \simeq \mathbb{Z}/2\mathbb{Z} \times \mathbb{Z}/2\mathbb{Z},$$
thus $G^* \subsetneq G$.

Exercise 5.18.
Let K be a field, $f(x)$ be irreducible and separable over K, $\alpha_1, \ldots, \alpha_n$ be the roots of f in an algebraic closure of K. Let $G = \text{Gal}(f, K)$, $H = \text{Gal}(f, K(\alpha))$ and $\phi \in G$. Consider the natural action of the group generated by ϕ on the set $\{\alpha_1, \ldots, \alpha_n\}$ and let

$$\{\alpha_1, \phi(\alpha_1), \ldots, \phi^{k_1-1}(\alpha_1)\}, \ldots, \{\alpha_s, \phi(\alpha_s), \ldots, \phi^{k_s-1}(\alpha_s)\},$$

be the orbits under this action, where k_i is the smallest positive integer such that $\phi^{k_i}(\alpha_i) = \alpha_i$. Let

$$\{H\sigma_1, H\sigma_1\phi, \ldots, H\sigma_1\phi^{m_1-1}\}, \ldots, \{H\sigma_r, H\sigma_r\phi, \ldots, H\sigma_r\phi^{m_r-1}\},$$

be the orbits of the elements of $(G/H)_d$ under the action of the same group as above, where m_i is the smallest positive integer such that $H\sigma_i\phi^{m_i} = H\sigma_i$. Show that $r = s$ and after changes of numbering if necessary, $m_i = k_i$ for all i.

Solution 5.18.
Let $\sigma \in G$, and m be a positive integer, then we have

$$\begin{aligned} H\sigma\phi^m = H\sigma &\Leftrightarrow \phi^m \in \sigma^{-1}H\sigma \\ &\Leftrightarrow \phi^m \in \text{Gal}(f, \sigma^{-1}(K)) \\ &\Leftrightarrow \phi^m(\sigma^{-1}(\alpha)) = \sigma^{-1}(\alpha). \end{aligned}$$

This shows that the orbits of $H\sigma$ and of $\sigma^{-1}(\alpha)$ under the action of the group generated by ϕ on the sets $(G/H)_d$ and $\{\alpha_1, \ldots, \alpha_n\}$ respectively, have the same cardinality and the result follows.

Exercise 5.19.
Let K be a field, $f(x) \in K[x]$ be irreducible and separable. Let $n = \deg f$, $\alpha_1, \ldots, \alpha_n$ be the roots of f in an algebraic closure of K and $G = \text{Gal}(f(x), K)$. Let $E = K\alpha_1 + \cdots + K\alpha_n$ be the K-vector space spanned by $\alpha_1, \ldots, \alpha_n$. Suppose that the α_i are labeled in order to satisfy the condition $\{\alpha_1, \ldots, \alpha_s\}$ is a basis of E over K with $s \leq n$. For any $\sigma \in G$ and any integer j, $1 \leq j \leq s$, write $\sigma(\alpha_j)$ in the form

$$\sigma(\alpha_j) = a_{1j}^\sigma \alpha_1 + a_{2j}^\sigma \alpha_2 + \cdots + a_{sj}^\sigma \alpha_s,$$

where the coefficients a_{ij}^σ belong to K. Consider the $s \times s$ matrix: $A(\sigma) = (a_{ij}^\sigma)$.

(1) Show that $A(\sigma) \in GL_s(K)$ for any $\sigma \in G$.
(2) Show that the map $\phi : \sigma \to A(\sigma)$ is an injective morphism of G into $GL_s(K)$.

(3) If G is the symmetric group and $\mathrm{Tr}_{K(\alpha)/K}(\alpha) \neq 0$, show that $s = n$.

Solution 5.19.

(1) Suppose that some linear combination of the column of $A(\sigma)$, with coefficients $\lambda_1, \ldots, \lambda_s$ in K, such that, $\lambda_1\sigma(\alpha_1) + \cdots + \lambda_s\sigma(\alpha_s) = 0$. Then $\sigma(\lambda_1\alpha_1 + \cdots + \lambda_s\alpha_s) = 0$. Therefore $\lambda\alpha_1 + \cdots + \lambda_s\alpha_s = 0$ and then $\lambda_1 = \cdots = \lambda_s = 0$. It follows that the columns of $A(\sigma)$ are linearly independent over K and then $\mathrm{Det}(A_\sigma) \neq 0$.

(2) Let $\sigma, \tau \in G$, then

$$\begin{aligned}\sum_{i=1}^{s} a_{ij}^{\sigma\circ\tau}\alpha_i &= \sigma\circ\tau(\alpha_j)\\ &= \sigma\left(\sum_{k=1}^{s} a_{kj}^{\tau}\alpha_k\right)\\ &= \sum_{k=1}^{s} a_{kj}^{\tau}\sigma(\alpha_k)\\ &= \sum_{k=1}^{s} a_{kj}^{\tau}\left(\sum_{i=1}^{s} a_{ik}^{\sigma}\alpha_i\right)\\ &= \sum_{k=1}^{s}\left(\sum_{k=1}^{s} a_{kj}^{\tau}a_{ik}^{\sigma}\right)\alpha_i.\end{aligned}$$

Therefore

$$a_{ij}^{\sigma\circ\tau} = \sum_{k=1}^{s} a_{kj}^{\tau}a_{ik}^{\sigma} = (a_{i1}^{\sigma}, \ldots, a_{is}^{\sigma})\begin{pmatrix} a_{1j}^{\tau}\\ a_{2j}^{\tau}\\ \cdots\\ a_{sj}^{\tau}\end{pmatrix},$$

which implies $A_{\sigma\circ\tau} = A_\sigma A_\tau$ and then $\phi(\sigma\circ\tau) = \phi(\sigma)\phi(\tau)$. It follows that ϕ is a morphism of groups. Let $\sigma \in G$. Suppose that $\sigma \in \mathrm{Ker}\,\phi$, then $a_{ij}^{\sigma} = 0$ if $i \neq j$ and $a_{ij}^{\sigma} = 1$ if $i = j$. Therefore $\sigma(\alpha_j) = \alpha_j$ for $j = 1, \ldots, s$. It follows that $\sigma = \mathrm{Id}_{K(\alpha_1,\ldots,\alpha_s)}$. Since $K(\alpha_1, \ldots, \alpha_s) = K(\alpha_1, \ldots, \alpha_n)$, then $\sigma = \mathrm{Id}_{K(\alpha_1,\ldots,\alpha_n)}$.

(3) Suppose that $G = \mathcal{S}_n$ and let $a_1, \ldots, a_n \in K$ such that

$$a_1\alpha_1 + \cdots + a_n\alpha_n = 0.$$

For any $i = 2, \ldots, n$, apply to this identity the transposition $(\alpha_1 \quad \alpha_i)$. We get the new identity

$$a_1\alpha_i + \cdots + a_{i-1}\alpha_{i-1} + a_i\alpha_1 + \cdots + a_n\alpha_n = 0.$$

Subtracting the second identity from the first, we obtain $(a_1 - a_i)(\alpha_1 - \alpha_i) = 0$, hence $a_1 = a_i$. Since $\alpha_1 + \cdots + \alpha_n \neq 0$, it follows that $\alpha_1, \ldots, \alpha_n$ are linearly independent over K, that is $s = n$.

Exercise 5.20.
Let K be a field of characteristic $p > 0$, E be a cyclic extension of K of degree p^e with $e \geq 2$ and let σ be a generator of $\mathrm{Gal}(E, K)$. Let F be the unique intermediate field $K \subset F \subset E$ of degree p^{e-1} over K.

(1) Show that there exists $\beta \in E$ such that $E = F(\beta)$, the minimal polynomial of β over F has the form $x^p - x - \delta$ and $\sigma^{p^{e-1}}(\beta) = \beta + 1$.
(2) Show that $E = K(\beta)$, $\sigma(\beta) = \beta + \alpha$, where $\alpha \in F$ satisfies the condition $\sigma(\delta) - \delta = \alpha^p - \alpha$ and $\mathrm{Tr}_{F/k}(\alpha) = 1$.

Solution 5.20.

(1) Since E/F is cyclic and $[E : F] = p$, there exists $\beta \in E$ such that $E = F(\beta)$ and the minimal polynomial of β over F has the form $f(x) = x^p - x - \delta$, [Lang (1965), Th. 11, Chap. 7.6]. The Galois group of E over F is generated by $\sigma^{p^{e-1}}$ and the roots of $f(x)$ are given by $\beta + i$ where $i \in \mathbb{Z}/p\mathbb{Z}$. Therefore $\sigma^{p^{e-1}}(\beta) = \beta + i_0$ where $i_0 \in \mathbb{Z}/p\mathbb{Z}$. Let $k_0 \in \mathbb{Z}/p\mathbb{Z}$ such that $k_0 i_0 = 1$, then

$$\sigma^{p^{e-1}}(k_0\beta) = k_0\sigma^{p^{e-1}}(\beta) = k_0\beta + 1,$$

hence replacing β by $k_0\beta$, we may suppose, if necessary, that $\sigma^{p^{e-1}}(\beta) = \beta + 1$.

(2) If $K(\beta) \neq E$, then $K(\beta)/K$ is cyclic of degree $\leq p^{e-1}$ hence $K(\beta) \subset F$, which is a contradiction. Therefore $E = K(\beta)$. Let $\alpha = \sigma(\beta) - \beta$. We show that $\alpha \in F$. For this it is necessary and sufficient to show that $\sigma^{p^{e-1}}(\alpha) = \alpha$. We have

$$\begin{aligned}\sigma^{p^{e-1}}(\alpha) &= \sigma(\sigma^{p^{e-1}}(\beta)) - \sigma^{p^{e-1}}(\beta)\\ &= \sigma(\beta + 1) - (\beta + 1)\\ &= \sigma(\beta) - \beta\\ &= \alpha,\end{aligned}$$

hence the claim is proved. Since $\beta^p - \beta - \delta = 0$, then

$$\sigma(\beta)^p - \sigma(\beta) - \sigma(\delta) = 0,$$

hence $(\beta + 1)^p - (\beta + 1) - \sigma(\delta) = 0$. Thus

$$\beta^p - \beta + \alpha^p - \alpha - \sigma(\delta) = 0.$$

Therefore $\alpha^p - \alpha - \sigma(\delta) = -\delta$, that is $\alpha^p - \alpha = \sigma(\delta) - \delta$. We have

$$\begin{aligned}
\mathrm{Tr}_{F/K}(\alpha) &= \sum_{k=1}^{p^{e-1}} \sigma^k(\alpha) \\
&= \sum_{k=1}^{p^{e-1}} \sigma^k(\sigma(\beta) - \beta) \\
&= \sum_{k=1}^{p^{e-1}} \sigma^{k+1}(\beta) - \sum_{k=1}^{p^{e-1}} \sigma^k(\beta) \\
&= \sigma^{p^{e-1}+1}(\beta) - \sigma(\beta) = \sigma(\beta + 1) - \sigma(\beta) \\
&= 1.
\end{aligned}$$

Exercise 5.21.

Let K be a algebraically closed field of characteristic $p \neq 2$ and m, n be positive integers. Let

$$f(x, y, t) = tx^m - 2x + 2 + (x^m - t)y^n \in K[x, y, t].$$

(1) Show that $f(x, y, t)$ is irreducible over K.

(2) Let $t_0 \in K$. Show that $f(x, y, t_0)$ is reducible over K if and only if $(t_0^2 + 2)^m - 2^n t_0 = 0$. If these equivalent conditions hold, show that $x - (t_0^2 + 2)/2$ is a factor of $f(x, y, t_0)$.

Solution 5.21.

(1) We have

$$f(x, y, t) = (x^m - y^n)t + x^m y^n - 2x + 2,$$

hence considering this polynomial as an element of $k(x, y)[t]$, it is of degree one, hence irreducible in this ring. Since

$$\gcd(x^m - y^n, x^m y^n - 2x + 2) = 1$$

this polynomial is irreducible in $k[x, y][t] = k[x, y, t]$.

(2) Necessity of the condition. Suppose first that, in $K(x)[y]$, we have $\gcd(t_0 x^m - 2x_0 + 2, x^m - t_0) = 1$, then $f(x, y, t)$ is reducible in $K(x)[y]$. Therefore $y^n - \frac{-t_0 x^m + 2x - 2}{x^m - t_0}$ is reducible over $K(x)$. By [Lang (1965), Theorem 16 VIII 9], there exists a prime number l such that

$$\begin{aligned}
&l \mid n \quad \text{and} \quad (-t_0 x^m + 2x - 2)/(x^m - t_0) \in K(x)^l \quad \text{or} \\
&4 \mid n \quad \text{and} \quad (-t_0 x^m + 2x - 2)/(x^m - t_0) \in -4K(x)^4.
\end{aligned}$$

In the first case we have

$$-t_0x^m + 2x + 2 \in K[x]^l \text{ and } x^m - t_0 \in K[x]^l.$$

The second condition implies $m = 0$ or $t_0 = 0$. But for $m = 0$ or for $t_0 = 0$ the first condition is not satisfied. In the second case, since K is algebraically closed, then

$$(-t_0x^m + 2x - 2)/(x^m - t_0) \in K(x)^2$$

and we are moved to the first case. We conclude that if

$$\gcd(t_0x^m - 2x + 2, x^m - t_0) = 1,$$

then $f(x, y, t_0)$ is irreducible over K. We have

$$\gcd(t_0x^m - 2x + 2, x^m - t_0) \neq 1$$

if and only if $\operatorname{Res}_x(t_0x^m - 2x + 2, x^m - t_0) = 0$. Straightforward computations show that

$$\operatorname{Res}_x(tx^m - 2x + 2, x^m - t_0) = (t^2 + 2)^m - 2^m t.$$

We conclude that if $f(x, y, t)$ is reducible over K, then

$$(t_0^2 + 2)^m - 2^m t_0 = 0.$$

Conversely this condition implies that there exists $d(x) \in K[x]\backslash K$ such that $d(x)|t_0x^m - 2x + 2$ and $d(x)|x^m - t_0$. Therefore

$$f(x, y, t) = d(x)\left(\frac{t_0x^m - 2x + 2}{d(x)} + \frac{x^m - t_0y^m}{d(x)}\right),$$

which implies that $f(x, y, t_0)$ is reducible over K. We have seen above that if $f_m(x, y, t_0)$ is reducible over K, then $\gcd(t_0x^m - 2x + 2, x^m - t_0) \neq 1$, hence these univariate polynomials have a common root, say u in K. We have $u^m = t_0$, hence $t_0^2 - 2u + 2 = 0$. Therefore $u = (t_0^2 + 2)/2$. It follows that $x - (t_0^2 + 2)/2$ is a factor of $f(x, y, t_0)$.

Exercise 5.22.
Let K be a field, L/K be a Galois extension, H be a subgroup of G and E be the invariant field of H. Show that $\operatorname{Aut}_K(E) \simeq (N_G(H))/H$.

Solution 5.22.
Consider the map $\theta : N_G(H) \to \operatorname{Aut}_K(E)$, defined by $\theta(\sigma) = \sigma_{|E}$. We first show that θ is well defined. Let $x \in E$ and $h \in H$, then

$$h(\sigma(x)) = \sigma(h_1(x)) = \sigma(x)$$

with $h_1 \in H$, hence $\sigma(x) \in E$. Clearly θ is a group homomorphism and is surjective. Let $\sigma \in N_G(H)$, then $\sigma \in \operatorname{Ker}\theta$ if and only if $\sigma_{|E} = Id_E$ if and only if $\sigma \in H$, hence $\operatorname{Ker}\theta = H$ and the result follows from the first isomorphism theorem of groups.

Exercise 5.23.
Let K be a field of characteristic $p \geq 0$, $n \geq 2$, be an integer. If $p > 0$, suppose that $p \nmid n$. Let ϵ be a primitive n-th root of unity in an algebraic closure of K. Let $f(x, y) \in K(x)[y]$. Show that $f(x, y) \in K(x^n)[y]$ if and only if $f(\epsilon x, y) = f(x, y)$.

Solution 5.23.
Necessity of the condition: Obvious.

Sufficiency of the condition.

First proof. Let $g(x, y) \in K(x, y)$. We show that $g(x, y) \in K(x^n, y)$ if and only if $g(\epsilon x, y) = g(x, y)$. Let G be the group of $K(\epsilon)$-automorphism of $K(\epsilon)(x, y)$ generated by the unique $K(\epsilon)$-automorphism σ such that $\sigma(x) = \epsilon x$ and $\sigma(y) = y$, then obviously G is of order n. Let $F \subset K(\epsilon)(x, y)$ be the invariant field of G, then by Artin's Theorem [Lang (1965), Th. 2, Chap. 8.1], $F/K(\epsilon)(x, y)$ is Galois and its Galois group is equal to G. We have $K(x^n, y) \subset F \subset K(x, y)$ and

$$[K(x, y) : K(x^n, y) = n = |G| = [K(x, y) : F],$$

hence $F = K(x^n, y)$ and the sufficiency of the condition follows.

Second proof. Write $f(x, y)$ in the form

$$f(x, y) = \sum_{i=0}^{m} (a_i(x)/b_i(x))y^i,$$

where $a_i(x), b_i(x) \in K[x]$ and $\gcd(a_i(x), b_i(x)) = 1$. Since $f(\epsilon x, y) = f(x, y)$, then $\frac{a_i(\epsilon x)}{b_i(\epsilon x)} = \frac{a_i(x)}{b_i(x)}$. It follows that

$$a_i(\epsilon, x) = \lambda_i a_i(x) \text{ and } b_1(\epsilon, x) = \lambda_i b_i(x),$$

where $\lambda_i \in K$ for any $i = 0, \ldots, m$. We deduce that

$$a_i(0) = \lambda_i a_i(0) \text{ and } b_i(0) = \lambda_i b_i(0).$$

Since $\gcd(a_i(x), b_i(x)) = 1$ then $a_i(0) \neq 0$ or $b_i(0) \neq 0$ thus $\lambda_i = 1$ for $i = 0, \ldots, m$. Fix $i \in \{0, \ldots, m\}$ and set $a_i(x) = c_q x^q + c_{q-1}x^{q-1} + \cdots + c_0$. Since $a_i(\epsilon x) = a_i(x)$ then for any $k \in \{0, \ldots, q\}$ we have $c_k \epsilon^k = c_k$. It follows that if $c_k \neq 0$, then $\epsilon^k = 1$, hence $n \mid k$. We conclude that $a_i(x) \in K[x^n]$ for $i = 0, \ldots, m$. The same method works for $b_i(x)$ and we may conclude that $b_i(x) \in K[x^n]$. Therefore $f(x, y) \in K(x^n)[y]$.

Exercise 5.24.

(1) Let $f(x) \in \mathbb{Q}[x]$ be irreducible. Suppose that $f(x)$ has a root $z = p + iq$, where $p, q \in \mathbb{R}, q \neq 0$ and $q^2 \in \mathbb{Q}$. Show that there exist $a \in \mathbb{Q}$ and $h(x) \in \mathbb{Q}[x]$ irreducible such that $f(x) = ah(x + iq)h(x - iq)$ and $h(p) = 0$.

(2) Let $f(x) \in \mathbb{Q}[x]$ of the degree $n \geq 3$ such that $f(x)$ has exactly two non real roots $z = p + iq$ and $\bar{z} = p - iq$. Show that $q \notin \mathbb{Q}$.

Solution 5.24.

(1) **First proof.** We may suppose that $f(x)$ is monic. Let $n = \deg f$ and $f(x) = \sum_{m=0}^{n} a_m x^m$, with $a_m \in \mathbb{Q}$ for $m = 0, \ldots, n-1$ and $a_n = 1$. We have

$$f(x+iq) = \sum_{m=0}^{n} a_m (x+iq)^m \quad \text{and}$$

$$(x+iq)^m = \sum_{k=0}^{m} \binom{m}{k} x^{m-k} (iq)^k.$$

If k is even, say $k = 2l$, then

$$(iq)^k = (iq)^{2l} = (-1)^l (q^2)^l \in \mathbb{Q}.$$

If k is odd, say $k = 2l+1$, then

$$(iq)^k = (iq)^{2l} \cdot iq = (-1)^l (q^2)^l (iq) := b_l \cdot iq,$$

where $b_l \in \mathbb{Q}$ hence

$$(x+iq)^m = F_m(x) + iqG_m(x),$$

where $F_m(x)$ and $G_m(x) \in \mathbb{Q}[x]$. Moreover $\deg F_m = m$ and $\deg G_m \leq m-1$. It follows that $f(x+iq)$ has the form $f(x+iq) = F(x) + iqG(x)$, where $F(x)$ and $G(x) \in \mathbb{Q}[x]$, $F(x)$ is monic, $\deg F = n$ and $\deg G \leq n-1$.

Since $0 = f(p+iq) = F(p) + iqG(p)$ and since $iq \in i\mathbb{Q}$ or $iq = i\sqrt{d}$, then $F(p) = G(p) = 0$, hence $F(x)$ and $G(x)$ have non trivial factor. Let $h(x) = \gcd(F(x), G(x))$ then $f(x+iq) = h(x)\big(F_1(x) + iqG_1(x)\big)$, where $F_1(x)$ and $G_1(x) \in \mathbb{Q}[x]$. Substitute $x - iq$ for x and obtain

$$f(x) = h(x-iq)F_1(x-iq) + iqG_1(x-iq).$$

By conjugation, we obtain

$$f(x) = h(x+iq)\big(F_1(x+iq) - iqG_1(x+iq)\big).$$

Hence

$$f^2(x) = h(x-iq)h(x+iq)(F_1(x-iq)+iqG_1(x-iq))(F_1(x+iq)-iq(x+iq)).$$

Let

$$A(x) = h(x-iq)h(x+iq) \text{ and}$$
$$B(x) = (F_1(x-iq) + iqG_1(x-iq))(F_1(x+iq) - iqG_1(x+iq))$$

then obviously, $A(x)$ and $B(x) \in \mathbb{Q}(iq)[x]$. Let σ be the unique automorphism of $\mathbb{Q}(iq)$ such that $\sigma(iq) = -iq$, then $A^\sigma = A$ and $B^\sigma = B$, hence $A(x)$ and $B(x) \in \mathbb{Q}[x]$ and then $F^2(x) = A(x)B(x)$. Let $D(x) = \gcd(f(x), A(x))$. Since any prime factor of $A(x)$ divides $f(x)$, then $D(x)$ is non-trivial. Moreover since $f(x)$ is irreducible, then $D(x) = f(x)$. Similarly we have $\gcd(f(x), B(x)) = f(x)$, hence $1 = \frac{A(X)}{f(x)}\frac{B(x)}{f(x)}$. Therefore

$$f(x) = A(x) = h(x - iq)h(x + iq).$$

Suppose that $h(x)$ is reducible over $\mathbb{Q}$ and let $h(y) = h_1(y)h_2(y)$ be non trivial factorization in $\mathbb{Q}[x, y]$, then $h(x + iq) = h_1(x + iq)h_2(x + iq)$ hence

$$f(x) = h_1(x - iq)h_1(x + iq)h_2(x - iq)h_2(x + iq),$$

contradicting the irreducibility of $f(x)$ over $\mathbb{Q}$.

Second proof. Let $h(x) = \mathrm{Irr}(p, \mathbb{Q})$ and let $g(x) = h(x+iq)h(x-iq)$. Let $H(x) = h(l(x))$, where $l(x) = x + iq$. We claim that $H(x)$ is irreducible over $\mathbb{Q}(iq)$. By contradiction, suppose that $H(x) = H_1(x)H_2(x)$ in $\mathbb{Q}(iq)[x]$, then

$$h(x) = H(l^{-1}(x)) = H_1(l^{-1}(x))H_2(l^{-1}(x)),$$

hence $h(x)$ is reducible over $\mathbb{Q}(iq)$. Consider the following diagram of fields.

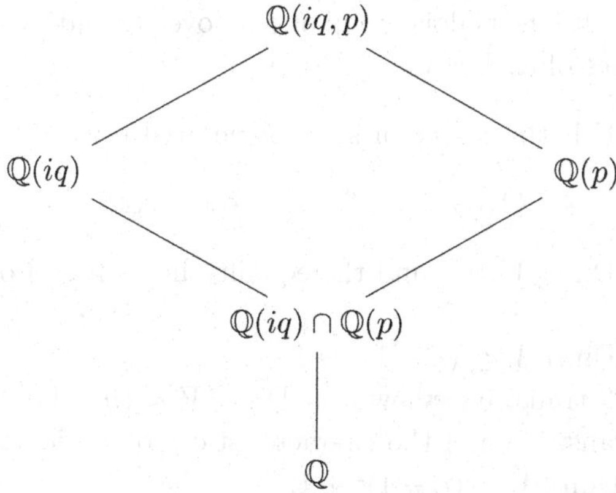

Since $\mathbb{Q}(p) \subset \mathbb{R}$ and $iq \notin R$, then $\mathbb{Q}(iq) \cap \mathbb{Q}(p) = \mathbb{Q}$. Since $\mathbb{Q}(iq)/\mathbb{Q}$ is Galois then $\mathbb{Q}(iq, p)/\mathbb{Q}(p)$ is Galois and their Galois groups are equal, see [Lang (1965), Chap. 8.1, Th. 4]. It follows that

$$[\mathbb{Q}(p) : \mathbb{Q}] = [\mathbb{Q}(p) : \mathbb{Q}(iq) \cap \mathbb{Q}(p)] = [\mathbb{Q}(iq, p) : \mathbb{Q}(iq)],$$

hence $H(x)$ is irreducible over $\mathbb{Q}(iq)$. Since $g(x) = \mathrm{N}_{\mathbb{Q}(iq)/\mathbb{Q}}(h(x+iq))$ and $\mathbb{Q}(iq)/\mathbb{Q}$ is normal and separable then, by **Exercise 4.2**, $g(x)$ is irreducible over $\mathbb{Q}$. We have

$$g(p+iq) = h(p+2iq)h(p) = 0,$$

hence $f(x)|g(x)$ in $\mathbb{Q}[x]$. Therefore $f(x) = g(x)$.

(2) Suppose that $q \in \mathbb{Q}$, then we may apply **(1)** and obtain:

$$f(x) = ah(x+iq)h(x-iq),$$

where $a \in \mathbb{Q}$, $h(x) \in \mathbb{Q}[x]$ irreducible and $h(p) = 0$. Let $\alpha_1, \ldots, \alpha_{n-2}$ be the real root of $f(x)$. Since

$$0 = f(\alpha_j) = ah(\alpha_j + iq)h(\alpha_j - iq),$$

then $h(\alpha_j + i_q) = 0$ or $h(\alpha_j - iq) = 0$ and then

$$h(\alpha_j + iq) = h(\alpha_j - iq) = 0.$$

It follows that $\frac{n}{2} = \deg h =$ which is the number of roots of $h(x)$ is $\geq 2(n-2)+1 = 2n-3$, that is $n \leq 2$, which is excluded by assumptions.

Exercise 5.25.

Let K be a field, α be algebraic, separable of degree n over K, $f(x) = \mathrm{Irr}(\alpha, K)$. Let G be the Galois group of $f(x)$ over K and $\alpha_1 = \alpha, \alpha_2, \ldots, \alpha_n$ be the conjugates of α. Let $V = \sum_{i=1}^{n} K(\alpha_i)$.

(1) Show that V is the K-vector space generated by

$$\{1, \alpha_1, \ldots, \alpha_1^{n-1}, \ldots, \alpha_n, \ldots, \alpha_n^{n-1}\}.$$

Show that $\mathrm{Dim}_K V \geq n$ and the equality holds if and only if $K(\alpha)/K$ is normal.

(2) Show that $\mathrm{Dim}_K V \leq (n-1)^2 + 1$.

(3) If G is not 2-transitive, show that $\mathrm{Dim}_K V < (n-1)^2 + 1$.

(4) If G is 2-transitive and the characteristic p of K does not divide $|G|$, show that $\mathrm{Dim}_K V = (n-1)^2 + 1$.

Solution 5.25.

(1) The statement about the generators of V is clear. Since $K(\alpha) \subset V$, then $\mathrm{Dim}_K V \geq \mathrm{Dim}_K K(\alpha) = n$. We have equality if and only if for any $i = 1, \ldots, n$, $\alpha_i \in K(\alpha)$, thus if and only if $K(\alpha)/K$ is normal.

(2) From **(1)**, we know a set of generators of V containing $1 + n(n-1)$ elements. Since for any $k = 1, \dots, n-1$, $\alpha_1^k + \cdots + \alpha_n^k \in K$, then $\alpha_n, \dots, \alpha_n^{n-1}$ are linear combinations of

$$1, \alpha_1, \dots, \alpha_1^{n-1}, \dots, \alpha_{n-1}, \dots, \alpha_{n-1}^{n-1}.$$

Therefore $\mathrm{Dim}_K V \le (n-1)^2 + 1$.

(3) Suppose that G is not 2-transitive, then the polynomial

$$f(x)/(x-\alpha) = (x-\alpha_2)\cdots(x-\alpha_n)$$

is reducible over $K(\alpha)$. Renumbering if necessary the roots of $f(x)$, we obtain a factorization over $K(\alpha)$ of the form:

$$f(x)/(x-\alpha) = \Big((x-\alpha_2)\cdots(x-\alpha_m)\Big)\Big((x-\alpha_{m+1})\cdots(x-\alpha_n)\Big)$$

for some integer $2 \le m \le n-1$. We deduce that for any $k = 1, \dots, n-1$,

$$\alpha_2^k + \cdots + \alpha_m^k \in K(\alpha) \text{ and } \alpha_{m+1}^k + \cdots + \alpha_n^k \in K(\alpha),$$

hence $\alpha_m, \dots, \alpha_m^{n-1}$ and $\alpha_n, \dots, \alpha_n^{n-1}$ are elements of the linear space generated by

$$\{1, \alpha_i^j, \text{ for } j = 1, \dots, n-1, \text{ and } i \ne m, n\}.$$

Therefore

$$\mathrm{Dim}_K V \le 1 + n(n-1) - 2(n-1) = 1 + (n-1)(n-2) < 1 + (n-1)^2.$$

(4) It is sufficient to show that $1, \alpha_1, \dots, \alpha_1^{n-1}, \dots, \alpha_{n-1}, \dots, \alpha_{n-1}^{n-1}$ are linearly independent over K. It is equivalent to show that if the polynomials $P_i(x) \in K[x]$, $i = 1, \dots, n-1$ satisfy the conditions: $\deg P_i \le n-1$, $P_i(0) = 0$ and $\sum_{i=1}^{n-1} P_i(\alpha_i) =: a \in K$, then $P_i(x) = 0$ for $i = 1, \dots, n-1$. For $i = 1, \dots, n$, let H_i be the subgroup of G fixing each element of $K(\alpha_i)$ and for $i \ne j$, let H_{ij} be the one which fixes each element of $K(\alpha_i, \alpha_j)$. Let

$$e = |H_{ij}| = [K(\alpha_1, \dots, \alpha_n) : K(\alpha_i, \alpha_j)].$$

Let $E = K(\alpha_1, \dots, \alpha_n)$. It is easy to verify the degree (resp. index) at each level as it is indicated in the following diagram of fields (resp. groups).

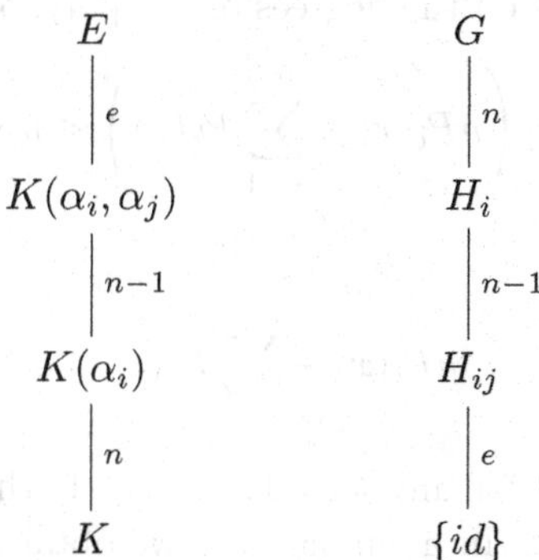

Now suppose that $\sum_{i=1}^{n-1} P_i(\alpha_i) =: a \in K$, where the polynomials $P_i(x)$ satisfy the conditions $\deg \leq n-1$ and $P_i(0) = 0$ for $i = 1, \ldots, n-1$. Then for any $\sigma \in H_1$, we have $\sum_{i=1}^{n-1} P_i(\sigma(\alpha_i)) =: a$. We may write this equation in the form $P_1(\alpha_1) + \sum_{i=2}^{n-1} P_i(\sigma(\alpha_i)) =: a$. Given j and l both distinct from 1, since $[E : K(\alpha_1, \alpha_j)] = e$, there exist e elements of H_1 which map α_j onto α_l. Therefore adding all these equations for $\sigma \in H_1$, we obtain

$$(n-1)eP_1(\alpha_1) + e\sum_{j=2}^{n}\sum_{s=2}^{n-1} P_s(\alpha_j) = (n-1)ea.$$

since $p \nmid |G|$, then $p \nmid e$. Therefore

$$(n-1)P_1(\alpha_1) + \sum_{j=2}^{n}\sum_{s=2}^{n-1} P_s(\alpha_j) = (n-1)a.$$

We may write this equation in the form

$$(n-1)P_1(\alpha_1) + P_2(\alpha_2) + \cdots + P_2(\alpha_n) + \cdots + P_{n-1}(\alpha_2) \\ + \cdots + P_{n-1}(\alpha_n) = (n-1)a.$$

Since for any $k \in \{2, \ldots, n-1\}$, $\sum_{j=1}^{n} P_k(\alpha_j) \in K$, then

$$(n-1)P_1(\alpha_1) - \sum_{k=2}^{n-1} P_k(\alpha_1) \in K.$$

We deduce that

$$nP_1(\alpha_1) - \sum_{k=1}^{n-1} P_k(\alpha_1) \in K.$$

The assumptions on the degrees of the polynomials P_k shows that

$$\deg\left(nP_1(x) - \sum_{k=1}^{n-1} P_k(x)\right) < n = \deg(\alpha_1).$$

Therefore

$$nP_1(x) - \sum_{k=1}^{n-1} P_k(x) = c \in K.$$

Since $P_k(0) = 0$ for any $k = 1, \ldots, n-1$, then $c = 0$ and $nP_1(x) - \sum_{k=1}^{n-1} P_k(x) = 0$. In a similar way we obtain the following equations

$nP_j(x)-\sum_{k=1}^{n-1} P_k(x)=0$ for $j=1,\dots,n-1$. We write these equations in the form

$$(n-1)P_1(x)-P_2(x)-\cdots-P_{n-1}(x)=0 \quad \text{(Eq 1)}$$
$$-P_1(x)+(n-1)P_2(x)-\cdots-P_{n-1}(x)=0 \quad \text{(Eq 2)}$$
$$\cdots\cdots\cdots\cdots\cdots\cdots\cdots\cdots=0$$
$$-P_1(x)-P_2(x)-\cdots+(n-1)P_{n-1}(x)=0. \quad \text{(Eq n-1)}$$

We obtain one more equation by adding these $n-1$ equations:

$$P_1(x)+P_2(x)+\cdots+P_{n-1}(x)=0. \quad \text{(Eq n)}$$

Adding the k-th equation to this last equation leads to $nP_k(x)=0$. Since $p\nmid n$, then $P_k(x)=0$ for $k=1,\dots,n-1$.

Exercise 5.26.

(1) Let L be a cyclic extension of degree 4 over $\mathbb{Q}$, M be its unique quadratic subfield. Show that $M\subset\mathbb{R}$.
(2) Let d be a square free integer such that $d\geq 2$ and $F=\mathbb{Q}(\sqrt{d})$. Let $a,b\in Z$ such that the number $\alpha=a+b\sqrt{d}$ is not a square in F and $K=F(\sqrt{\alpha})$. Show that $K/\mathbb{Q}$ is normal if and only if $a^2-db^2\in F^2$.
(3) Keep the assumptions of (2) and suppose that $a^2-db^2\in F^2$. Show that $a^2-db^2=y^2$ or $a^2-db^2=dy^2$ with $y\in\mathbb{Z}$. Show that in the first (resp. second) case $\mathrm{Gal}(K,\mathbb{Q})\simeq\mathbb{Z}/2\mathbb{Z}\times\mathbb{Z}/2\mathbb{Z}$ (resp. $\mathrm{Gal}(K,\mathbb{Q})\simeq\mathbb{Z}/4\mathbb{Z}$).
(4) Let E be a number field of degree 4 over $\mathbb{Q}$. Show that $E/\mathbb{Q}$ is cyclic if and only if the following conditions hold.
   (a) There exist a square free integer $d\geq 2$ and $s,q,r\in\mathbb{Z}$ such that $d=(\frac{s}{r})^2+(\frac{q}{r})^2$.
   (b) $E=\mathbb{Q}(\sqrt{rd+s\sqrt{d}})$.

Solution 5.26.

(1) If $L\subset\mathbb{R}$, the proof is obvious. Suppose next that $L\not\subset\mathbb{R}$. Let σ be the restriction of the complex conjugation to L, then σ is an automorphism of L, i.e. $\sigma\in\mathrm{Gal}(L,\mathbb{Q})$. Since this group is cyclic of order 4, it contains a unique subgroup of order 2, namely $H=\mathrm{Gal}(L,M)$. Moreover since σ is of order 2, then $H=\langle\sigma\rangle$. Therefore $M=\mathrm{Inv}(\langle\sigma\rangle)\subset\mathbb{R}$.
(2) Let $\beta=a-b\sqrt{d}$. It is clear that the conjugates of $\sqrt{\alpha}$ are the followings: $\sqrt{\alpha},-\sqrt{\alpha},\sqrt{\beta},-\sqrt{\beta}$. Suppose that $a^2-db^2=\gamma^2\in F^2$. We have $\alpha\beta=\gamma^2$, hence $\sqrt{\alpha}\sqrt{\beta}=\pm\gamma$. It follows that $\sqrt{\beta}=\pm\gamma/\sqrt{\alpha}\in K$, which shows that $K/\mathbb{Q}$ is normal. Since $\chi(\mathbb{Q})=0$, then $K/\mathbb{Q}$ is Galois.

Suppose now that $K/\mathbb{Q}$ is Galois. Then $\sqrt{\beta} \in K$. Set $\sqrt{\beta} = A + B\sqrt{\alpha}$, where $A, B \in F$. We deduce that

$$A^2 + B^2\alpha + 2AB\sqrt{\alpha} = a - b\sqrt{d} \in F.$$

Therefore $A = 0$ or $B = 0$. If $B = 0$, then $A^2 = a - b\sqrt{d}$. Applying the automorphism $\tau : F \to F$ defined by $\tau(x + y\sqrt{d}) = x - y\sqrt{d}$, we obtain $a + b\sqrt{d} = \tau(A)^2 \in F^2$, which is a contradiction. Suppose next that $A = 0$. We have

$$a - b\sqrt{d} = B^2\alpha = B^2(a + b\sqrt{d}),$$

hence

$$B^2 = \frac{a - b\sqrt{d}}{a + b\sqrt{d}} = \frac{a^2 - db^2}{(a + b\sqrt{d})^2}.$$

This proves $a^2 - db^2 \in F^2$.

(3) Since $a^2 - db^2 \in F^2$, set $a^2 - db^2 = (c + e\sqrt{d})^2$, where $c, d \in \mathbb{Q}$. Then $a^2 - db^2 = c^2 + de^2 + 2ce\sqrt{d}$. We conclude that $c = 0$ or $e = 0$. If $e = 0$, then $a^2 - db^2 = c^2$, which represents the first case mentioned in the conclusion of **(3)**. Suppose next that $c = 0$. We have $a^2 - db^2 = de^2$, which represents the second case. We now prove the assertion relative to the Galois group. Let $\sigma_1, \sigma_2, \sigma_3, \sigma_4$ be the elements of the Galois group of K over $\mathbb{Q}$ such that

$$\sigma_1 = Id_K,\ \sigma_2(\sqrt{\alpha}) = -\sqrt{\alpha},\ \sigma_3(\sqrt{\alpha}) = \sqrt{\beta} \text{ and } \sigma_3(\sqrt{\alpha}) = -\sqrt{\beta}.$$

It is clear that σ_2 is of order 2, hence $K/\mathbb{Q}$ is cyclic if and only if σ_3 is of order 4. We have $\sigma_3(\alpha) = (\sigma_3(\sqrt{\alpha}))^2 = \beta$. We deduce that $\sigma_3(\sqrt{d}) = -\sqrt{d}$. We have $\sqrt{\alpha}\sqrt{\beta} = \gamma \in F$, hence

$$\begin{aligned}\sigma_3^2(\sqrt{\alpha}) &= \sigma_3(\sqrt{\beta})\\ &= \sigma_3(\gamma/\sqrt{\alpha})\\ &= \sigma_3(\gamma)/\sigma_3(\sqrt{\alpha})\\ &= \sigma_3(\gamma)/\sqrt{\beta}\\ &= (\sigma_3(\gamma)/\gamma)\sqrt{\alpha}.\end{aligned}$$

Therefore σ_3 is of order 4 or 2 according to $\sigma_3(\gamma)/\gamma = -1$ or $\sigma_3(\gamma)/\gamma = 1$. We deduce that $\text{Gal}(K, \mathbb{Q}) \simeq \mathbb{Z}/4\mathbb{Z})$ if and only if $\sigma_3(\gamma) = -\gamma$, which is equivalent to $\gamma = y\sqrt{d}$ with $y \in \mathbb{Z}$ and then equivalent to $a^2 - db^2 = dy^2$. Similarly we have $\text{Gal}(K, \mathbb{Q}) \simeq \mathbb{Z}/2\mathbb{Z} \times \mathbb{Z}/2\mathbb{Z}$ if and only if $\sigma_3(\gamma) = \gamma$, which is equivalent to $\gamma = y$ with $y \in \mathbb{Z}$ and then equivalent to $a^2 - db^2 = y^2$.

(4) • **Necessity of the conditions.**
By **(2)** and **(3)**, there exist $a, b \in \mathbb{Z}$ such that $E = \mathbb{Q}(\sqrt{a + b\sqrt{d}})$, for some integer $d \geq 2$ and some integers, a, b such that $a+b\sqrt{d} \notin \mathbb{Q}(\sqrt{d})$ and $a^2 - db^2 = du^2$ with $u \in \mathbb{Z}$. This last relation implies that $d \mid a$. Set $a = da_1$, then $da_1^2 = b^2 + u^2$. We deduce that $d = (b/a_1)^2 + (u/a_1)^2$. If we replace b by s, u by q and a by rd, we obtain the proof of (b) and (c). Notice that the identity $d = (\frac{s}{r})^2 + (\frac{q}{r})^2$ shows that d is a sum of two squares of integers. Therefore any prime divisor of d is equal to 2 or congruent to 1 modulo 4.

• **Sufficiency of the conditions.**
We must show that $rd + s\sqrt{d} \notin \mathbb{Q}(\sqrt{d})^2$ and $r^2d^2 - s^2d = dy^2$ with $y \in \mathbb{Q}$. Since $d = (\frac{s}{r})^2 + (\frac{q}{r})^2$, then $r^2d - s^2 = q^2$ and we obtain the second claim by multiplying by d. Suppose that $rd+s\sqrt{d} \in \mathbb{Q}(\sqrt{d})^2$, then $(rd + s\sqrt{d})(rd - s\sqrt{d}) \in \mathbb{Q}^2$, hence $r^2d^2 - s^2d = y^2$ with $y \in \mathbb{Q}$ which contradicts the equation $r^2d - s^2 = q^2$.

Exercise 5.27.
Let p be a prime number, K be a field, $K(\alpha)$ and $K(\beta)$ be Galois extensions of K and N be a normal p-subgroup of $\mathrm{Gal}(K(\alpha, \beta), K)$ such that every element of order p in N fixes α or β. Show that the order of N divides $[K(\alpha\beta) : K]$.

Solution 5.27.
Let $G = \mathrm{Gal}(K(\alpha, \beta) : K)$, $\gamma = \alpha\beta$ and $f(x) = \mathrm{Irr}(\gamma, K)$. Obviously N acts in the set of roots of $f(x)$. Let $\gamma^{'}$ be any root of $f(x)$. We show that the orbits, under the action of N, of γ and $\gamma^{'}$ have the same cardinal. Since γ and $\gamma^{'}$ are conjugate over K, there exists $\sigma \in G$ such that $\sigma(\gamma) = \gamma^{'}$. For any $n \in N$, we have

$$\sigma(n(\gamma)) = n^{'}(\sigma(\gamma)) = n^{'}(\gamma^{'})$$

for some $n^{'} \in N$. Therefore, σ induces an injective map from $Orb(\gamma)$ into $Orb(\gamma^{'})$. Clearly, σ^{-1} induces an injective map form $Orb(\gamma^{'})$ into $Orb(\gamma)$. We conclude that all the orbits under the action of N have the same cardinality. Thus, to obtain the stated result, it is sufficient to prove $|Orb(\gamma)| = (N : 1)$. This will follow if no non trivial element of N fixes γ. Suppose that there exists $n \in N$, $n \neq Id$ such that $n(\gamma) = \gamma$. Let p^e, with $e \geq 1$, be the order of n. We may suppose that $e = 1$, otherwise $n^{p^{e-1}}$ has order p and fixes γ. Our assumptions on the elements of order p of N implies $n(\alpha) = \alpha$ or $n(\beta) = \beta$. If $n(\alpha) = \alpha$, then $\alpha\beta = \gamma = n(\gamma) =$

$n(\alpha)n(\beta) = \alpha n(\beta)$, hence $n(\beta) = \beta$, thus $n = Id$ which is a contradiction. We get the same contradiction if we suppose that $n(\beta) = \beta$.

Remark. We give a second proof of the claim that all the orbits under the action of N have the same cardinality. Let $H = N \cap \mathrm{Gal}(K(\alpha, \beta), K(\gamma))$. We first show that $Orb(\gamma) = \{n_1(\gamma), \ldots, n_s(\gamma)\}$, where $\{n_1, \ldots, n_s\}$ is a complete set of representatives of the elements of N modulo H. Let $n_1, n_2 \in N$, then

$$\begin{aligned} n_1(\gamma) = n_2(\gamma) &\Leftrightarrow n_2^{-1}n_1(\gamma) = \gamma \\ &\Leftrightarrow n_2^{-1}n_1 \in H \\ &\Leftrightarrow n_1H = n_2H. \end{aligned}$$

We conclude that $|Orb(\gamma)| = (N : H)$.

Let $\gamma^{'} = \sigma(\gamma)$ be a conjugate of γ, then as above, we have $|Orb\gamma^{'}| = (N : \sigma H \sigma^{-1}) = (N : H)$. Therefore all the orbits have the same cardinality.

Exercise 5.28.
Let K be a field.

(1) Let G be a finite group. Show that there exist fields E and F such that $K \subset F \subset E$, E/F is Galois and $\mathrm{Gal}(E, F) \simeq G$.
(2) Let n be a positive integer. Show that there exists an extension J of K which has no finite extension of degree k with $1 \leq k \leq n$.

Solution 5.28.

(1) Let $m = |G|$ and let $x_1, \ldots, x_m$ be algebraically independent variables over K and let σ_i for $i = 1, \ldots, m$ be the elementary symmetric function of degree i of the x_j. Then $K(x_1, \ldots, x_m)$ is a Galois extension of $k(\sigma_1, \ldots, \sigma_m)$ whose Galois group is isomorphic to $\mathcal{S}_m$. Embedding G in $\mathcal{S}_{\Updownarrow}$, we may consider G as a subgroup of $\mathcal{S}_m$. Let F be the invariant field of G, then $K \subset F \subset K(x_1, \ldots, x_m)$, $K(x_1, \ldots, x_m)/F$ is Galois and $\mathrm{Gal}(K(x_1, \ldots, x_m), F) \simeq G$. The proof is compete if we let $E = K(x_1, \ldots, x_m)$.
(2) Let m be an integer such that $n! < m!/2$. Let $G = \mathcal{A}_m$ and let E and F be the fields determined by **(1)** and satisfying the conditions: $K \subset F \subset E$, E/F is Galois and $\mathrm{Gal}(E, F) \simeq \mathcal{A}_m$. Let Ω be an algebraic closure of F containing E. Consider the set $\mathcal{L}$ of subfields of Ω containing F such that $L \cap E = F$. These conditions on L are equivalent to LE/L is Galois and $\mathrm{Gal}(LE, L) \simeq \cap\mathcal{A}_m$. The following diagram may be useful.

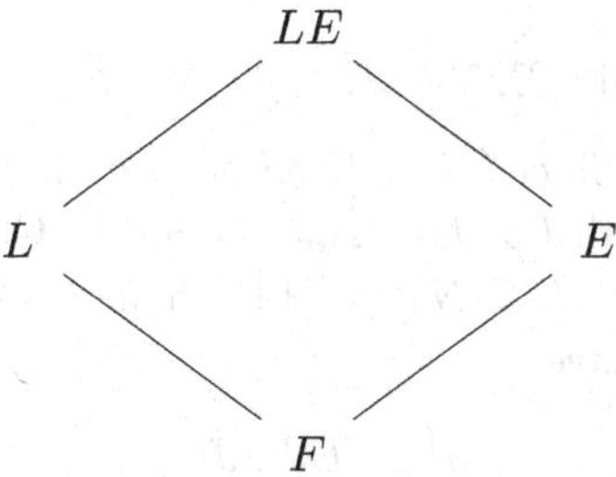

The set $\mathcal{L}$ is non empty since it contains F. Moreover $\mathcal{L}$ is ordered by the inclusion. Any family of totally ordered elements of $\mathcal{L}$ has a greatest element (which is the union of the sets in the family). It follows that $\mathcal{L}$ is an inductive set and then Zorn's lemma implies that it has a maximal element say J. Suppose now that J has an extension of degree k with $1 \leq k \leq n$ and let $N \subset \Omega$ be its normal closure over J. We may visualize the lattice of fields trough the following diagram.

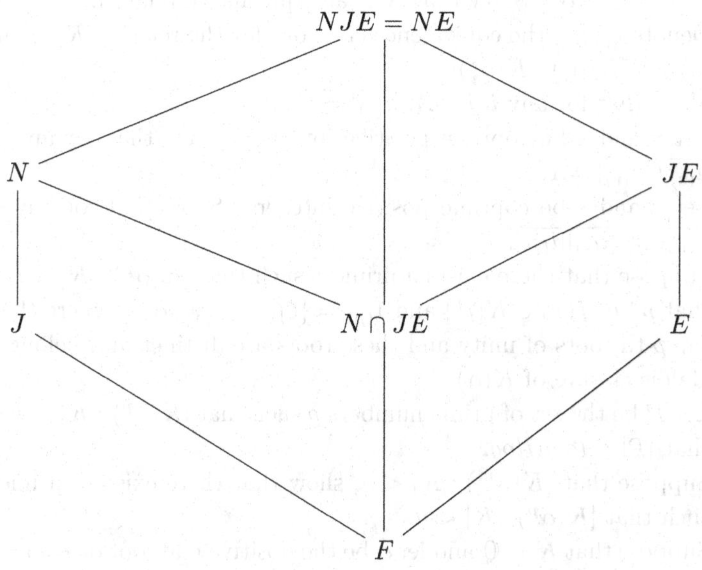

Since $J \in \mathcal{L}$, then $\mathrm{Gal}(JE, J) \simeq \mathcal{A}_m$. By [Lang (1965), Th. 4, Chap. 8.1], $JE/(N \cap JE)$ is Galois, NE/N is also Galois and $\mathrm{Gal}(JE, N \cap JE) \simeq \mathrm{Gal}(NE, N)$. Since J is maximal in $\mathcal{L}$ then $N \in \mathcal{L}$.

Since

$$\mathrm{Gal}(JE, N \cap JE) \subsetneq \mathrm{Gal}(JE, J) = \mathcal{A}_m,$$

then $J \subsetneq N \cap JE$. Since $(N \cap JE)/J$ is normal, then $\mathrm{Gal}(JE, N \cap JE)$ is distinguished in $\mathrm{Gal}(JE, J) = \mathcal{A}_m$. Therefore $\mathrm{Gal}(JE, N \cap JE) = \{\mathrm{Id}\}$. It follows that $\mathrm{Gal}(NE, N) = \{\mathrm{Id}\}$, hence $NE = N$ which implies $E \subset N$. Now we have

$$[N : J] \geq [EJ : J] = \frac{m!}{2} > n!$$

and this contradicts the fact that N is the normal closure over J of some extension L/J with $[L : J] = k$ and $1 \leq k \leq n$.

Exercise 5.29.
Let K be a field, Ω be an algebraic closure of K, $\alpha \in \Omega$ be separable of degree n over K. Let $\alpha_1, \alpha_2, \ldots, \alpha_n$ be the conjugates of α, where $\alpha_1 = \alpha$.

(1) Let $a \in K^\star$ and β be a root of $f(x)$. If $a\alpha$ is a root of $f(x)$, show that $a\beta$ is a root of $f(x)$.
(2) Let $s \geq 1$ be an integer. Define in $A = \{1, \ldots, n\}$ the relation $\mathcal{R}$ by i $\mathcal{R}$ j if $\alpha_i^s = \alpha_j^s$. Show that $\mathcal{R}$ is an equivalence relation.
(3) Denote by $\overline{i_{(s)}}$ the equivalence class of i for the relation $\mathcal{R}$. Show that $|\overline{i_{(s)}}| = [K(\alpha_i) : K(\alpha_i^s)]$.
(4) Show that for any $i, j \in A$, $|\overline{i_{(s)}}| = |\overline{j_{(s)}}|$.
(5) Let r and s be coprime positive integers. Show that for any $i \in A$, $|\overline{i_{(s)}} \cap \overline{i_{(r)}}| = 1$.
(6) Let r and s be coprime positive integers. Show that for any $i \in A$, $|\overline{i_{(rs)}}| \geq |\overline{i_{(s)}}||\overline{i_{(r)}}|$.
(7) Suppose that there exists a prime p such that $[K(\alpha^p) : K] < n$. Show that $p \mid n$, $f(x) \in K[x^p]$ and $\overline{1_{(p)}} = \{\zeta_1\alpha_1, \ldots, \zeta_p\alpha_1\}$, where the ζ_i are the p-th roots of unity and these roots are distinct and belong to the Galois closure of $K(\alpha)$.
(8) Let $\mathcal{P}$ be the set of prime numbers p such that $[K(\alpha^p) : K] < n$. Show that $|\mathcal{P}| \leq logn/log2$.
(9) Suppose that $[K(\alpha^m) : K] < n$, show that there exists a prime $p \mid m$ such that $[K(\alpha^p) : K] < n$.
(10) Suppose that $K = \mathbb{Q}$ and let α be the positive real root of $f(x) = x^6 - 2$ and p be a prime number.
    (a) Show that $[\mathbb{Q}(\alpha^p) : \mathbb{Q}] < 6$ if and only if $p = 2$ or $p = 3$. Deduce that for any positive integer m, $[\mathbb{Q}(\alpha^m) : \mathbb{Q}] < 6$ if and only if $m \equiv 0 \pmod 2$ or $m \equiv 0 \pmod 3$.

(b) Let $\beta = \sum_{i=0}^{5} a_i\alpha^i$ be a primitive element of $\mathbb{Q}(\alpha)$ over $\mathbb{Q}$. Show that $[\mathbb{Q}(\beta^p) : \mathbb{Q}] < 6$ if and only if $p = 2$ (resp.$p = 3$) and the a_i satisfy the following equations

$$\begin{aligned} a_0a_1 + 2a_2a_5 + 2a_3a_4 &= 0 \\ a_0a_3 + a_1a_2 + 2a_4a_5 &= 0 \\ a_0a_5 + a_1a_4 + a_2a_3 &= 0 \quad \text{(resp.} \\ b_1a_0 + b_0a_1 + 2b_5a_2 + 2b_4a_3 + 2b_3a_4 + 2b_2a_5 &= 0 \\ b_2a_0 + b_1a_1 + b_0a_2 + 2b_5a_3 + 2b_4a_4 + 2b_3a_5 &= 0 \\ b_4a_0 + b_3a_1 + b_2a_2 + b_1a_3 + b_0a_4 + 2b_5a_5 &= 0 \\ b_5a_0 + b_4a_1 + b_3a_2 + b_2a_3 + b_1a_4 + b_0a_5 &= 0). \end{aligned}$$

Here the b_i are given by

$$\begin{aligned} b_0 &= a_0^2 + 2a_3^2 + 4a_1a_5 + 4a_2a_4 \\ b_1 &= 2a_0a_1 + 4a_2a_5 + 4a_3a_4 \\ b_2 &= a_1^2 + 2a_4^2 + 2a_0a_2 + 4a_3a_5 \\ b_3 &= 2a_0a_3 + 2a_1a_2 + 4a_4a_5 \\ b_4 &= a_2^2 + 2a_5^2 + 2a_0a_4 + 2a_1a_3 \\ b_5 &= 2a_0a_5 + 2a_1a_4 + 2a_2a_3. \end{aligned}$$

Solution 5.29.

(1) Let $q(x)$ and $r(x) \in K[x]$ such that $\deg r(x) < n$ and

$$f(ax) = f(x)q(x) + r(x),$$

then obviously $q(x) = a^n$. Substituting α for x in the preceding identity, leads to $r(\alpha) = 0$, hence $r(x) = 0$. Thus $f(ax) = a^n f(x)$. It follows that $f(a\beta) = 0$.

(2) Easy.

(3) Let

$$\Sigma = \{\sigma : K(\alpha_i) \to \overline{K}, \sigma \,\text{is a}\, K(\alpha_i^s)\text{-embedding}\}$$

and consider the map $\phi : \overline{i_{(s)}} \to \Sigma$ such that for any $j \in \overline{i_{(s)}}$, we have $\phi(j)$ is the unique K-embedding of $K(\alpha_i)$ into $\overline{K}$ satisfying $\phi(j)(\alpha_i) = \alpha_j$. We verify that $\phi(j)$ fixes α_i^s. We have $\phi(j)(\alpha_i^s) = \alpha_j^s = \alpha_i^s$, hence the result. We show that ϕ is one to one. Suppose that $\phi(j) = \phi(k)$, then

$$\alpha_j = \phi(j)(\alpha_i) = \phi(k)(\alpha_i) = \alpha_k,$$

hence $j = k$. We show that ϕ is onto. Let $\sigma : K(\alpha_i) \to \overline{K}$ be a $K(\alpha_i^s)$-embedding such that $\sigma(\alpha_i) = \alpha_j$, then $\alpha_i^s = \sigma(\alpha_i^s) = \alpha_j^s$, hence $j \in \overline{i_{(s)}}$.

(4) We define an injective map from $\overline{i_{(s)}}$ into $\overline{j_{(s)}}$. Let $f(x) = \mathrm{Irr}(\alpha, K)$, $G = \mathrm{Gal}(f(x), K)$. Since G is transitive, then there exists $\sigma \in G$ such that $\sigma(\alpha_i) = \alpha_j$. Let $k \in \overline{i_{(s)}}$, then $\alpha_k^s = \alpha_i^s$, hence $\sigma(\alpha_k)^s = \alpha_j^s$. Therefore σ induces an injective map from $\overline{i_{(s)}}$ into $\overline{j_{(s)}}$. By symmetry there exists an injective map from the later into the former. Therefore the two sets have the same cardinality.

(5) Obviously $i \in \overline{i_{(s)}} \cap \overline{i_{(r)}}$. Let k and $l \in \overline{i_{(s)}} \cap \overline{i_{(r)}}$, then$\alpha_k^r = \alpha_l^r$ and $\alpha_k^s = \alpha_l^s$. By Bezout's identity, there exist u and v such that $ur + vs = 1$. We deduce that

$$\alpha_k = \alpha_k^{ur}\alpha_k^{vs} = \alpha_l^{ur}\alpha_l^{vs} = \alpha_l,$$

hence the result.

(6) By **(2)**, it is sufficient to show that $\cup_{j\in\overline{i_{(r)}}}\overline{j_{(s)}} \subset \overline{i_{(rs)}}$ and that this union is disjoint. Let $j \in \overline{i_{(r)}}$ and $k \in \overline{j_{(s)}}$, then $\alpha_j^r = \alpha_i^r$ and $\alpha_k^s = \alpha_j^s$, hence $\alpha_k^{rs} = \alpha_j^{rs} = \alpha_i^{rs}$, which means $k \in \overline{i_{(rs)}}$. Suppose that there exist j_1 and $j_2 \in \overline{i_{(r)}}$ such that $\overline{j_{1(s)}} = \overline{j_{2(s)}}$, then j_1 and $j_2 \in \overline{j_{1(r)}} \cap \overline{j_{1(s)}}$, hence by **(5)**, $j_1 = j_2$.

(7) By **(3)**, we conclude that $[K(\alpha^p) : K] < n$ if and only if there exists a root β of $f(x)$ such that $\beta \neq \alpha$ and $\beta^p = \alpha^p$. We use this β in our proof. We have $\beta = \zeta\alpha$, where ζ is a primitive p-th root of unity. As in (1), we conclude that $f(\zeta x) = \zeta^n f(x)$. Let $f(x) = \sum_{i=0}^n a_i x^i$, where $a_i \in K$ and $a_n = 1$, then

$$\zeta^n \sum_{i=0}^{n} \zeta^{i-n} a_i x^i = \zeta^n \sum_{i=0}^{n} a_i x^i.$$

It follows that for any i for which $a_i \neq 0$, we have $\zeta^{i-n} = 1$, that is $i \equiv n \pmod{p}$. In particular $0 \equiv n \pmod{p}$ and then $i \equiv 0 \pmod{p}$ for all the indices i such that $a_i \neq 0$. We conclude that $f(x) \in K[x^p]$. For any $i = 1, \dots, p$, we have $(\zeta_i\alpha_1)^p = \alpha_1^p$ hence $\zeta_i\alpha_1 \in \overline{1_{(p)}}$. Conversely if $\alpha_i \in \overline{1_{(p)}}$, then $\alpha_i^p = \alpha_1^p$, thus $\alpha_i = \eta\alpha_1$ for some p-th root of unity η. Since for any $i = 1, \dots, p$, $\zeta_i = \alpha_1/\alpha_j$ for some $j \in \{1, \dots, n\}$, then the Galois closure of $K(\alpha)$ contains the p-th roots of unity and these roots are distinct since $f(x)$ is separable over K.

(8) Notice that by the definition of $\mathcal{P}$, for any $p \in \mathcal{P}$, we have $|\overline{i_{(p)}}| \geq 2$. Let B be any finite subset of $\mathcal{P}$, then by (4) and (6), we have

$$2^{|B|} \leq \prod_{p\in B} |\overline{i_{(p)}}| \leq |\overline{i_{(\prod_{p\in B} p)}}| \leq n.$$

We deduce that $|B| \leq \log n/\log 2$, hence $\mathcal{P}$ is finite and $|\mathcal{P}| \leq (\log n)/(\log 2)$.

Remark. Since any element of $\mathcal{P}$ is a divisor of n, then $|\mathcal{P}| \leq r$, where r is the number of prime factors of n. Write n in the form $n = \prod_{i=1}^{r} p_i^{e_i}$. Then

$$\log n = \sum_{i=1}^{r} e_i \log p_i \geq \sum_{i=1}^{r} e_i \log 2 \geq r \log 2,$$

hence the result.

(9) Let $k = |\overline{1_{(m)}}|$, then the assumptions implies that $k \geq 2$. We may suppose that $\overline{1_{(m)}} = \{1, \ldots, k\}$. For any $i \in \overline{1_{(m)}}$, let $\zeta_i = \alpha_1/\alpha_i$, then obviously ζ_i is a m-th root of unity. Let d be the least common multiple of the orders of these ζ_i, then $d \mid m$. Let U_m be the set of m-th roots of unity. We show that $H := \{\zeta_1, \ldots, \zeta_k\}$ is a subgroup of U_m. Let ζ_i and ζ_j be elements of H, then $\zeta_i^{-1}\alpha_1$ and $\zeta_j^{-1}\alpha_1$ are roots of $f(x)$. By **(1)**, we conclude that $\zeta_j^{-1}\zeta_i^{-1}\alpha_1$ is a root say α_t of $f(x)$. It follows that $\alpha_1/\alpha_t = \zeta_j\zeta_i$. Thus $t \in \overline{1_{(m)}}$ and then $\zeta_j\zeta_i \in H$. Now H being a subgroup of the cyclic group U_m is itself cyclic of order k. By definition of d it follows that $k = d$. Since $k \geq 2$, then k has at least a prime divisor p. Therefore H contains an element of order p, say ζ_{i_0} with $i_0 \in \{1, \ldots, k\}$. We deduce that $\alpha_1 = \zeta_{i_0}\alpha_{i_0}$. Thus $\alpha_1^p = \alpha_{i_0}^p$. Since $\alpha_1 \neq \alpha_{i_0}$, then $[K(\alpha^p) : K] < n$.

(10)(a) Suppose that $[\mathbb{Q}(\alpha^p) : \mathbb{Q}] < 6$, then by (7), $p \mid 6$ so that $p = 2$ or $p = 3$. The roots of $f(x)$ are given by α, $-\alpha$, $j\alpha$, $-j\alpha$, $j^2\alpha$ and $-j^2\alpha$, where j is a primitive cube root of unity. Therefore we have the following relations between α and some of its conjugates

$$\alpha^2 = (-\alpha)^2 \quad \text{and} \quad \alpha^3 = (j\alpha)^3 = (j^2\alpha)^3.$$

This shows that $[\mathbb{Q}(\alpha^2) : \mathbb{Q}] = 3$ and $[\mathbb{Q}(\alpha^3) : \mathbb{Q}] = 2$. We deduce that

$$\begin{aligned}
[\mathbb{Q}(\alpha^m) : \mathbb{Q}] < 6 &\iff \text{There exists a prime} \quad p \mid m\, [\mathbb{Q}(\alpha^p) : \mathbb{Q}] < 6 \\
&\iff m \equiv 0 \pmod 2 \quad \text{or} \quad m \equiv 0 \pmod 3.
\end{aligned}$$

(b) Let $\sigma_1, \ldots, \sigma_6 : \mathbb{Q}(\alpha) \to \mathbb{C}$ be the distinct embeddings. We suppose that they are labeled in the following way:

$$\begin{aligned}
&\sigma_1(\alpha) = \alpha, \quad \sigma_2(\alpha) = -\alpha, \quad \sigma_3(\alpha) = j\alpha, \\
&\sigma_4(\alpha) = -j\alpha, \quad \sigma_5(\alpha) = j^2\alpha, \quad \text{and} \quad \sigma_6(\alpha) = -j^2\alpha.
\end{aligned}$$

Let $\gamma = \sum_{i=0}^{5} c_i\alpha^i$ be an element of $\mathbb{Q}(\alpha)$. We look at the conditions $\sigma_i(\gamma) = \gamma$ for $i = 2, \ldots, 6$. We have

$$\begin{aligned}\sigma_2(\gamma) = \gamma &\Longleftrightarrow c_0 - c_1\alpha + c_2\alpha^2 - c_3\alpha^3 + c_4\alpha^4 - c_5\alpha^5 = \sum_{i=0}^{5} c_i\alpha^i\\ &\Longleftrightarrow c_1 = c_3 = c_5 = 0\\ &\Longleftrightarrow \gamma \in \mathbb{Q}(\alpha^2),\end{aligned}$$

$$\begin{aligned}\sigma_3(\gamma) = \gamma &\Longleftrightarrow c_0 + c_1j\alpha + c_2j^2\alpha^2 + c_3\alpha^3 + c_4j\alpha^4 + c_5j^2\alpha^5 = \sum_{i=0}^{5} c_i\alpha^i\\ &\Longleftrightarrow c_0 + c_1j\alpha + c_2(-1-j)\alpha^2 + c_3\alpha^3 + c_4j\alpha^4 + c_5(-1-j)\alpha^5\\ &= \sum_{i=0}^{5} c_i\alpha^i\\ &\Longleftrightarrow c_0 - c_2\alpha^2 + c_3\alpha^3 - c_5\alpha^5 + c_1j\alpha - c_2j\alpha^2 + c_4j\alpha^4 - c_5j\alpha^5\\ &= \sum_{i=0}^{5} c_i\alpha^i\\ &\Longleftrightarrow c_1 = c_2 = c_4 = c_5 = 0\\ &\Longleftrightarrow \gamma \in \mathbb{Q}(\alpha^3),\end{aligned}$$

$$\begin{aligned}\sigma_4(\gamma) = \gamma &\Longleftrightarrow c_0 - c_1j\alpha + c_2j^2\alpha^2 - c_3\alpha^3 + c_4j\alpha^4 - c_5j^2\alpha^5 = \sum_{i=0}^{5} c_i\alpha^i\\ &\Longleftrightarrow c_0 - c_1j\alpha + c_2(-1-j)\alpha^2 - c_3\alpha^3 + c_4j\alpha^4 - c_5(-1-j)\alpha^5\\ &= \sum_{i=0}^{5} c_i\alpha^i\\ &\Longleftrightarrow c_0 - c_2\alpha^2 - c_3\alpha^3 + c_5\alpha^5 - c_1j\alpha - c_2j\alpha^2 + c_4j\alpha^4 + c_5j\alpha^5\\ &= \sum_{i=0}^{5} c_i\alpha^i\\ &\Longleftrightarrow c_1 = c_2 = c_3 = c_4 = c_5 = 0\\ &\Longleftrightarrow \gamma \in \mathbb{Q},\end{aligned}$$

$$
\begin{aligned}
\sigma_5(\gamma) = \gamma &\iff c_0 + c_1 j^2\alpha + c_2 j\alpha^2 + c_3\alpha^3 + c_4 j^2\alpha^4 + c_5 j\alpha^5 = \sum_{i=0}^{5} c_i\alpha^i \\
&\iff c_0 + c_1(-1-j)\alpha + c_2 j\alpha^2 + c_3\alpha^3 + c_4(-1-j)\alpha^4 + c_5 j\alpha^5 \\
&= \sum_{i=0}^{5} c_i\alpha^i \\
&\iff c_0 - c_1\alpha + c_3\alpha^3 - c_4\alpha^4 - c_1 j\alpha + c_2 j\alpha^2 - c_4 j\alpha^4 + c_5 j\alpha^5 \\
&= \sum_{i=0}^{5} c_i\alpha^i \\
&\iff c_1 = c_2 = c_4 = c_5 = 0 \\
&\iff \gamma \in \mathbb{Q}(\alpha^3),
\end{aligned}
$$

and

$$
\begin{aligned}
\sigma_6(\gamma) = \gamma &\iff c_0 - c_1 j^2\alpha + c_2 j\alpha^2 - c_3\alpha^3 + c_4 j^2\alpha^4 - c_5 j\alpha^5 \\
&= \sum_{i=0}^{5} c_i\alpha^i \\
&\iff c_0 - c_1(-1-j)\alpha + c_2 j\alpha^2 - c_3\alpha^3 + c_4(-1-j)\alpha^4 - c_5 j\alpha^5 \\
&= \sum_{i=0}^{5} c_i\alpha^i \\
&\iff c_0 - c_1\alpha - c_3\alpha^3 - c_4\alpha^4 + c_1 j\alpha + c_2 j\alpha^2 - c_4 j\alpha^4 - c_5 j\alpha^5 \\
&= \sum_{i=0}^{5} c_i\alpha^i \\
&\iff c_1 = c_2 = c_3 = c_4 = c_5 = 0 \\
&\iff \gamma \in \mathbb{Q}.
\end{aligned}
$$

In these equivalences, we have used the fact that $\{1, \alpha, \ldots, \alpha^5\}$ (resp. $\{1, \alpha, \ldots, \alpha^5, j, j\alpha, \ldots, j\alpha^5\}$ is a basis of $\mathbb{Q}(\alpha)$ (resp. $\mathbb{Q}(\alpha, j)$) over $\mathbb{Q}$. By (9), we conclude that $[\mathbb{Q}(\beta^p) : \mathbb{Q}] < 6$ if and only if $p = 2$ and there exists $i \in \{2, \ldots, 6\}$ such that $\sigma_i(\beta)^2 = \beta^2$ or $p = 3$ and there exists $i \in \{2, \ldots, 6\}$ such that $\sigma_i(\beta)^3 = \beta^3$. Using the above equivalences, we conclude that

$$
\begin{aligned}
\sigma_i(\beta)^2 = \beta^2 &\iff \sigma_i(\beta^2) = \beta^2 \\
&\iff i = 2 \\
&\iff \beta^2 \in \mathbb{Q}(\alpha^2).
\end{aligned}
$$

Since

$$\begin{aligned}\beta^2 = a_0^2 + 2a_3^2 + 4a_1a_5 + 4a_2a_4 \\ + \alpha(2a_0a_1 + 4a_2a_5 + 4a_3a_4) \\ + \alpha^2(a_1^2 + 2a_4^2 + 2a_0a_2) \\ + \alpha^3(2a_0a_3 + 2a_1a_2 + 4a_4a_5) \\ + \alpha^4(2a_5^2 + 2a_0a_4 + 2a_1a_3) \\ + \alpha^5(2a_0a_5 + 2a_1a_4 + 2a_2a_3),\end{aligned}$$

then $\sigma_i(\beta)^2 = \beta^2$ if and only if β satisfies the system of equations in the a_j given in the statement, which is obtained by equating to 0 the coefficients of α, α^3 and α^5 in this expression of β^2. Similarly we have

$$\begin{aligned}\sigma_i(\beta)^3 = \beta^3 &\Longleftrightarrow \sigma_i(\beta^3) = \beta^3 \\ &\Longleftrightarrow i = 3 \quad \text{or} \quad i = 5 \Longleftrightarrow \beta^3 \in \mathbb{Q}(\alpha^3).\end{aligned}$$

We may conclude the proof on expressing β^3 in the form $\beta^3 = \sum_{i=0}^{5} c_i\alpha^i$ and then equating to 0 the coefficients c_1, c_2, c_4 and c_5.

Remark. For example, any primitive element of $\mathbb{Q}(\alpha)$ of the form $\beta = a_3\alpha^3 + a_5\alpha^5$ (resp. $\beta = a_1\alpha + a_4\alpha^4$) satisfies the condition $[\mathbb{Q}(\beta^2) : \mathbb{Q}] = 3$ (resp. $[\mathbb{Q}(\beta^3) : \mathbb{Q}] = 2$).

Exercise 5.30.

Two complex numbers α and β are said to be equivalent if there exists $(A, B, C, D) \in \mathbb{Z}^4$ such that $\beta = (A\alpha + B)/(C\alpha + D)$ with $AD - BC = \pm 1$.

(1) Let $(a, b, c) \in \mathbb{Z}^3$ such that $a \neq 0$ and the polynomial $f(x) = ax^2 + bx + c$ is irreducible over $\mathbb{Z}$. Show that the roots of this polynomial are equivalent if and only if, there exists $(x, y) \in \mathbb{Z}^2$ such that

$$a^2x^2 + (b^2 - 2ac)xy + c^2y^2 = \pm b^2. \qquad \text{(Eq 1)}$$

(2) Show that the equivalent conditions in **(1)** hold in the following cases.

 (i) $a = \pm c$.
 (ii) $a \mid b$.
 (iii) $c \mid b$.

(3) Let $(a, b, c, d) \in \mathbb{Z}^4$ such that $a \neq 0$ and the polynomial $g(x) = ax^3 + bx^2 + cx + d$ is irreducible over $\mathbb{Z}$.

 (a) Suppose that $g(x)$ has two distinct equivalent roots. Show that $\text{Disc}(g)$ is a perfect square.

(b) Let $h(x) = 8x^3 - 24x^2 + 18x - 1$ and γ be a root of $h(x)$. Show that $\mathrm{Disc}(h) = 3^4 \cdot 2^6$ and the list of the roots of $h(x)$ is given by $\gamma, \gamma' = \frac{6\gamma-7}{4\gamma-4}, \gamma'' = \frac{4\gamma-7}{4\gamma-6}$. Conclude that the converse of (a) is false.

(c) Suppose that $\mathrm{Disc}(g)$ is a perfect square and let α, β be two distinct roots of $g(x)$. Show that there exists $(A, B, C, D) \in \mathbb{Z}^4$ such that $\beta = (A\alpha + B)/(C\alpha + D)$ with $AD - BC = (A + D)^2$ and $AD - BC \neq 0$.

(4) Let $(a, b, c, d) \in \mathbb{Z}^4$ such that $a \neq 0$ and the polynomial $g(x) = ax^3 + bx^2 + cx + d$ is irreducible over $\mathbb{Z}$ and let α be a root of $g(x)$. Suppose that $\mathrm{Disc}(g)$ is a perfect square. Show that there exists $(u, v, w) \in \mathbb{Z}^3$ such that the roots of the polynomial $\hat{g}(x) := \mathrm{Irr}((u\alpha + v)/w, \mathbb{Q})$ are equivalent.

(5) Consider the polynomial $h(x)$ defined in **(3), (b)**. Determine a polynomial $\hat{h}(x)$ whose roots are equivalent and generate the field $\mathbb{Q}(\gamma)$.

(6) Let $(a, b, c, d) \in \mathbb{Z}^4$ such that $a \neq 0$ and the polynomial $g(x) = ax^3 + bx^2 + cx + d$ is irreducible over $\mathbb{Z}$ and let α be a root of $g(x)$. Suppose that $Dis(g)$ is a perfect square. Let $\beta = (A\alpha + B)/(C\alpha + D)$, with $(A, B, C, D) \in \mathbb{Z}^4$, $\gcd(A, B, C, D) = 1$, be a root of $g(x)$ distinct from α. Show that any prime factor of $AD - BC$ is a divisor of $a\,\mathrm{Disc}(g)$.

Solution 5.30.

(1) Let $\alpha = -(b + \sqrt{d})/2a$ and $\beta = -(b - \sqrt{d})/2a$ be the roots of $f(x)$, where $d = b^2 - 4ac$. Suppose that $\beta = (A\alpha + B)/(C\alpha + D)$ with $(A, B, C, D) \in \mathbb{Z}^4$ and $AD - BC = \pm 1$. Then

$$A\alpha + B = C\alpha\beta + D\beta = Cc/a + D\beta,$$

hence $(A + D)\alpha + B = Cc/a + D(\alpha + \beta)$. Therefore

$$a(A + D)\alpha + aB = Cc - bD.$$

Since $1, \alpha$ is a basis of $\mathbb{Q}(\alpha)$ over $\mathbb{Q}$, then $A + D = 0$ and $aB = Cc - bD$. We have

$$A^2 + BC = A(-D) + BC = \mp 1,$$

hence

$$\mp b^2 = b^2BC + b^2A^2 = b^2BC + (aB - cC)^2 = a^2B^2 + (b^2 - 2ac)BC + c^2C^2.$$

This implies that $x = B, y = C$ is an integral solution of the equation

$$a^2x^2 + (b^2 - 2ac)xy + c^2y^2 = \mp b^2.$$

Conversely, suppose that

$$a^2x^2 + (b^2 - 2ac)xy + c^2y^2 = \pm b^2$$

with $(x, y) \in \mathbb{Z}^2$. If $b = 0$, then $\beta = -\alpha$ which implies that α and β are equivalent. Suppose that $b \neq 0$, then $b^2 \mid (ax - cy)^2$. Let $z \in \mathbb{Z}$ such that $ax - cy = bz$, then $\pm b^2 = b^2xy + b^2z^2$, hence $z^2 + xy = \pm 1$. We have $x - \frac{c}{a}y = \frac{b}{a}z$, hence $x - \alpha\beta y = -(\alpha + \beta)z$. Therefore $\beta = \frac{z\alpha + x}{y\alpha - z}$. We have proved that $z^2 + xy = \pm 1$, hence α and β are equivalent.

(2) (i) If $a = \epsilon c$, then **(Eq 1)** becomes:

$$c^2x^2 + (b^2 - \epsilon 2c^2)xy + c^2y^2 = \pm b^2$$

or

$$b^2xy + (cx - \epsilon cy)^2 = \pm b^2.$$

Clearly $x = 1$, $y = \epsilon$ is a solution of this equation.

(ii) If $a \mid b$, then $(b/a, 0)$ is a solution of **(Eq 1)**.

(iii) If $c \mid b$, then $(0, b/c)$ is a solution of **(Eq 1)**.

(3)(a) Let α and β be two distinct equivalent roots of $g(x)$. Let $(A, B, C, D) \in \mathbb{Z}^4$ such that $\beta = (A\alpha + B)/(C\alpha + D)$ with $AD - BC = \pm 1$. The third root γ of $g(x)$ is given by $\gamma = -b/a - \alpha - \beta$, hence $\mathbb{Q}(\alpha, \beta, \gamma) = \mathbb{Q}(\alpha)$. Therefore $\mathrm{Gal}(g(x), \mathbb{Q}) \simeq \mathbb{Z}/3\mathbb{Z} \simeq \mathcal{A}_3$. It follows that $\mathrm{Disc}(g)$ is a square in $\mathbb{Z}$.

(b) The computation of the discriminant of $h(x)$ may be done by using the formula given in [Lang (1965), Exercise 11, Chap. 5]. It is possible to obtain the same result by using the following general formula for discriminants. Let $F(x)$ be a polynomial of degree n and let a be its leading coefficient, then

$$\mathrm{Disc}(F) = (-1)^{n(n-1)/2} a^{n-2} \prod_{i=1}^{n} F'(\theta_i),$$

where $\theta_1, \ldots, \theta_n$ denote the roots of $F(x)$. Moreover, if $F(x)$ is irreducible over K, then

$$\mathrm{Disc}(F) = (-1)^{n(n-1)/2} a^{n-2}\, \mathrm{N}_{\mathbb{Q}(\theta)/\mathbb{Q}}(F'(\theta)).$$

Since $g'(x) = 6(4x^2 - 8x + 3)$, then

$$\mathrm{Disc}(g) = -2^6.3^3\, \mathrm{N}_{\mathbb{Q}(\gamma)/\mathbb{Q}}(4\gamma^2 - 8\gamma + 3).$$

Set $\delta = 4\gamma^2 - 8\gamma + 3$. We compute the characteristic polynomial $p(x)$ of δ. We have

$$\delta = 4\gamma^2 - 8\gamma + 3$$
$$\gamma\delta = 4\gamma^2 - 6\gamma + \frac{1}{2}$$
$$\gamma^2\delta = 6\gamma^2 - \frac{17}{2}\gamma + \frac{1}{2},$$

hence

$$p(x) = \begin{vmatrix} 3-x & -8 & 4 \\ 1/2 & -(6+\gamma) & 4 \\ 1/2 & -17/2 & 6-\gamma \end{vmatrix} = x^3 + 19x^2 + 3.$$

It follows that $\mathrm{N}_{\mathbb{Q}(\gamma)/\mathbb{Q}}(\delta) = -3$ and then $\mathrm{Disc}(g) = 2^6.3^4$.
On the other hand, let $u(x) = \frac{6x-7}{4x-4}$, then it is easy to verify the following identities:

$$u^{(2)}(x) := u(u(x)) = \frac{4x-7}{4x-6},$$

$\gamma' = u(\gamma)$ and $\gamma'' = u^{(2)}(\gamma)$. Moreover straightforward computations lead to the identity: $(4x-4)^3 h(u(x)) = -8h(x)$. This shows that $u(\gamma)$ and $u^{(2)}(\gamma)$ are roots of $h(x)$. Therefore $\gamma, \gamma', \gamma''$ are equivalent but here, although the discriminant is a perfect square, we have $AD - BC = 4 \neq \pm 1$ for $u(x)$ and for $u^{(2)}(x)$. This shows that the converse of (a) is false.

(c) Let α, β, γ be the roots of $g(x)$. Since $\mathrm{Disc}(g)$ is a square, then $\mathbb{Q}(\alpha)/\mathbb{Q}$ is Galois, cyclic. Therefore $\beta \in \mathbb{Q}(\alpha)$. Let $a_0, a_1, a_2 \in \mathbb{Q}$ such that $\beta = a_0 + a_1\alpha + a_2\alpha^2$. We distinguish the cases $a_2 = 0$ and $a_2 \neq 0$.

- $a_2 \neq 0$. The equation satisfied by α may be written in the form $\alpha^3 = b_0 + b_1\alpha + b_2\alpha^2$. Using this equation, it is easy to show the equivalence of the following propositions
  (P_1): There exists $(A, B, C, D) \in \mathbb{Z}^4$ such that $(C, D) \neq (0, 0)$ and $\beta = (A\alpha + B)/(C\alpha + D)$.
  (P_2): The linear system of equations

$$-B + a_2b_0C + a_0D = 0$$
$$-A + (a_0 + a_2b_1)C + a_1D = 0$$
$$(a_1 + a_2b_2)C + a_2D = 0,$$

  has a solution $(A, B, C, D) \in \mathbb{Q}^4$ such that $(C, D) \neq (0, 0)$.

We will prove (P_2). The matrix of the system is given by

$$M = \begin{pmatrix} 0 & -1 & a_2b_0 & a_0 \\ -1 & 0 & a_0 + a_2b_1 & a_1 \\ 0 & 0 & a_1 + a_2b_2 & a_2 \end{pmatrix}.$$

The determinant of the matrix extracted from M by suppressing its last column is equal to $-(a_2b_2 + a_1)$.

- $a_1 \neq -a_2b_2$. In this case we may compute A, B, C in terms of D. If we choose $D \in \mathbb{Q} \setminus \{0\}$, then we find $A, B, C \in \mathbb{Q}$ such that (A, B, C, D) is a solution of the system.
- $a_1 = -a_2b_2$. The third equation of the system shows that $a_2D = 0$. If $a_2 = 0$, then $a_1 = 0$ and then $\beta = a_0 \in \mathbb{Q}$, which is a contradiction. Therefore $D = 0$ and then solving the first two equations gives $A = (a_0 + a_2b_1)C$ and $B = a_2b_0C$. Therefore $\beta = \frac{(a_0+a_2b_1)\alpha+a_2b_0}{\alpha}$. If we multiply the numerator and denominator of this fraction by a same integer, we obtain a similar relation with integral coefficients.

To end the proof of **(c)**, we must show that $AD - BC = (A + D)^2$ and $AD - BC \neq 0$. Since $g(\beta) = 0$, then $g(x) \mid (Cx + D)^3 g(u(x))$ in $\mathbb{Q}[x]$, where $u(x) = \frac{Ax+B}{Cx+D}$. It follows that there exists $r \in \mathbb{Q}$ such that

$$(Cx + D)^3 g(u(x)) = rg(x).$$

From this identity it follows that $u(\beta)$ is a root of $g(x)$. We discuss the possible values of $u(\beta)$. If $u(\beta) = \beta$, we find that $u(x) = x$, which is a contradiction. Suppose that $u(\beta) = \alpha$. Since $u^{(2)}(x) = \frac{(A^2+BC)x+B(A+D)}{C(A+D)x+D^2+BC}$, then

$$C(A + D) = D^2 - A^2 = B(A + D) = 0,$$

hence $A + D = 0$ or $B = C = A - D = 0$. Clearly, the second possibility is impossible, hence $A + D = 0$. This implies that $u^{(2)}(x) = x$. Consider the value of $u(\gamma)$. If $u(\gamma) = \alpha$, then $u^{(2)}(\gamma) = u(\alpha) = \beta$, which is a contradiction. Obviously me may reject the possibility $u(\gamma) = \gamma$. If $u(\gamma) = \beta$, then $u^{(2)}(\gamma) = u(\beta) = \alpha$, which is a contradiction. We conclude that the only possibility for $u(\beta)$ is $u(\beta) = \gamma$. Therefore $u^{(3)}(\alpha) = \alpha$. Since

$$u^{(3)}(x) = \frac{(A^3 + 2ABC + BCD)x + B(A^2 + BC + D^2 + AD)}{C(A^2 + AD + D^2 + BC)x + D^3 + 2BCD + ABC},$$

then

$$B(A^2 + BC + D^2 + AD) = 0,$$
$$C(A^2 + AD + D^2 + BC) = 0 \quad \text{and}$$
$$(A - D)(A^2 + AD + D^2 + BC) = 0.$$

If $B = C = A - D = 0$, then $u^{(2)}(x) = x$, which is a contradiction. Hence

$$(A + D)^2 - (AD - BC) = A^2 + BC + D^2 + AD = 0.$$

It follows that $u^{(3)}(x) = x$ and $(A+D)^2 = (AD-BC)$. This implies that $AD - BC \neq 0$.

(4) Let β be a root of $g(x)$ distinct from α. According to **(3)**, **(c)**, there exists $(A, B, C, D) \in \mathbb{Z}^4$ such that $\beta = (A\alpha + B)/(C\alpha + D)$ with $AD - BC = (A + D)^2$ and $AD - BC \neq 0$. We may suppose that $\gcd(A, B, C, D) = 1$. If $AD - BC = \pm 1$, then the result is trivial. If not, we make a finite number of transformations on the generator of the field $\mathbb{Q}(\alpha)$.

First kind of transformations. Suppose that $|AD - BC| \geq 2$ and $p \mid \gcd(A, D)$, where p is a prime number. Since $BC \equiv AD \pmod{p^2}$, then $p^2 \mid B$ or $p^2 \mid C$. If $p^2 \mid B$, then

$$\beta/p = \frac{(A/p)(\alpha/p) + (B/p^2)}{C(\alpha/p) + (D/p)} = \frac{A_1(\alpha/p) + B_1}{C_1(\alpha/p) + D_1},$$

where

$$A_1 = A/p, \quad B_1 = B/p^2, \quad C_1 = C \quad \text{and} \quad D_1 = D/p.$$

We have

$$|A_1D_1 - B_1C_1| = |(AD - BC)/p^2| < |AD - BC|,$$

hence replacing α by $\alpha' = \alpha/p$, we reduce $AD - BC$. If $p^2 \mid C$, then

$$p\beta = \frac{(A/p)(p\alpha) + (B)}{(C/p^2)(p\alpha) + (D/p)} = \frac{A_1(p\alpha) + B_1}{C_1(p\alpha) + D_1},$$

where

$$A_1 = A/p, \quad B_1 = B, \quad C_1 = C/p^2 \quad \text{and} \quad D_1 = D/p.$$

We have

$$|A_1D_1 - B_1C_1| = |(AD - BC)/p^2| < |AD - BC|,$$

hence replacing α by $\alpha' = p\alpha$, we reduce $AD - BC$. We repeat these transformations of the first kind until $|AD - BC| = 1$ or $|AD - BC| \geq$

2 and $\gcd(A, D) = 1$. If $|AD - BC| = 1$, then we have found our generating element of $\mathbb{Q}(\alpha)$. If $|AD - BC| \geq 2$ and $\gcd(A, D) = 1$, we make a second kind transformation, hereafter described.

Second kind of transformations. Here we suppose that $|AD - BC| \geq 2$ and $\gcd(A, D) = 1$. For any $k \in \mathbb{Z}$, we have

$$\beta + k = \frac{(A + Ck)(\alpha + k) + B + kD - kA - k^2C}{C(\alpha + k) + D - kC} = \frac{A_1\alpha + B_1}{C_1\alpha + D_1},$$

where

$$A_1 = A + Ck, \quad B_1 = B + kD - kA - k^2C,$$
$$C_1 = C \quad \text{and} \quad D_1 = D - kC.$$

One may verify that $A_1D_1 - B_1C_1 = AD - BC$. Let p be a prime number such that $p \mid AD - BC$. Since $AD - BC = (A + D)^2$ and $\gcd(A, D) = 1$, then $p \nmid A$, $p \nmid D$ and $p \nmid BC$. Choose $k \in \mathbb{Z}$ such that $p \mid A_1$, that is $k \equiv -A/C \pmod{p}$. Since $-A/C \equiv D/C \pmod{p}$, then $p \mid D_1$. At this stage we have $|A_1D_1 - B_1C_1| \geq 2$ and $\gcd(A_1, D_1) \equiv 0 \pmod{p}$, so that we can use the transformations of the first kind to reduce $|A_1D_1 - B_1C_1|$. It is clear that after a finite number of transformations, we reach a stage for which $AD - BC| = 1$. The composition of all the transformations used in the process will produce a transformation on α of the form $T(\alpha) = (u\alpha + v)/w$.

(5) We have $\gamma' = \frac{6\gamma-7}{4\gamma-4}$. Here

$$AD - BC = 4,\ A = 6,\ D = -4,\ \gcd(A, D) = 2,\ B = -7 \text{ and } C = 4,$$

hence we can apply a transformation of the first kind on replacing γ by $\hat{\gamma} = 2\gamma$. We obtain the following identity: $2\gamma' = \frac{3(2\gamma)-7}{(2\gamma)-2} = \frac{3\hat{\gamma}-7}{\hat{\gamma}-2}$. Here $AD - BC = 1$ so that we have reached the final primitive element, namely 2γ. The polynomial $\hat{h}(x)$ is given by $\hat{h}(x) = x^3 - 6x^2 + 9x - 1$.

(6) Let p be a prime factor of $AD - BC$. Suppose that $p \nmid a$. We will show that $p \mid \mathrm{Disc}(g)$, that is $g(x)$ is inseparable over $\mathbb{Q}$. Let α and β be two roots of $g(x)$ related by the identity $\beta = (A\alpha + B)/(C\alpha + D)$ with $(A, B, C, D) \in \mathbb{Z}^4$,

$$\gcd(A, B, C, D) = 1,\ AD - BC = (A + D)^2 \text{ and } AD - BC \neq 0.$$

Set $u(x) = (Ax + B)/(Cx + D)$, then $u^{(2)}(x) = (A_2x + B_2)/(C_2x + D_2)$, where

$$A_2 = A^2 + BC,\ B_2 = B(A + D),\ C_2 = C(A + D) \text{ and } D_2 = D^2 + BC.$$

We first claim that $(C, D) \not\equiv (0,0) \pmod p$ and $(C_2, D_2) \not\equiv (0,0) \pmod p$. We prove the claim only for (C, D). The proof for the other is similar. Since $g(x) \mid (Cx+D)^3 g(u(x))$, then

$$(Cx+D)^3 g(u(x)) = (r/s)g(x),$$

where $r, s \in \mathbb{Z}$, $\gcd(r,s)=1$ and $rs \neq 0$. We may write this identity in the form

$$\begin{aligned} &s\Big(a(Ax+B)^3 + b(Ax+B)^2(Cx+D) \\ &+ c(Ax+B)(Cx+D)^2 + d(Cx+D)^3\Big) \\ &= rg(x). \end{aligned}$$

If $p \mid s$, then $g(x) \equiv 0 \pmod p$, which is a contradiction. Therefore $s \not\equiv 0 \pmod p$. Suppose by contradiction that $C \equiv 0 \pmod p$ and $D \equiv 0 \pmod p$, then $A \equiv 0 \pmod p$ and by the preceding identity, we conclude that $saB^3 \equiv rg(x) \pmod p$. Since none of the integers s, a, B is 0 modulo p, then $r \not\equiv 0 \pmod p$. We look at this congruence as an identity in $\mathbb{F}_p[x]$. While saB^3 is a constant, $rg(x)$ is a polynomial of degree 3. Therefore, we have reached a contradiction and our claim is established. From the relations

$$\begin{aligned} \alpha + u(\alpha) + u^{(2)}(\alpha) &= -b/a, \\ \alpha u(\alpha) + \alpha u^{(2)}(\alpha) + u(\alpha)u^{(2)}(\alpha) &= c/a \quad \text{and} \\ \alpha u(\alpha)u^{(2)}(\alpha) &= -d/a, \end{aligned}$$

we deduce the following identities

$$\begin{aligned} &(Cx+D)(C_2x+D_2)x \\ &\quad + (C_2x+D_2)(Ax+B) \\ &\quad + (Cx+D)(A_2x+B_2) \\ &= -b/a(Cx+D)(C_2x+D_2) + \lambda_1 g(x), \end{aligned}$$

$$\begin{aligned} &x(C_2x+D_2)(Ax+B) \\ &\quad + x(Cx+D)(A_2x+B_2) \\ &\quad + (Ax+B))(A_2x+B_2) \\ &= (c/a)(Cx+D)(C_2x+D_2) + \lambda_2 g(x), \end{aligned}$$

and

$$x(Ax+B))(A_2x+B_2) = -(d/a)(Cx+D)(C_2x+D_2) + \lambda_3 g(x),$$

where $\lambda_1, \lambda_2, \lambda_3 \in \mathbb{Q}$. Indeed λ_1 has the form$\lambda_1 = u_1/a^k$, where $u_1 \in \mathbb{Z}$ and k is a non negative integer. To prove this let $S = \{a^m, m \in \mathbb{N}\}$, then S is a multiplicative subset of $\mathbb{Z}$. Let $R = S^{-1}\mathbb{Z}$ be the ring of fractions of $\mathbb{Z}$ for this multiplicative subset. Since a is invertible in R, we can perform the Euclidean division in $R[x]$ of

$$\begin{aligned}&(Cx + D)(C_2x + D_2)x\\&\quad + (C_2x + D_2)(Ax + B)\\&\quad + (Cx + D)(A_2x + B_2)\\&\quad - b/a(Cx + D)(C_2x + D_2)\end{aligned}$$

by $g(x)$ and obtain a quotient $q_1(x)$ and a remainder $r_1(x)$ both in $R[x]$. The quotient and the remainder being unique in $\mathbb{Q}[x]$, we conclude that $r_1(x) = 0$ and $\lambda_1 \in R$. The proof for λ_2 and λ_3 is similar and will be omitted. This result allows us to reduce modulo p the three identities and obtain identities in $\mathbb{F}_p[x]$. Let $\bar{g}(x)$ be the reduction of $g(x)$ modulo p and let ρ be one of its roots in an algebraic closure of $\mathbb{F}_p$. In each of these identities, we substitute ρ for x. This is possible because $a \not\equiv 0 \pmod p$, $(C, D) \neq (0, 0) \pmod p$ and $(C_2, D_2) \neq (0, 0) \pmod p$. For example the first identity becomes: $\rho + u(\rho) + u^{(2)}(\rho) = -\bar{b}/\bar{a}$. This implies that $\rho, u(\rho), u^{(2)}(\rho)$ is the complete list of the roots of $\bar{g}(x)$ in an algebraic closure of $\mathbb{F}_p$. Since $AD - BC \equiv 0 \pmod p$, then the determinant $\begin{vmatrix} \bar{A} & \bar{B} \\ \bar{C} & \bar{D} \end{vmatrix}$ is 0 in $\mathbb{F}_p$. It follows that there exist $\mu_1, \mu_2 \in \mathbb{F}_p$, not both 0, such that $\mu_1(\bar{A}, \bar{B}) + \mu_2(\bar{C}, \bar{D}) = (0, 0)$. Since $(\bar{C}, \bar{D}) \neq (0, 0)$, then $\mu_1 \neq 0$. Therefore $\bar{A} = (\mu_2/\mu_1)\bar{C}$ and $\bar{B} = (\mu_2/\mu_1)\bar{D}$. It follows that $\overline{u}(x) = \mu_2/\mu_1$ and then $\overline{u}(\rho) = \overline{u^{(2)}}(\rho) = \mu_2/\mu_1$. This shows that $\bar{g}(x)$ is not separable over $\mathbb{F}_p$.

Exercise 5.31.

Let p be a prime number, r be a positive integer, $q = p^r$, $E = \mathbb{F}_q(t)$ and $F = \mathbb{F}_q(t^q - t)$.

(1) For any $\alpha \in \mathbb{F}_q$, let σ_α be the unique $\mathbb{F}_q$-automorphism of $\mathbb{F}_q(t)$ such that $\sigma_\alpha(t) = t + \alpha$. Let $G = \{\sigma_\alpha, \alpha \in \mathbb{F}_q\}$. Show that G is a subgroup of $\mathrm{Aut}_{\mathbb{F}_q}(\mathbb{F}_q(t))$ isomorphic to $(\mathbb{F}_q, +)$. Show that $\mathrm{Inv}(G) = F$. Deduce that E is a Galois extension of F of degree q and $\mathrm{Gal}(E, F) = G \simeq (\mathbb{Z}/p\mathbb{Z})^r$.

(2) Determine all of the fields M such that $F \subset M \subset \mathbb{F}_q(t)$ and $[\mathbb{F}_q(t) : M] = p$. For any of these fields M, show that $M = \mathbb{F}_q(t^p - t\alpha^{p-1})$ for some $\alpha \in \mathbb{F}_q^*$.

(3) Let $\alpha \in \mathbb{F}_q^*$ be the element determined in **(2)**. Show that we may express explicitly $t^q - t$ in the form:

$$t^q - t = \sum_{i=1}^{r} a_i \left(t^{p^{r+1-i}} - (t\alpha^{p-1})^{p^{r-i}}\right), \qquad \text{(Eq 1)}$$

with $a_r = 1$.

(4)(a) For any $k \in \{1, ..., r\}$ show that there exist $\binom{r}{k}$ fields L_k such that $\mathbb{F}_q(t^q - t) \subset L_k \subset \mathbb{F}_q(t)$ and $[\mathbb{F}_q(t) : L_k] = p^k$.

(b) Let $\alpha_1, \ldots, \alpha_k$ be linearly independent elements of $\mathbb{F}_q$ over $\mathbb{F}_1(t)$. Let H_k be the subgroup of G generated by $\sigma_{\alpha_1}, \ldots, \sigma_{\alpha_k}$. Show that $H_k = \{\sigma_\beta, \text{ for } \beta \sum_{i=1}^k b_i\alpha_i, b_i \in \mathbb{F}_p\}$. Deduce that $[\mathbb{F}_q(t) : \mathrm{Inv}(H_k)] = p^k$.

(c) Show that there exists a unique polynomial $g_k(t) \in \mathbb{F}_q[t]$, of the form $g_k(t) = \sum_{i=0}^k a_i t^{p^i}$, where $a_t \in \mathbb{F}_q$, $a_k = 1$, and satisfying the condition $\mathrm{Inv}(H_k) = \mathbb{F}_q(g_k(t))$.

(5) Suppose that $r = 3$ ie $q = p^3$. Find the list of fields L such that $\mathbb{F}_q(t^q - t) \subset L \subset \mathbb{F}_q$, by giving for each L a polynomial $g(t) \in \mathbb{F}_q[t]$ such that $L = \mathbb{F}_q(g(t))$. Draw the diagram of these fields.

Solution 5.31.

(1) Let $\phi : \mathbb{F}_q \to Aut_{\mathbb{F}_q}(\mathbb{F}_q(t))$ be the map defined by $\phi(\alpha) = \sigma_\alpha$. Clearly ϕ is a morphism of the group $(\mathbb{F}_q, +)$ into the group $\mathrm{Aut}_{\mathbb{F}_q}(\mathbb{F}_q(t), \circ)$. We obviously have $\mathrm{Ker}\,\phi = \{0\}$ and $\mathrm{Im}\,\phi = G$. Therefore ϕ induces an isomorphism of $\mathbb{F}_q$ onto G. Let K be the invariant field of G. For any $\alpha \in \mathbb{F}_q$, we have

$$\sigma_\alpha(t^q - t) = (t + \alpha)^q - (t + \alpha) = t^q + \alpha^q - t - \alpha = t^q - t,$$

hence $F \subset K$. Consider the following inclusion: $F \subset K \subset \mathbb{F}_q(t)$. Clearly we have $[\mathbb{F}_q(t) : F] = q$. By Artin's Theorem [Lang (1965), Th. 2, Chap. 8.1], $\mathbb{F}_q(t)/K$ is Galois with Galois group equal to G and $[\mathbb{F}_q(t) : K] = |G| = q$. Therefore $K = F = \mathbb{F}_q(t^q - t)$. We conclude that E is a Galois extension of F of degree q and its Galois group is equal to G. Since G is isomorphic to $\mathbb{F}_q$, then G is isomorphic to $(\mathbb{Z}/p\mathbb{Z})^r$.

(2) By the fundamental theorem of Galois [Lang (1965), Th. 1 and Cor., Chap. 8.1], M is the invariant field of a subgroup H of G of order p. Since G contains r such subgroups, there are exactly r fields M. Let $\alpha \in \mathbb{F}_q^*$ and let σ_α be the $\mathbb{F}_q$-automorphism of $\mathbb{F}_q(t)$ defined in **(1)**, then σ_α generates a subgroup H of order p of G. Moreover any subgroup of G of order p arises in this way. Let $\beta \in \mathbb{F}_q^*$, then β generates the

same subgroup as α if and only if $\beta = \lambda\alpha$ with $\lambda \in \mathbb{F}_p^*$. We show that $\mathrm{Inv}(H) = \mathbb{F}_q(t^p - t\alpha^{p-1})$. We have

$$\sigma_\alpha(t^p - t\alpha^{p-1}) = (t+\alpha)^p - (t+\alpha)\alpha^{p-1} = t^p + \alpha^p - t\alpha^{p-1} - \alpha^p = t^p - t\alpha^{p-1},$$

hence $t^p - t\alpha^{p-1} \in \mathrm{Inv}(H)$. We have

$$\mathbb{F}_q(t^p - t\alpha^{p-1}) \subset \mathrm{Inv}(H) \subset \mathbb{F}_q(t).$$

By Artin's Theorem [Lang (1965), Th. 2, Chap. 8.1],

$$H = \mathrm{Gal}(\mathbb{F}_q(t), \mathrm{Inv}\, H),$$

hence

$$p = (H : 1) = [\mathbb{F}_q(t) : \mathrm{Inv}(H)].$$

On the other hand,

$$[\mathbb{F}_q(t) : \mathbb{F}_q(t^p - \alpha t^{p-1})] = p.$$

We deduce that $M = \mathrm{Inv}(H) = \mathbb{F}_q(t^p - t\alpha^{p-1})$.

(3) We may write **(Eq 1)** in the form:

$$\begin{aligned} t^q - t = t^{p^r} - t^{p^{r-1}}(\alpha^{p-1})^{p^{r-1}} &+ a_2(t^{p^{r-1}} - t^{p^{r-2}}(\alpha^{p-1})^{p^{r-2}}) + \cdots \\ &+ a_{k-1}(t^{p-k+2} - t^{p^{r-k+1}}(\alpha^{p-1})^{p^{r-k+1}}) \\ &+ a_k(t^{p^{r-k+1}} - t^{p^{r-k}}(\alpha^{p-1})^{p^{r-k}}) + \cdots + a_r(t^p - t\alpha^{p-1}), \end{aligned}$$

hence (1) is equivalent to

$$\begin{aligned} a_2 &= (\alpha^{p-1})^{p^{r-1}} = \beta^{p^{r-1}}, \\ a_k &= (\alpha^{p-1})p^{r-k+1}a_{k-1}, \quad \text{for} \quad k = 2, \ldots, r \quad \text{and} \\ a_r\alpha^{p-1} &= a_r\beta = 1, \end{aligned}$$

where $\beta = \alpha^{p-1}$. Using the first $r-1$ equation, it is easy to prove by induction that $a_k = \beta^{p^{r-1}+p^{r-2}+\cdots+p^{r-(k-1)}}$. In particular, we have $a_r = \beta^{p^{r-1}+p^{r-2}+\cdots+p}$ and then

$$a_r\beta = \beta^{p^{r-1}+\cdots+p+1} = \beta^{(p^r-1)/(p-1)} = \alpha^{p^r-1} = 1$$

and the last equation is satisfied.

(4)(a) Since $G \simeq (\mathbb{Z}/p\mathbb{Z})^r$, then any subgroup of G is of order p^k with $k = 0, 1, \ldots, r$. The number of subgroups H_k of order p^k is equal to $\binom{r}{k}$. By Galois theorem relative to the correspondence between subgroups and subfields, there are exactly $\binom{r}{k}$ fields L_k such that $F \subset L_k \subset \mathbb{F}_q(t)$ and $[\mathbb{F}_q(t) : L_k] = p^k$.

(b) Let $\sigma_\beta \in G$, then

$$\begin{aligned}\sigma_\beta \in H_k &\Leftrightarrow \text{there exist } b_1,\dots,b_k \in \mathbb{N}, \sigma_\beta = \sigma_{\alpha_1}^{b_1}\circ\sigma_{\alpha_2}^{b_2}\circ\cdots\circ\sigma_{\alpha_k}^{b_k}\\ &\Leftrightarrow \text{there exist } b_1,\dots,b_k \in \mathbb{N}, \sigma_\beta = \sigma_{b_1\alpha_1+\cdots+b_k\alpha_k}\\ &\Leftrightarrow \text{there exist } b_1,\dots,b_k \in \mathbb{N}, t+\beta = t+b_1\alpha_1+\cdots+b_k\alpha_k\\ &\Leftrightarrow \text{there exist } b_1,\dots,b_k \in \mathbb{N}, \beta = b_1\alpha_1+\cdots+b_k\alpha_k.\end{aligned}$$

Since $\mathbb{F}_q(t)$ is Galois over $\operatorname{Inv} H_k$, then $[\mathbb{F}_q(t) : \operatorname{Inv} H_k] = |H_k| = p^k$.

(c) Let $g_k(t) = \sum_{i=0}^{k} a_i t^{p^i} \in \mathbb{F}_q[t]$ with $a_k = 1$. Then

$$\begin{aligned}g_k(t) \in \operatorname{Inv}(H) &\Leftrightarrow \sigma_{\alpha_i}(g_k(t)) = g_k(t) \quad \text{for} \quad j = 1,\dots,k\\ &\Leftrightarrow \sum_{i=0}^{k} a_i(t^{p^i} + \alpha_j^{p^i}) = \sum_{i=0}^{k} a_i(t^{p^i}) \quad \text{for} \quad j=1,\dots,k\\ &\Leftrightarrow \sum_{i=0}^{k} a_i\alpha_j^{p^i} = 0 \quad \text{for} \quad j=1,\dots,k\\ &\Leftrightarrow a_0\alpha_1 + a_1\alpha_1^p + \cdots + a_{k-1}\alpha_1^{p^{k-1}} = -\alpha_1^{p^k}\\ &\Leftrightarrow \dots\dots\dots\dots\dots\dots\dots\dots\\ &\quad a_0\alpha_k + a_1\alpha_k^p + \cdots + a_{k-1}\alpha_k^{p^{k-1}} = -\alpha_k^{p^k}.\end{aligned}$$

Therefore the polynomial $g_k(t)$ has the given form and satisfies the condition $g_k(t) \in \operatorname{Inv}(H_k)$ if and only if the above system has a solution $(a_0, a_1, \dots, a_{k-1}) \in (\mathbb{F}_q)^k$. The determinant of this system is given by

$$D = \begin{vmatrix} \alpha_1\alpha_1^p & \cdots & \alpha_1^{p^{k-1}}\\ \alpha_2\alpha_2^p & \cdots & \alpha_2^{p^{k-1}}\\ \dots\dots & \cdots & \dots\dots\\ \alpha_k\alpha_k^p & \cdots & \alpha_k^{p^{k-1}}\end{vmatrix}.$$

By [Lang (1965), Coro. 2, Chap. 8.5], $D \neq 0$, hence $a_0, a_1, \dots, a_{k-1}$ are uniquely determined. This implies that there exists one and only one polynomial $g_k(t) \in \mathbb{F}_q[t]$ having the given shape and belonging to $\operatorname{Inv}(H_k)$.

(5) We begin with the fields $\mathbb{F}_q(t^q - t)$ and $\mathbb{F}_q(t)$ of degree 1 and q respectively over F. The degree of any other intermediate field L is a divisor of $q = p^3$, hence equal to p or p^2.

(a) Intermediate fields L such that $[L : F] = p$. Let α be a primitive element of $\mathbb{F}_q$ over $\mathbb{F}_p$ (i.e. a root of a monic polynomial $f(x) \in \mathbb{F}_p[x]$ of degree 3). The group G contains 3 subgroups of order p, namely the subgroups generated by $\sigma_1, \sigma_\alpha, \sigma_{\alpha^2}$ respectively. Therefore there are 3 fields L such that $\mathbb{F}_q(t^q - t) \subset L \subset \mathbb{F}_q(t)$ and $[\mathbb{F}_q(t) : L] = p$, namely

$$L = \mathbb{F}_q(t^p - t) \text{ or } L = \mathbb{F}_q(t^p - t\alpha^{p-1}) \text{ or } L = \mathbb{F}_q(t^p - t\alpha^{2(p-1)}).$$

(b) Intermediate fields L such that $[L : F] = p^2$. Such fields are invariant fields of subgroups of G of order p^2. There are $\binom{3}{2} = 3$ such subgroups namely the subgroups generated by $\{\sigma_1, \sigma_\alpha\}, \{\sigma_1, \sigma_{\alpha^2}\}$ and $\{\sigma_\alpha, \sigma_{\alpha^2}\}$ respectively.

- Invariant field of the subgroup generated by $\{\sigma_1, \sigma_\beta\}$ where $\beta = \alpha$ or $\beta = \alpha^2$. We compute the coefficients $a_1, a_2 \in \mathbb{F}_q$ of $g(t) = t^{p^2} + a_1 t^p + a_2 t$ such that $g(t+1) = g(t)$ and $g(t+\beta) = g(t)$. These condition lead to the equations

$$1 + a_1 + a_2 = 0 \text{ and } \beta^{p^2} + a_1\beta^p + a_2\beta = 0,$$

that is

$$a_1 + a_2 = -1, \text{and } a_1\beta^p + a_2\beta = -\beta^{p^2}.$$

The determinant of this system of linear equations is equal to $D = \begin{vmatrix} 1 & 1 \\ \beta^p & \beta \end{vmatrix} = \beta - \beta^p$, hence non zero. We deduce that

$$a_1 = \begin{vmatrix} -1 & 1 \\ -\beta^{p^2} & \beta \end{vmatrix} / (\beta - \beta^p) = -\frac{\beta^{p^2-1} - 1}{\beta^{p-1} - 1}$$

$$= -((\beta^{p-1})^p + \cdots + \beta^{p-1} + 1),$$

$$a_2 = -1 - a_1 = (\beta^{p-1})^p + \cdots + \beta^{p-1}.$$

Therefore

$$L = \mathbb{F}_q(t^{p^2} + a_1 t^p - (1 + a_1)t)$$

with $a_1 = -(\beta^{p-1})^p - \cdots - \beta^{p-1} - 1$.

- Invariant field of the subgroup generated by $\{\sigma_\alpha, \sigma_{\alpha^2}\}$. We compute the coefficients $b_1, b_2 \in \mathbb{F}_q$ of $g(t) = t^{p^2} + b_1 t^p + b_2 t$ such that $g(t+\alpha) = g(t)$ and $g(t+\alpha^2) = g(t)$. We find the equations: $b_1\alpha^p + b_2\alpha = -\alpha^{p^2}$ and $b_1\alpha^{2p} + b_2\alpha^2 = -\alpha^{2p^2}$. The determinant of this system is given by $D = \begin{vmatrix} \alpha^p & \alpha \\ \alpha^{2p} & \alpha^2 \end{vmatrix} = \alpha^{p+2} - \alpha^{2p+1}$, hence

$$
\begin{aligned}
b_1 &= \begin{vmatrix} -\alpha^{p^2} & \alpha \\ -\alpha^{2p^2} & \alpha^2 \end{vmatrix} /(\alpha^{p+2} - \alpha^{2p+1}) \\
&= -\frac{\alpha^{2p^2+1} - \alpha^{p^2+2}}{\alpha^{2p+1} - \alpha^{p+2}} \\
&= -\frac{\alpha^{p^2+2}(\alpha^{p^2-1} - 1)}{\alpha^{p+2}(\alpha^{p-1} - 1)} \\
&= -\alpha^{p^2-p}((\alpha^{p-1})^p + \cdots + (\alpha^{p-1}) + 1) \\
&= -(\alpha^{p-1})^p[(\alpha^{p-1})^p + \cdots + \alpha^{p-1} + 1] \\
&= -((\alpha^{p-1})^{2p} + \cdots + (\alpha^{p-1})^{p+1} + (\alpha^{p-1})^p)
\end{aligned}
$$

and

$$
\begin{aligned}
b_2 &= \begin{vmatrix} \alpha^p & -\alpha^{p^2} \\ \alpha^{2p} & -\alpha^{2p^2} \end{vmatrix} /(\alpha^{p+2} - \alpha^{2p+1}) \\
&= \frac{\alpha^{p^2+2p} - \alpha^{2p^2+p}}{\alpha^{p+2} - \alpha^{2p+1}} \\
&= \frac{\alpha^{2p^2+p} - \alpha^{p^2+2p}}{\alpha^{2p+1} - \alpha^{p+2}} \\
&= \frac{\alpha^{p^2+2p}(\alpha^{p^2-p} - 1)}{\alpha^{p+2}(\alpha^{p-1} - 1)} \\
&= \alpha^{p^2+p-2}((\alpha^{p-1})^{p-1} + \cdots + (\alpha^{p-1}) + 1) \\
&= (\alpha^{p-1})^{p+2}((\alpha^{p-1})^{p-1} + \cdots + \alpha^{p-1} + 1) \\
&= (\alpha^{p-1})^{2p+1} + \cdots + (\alpha^{p-1})^{p+3} + (\alpha^{p-1})^{p+2}.
\end{aligned}
$$

We now draw the diagram of intermediate fields.

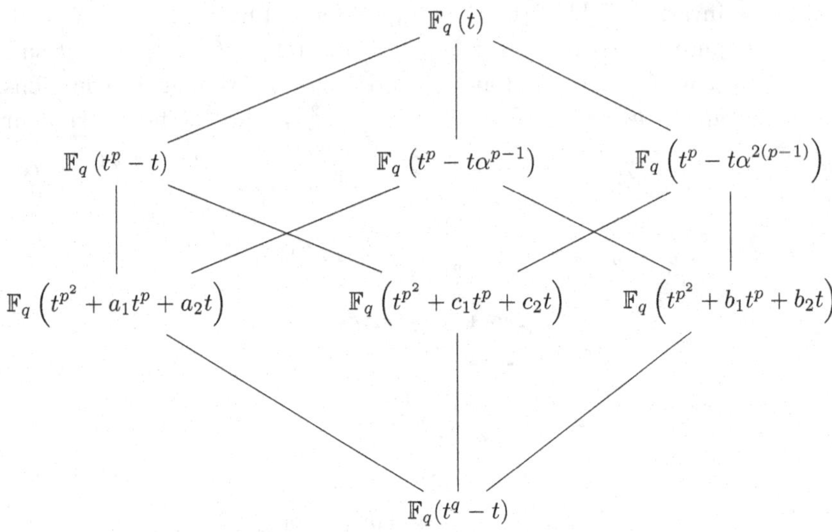

Here

$$a_1 = -\sum_{i=0}^{p} \left(\alpha^{p-1}\right)^i, \quad a_2 = \sum_{i=1}^{p} \left(\alpha^{p-1}\right)^i,$$

$$c_1 = -\sum_{i=0}^{p} \left(\alpha^{2(p-1)}\right)^i, \quad c_2 = \sum_{i=1}^{p} \left(\alpha^{2(p-1)}\right)^i \quad \text{and}$$

$$b_1 = -\sum_{i=p}^{2p} \left(\alpha^{p-1}\right)^i, \quad b_2 = \sum_{i=p+2}^{2p+1} \left(\alpha^{p-1}\right)^i.$$

Exercise 5.32.

Let k and n be positive integers such that $1 \leq k \leq n$. A group G acting on a set Ω of cardinality n is said to be k-transitive if given two ordered subsets of cardinality k of Ω, $\{a_1, \ldots, a_k\}$ and $\{b_1, \ldots, b_k\}$, there exists $\sigma \in G$ such that $\sigma(a_i) = b_i$ for $i = 1, \ldots, k$. Denote by $G_{\{a_1,\ldots,a_k\}}$ be the subgroup of G whose elements are the $\sigma \in G$ such that $\sigma(a_i) = a_i$ for $i = 1, \ldots, k$.

(1) Show that the following propositions are equivalent.
   (i) G is k-transitive.
   (ii) There exists a subset $\{a_1, \ldots, a_k\}$ of cardinality k of Ω satisfying the condition: for any subset $\{b_1, \ldots, b_k\}$ of cardinality k of Ω, there exists $\sigma \in G$ such that $\sigma(a_i) = b_i$ for $i = 1, \ldots, k$.
(2) Let $1 \leq l < k$. Show that the following propositions are equivalent.

(a) G is k-transitive.

(b) G is l-transitive and for any $\{a_1, \ldots, a_l\} \subset \Omega$, $G_{\{a_1,\ldots,a_l\}}$ is $(k-l)$-transitive on $\Omega \setminus \{a_1, \ldots, a_l\}$.

(c) G is l-transitive and there exists $\{a_1, \ldots, a_l\} \subset \Omega$ such that $G_{\{a_1,\ldots,a_l\}}$ is $(k-l)$-transitive on $\Omega \setminus \{a_1, \ldots, a_l\}$.

(3) Let K be a field, $f(x)$ be a separable polynomial of degree n with coefficients in K. Let Ω be the set of roots of $f(x)$ in an algebraic closure of K and $G = \mathrm{Gal}(f(x), K)$. Let k be an integer such that $2 \le k \le n$. Show that the following propositions are equivalent.

(d) G is k-transitive.

(e) $f(x)$ is irreducible over K and for any integer $2 \le l \le k$ and any subset $\{\alpha_1, \ldots, \alpha_{l-1}\}$ of cardinality $l-1$ of Ω, the polynomial $f_l(x) = f(x)/\prod_{i=1}^{l-1}(x-\alpha_i)$ is irreducible over $K(\alpha_1, \ldots, \alpha_{l-1})$.

(h) $f(x)$ is irreducible over K and and there exists a subset of Ω, $\{\alpha_1, \ldots, \alpha_{k-1}\}$ of cardinality $k-1$ such that for any integer $2 \le l \le k$, the polynomial $f_l(x) = f(x)/\prod_{i=1}^{l-1}(x-\alpha_i)$ is irreducible over $K(\alpha_1, \ldots, \alpha_{l-1})$.

(4) A subset A of Ω is called a non trivial block of G if $2 \le |A| < |\Omega|$ and for any $\sigma \in G$, we have $\sigma(A) = A$ or $\sigma(A) \cap A = \emptyset$.
Show that if $G \ne 1$ and G is intransitive, then it has a non trivial block.

(5) The group G is said to be imprimitive if there exists a partition of Ω of the form $\Omega = \Omega_1 \cup \cdots \cup \Omega_r$, satisfying the following conditions.

- $2 \le |\Omega_i| < |\Omega|$ for $i = 1, \ldots, r$.
- For any $i \in \{1, \ldots, r\}$, there exists $j \in \{1, \ldots, r\}$ such that $\sigma(\Omega_i) = \Omega_j$.

Show that if G is imprimitive, then it has a non trivial block. Show that the converse holds if G is transitive.

(6) Let $f(x) \in K[x]$ be irreducible and separable. Show that the following propositions are equivalent.

(u) G is imprimitive.

(v) For any root α of $f(x)$ there exists a field F such that $K \subsetneq F \subsetneq K(\alpha)$.

(w) There exists a root α of $f(x)$ and a field F such that $K \subsetneq F \subsetneq K(\alpha)$

(7) Let G be an imprimitive group. If G is k-transitive, show that $k = 1$.

Solution 5.32.

(1) • $(i) \Rightarrow (ii)$. Obvious.

- $(ii) \Rightarrow (i)$. Let $\{c_1,\dots,c_k\}$ and $\{b_1,\dots,b_k\}$ be two subsets of Ω of cardinality k. By (ii), there exist σ_1 and $\sigma_2 \in G$ such that $\sigma_1(a_i) = c_i$ and $\sigma_2(a_i) = b_i$ for $i = 1,\dots,k$. Let $\sigma = \sigma_1\sigma_2^{-1}$. Then $\sigma(b_i) = c_i$ for $i = 1,\dots,k$.

(2)
- $(a) \Rightarrow (b)$. That G is l-transitive is obvious. Fix a subset $\{a_1,\dots,a_l\}$ of Ω of cardinality l and let $\{b_1,\dots,b_{k-l}\}$ and $\{c_1,\dots,c_{k-l}\}$ be two subsets of $\Omega \setminus \{a_1,\dots,a_l\}$. Consider the subsets of Ω of cardinality k,
$$B = \{a_1,\dots,a_l,b_1,\dots,b_{k-l}\} \text{ and } C = \{a_1,\dots,a_l,b_1,\dots,b_{k-l}\}.$$
By **(a)**, there exists $\sigma \in G$ such that $\sigma(a_i) = a_i$ and $\sigma(b_j) = c_j$ for $i = 1,\dots,l$ and $j = 1,\dots,k-l$. This σ belongs to $G_{\{a_1,\dots,a_l\}}$ and maps b_j onto c_j for $j = 1,\dots,k-l$.
- $(b) \Rightarrow (c)$. Obvious.
- $(c) \Rightarrow (a)$. Let $\{b_1,\dots,b_k\}$ and $\{c_1,\dots,c_k\}$ be two subsets of Ω of cardinality k. Let σ_1 and $\sigma_2 \in G$ such that $\sigma_1(b_i) = a_i$ and $\sigma_2(c_i) = a_i$ for $i = 1,\dots,l$. Let $\tau \in G_{\{a_1,\dots,a_l\}}$ such that $\tau(\sigma_1(b_i)) = \sigma_2(c_i)$ for $i = l+1,\dots,k$. Then $\sigma := \sigma_2^{-1}\tau\sigma_1$ maps b_i onto c_i for $i = 1,\dots,k$.

(3)
- $(d) \Rightarrow (e)$. We use the well known result that a separable polynomial over a given field is irreducible if and only if its Galois group acts transitively on its roots. Set $f_1(x) = f(x)$. Fix $l \in \{2,\dots,k\}$ and $\{\alpha_1,\dots,\alpha_{l-1}\}$ be a subset of Ω. Suppose that $f_{l-1}(x)$ is irreducible over $K(\alpha_1,\dots,\alpha_{l-2})$. To show that $f_l(x)$ is irreducible over $K(\alpha_1,\dots,\alpha_{l-1})$, it is sufficient to prove that $\mathrm{Gal}(f_l(x), K(\alpha_1,\dots,\alpha_{l-1}))$ is transitive on $\Omega \setminus \{\alpha_1,\dots,\alpha_{l-1}\}$. This assertion is true by **(2)**.
- $(e) \Rightarrow (h)$. Obvious.
- $(h) \Rightarrow (d)$. Since $f(x)$ is irreducible, then G is transitive. Since $f(x)/(x-\alpha_1)$ is irreducible over $K(\alpha_1)$, then G_{α_1} is transitive on $\Omega \setminus \{\alpha_1\}$. It follows, from **(2)**, that G is 2-transitive on Ω. Suppose that G is $(l-1)$-transitive. Since $f_l(x)$ is irreducible over $K(\alpha_1,\dots,\alpha_{l-1})$, then $G_{\{\alpha_1,\dots,\alpha_{l-1}\}}$ is transitive over $\Omega \setminus \{\alpha_1,\dots,\alpha_{l-1}\}$. By **(2)**, G is l-transitive.

(4) Since $G \neq 1$, there exists $\alpha \in \Omega$ and $\sigma \in G$ such that $\sigma(\alpha) = \alpha$. Let $A = \{\sigma(\alpha), \sigma \in G\}$, then $|A| \geq 2$. On the other hand, since G is transitive, there exists $\theta \in \Omega$ such that $\theta \notin A$. Thus $|A| < |\Omega|$. Obviously $\sigma(A) = A$ for any $\sigma \in G$. Therefore A is a non trivial block of G.

(5) Suppose that G is imprimitive and let $A = \Omega_1$. Since for any $\sigma \in G$, $\sigma(\Omega_1) = \Omega_j$, then $\sigma(A) = A$ or $\sigma(A) \cap A = \emptyset$. To prove the reverse

implication, when G is transitive, let A satisfying one of the preceding conditions. Let $A_1 = A, \dots, A_r$ be the distinct values of $\sigma(A)$ for $\sigma \in G$. Let $b \in \Omega$. Let $a \in A$, since G is transitive, let $\sigma \in G$ such that $\sigma(a) = b$, then $b \in \sigma(A)$, thus $\cup A_i = \Omega$. It is easy to verify that $A_i \cap A_j = \emptyset$ for $i \neq j$.

(6) • $(u) \Rightarrow (v)$. Let $\Omega = \Omega_1 \cup \cdots \cup \Omega_r$ be a partition of Ω satisfying the condition of imprimitivity. Let α be a root of $f(x)$, which we may suppose that it belongs to Ω_1. Let

$$L = K(\alpha_1, \dots, \alpha_n),\ H = \mathrm{Gal}(L, K(\alpha)),\ J = \{\sigma \in G, \sigma(\Omega_1) = \Omega_1\}.$$

Clearly J is a subgroup of G containing H. Let $F = \mathrm{Inv}(J)$, then $K \subset F \subset K(\alpha)$. It remains to show that these inclusions are strict. Let $\beta \in \Omega_1$ such that $\beta \neq \alpha$ and let $\sigma \in G$ such that $\sigma(\alpha) = \beta$. Then $\sigma \in J$ and $\sigma \notin H$, thus $J \neq H$. Let $\alpha_2 \in \Omega_2$ and let $\sigma \in G$ such that $\sigma(\alpha) = \alpha_2$, then $\sigma \in G \setminus J$.

• $(v) \Rightarrow (w)$. Obvious.

• $(w) \Rightarrow (u)$. Let $\bar{K}$ be an algebraic closure of K. Let $m = [F : K]$ and $\sigma_1, \dots, \sigma_m$ be the distinct K-embeddings of F into $\bar{K}$. For any $i \in \{1, \dots, m\}$ let σ_{ij}, $j = 1, \dots, n/m$ be the extensions of σ_i to $K(\alpha)$. Let $\Omega_i := \{\sigma_{ij}(\alpha), j = 1, \dots, n/m\}$. Since the σ_{ij} for $i \in \{1, \dots, m\}$ and $j \in \{1, \dots, n/m\}$ represent the list of all the K-embeddings of $K(\alpha)$ into $\bar{K}$, then $\Omega = \cup_{i=1}^{n/m} \Omega_i$. Since the inclusions of fields are strict, then $2 \leq |\Omega_i| < n$ for $i = 1, \dots, n/m$. If $\Omega_i \cup \Omega_h \neq \emptyset$, then $\sigma_{ij}(\alpha) = \sigma_{hl}(\alpha)$ for some j and l. It follows that $\sigma_{ij} = \sigma_{hl}$, hence $i = h$, which implies $\Omega_i = \Omega_h$.

(7) Suppose that G is k-transitive and by contradiction that $k \geq 2$. In particular G is 2-transitive. Let $\Omega = \Omega_1 \cup \cdots \cup \Omega_r$ be a partition of Ω satisfying the condition of imprimitivity. Let $a_1, b_1 \in \Omega_1$ and $a_2 \in \Omega_2$, then there exists $\sigma \in G$ such that $\sigma(a_1) = a_1$ and $\sigma(b_1) = a_2$. This is a contradiction since we have $\sigma(\Omega_1) \neq \Omega_1$ and $\Omega_1 \cap \sigma(\Omega_1) \neq \emptyset$.

Chapter 6

Finite fields

Exercise 6.1.
Let p be a prime number and n be a positive integer. Show that in $\mathbb{Z}[x]$, we have $(1+x+\cdots+x^{p-1})^n \equiv (1-x)^{n(p-1)} \pmod{p}$.

Solution 6.1.
We do computations in $\mathbb{F}_p(x)$. We have

$$(1+x+\cdots+x^{p-1})^n = \left(\frac{1-x^p}{1-x}\right)^n = \left(\frac{(1-x)^p}{1-x}\right)^n = (1-x)^{n(p-1)},$$

hence the result.

Exercise 6.2.
Let q be a prime power. Let $f(x)$ be a polynomial of degree 3 with coefficients in $\mathbb{F}_q$.

(1) If $q > 3$, show that the equation $y^2 = f(x)$ has at least one solution $(x,y) \in \mathbb{F}_q^2$.
(2) Show that the above conclusion does not hold for $q = 3$.

Solution 6.2.

(1) If f has a root in $\mathbb{F}_q$, then the claim is obvious, so may suppose that f has no root in $\mathbb{F}_q$. Suppose by contradiction that for any $a \in \mathbb{F}_q$, $f(a) \notin \mathbb{F}_q^{\star 2}$ then $f(a)^{(q-1)/2} = -1$ for any $a \in \mathbb{F}_q$, hence $f(a)^r + 1 = 0$ for any $a \in \mathbb{F}_q$, where $r = (q-1)/2$. It follows that

$$f(x)^r + 1 = (x^q - x)g(x), \qquad \text{(Eq 1)}$$

where $g(x) \in \mathbb{F}_q[x]$ and $\deg g = 3r - q = 3r - (2r+1) = r - 1$. Differentiating this equation, we obtain:

$$rf'(x)f(x)^{r-1} = g'(x)(x^q - x) - g(x). \qquad \text{(Eq 2)}$$

Multiply equations **(Eq 1)** and **(Eq 2)** by $rf'(x)$ and $f(x)$ respectively and obtain equations (1′) and (2′) respectively. Subtract (2′) from (1′) and obtain:

$$rf'(x) = rf'(x)(x^q - x)g(x) - f(x)g'(x)(x^q - x) + f(x)g(x),$$

hence

$$rf'(x) - f(x)g(x) = (x^q - x)(rf'(x)g(x) - f(x)g'(x)).$$

Equating the degrees of both sides of this equation, we obtain

$$3 + \deg g = r + 2 \geq q = 2r + 1$$

that is $r \leq 1$ thus $q \leq 3$, which is a contradiction.

(2) Let $f(x) = x^3 - x - 1$, then for any $a \in \mathbb{F}_3$, we have $f(a) = -1$, so that $f(a)$ is never a square in $\mathbb{F}_3$.

Exercise 6.3.
Let K be a field of characteristic $p > 0$ and d be a positive integer.

(1) Let $K_d = \{x_1^d + \cdots + x_n^d, n \geq 1, x_i \in K\}$. Show that K_d is a subfield of K.
(2) Let $K^d = \{x^d, x \in K\}$. Show that K_d is the smallest subfield of K containing K^d.
(3) Determine K_d for any $d \geq 1$ if $K = \mathbb{F}_{3^4}$.
(4) Suppose that $K = \mathbb{F}_q$. Let F be a subfield of K. Show that there exists $d \geq 1$ such that $F = K_d$.

Solution 6.3.

(1) Let $y = x_1^d + \cdots + x_n^d$ and $z = u_1^d + \cdots + u_m^d$ be elements of K_d, then clearly $y+z$ and $yz \in K^d$. Moreover, we have $-1 = p-1 = 1^d + \cdots + 1^d$ ($p-1$ times), hence $-1 \in K_d$ and then $-y = -1 \cdot y \in K_d$. Let $y \in K_d$ such that $y \neq 0$, then

$$y^{-1} = y^{d-1} \cdot (y^{-1})^d = y \cdots y \cdot (y^{-1})^d \in K_d.$$

We conclude that K_d is a subfield of K.

(2) By **(1)**, K_d is a subfield of K. Let F be a subfield of K containing K^d, then it contains any finite sum of elements of K^d. Therefore $F \supset K_d$. We conclude that K_d is the smallest subfield of K containing K^d.

(3) The subfields of K are: $\mathbb{F}_3$, $\mathbb{F}_{3^2}$ and $\mathbb{F}_{3^4}$. Let $\delta = \gcd(d, 3^4 - 1) = \gcd(d, 2^4.5)$. We examine several cases.

- $\delta = 1$. In this case $K^d \backslash \{0\}$ is a subgroup of K^* of order $\frac{3^4-1}{\delta} = 80$. Therefore, according to **(2)** and the list of subfields of K, $K_d = \mathbb{F}_{3^4}$.

- $\delta = 2$. Here $K^d\backslash\{0\}$ is a subgroup of K^* of order $\frac{3^4-1}{2} = 40$. Therefore, for the same reason as above, $K_d = \mathbb{F}_{3^4}$.
- $\delta = 4$. Here $K^d\backslash\{0\}$ is a subgroup of K^* of order $\frac{3^4-1}{4} = 20$, hence $K_d = \mathbb{F}_{3^4}$.
- $\delta = 5$. Here $K^d\backslash\{0\}$ is a subgroup of K^* of order $\frac{3^4-1}{5} = 16$, hence $K_d = \mathbb{F}_{3^4}$.
- $\delta = 10$. Here $K^d\backslash\{0\}$ is a subgroup of K^* of order $\frac{3^4-1}{10} = 8$, hence $K_d = \mathbb{F}_{3^2}$.
- $\delta = 16$. Here $K^d\backslash\{0\}$ is a subgroup of K^* of order $\frac{3^4-1}{16} = 5$, hence $K_d = \mathbb{F}_{3^4}$, because if a a subfield F of K contains K^d, then $K^d\backslash\{0\}$ is a subgroup of F^*. In this case we may exclude the possibility $F = \mathbb{F}_{3^2}$.
- $\delta = 8$. Here $K^d\backslash\{0\}$ has order $\frac{3^4-1}{8} = 10$, hence $K_d = \mathbb{F}_{3^4}$.
- $\delta = 20$. Here $K^d\backslash\{0\}$ has order $\frac{3^4-1}{20} = 4$, hence $K_d = \mathbb{F}_{3^2}$.
- $\delta = 40$. Here $K^d\backslash\{0\}$ has order $\frac{3^4-1}{40} = 2$, hence $K_d = \mathbb{F}_{3}$.
- $\delta = 80$. Here $K^d\backslash\{0\}$ has order 1, hence $K_d = \mathbb{F}_3$.

(4) Let e be the order of F^*, then $F^* = \{x^{\frac{q-1}{e}}, x \in \mathbb{F}_q^*\}$. Therefore $F = \mathbb{F}_{q^d}$, with $d = \frac{q-1}{e}$. Since F is a field then $F = K_d$.

Exercise 6.4.
Let p be a prime number, m and n be positive integers such that $m \geq 2$.

(1) If $m \equiv 0 \pmod p$ or $m \equiv 1 \pmod p$, show that the polynomial $f(x) = x^m + x + 1$ is separable over $\mathbb{F}_p$.
(2) Let $g(x) = x^{p^n} + x + 1$ and let $g_1(x), \ldots, g_r(x)$ be its irreducible factors in $\mathbb{F}_p[x]$ of degree $d_1, \ldots, d_r$ respectively. Show that $\text{lcm}(d_1, \ldots, d_r) = 2n$. Deduce that if ($p = 2$ and $n \geq 3$) or $p \geq 3$, then $g(x)$ is reducible over $\mathbb{F}_p$.
(3) Let $h(x) = x^{p^n+1} + x + 1$ and let $h_1(x), \ldots, h_s(x)$ be its irreducible factors in $\mathbb{F}_p[x]$ of degree $e_1, \ldots, e_s$ respectively. Show that $\text{lcm}(e_1, \ldots, e_s) = 3n$. Deduce that if ($p = 2$ and $n \notin \{1, 3\}$) or $p \geq 3$, then $h(x)$ is reducible over $\mathbb{F}_p$.
(4) Show that $r \geq p^n/(2n)$ and $s \geq (p^n + 1)/(3n)$.

Solution 6.4.

(1) We have $f'(x) = mx^{m-1} + 1$. If $m \equiv 0 \pmod p$, then $f'(x) = 1$ which implies that $f(x)$ is separable over $\mathbb{F}_p$. If $m \equiv 1 \pmod p$, then $f'(x) = x^{m-1} + 1$. Suppose that $f(x)$ and $f'(x)$ have a common root say α, then $\alpha^{m-1} = -1$ and $\alpha^m = -\alpha - 1$. It follows that $\alpha^m = -\alpha$

and $\alpha^m = -\alpha - 1$, which is a contradiction. Therefore $f(x)$ is separable over $\mathbb{F}_p$.

(2) Let I be the ideal generated by $g(x)$ and let I_j, for $j = 1, \dots, r$, be the ideal generated by $g_j(x)$ in $\mathbb{F}_p[x]$. Since by (1), $g(x)$ is separable over $\mathbb{F}_p$, we may consider the following sequence of rings isomorphisms

$$\mathbb{F}_p[x]/I \xrightarrow{\psi} \prod_{j=1}^{r}(\mathbb{F}_p[x]/I_j) \xrightarrow{\hat{\phi}} \prod_{j=1}^{r}(\mathbb{F}_p[x]/I_j) \xrightarrow{\psi^{-1}} \mathbb{F}_p[x]/I,$$

defined by

$$\psi(a(x) + I) = (a(x) + I_1, \dots, a(x) + I_r),$$

ψ^{-1} is the inverse of ψ and

$$\hat{\phi}(a_1(x) + I_1, \dots, a_r(x) + I_r) = (a_1(x^p) + I_1, \dots, a_r(x^p) + I_r).$$

Let $\phi = \psi^{-1} \circ \hat{\phi} \circ \psi$, then $\phi(a(x) + I) = a(x^p) + I$. Let k be a positive integer, then we have

$$\begin{aligned} \phi^k(x + I) = x + I &\Longleftrightarrow \hat{\phi}^k(x + I_1, \dots, x + I_r) = (x + I_1, \dots, x + I_r) \\ &\Longleftrightarrow x^{p^k} + I_j = x + I_j \quad \text{for} \quad j = 1, \dots, r \\ &\Longleftrightarrow \deg g_j \mid k \quad \text{for} \quad j = 1, \dots, r \\ &\Longleftrightarrow \operatorname{lcm}(d_1, \dots, d_r) \mid k. \end{aligned}$$

Let $d = \operatorname{lcm}(d_1, \dots, d_r)$. We have

$$\phi^n(x + I) = x^{p^n} + I = -x - 1 + I \quad \text{and} \quad \phi^{2n}(x + I) = x + I,$$

hence by the above equivalences, $d \mid 2n$. Since $\phi^n(x + I) = -x - 1 + I$, then $d \neq n$. Suppose that $d < n$. From the identity $x^{p^d} + I = x + I$, we conclude that $x^{p^d} - x \equiv 0 \pmod{g(x)}$, which is a contradiction. Suppose now that $n < d < 2n$. Then $d = 2n'$ with $n' \mid n$ and $0 < n' < n$. Set $d = n + t$, where $0 < t < n$. From the identity $x^{p^d} + I = x + I$, we conclude that

$$\begin{aligned} x + I = x^{p^d} + I &= x^{p^{n+t}} + I \\ &= (x^{p^n})^{p^t} + I = (-x - 1)^{p^t} + I \\ &= -x^{p^t} - 1 + I. \end{aligned}$$

We deduce that $x^{p^t} + x + 1 \equiv 0 \pmod{g(x)}$, which is a contradiction. Therefore $d = 2n$.

Suppose that $g(x)$ is irreducible over $\mathbb{F}_p$, then $r = 1$, $d = d_1 = p^n = 2n$. From this it is obvious that $p = 2$ and $n = 2^j$. We deduce that $2^j =$

$j+1$. Using arguments from analysis one may show that $j \in \{0,1\}$ and then $n \in \{1,2\}$. It is possible to get the same conclusion by arguing that 2^j is equal to the number of subsets of a set, say E, containing j elements. Since the empty set and the sets $\{e\}$ for $e \in E$ are such subsets and their number is equal to $j+1$, then $j < 2$.

(3) We use the same proof as in **(2)**. Making the necessary changes, mutadis mutandis, we keep the same notations as in **(2)**. Here we have

$$\begin{aligned}\phi^n(x+I) &= x^{p^n} + I = -1 - 1/x + I,\\ \phi^{2n}(x+I) &= -1/(x+1) + I \quad \text{and}\\ \phi^{3n}(x+I) &= x + I,\end{aligned}$$

hence by the above equivalences, $\operatorname{lcm}(e_1, \ldots, e_s) \mid 3n$. Since $\phi^n(x+I) = -x - 1 + I$, then $d \neq n$. Suppose that $d < n$. From the identity $x^{p^d} + I = x + I$, we conclude that $x^{p^d} - x \equiv 0 \pmod{h(x)}$, which is a contradiction. Suppose now that $n < d < 2n$. Then $d = 3n'$ with $n' \mid n$ and $0 < n' < n$. Set $d = n + t$, where $0 < t < n$. From the identity $x^{p^d} + I = x + I$, we conclude that

$$\begin{aligned}x + I = x^{p^d} + I &= x^{p^{n+t}} + I\\ &= (x^{p^n})^{p^t} + I = (-1 - 1/x)^{p^t} + I\\ &= -1 - (1/x)^{p^t} + I.\end{aligned}$$

We deduce that $x^{p^t+1} + x + 1 \equiv 0 \pmod{h(x)}$, which is a contradiction. From the computation of $\phi^{2n}(x+I)$, it is seen that $d \neq 2n$. Suppose that $2n < d < 3n$. Set $d = 2n + u$, where $0 < u < n$, then

$$\begin{aligned}x + I = x^{p^d} + I &= x^{p^{2n+u}} + I\\ &= (x^{p^{2n}})^{p^u} + I = (\frac{-1}{x+1})^{p^u} + I\\ &= \frac{-1}{x^{p^u} + 1} + I.\end{aligned}$$

We deduce that $x^{p^u+1} + x + 1 \equiv 0 \pmod{h(x)}$, which is a contradiction. Therefore $d = 3n$.

Suppose that $h(x)$ is irreducible over $\mathbb{F}_p$, then $r = 1$, $d = d_1 = p^n + 1 = 3n$. Using the function defined by $F(x) = p^x - 3x + 1$, it is seen that this equality holds only if $p = 2$, $n = 1$ or $n = 3$.

(4) Since $d = \operatorname{lcm}(d_1, \ldots, d_r)$, then $d_i \leq d$ for any $i \in \{1, \ldots, r\}$. It follows from **(2)**, that $p^n = \sum_{i=1}^{r} d_i \leq 2nr$, thus $r \geq p^n/(2n)$. The proof for s is similar and will be omitted.

Exercise 6.5.
Let p be a prime number, $q = p^n$ and $L(x) = \sum_{i=0}^m a_i x^{q^i} \in \mathbb{F}_q[x]$. Show that for any $\alpha \in \mathbb{F}_{q^n}$, $L(\mathrm{Tr}_{\mathbb{F}_{q^n}/\mathbb{F}_q}(\alpha)) = \mathrm{Tr}_{\mathbb{F}_{q^n}/\mathbb{F}_q}(L(\alpha))$.

Solution 6.5.
We have

$$\begin{aligned}
L(\mathrm{Tr}_{\mathbb{F}_{q^n}/\mathbb{F}_q}(\alpha)) &= L(\alpha + \alpha^q + \cdots + \alpha^{q^{n-1}}) \\
&= \sum_{i=0}^m a_i(\alpha + \alpha^q + \cdots + \alpha^{q^{n-1}})^{q^i} \\
&= \sum_{i=0}^m a_i(\alpha^{q^i} + \alpha^{q^{i+1}} + \cdots + \alpha^{q^{i+n-1}}) \\
&= \sum_{i=0}^m a_i\alpha^{q^i} + \sum_{i=0}^m a_i\alpha^{q^{i+1}} + \cdots + \sum_{i=0}^m a_i\alpha^{q^{i+n-1}} \\
&= \sum_{i=0}^m a_i\alpha^{q^i} + \left(\sum_{i=0}^m a_i\alpha^{q^i}\right)^q + \cdots + \left(\sum_{i=0}^m a_i\alpha^{q^i}\right)^{q^{n-1}} \\
&= L(\alpha) + L(\alpha)^q + \cdots + L(\alpha)^{q^{n-1}} \\
&= \mathrm{Tr}_{\mathbb{F}_{q^n}/\mathbb{F}_q}(L(\alpha)),
\end{aligned}$$

hence the result.

Exercise 6.6.
Let q be a prime power, n be a positive integer and $x \in \mathbb{F}_{q^n}$. Show that $x \in \mathbb{F}_{q^n}^2$ if and only if $\mathrm{N}_{\mathbb{F}_{q^n}/\mathbb{F}_q}(x) \in \mathbb{F}_q^2$.

Solution 6.6.
If $x = 0$, the equivalence is obvious. From now on, we suppose that $x \neq 0$. Since the norm map from $\mathbb{F}_{q^n}$ into $\mathbb{F}_q$ is surjective, then the proposition

(P) For $x \in \mathbb{F}_{q^n}^*, x \in \mathbb{F}_{q^n}^{*2}$ if and only if $\mathrm{N}_{\mathbb{F}_{q^n}/\mathbb{F}_q}(x) \in \mathbb{F}_q^{*2}$

is equivalent to the proposition

(Q) $\mathrm{N}_{\mathbb{F}_{q^n}/\mathbb{F}_q}(\mathbb{F}_{q^n}^{*2}) = \mathbb{F}_q^{*2}$,

so we prove **(Q)**.

In the sequel, we omit the reference to the index $\mathbb{F}_{q^n}/\mathbb{F}_q$ in the norm and will write $N(x)$ for the norm of x. Obviously we have $\mathrm{N}_{\mathbb{F}_{q^n}/\mathbb{F}_q}(\mathbb{F}_{q^n}^{*2}) \subset \mathbb{F}_q^{*2}$. To get equality of these two sets, we show that they have the same cardinality. It is easy to show that $|\mathbb{F}_q^{*2}| = (q-1)/2$. For the other set consider the norm maps $N_1 : \mathbb{F}_{q^n}^{\star} \to \mathbb{F}_q^*$ and $N_2 : \mathbb{F}_{q^n}^{\star 2} \to \mathbb{F}_q^{*2}$ which are morphisms of groups. Since N_1 is surjective then $|\operatorname{Ker} N_1| = (q^n-1)/(q-1)$.

We have

$$\operatorname{Im} N_2 = N_2(\mathbb{F}_{q^n}^{*2}) \simeq \mathbb{F}_{q^n}^{*2} / \operatorname{Ker} N_2 = \mathbb{F}_{q^n}^{*2}/(\operatorname{Ker} N_1 \cap \mathbb{F}_{q^n}^{*2}),$$

hence

$$\begin{aligned} |N(\mathbb{F}_{q^n}^{*2})| &= |N_2(\mathbb{F}_{q^n}^{*2})| \\ &= |\mathbb{F}_{q^n}^{*2}|/|\operatorname{Ker} N_1 \cap \mathbb{F}_{q^n}^{*2}| \\ &\geq |\mathbb{F}_{q^n}^{*2}|/|\operatorname{Ker} N_1| \\ &= \frac{q^n-1}{2}/\frac{q^n-1}{q-1} \\ &= (q-1)/2. \end{aligned}$$

We conclude that $N(\mathbb{F}_{q^n}^{*2}) = \mathbb{F}_q^{*2}$.

Exercise 6.7.
Let $f(x) = x^2 + ax + b \in \mathbb{F}_q[x]$. Suppose that $f(x)$ is irreducible over $\mathbb{F}_q$ and that $a \neq 0$ if q is odd. Let α be a root of $f(x)$ in an algebraic closure of $\mathbb{F}_q$. Show that $\{\alpha, \alpha^q\}$ is a basis of $\mathbb{F}_q(\alpha)$ over $\mathbb{F}_q$.

Solution 6.7.
Suppose that $\lambda\alpha + \mu\alpha^q = 0$, where $\lambda, \mu \in \mathbb{F}_q$. Suppose that $\mu \neq 0$, then $\lambda \neq 0$ and $\alpha^{q-1} = -\lambda/\mu \in \mathbb{F}_q^\star$, hence $\alpha^{(q-1)^2} = 1$. It follows that the order of α in the multiplicative group $\mathbb{F}_q(\alpha)^\star$ divides $\gcd((q-1)^2, q^2-1)$. We have

$$\gcd((q-1)^2, q^2-1) = \begin{cases} q-1 & \text{if } q \text{ is even} \\ 2(q-1) & \text{if } q \text{ is odd} \end{cases}.$$

If q is odd then $\alpha^{2(q-1)} = 1$, hence $\alpha^2 \in \mathbb{F}_q$, which is excluded by our assumptions. If q is even, then $\alpha^{q-1} = 1$. Therefore $\alpha \in \mathbb{F}_q$ which is a contradiction. It follows that $\mu = 0$, then $\lambda = 0$, which implies that $\{\alpha, \alpha^q\}$ is a basis of $\mathbb{F}_q(\alpha)$ over $\mathbb{F}_q$.

Exercise 6.8.
Let p be a prime number, q be a positive power of p, n be a positive integer and $\alpha \in \mathbb{F}_{q^n}$ be an element generating a normal basis of $\mathbb{F}_{q^n}$ over $\mathbb{F}_q$. Suppose that there exists $a \in \mathbb{F}_q$ such that $\alpha + a$ does not generate a normal basis of $\mathbb{F}_{q^n}$ over $\mathbb{F}_q$. Show that $p \nmid n$ and $a = -\operatorname{Tr}_{\mathbb{F}_{q^n}/\mathbb{F}_q}(\alpha)/n$.

Solution 6.8.
Since $\alpha + a$ does not generate a normal basis of $\mathbb{F}_{q^n}$ over $\mathbb{F}_q$, then there exists a positive integer m, $1 \leq m \leq n-1$ and $a_0, a_1, \ldots, a_{m-1} \in \mathbb{F}_q$ such that

$$a_0(\alpha+a) + a_1(\alpha+a)^q + \cdots + a_{m-1}(\alpha+a)^{q^{m-1}} + (\alpha+a)^{q^m} = 0.$$

We deduce that

$$a_0\alpha + a_1\alpha^q + \cdots + a_{m-1}\alpha^{q^{m-1}} + \alpha^{q^m} + a(a_0 + \cdots + a_{m-1} + 1) = 0.$$

We write this equation in the form:

$$a_0\alpha + a_1\alpha^q + \cdots + a_{m-1}\alpha^{q^{m-1}} + \alpha^{q^m} = -a(a_0 + \cdots + a_{m-1} + 1).$$

Using the trace map, we obtain

$$\begin{aligned} a_0 \operatorname{Tr}_{\mathbb{F}_{q^n}/\mathbb{F}_q}(\alpha) &+ a_1 \operatorname{Tr}_{\mathbb{F}_{q^n}/\mathbb{F}_q}(\alpha^q) \\ &+ \cdots + a_{m-1} \operatorname{Tr}_{\mathbb{F}_{q^n}/\mathbb{F}_q}\left(\alpha^{q^{m-1}}\right) \\ &+ \operatorname{Tr}_{\mathbb{F}_{q^n}/\mathbb{F}_q}\left(\alpha^{q^m}\right) \\ &= -na(a_0 + \cdots + a_{m-1} + 1). \end{aligned}$$

Since conjugate elements in a given extension have the same trace, then

$$\operatorname{Tr}_{\mathbb{F}_{q^n}/\mathbb{F}_q}(\alpha)(a_0 + \cdots + a_{m-1} + 1) = -na(a_0 + \cdots + a_{m-1} + 1).$$

Since α generates a normal basis, then $a_0 + \cdots + a_{m-1} + 1 \neq 0$ and $\operatorname{Tr}_{\mathbb{F}_{q^n}/\mathbb{F}_q}(\alpha) \neq 0$. It follows that $p \nmid n$ and $a = -\operatorname{Tr}_{\mathbb{F}_{q^n}/\mathbb{F}_q}(\alpha)/n$.

Exercise 6.9.
Let $\sigma : \mathbb{F}_{q^n} \to \mathbb{F}_{q^n}$ be the Frobenius automorphism and let $f(x)$ (resp. $g(x)$) be the characteristic (resp. minimal) polynomial of the $\mathbb{F}_q$-linear map σ. Show that $f(x) = g(x) = x^n - 1$.

Solution 6.9.
For any $\alpha \in \mathbb{F}_{q^n}$, we have $\sigma^n(\alpha) = \alpha^{q^n} = \alpha$, hence $\sigma^n = \operatorname{Id}_{\mathbb{F}_{q^n}}$. It follows that $f(x) = x^n - 1$. It is known that $g(x)$ is monic and $g(x) \mid f(x)$. Suppose that $g(x) \neq x^n - 1$, then the degree m of $g(x)$ is at most equal to $n-1$. Set $g(x) = x^m + a_{m-1}x^{m-1} + \cdots + a_0$, where $a_0, \ldots, a_{m-1} \in \mathbb{F}_q$. For any $\alpha \in \mathbb{F}_{q^n}$, we have $g(\sigma)(\alpha) = 0$, hence

$$\sigma^m + a_{m-1}\sigma^{m-1} + \cdots + a_0 Id_{\mathbb{F}_{q^n}} = 0.$$

It follows that any $\alpha \in \mathbb{F}_{q^n}$ is a root of the polynomial

$$h(x) = x^{q^m} + a_{m-1}x^{q^{m-1}} + \cdots + a_0 x.$$

This fact is impossible since the number of roots of this polynomial, in an algebraic closure of $\mathbb{F}_q$, is at most equal to q^m. We conclude that $g(x) = x^n - 1$.

Exercise 6.10.

(1) Let E be a vector space of finite dimension over the field K and let $T : E \to E$ be a linear map. Let $v \in E$ and

$$I = \{f(x) \in K[x], f(T)(v) = 0\}.$$

Show that I is a principal ideal of $K[x]$ generated by some monic polynomial which will be denoted by $M_{(T,v)}(x)$. Show that this polynomial divides the characteristic polynomial and the minimal polynomial of T over K.

(2) Let p be a prime number, r be a positive integer and $q = p^r$. Let n be a positive integer, α be an element of the field $\mathbb{F}_{q^n}$ and σ be the Frobenius automorphism of this field. Show that α generates a normal basis of $\mathbb{F}_{q^n}$ over $\mathbb{F}_q$ if and only if $M_{(\sigma,\alpha)}(x) = x^n - 1$.
Hint. One may use **Exercise 6.9**.

(3) Write n in the form $n = p^e t$, where e and t are positive integers and $\gcd(p, t) = 1$. Let $x^n - 1 = (f_1(x) \cdots f_r(x))^{p^e}$ be the factorization of $x^n - 1$ into a product of irreducible factors over $\mathbb{F}_q$. Let $\phi_i(x) = (x^n - 1)/f_i(x)$. Show that α generates a normal basis over $\mathbb{F}_q$ if and only if $\phi_i(\sigma)(\alpha) \neq 0$ for $i = 1, \dots, r$.

(4) If moreover $t = 1$, show that α generates a normal basis over $\mathbb{F}_q$ if and only if $\mathrm{Tr}_{\mathbb{F}_{q^n}/\mathbb{F}_q}(\alpha) \neq 0$.

Solution 6.10.

(1) Clearly I is an ideal of the ring $K[x]$. Since this ring is principal, then I is principal generated by some polynomial, which may be supposed to be monic. Since the characteristic polynomial and the minimal polynomial of T over K belong to I, then they are divisible by $M_{(T,v)}(x)$.

(2) Suppose that $M_{(\sigma,\alpha)}(x) \neq x^n - 1$. By (1) and by **Exercise 6.9**, $x^n - 1 \in I$, hence $M_{(\sigma,\alpha)}(x) \mid x^n - 1$. Therefore the degree m of $g(x) := M_{(\sigma,\alpha)}(x)$ satisfies the condition $m < n$. Set $g(x) = x^m + a_{m-1}x^{m-1} + \cdots + a_0$, where $a_0, \dots, a_{m-1} \in \mathbb{F}_q$, then

$$\sigma^m + a_{m-1}\sigma^{m-1} \cdots + a_0 Id_{\mathbb{F}_{q^n}}(\alpha) = 0.$$

Therefore

$$\alpha^{q^m} + a_{m-1}\alpha^{q^{m-1}} \cdots + a_0\alpha = 0.$$

We conclude that α does not generate a normal basis of $\mathbb{F}_{q^n}$ over $\mathbb{F}_q$.

Suppose that α does not generate a normal basis of $\mathbb{F}_{q^n}$ over $\mathbb{F}_q$, then there exist a positive integer m, $m \leq n-1$ and $a_0, \ldots, a_m \in \mathbb{F}_q$, with $a_m \neq 0$ such that $a_0\alpha + a_1\alpha^q + \cdots + a_m\alpha^{q^m} = 0$. Let

$$g(x) = a_0 + a_1x + \cdots + a_mx^m,$$

then $g(\sigma)(\alpha) = 0$. Therefore $M_{(\sigma,\alpha)}(x) \mid g(x)$ and then $M_{(\sigma,\alpha)}(x) \neq x^n - 1$.

(3) Since, by **Exercise 6.9**, $M_{(\sigma,\alpha)}(x) \mid x^n - 1$, then the equivalence follows from **(2)**.

(4) If $t = 1$, that is

$$x^n - 1 = x^{p^e} - 1 = (x-1)^{p^e},$$

then $r = 1$, $f_1(x) = x - 1$ and

$$\phi_1(x) = (x^n - 1)/(x-1) = x^{n-1} + \cdots + x + 1,$$

hence by **(3)**, α generates a normal basis over $\mathbb{F}_q$ if and only if $\mathrm{Tr}_{\mathbb{F}_{q^n}/\mathbb{F}_q}(\alpha) \neq 0$.

Exercise 6.11.
For any finite family $\{u_1(X_1, \ldots, X_n), \ldots, u_r(X_1, \ldots, X_n)\}$ of polynomials with coefficients in $\mathbb{F}_q$, we denote by $V(u_1, \ldots, u_n)$ the subset of $\mathbb{F}_q^n$,

$$V(u_1, \ldots, u_n) = \{(x_1, \ldots, x_n) \in \mathbb{F}_q^n, u_i(x_1, \ldots, x_n) = 0, \text{ for } i = 1, \ldots, r\}.$$

Let $f_1(X_1, \ldots, X_n), \ldots, f_r(X_1, \ldots, X_n) \in \mathbb{F}_q[X_1, \ldots, X_n]$ and let

$$F(X_1, \ldots, X_n) = (1 - f_1^{q-1}) \cdots (1 - f_r^{q-1}) - 1.$$

Show that $V(f_1, \ldots, f_r) = V(F)$.

Solution 6.11.
Set $\vec{X} = (X_1, \ldots, X_n)$ and $\vec{x} = (x_1, \ldots, x_n)$. Let $\vec{x} \in V(f_1, \ldots, f_n)$, then clearly $F(\vec{x}) = 0$, hence $\vec{x} \in V(F)$. Let $\vec{x} \in V(F)$. Suppose that there exists $j \in \{1, \ldots, r\}$ such that $f_j(\vec{x}) \neq 0$, then $f_j(\vec{x})^{q-1} = 1$, hence $F(\vec{x}) = -1$. Therefore we reach a contradiction. We conclude that $\vec{x} \in V(f_1, \ldots, f_r)$.

Exercise 6.12.
Let q be a prime power, $N = q - 1$ and $a \in \mathbb{F}_q$. Let

$$\delta_a(X) = \begin{cases} 1 - X^N & \text{if } a = 0 \\ -\sum_{k=1}^{N}(X/a)^k & \text{if } a \neq 0 \end{cases}.$$

(1) Let $x \in \mathbb{F}_q$. Show that $\delta_a(x) = \begin{cases} 1 & \text{if } x = a \\ 0 & \text{if } x \neq a \end{cases}$.

(2) Let $(a_1^1, \ldots, a_1^n), \ldots, (a_M^1, \ldots, a_M^n)$ be M distinct tuples of elements of $\mathbb{F}_q$ and let $b_1, \ldots, b_M \in \mathbb{F}_q$. Consider the polynomial

$$f(X_1, \ldots, X_n) = \sum_{i=1}^{M} b_i \prod_{j=1}^{n} \delta_{a_i^j}(X_j).$$

Show that $f(X_1, \ldots, X_n) \in \mathbb{F}_q[X_1, \ldots, X_n]$ and for any $k \in \{1, \ldots, M\}$, $f(a_k^1, \ldots, a_k^n) = b_k$.

Solution 6.12.

(1) If $a = 0$, then $\delta_a(X) = 1 - X^N$, hence $\delta_a(0) = 1$ and for $x \neq 0$, $\delta_a(x) = 1 - x^{q-1} = 0$. If $a \neq 0$, then $\delta_a(X) = -\sum_{k=1}^{N}(X/a)^k$, hence

$$\delta_a(a) = -\sum_{k=1}^{N} 1^k = -(q-1) = 1$$

and for $x \neq a$,

$$\delta_a(x) = -\sum_{k=1}^{N} y^k = -y(y^{q-1} - 1)/(y - 1),$$

where $y = x/a$. Therefore $\delta_a(x) = 0$.

(2) Clearly, $f(X_1, \ldots, X_n) \in \mathbb{F}_q[X_1, \ldots, X_n]$. Let $k \in \{1, \ldots, M\}$. To get the result, we show that $\prod_{j=1}^{n} \delta_{a_k^j}(a_k^j) = 1$ and $\prod_{j=1}^{n} \delta_{a_i^j}(a_k^j) = 0$ for $i \neq k$. These claims are true by **(1)** and the proof is complete.

Exercise 6.13.

Let n be a positive integer, p be a prime number and $q = p^r$, where r is positive integer. Let $\bar{n} \in \{1, \ldots, q-1\}$ be the unique integer such that $n \equiv \bar{n} \pmod{q-1}$.

(1) Let $(a, b) \in \mathbb{F}_q^2$ such that $a \neq 0$. Show that

$$(aX + b)^n \equiv (aX + b)^{\bar{n}} \pmod{(X^q - X)\mathbb{F}_q[X]}.$$

(2) Deduce that for any $j \in \{1, \ldots, q-1\}$ we have

$$\sum_{\substack{j \leq m \leq n \\ m \equiv j \pmod{q-1}}} \binom{n}{m} \equiv \binom{\bar{n}}{j} \pmod{p}.$$

Solution 6.13.

(1) It is equivalent to prove that the functions

$$f(x) = (ax+b)^n \text{ and } g(x) = (ax+b)^{\bar{n}}$$

take the same values point wise in $\mathbb{F}_q$. This assertion is trivial for $x = -b/a$. Set $n = (q-1)m + \bar{n}$, then for any $x \neq -b/a$, we have

$$f(x) = (ax+b)^{(q-1)m}(ax+b)^{\bar{n}} = (ax+b)^{\bar{n}} = g(x).$$

(2) On one hand, by **(1)**, we have

$$(X+1)^n \equiv (X+1)^{\bar{n}} \pmod{(X^q - X)\mathbb{F}_q[X]}.$$

On the other hand, we have

$$(X+1)^n = \sum_{m=0}^{n} \binom{n}{m} X^m = 1 + \sum_{j=1}^{q-1} \left(\sum_{\substack{j \le m \le n \\ m \equiv j \pmod{q-1}}} \binom{n}{m} X^m \right),$$

hence

$$(X+1)^n \equiv 1 + \sum_{j=1}^{q-1} \left(\sum_{\substack{j \le m \le n \\ m \equiv j \pmod{q-1}}} \binom{n}{m} \right) X^j \pmod{(X^q - X)\mathbb{F}_q[X]}.$$

By equating the coefficients of X^j in the right sides of the two congruence equations, we obtain the result.

Exercise 6.14.
Let a and $b \in \mathbb{F}_q^\star$. Show that $\sum_{j=0}^{q-1} a^{q-1-j} b^j = 1$.

Solution 6.14.
If $b = a$, then

$$\sum_{j=0}^{q-1} a^{q-1-j} b^j = \sum_{j=0}^{q-1} a^{q-1} = \sum 1 = 0.$$

If $b \neq a$, then

$$\sum_{j=0}^{q-1} a^{q-1-j} b^j = b^{q-1} \sum_{j=0}^{q-1} (a/b)^j = b^{q-1} \frac{1 - (a/b)^q}{1 - a/b} = 1.$$

Exercise 6.15.
Let $f(x) \in \mathbb{F}_q[x]$. Show that the following propositions are equivalent.

(i) For any $a \in \mathbb{F}_q$, $f(x+a) = f(x)$.
(ii) $f(x)$ has the form $f(x) = \sum_{k=0}^{m} c_k(x^q - x)^k$ with $c_k \in \mathbb{F}_q$.

Solution 6.15.

- $(i) \Rightarrow (ii)$. The representation of $f(x)$ in base $x^q - x$ [Lang (1965), Th. 9, Chap. 5.5] has the form $f(x) = \sum_k c_k(x)(x^q - x)^k$, where $c_k(x) \in \mathbb{F}_q[x]$ and $\deg c_k < q$. From the hypothesis and from the uniqueness of this representation, we conclude that for any $k = 0, \ldots, m$, and any $a \in \mathbb{F}_q$, $c_k(x+a) = c_k(x)$, hence $c_k(x+a) - c_k(x) = 0$ for any k and any a. It follows that $c_k(x+y) = c_k(x)$, which implies $c_k(x) \in \mathbb{F}_q$.
- $(ii) \Rightarrow (i)$. Easy.

Exercise 6.16.
Let p be a prime number, $q = p^r$, $f(x) \in \mathbb{F}_q[x]$ be a monic polynomial of degree d and let $E = \mathbb{F}_q[x]/(f(x))$. Recall that the trace map, $\mathrm{Tr} : E \to \mathbb{F}_q$, defined by $\mathrm{Tr}(\alpha) = \alpha + \alpha^q + \cdots + \alpha^{q^{d-1}}$ is linear over $\mathbb{F}_q$.

(1) If f is irreducible, show that the trace map is surjective.
(2) Suppose that f is reducible over $\mathbb{F}_q$.

(a) If $f = g^k$, where g is irreducible over $\mathbb{F}_q$ and $k \geq 2$, show that there exists $e \in E \setminus \{0\}$, such that $e^2 = 0$ and $\mathrm{Tr}(e) = e$.
(b) If $f = f_1^{e_1} \cdots f_r^{e_r}$, where $r \geq 2$ and the f_i are irreducible over $\mathbb{F}_q$, show that there exists $e \in E \setminus \{0, 1\}$, such that $e^2 = e$ and $\mathrm{Tr}(e) = de$.
(c) Suppose that $p \nmid d$. Deduce that $\mathbb{F}_q \subset \mathrm{Im}(\mathrm{Tr})$ and that this inclusion is strict.

Solution 6.16.

(1) It is sufficient to prove that the image of the map is not trivial. By contradiction suppose that for any $\alpha \in E$, $\alpha + \alpha^q + \cdots + \alpha^{q^{d-1}} = 0$, then the equation $x + x^q + \cdots + x^{q^{d-1}} = 0$ has q^d solutions, which is impossible.
(2)(a) Let $e = \overline{(g(x))^h}$, where

$$h = \begin{cases} k/2 & \text{if} \quad k \quad \text{is even} \\ (k+1)/2 & \text{if} \quad k \quad \text{is odd} \end{cases},$$

then $e \neq 0$ and $e^2 = 0$. We deduce that for any $j \geq 2$, $e^j = 0$, hence $\mathrm{Tr}(e) = e$.

(b) Let $u(x) \in \mathbb{F}_q[x]$ such that

$$u(x) \equiv 1 \pmod{f_1^{e_1}(x)} \text{ and } u(x) \equiv 0 \pmod{f_2^{e_2}(x) \cdots f_r^{e_r}(x)},$$

then

$$u^2 - u = u(u-1) \equiv 0 \pmod{f(x)}.$$

Let $e = \overline{u(x)}$, then clearly $e \neq 0$, $e \neq 1$ and $e^k = e$ for any integer $k \geq 1$. Therefore $\mathrm{Tr}(e) = de$.

(c) Since $p \nmid d$, then $1/d \in \mathbb{F}_q$. We deduce that $1 = \mathrm{Tr}(1/d) \in \mathrm{Im}(Tr)$, hence $\mathbb{F}_q \subset \mathrm{Im}(\mathrm{Tr})$. In the case (a), we have $\mathrm{Tr}(e) = e$, $e \neq 0$ and $e^2 = 0$, hence $e \notin \mathbb{F}_q$. In the case (b), we have $e \neq 0$ and $e^2 = e$, hence e is a 0 divisor and then $e \notin \mathbb{F}_q$. But in both cases we have proved that $e \in \mathrm{Im}(\mathrm{Tr})$, hence the inclusion is strict.

Exercise 6.17.
Show that the polynomial $f(x) = x^n + x + 1$ is reducible over $\mathbb{F}_2$ if and only if there exists a non constant polynomial $g(x) \in \mathbb{F}_2[x]$ of degree at most $n-1$ such that $f(x) \mid g(x^2) - g(x)$.

Solution 6.17.
We prove the necessity of the condition. Let $f(x) = f_1(x)f_2(x)$ be a non trivial factorization of $f(x)$ in $\mathbb{F}_2[x]$ with $\gcd(f_1(x), f_2(x)) = 1$. Let $g(x) \in \mathbb{F}_2[x]$ be the unique polynomial of degree $k < n$ (determined by the Chinese remainder theorem) such that $g(x) \equiv 0 \pmod{f_1(x)}$ and $g(x) \equiv 1 \pmod{f_2(x)}$. Then we have $g(x)^2 - g(x) \equiv 0 \pmod{f(x)}$, hence

$$g(x^2) - g(x) \equiv 0 \pmod{f(x)}.$$

The sufficiency of the condition is obvious.

Exercise 6.18.
Let m be a positive integer, q be a prime power and $a \in \mathbb{F}_q^*$. Show that there exists $f(x) \in \mathbb{F}_q[x]$ monic, irreducible of degree m such that $f(0) = a$.

Solution 6.18.
Since the result is obvious if $m = 1$, then we may suppose that $m \geq 2$. Let α be a generator of the group $\mathbb{F}_{q^m}^*$. We have

$$\mathrm{N}_{\mathbb{F}_{q^m}/\mathbb{F}_q}(\alpha) = \alpha\alpha^q \cdots \alpha^{q^{m-1}} = \alpha^{1+q+\cdots+q^{m-1}} = \alpha^{(q^m-1)/(q-1)}.$$

Let $c = \mathrm{N}_{\mathbb{F}_{q^m}/\mathbb{F}_q}(\alpha)$, then $c \in \mathbb{F}_{q^*}$ and c is a generator of this group. It follows that there exists $k \in \{0, \ldots, q-2\}$ such that $a = c^k$. Let $\mathbb{F}_{q^d}$ be the field generated by α^k over $\mathbb{F}_q$, then $d | m$. Suppose that $d < m$

then $(\alpha^k)^{q^d-1} = 1$, hence $\alpha^{(q^d-1)k} = 1$, thus $k(q^d - 1) \equiv 0 \mod q^m - 1$ contradicting the fact that $k \leq q - 2$. We conclude that $d = m$, that is α^k is a primitive element of $\mathbb{F}_{q^m}$ over $\mathbb{F}_q$. Clearly $(-1)^m\alpha^k$ is also a primitive element. Let $g(x) = \mathrm{Irr}((-1)^m\alpha^k, \mathbb{F}_q)$, then $\deg g = m$ and we have

$$\begin{aligned} g(0) &= (-1)^m \mathrm{N}_{\mathbb{F}_{q^m}/\mathbb{F}_q}((-1)^m\alpha^k) \\ &= (-1)^m(-1)^{m^2}(N_{\mathbb{F}_{q^m}/\mathbb{F}_q}(\alpha))^k \\ &= (-1)^{m^2+m}c^k = c^k \\ &= a. \end{aligned}$$

Exercise 6.19.
Let m, n be positive integers and q be a prime power. Let $\alpha \in \mathbb{F}_{q^{mn}}$ be an element generating a normal basis over $\mathbb{F}_q$.

(1) Show that $\mathrm{Tr}_{\mathbb{F}_{q^{mn}}/\mathbb{F}_{q^m}}(\alpha)$ generates a normal basis of $\mathbb{F}_{q^m}$ over $\mathbb{F}_q$.
(2) Show that the preceding result no longer holds if the trace is replaced by the norm.

Solution 6.19.

(1) Let $\gamma = \mathrm{Tr}_{\mathbb{F}_{q^{mn}}/\mathbb{F}_{q^m}}(\alpha)$. Then
$$\gamma = \sum_{j=0}^{n-1} \alpha^{(q^m)^j} = \sum_{j=0}^{n-1} \alpha^{q^{mj}}.$$
For $k = 0, \ldots, m-1$, we have $\gamma^{q^k} = \sum_{j=0}^{n-1} \alpha^{q^{mj+k}}$. Suppose that $\sum_{k=0}^{m-1} a_k\gamma^{q^k} = 0$, where the coefficients a_k belong to $\mathbb{F}_q$. Then $\sum_{k=0}^{m-1}\sum_{j=0}^{n-1} a_k\alpha^{q^{mj+k}} = 0$. The left side of this identity is a linear combination, with coefficients in $\mathbb{F}_q$ of the family $\{\alpha^{q^i}, i = 0, \ldots, mn-1\}$ and each α^{q^i} appears one and only one time. Hence $a_k = 0$ for $k = 0, \ldots, m-1$. Therefore, $\gamma, \gamma^q, \ldots, \gamma^{q^{m-1}}$ are linearly independent over $\mathbb{F}_q$, and then they constitute a normal basis of $\mathbb{F}_{q^m}$ over $\mathbb{F}_q$.
(2) Let α be a root of $f(x) = x^4 + x^3 + x^2 + x + 1$ in an algebraic closure of $\mathbb{F}_2$. It is easy to verify the irreducibility of this polynomial over $\mathbb{F}_2$, therefore $\mathbb{F}_2(\alpha) = \mathbb{F}_{2^4}$. The conjugates of α over $\mathbb{F}_2$ are
$$\alpha,\ \alpha^2,\ \alpha^4 = \alpha^3 + \alpha^2 + \alpha + 1 \text{ and } \alpha^8 = \alpha^3.$$
It is easy to verify that these elements are linearly independent over $\mathbb{F}_2$, hence they form a normal basis of $\mathbb{F}_{2^4}$ over $\mathbb{F}_2$. On the other hand, we have
$$\mathrm{N}_{\mathbb{F}_{2^4}/\mathbb{F}_{2^2}}(\alpha) = \alpha\alpha^4 = \alpha^5 = 1,$$
hence this norm does not generate a normal basis of $\mathbb{F}_{2^2}$ over $\mathbb{F}_2$.

Exercise 6.20.
Let p be a prime number, q be a power of p, $f(x) = x^m - ax - b$ be an irreducible polynomial with coefficients in $\mathbb{F}_q$ and let α be a root of $f(x)$ in an algebraic closure of $\mathbb{F}_q$.

(1) Show that for any $1 \leq k \leq m-2$, $\mathrm{Tr}_{\mathbb{F}_q(\alpha)/\mathbb{F}_q}(\alpha^k) = 0$.
(2) Deduce that if $g(\alpha)$ generates a normal basis of $\mathbb{F}_q(\alpha)$ over $\mathbb{F}_q$, where $g(x) \in \mathbb{F}_q[x]$, $1 \leq \deg g \leq m-2$, then $m \not\equiv 0 \pmod p$ and $g(0) \neq 0$.

Solution 6.20.

(1) Let $\alpha_1 = \alpha, \alpha_2, \ldots, \alpha_m$ be the roots of $f(x)$ in an algebraic closure of $\mathbb{F}_q$. For any $k \in \{1, \ldots, m\}$ let σ_k be the elementary symmetric function of the α_i of degree k and let $s_k = \sum_{i=1}^m \alpha_i^k$. Newton's formula [Small (1991), Prop. 2.3, Chap. 2] reads for $k = 1, \ldots, m$:
$$s_k - \sigma_1 s_{k-1} + \cdots + (-1)^{k-1}\sigma_{k-1}s_1 + (-1)^k k\sigma_k = 0.$$
By induction it is seen that $s_k = 0$ for $k = 1, \ldots, m-2$. Therefore $\mathrm{Tr}_{\mathbb{F}_q(\alpha)/\mathbb{F}_q}(\alpha^k) = 0$.
(2) Let $g(x) = a_0 + a_1 x + \cdots + a_{m-2}x^{m-2}$ be a polynomial of degree at most $m-2$ with coefficients in $\mathbb{F}_q$. Suppose that $m \equiv 0 \pmod p$ or $g(0) = a_0 = 0$, then we have
$$r_{\mathbb{F}_q(\alpha)/\mathbb{F}_q}(g(\alpha)) = ma_0 + a_1 r_{\mathbb{F}_q(\alpha)/\mathbb{F}_q}(\alpha) + \cdots + a_{m-2} r_{\mathbb{F}_q(\alpha)/\mathbb{F}_q}(\alpha^{m-2})$$
hence $g(\alpha), g(\alpha)^q, \ldots, g(\alpha)^{q^m - 1}$ are linearly dependent over $\mathbb{F}_q$. It follows that $g(\alpha)$ does not generate a normal basis of $\mathbb{F}_q(\alpha)$ over $\mathbb{F}_q$.

Exercise 6.21.
Let q be a prime power, E be the set of maps $f : \mathbb{F}_q^n \to \mathbb{F}_q$. Let
$$I = \{g \in \mathbb{F}_q[x_1, \ldots, x_n],\, g(a_1, \ldots, a_n) = 0 \text{ for any } (a_1, \ldots, a_n) \in \mathbb{F}_q^n\}.$$

(1) Let $\psi : \mathbb{F}_q[x_1, \ldots, x_n] \to E$ be the map such that $\psi(g(x_1, \ldots, x_n))$ is the map of $\mathbb{F}_{q^n}$ into $\mathbb{F}_q$ induced by g. Show that ψ is morphism of $\mathbb{F}_q$-algebras of $\mathbb{F}_q[x_1, \ldots, x_n]$ onto E. Deduce $\mathbb{F}_q[x_1, \ldots, x_n]/I \simeq E$.
(2) Let $F_1(x_1, \ldots, x_n), \ldots, F_r(x_1, \ldots, x_n) \in \mathbb{F}_q[x_1, \ldots, x_n]$,
$$V = \{(a_1, \ldots, a_n) \in \mathbb{F}_q^n, F_i(a_1, \ldots, a_n) = 0 \text{ for any } i = 1, \ldots, r\}$$
and
$$J = \{F \in \mathbb{F}_q[x_1, \ldots, x_n], F(a_1, \ldots, a_n) = 0, \text{ for any } (a_1, \ldots, a_n) \in V\}.$$
Show that
$$J = I + F_1\mathbb{F}_q[x_1, \ldots, x_n] + \cdots + F_r\mathbb{F}_q[x_1, \ldots, x_n].$$

Solution 6.21.

(1) We omit the proof of every thing except that ψ is onto.

For any $a = (a_1, \ldots, a_n) \in \mathbb{F}_q^n$, let $f_a(x) = \begin{cases} 1 & \text{if } x = a \\ 0 & \text{if not} \end{cases}$. We first show that the set $\{f_a, a \in \mathbb{F}_q^n\}$ is a basis of E. Suppose that

$$\lambda_1 f_{a^{(1)}} + \cdots + \lambda_m f_{a^{(m)}} = 0, \tag{6.1}$$

where $\lambda_1, \ldots, \lambda_m \in \mathbb{F}_q$ and $a^{(1)}, \ldots, a^{(m)}$ are distinct elements of $\mathbb{F}_q^n$, then for any $j \in \{1, \ldots, m\}$, $\lambda_j = (\lambda_1 f_{a^{(1)}} + \cdots + \lambda_m f_{a(m)})(a^{(j)}) = 0$, hence the set is free over $\mathbb{F}_q$.

Let $f \in E$, then one easily verifies that $f = \sum f(a) f_a$, which proves that the given set generates E over $\mathbb{F}_q$. Now to prove the surjectivity of ψ, it is sufficient to prove that for any $a \in \mathbb{F}_q^n$, f_a is a image of some polynomial, i.e., f_a is a polynomial function. Let $a = (a_1, \ldots, a_n) \in \mathbb{F}_q^n$ and

$$F_a(x_1, \ldots, x_n) = (1 - (x_1 - a_1)^{q-1}) \cdots (1 - (x_n - a_n)^{q-1}),$$

then clearly $\psi(F_a) = f_a$.

(2) Obviously we have $I + \sum F_i \mathbb{F}_q[x_1, \ldots, x_n] \subset J$. Let $H \in J$ and let

$$F(x_1, \ldots, x_n) = 1 - (1 - F_1(x_1, \ldots, x_n)^{q-1}) \cdots (1 - F_r(x_1, \ldots, x_n)^{q-1}).$$

One verifies that

$$F(a_1, \ldots, a_n) = \begin{cases} 0 & \text{if } (a_1, \cdots, a_n) \in V \\ 1 & \text{if not} \end{cases}$$

and $F(x_1, \ldots, x_n) \in \sum F_i \mathbb{F}_q[x_1, \ldots, x_n]$. Let $G = H - HF$. If $(a_1, \ldots, a_n) \in V$, then $G(a_1, \ldots, a_n) = 0$ and if $(a_1, \ldots, a_n) \notin V$, then

$$G(a_1, \ldots, a_n) = H(a_1, \ldots, a_n) - H(a_1, \ldots, a_n) = 0,$$

hence $G(a_1, \ldots, a_n) = 0$.

It follows that $G \in I$ and

$$H = G + HF \in I + \sum_{i=1}^{r} F_i \mathbb{F}_q[x_1, \ldots, x_n].$$

Chapter 7

Permutation polynomials

Exercise 7.1.
Let $r \geq 2$ be an integer and $f(x)$ be a polynomial of degree n with coefficients in $\mathbb{F}_{q^r}$. Suppose that $f(\mathbb{F}_q) \subseteq \mathbb{F}_q$.

(1) If $n \leq q-1$, show that $f(x) \in \mathbb{F}_q[x]$.
(2) If $n \geq q$, let $g(x) \in \mathbb{F}_{q^r}[x]$ be the unique polynomial such that $\deg g \leq q-1$ and $f(x) \equiv g(x) \pmod{x^q - x}$. Show that $g(x) \in \mathbb{F}_q[x]$.

Solution 7.1.

(1) Set $f(x) = a_0 + a_1 x + \cdots + a_{q-1}x^{q-1}$, where $a_i \in \mathbb{F}_{q^r}$ for $i = 0, \ldots, q-1$ and let $\xi_0, \ldots, \xi_{q-1}$ be the distinct elements of $\mathbb{F}_q$ with $\xi_0 = 0$. We may write the equations $f(\xi_i) = a_0 + a_1\xi_i + \cdots + a_{q-1}\xi_i^{q-1}$ in the form

$$\begin{pmatrix} f(\xi_0) \\ f(\xi_1) \\ \cdots \\ f(\xi_{q-1}) \end{pmatrix} = A \begin{pmatrix} a_0 \\ a_1 \\ \cdots \\ a_{q-1} \end{pmatrix},$$

where

$$A = \begin{pmatrix} 1 & 0 & \cdots & 0 \\ 1 & \xi_1 & \cdots & \xi_1^{q-1} \\ \cdots & \cdots & \cdots & \cdots \\ 1 & \xi_{q-1} & \cdots & \xi_{q-1}^{q-1} \end{pmatrix}.$$

We have

$$\operatorname{Det} A = \xi_1 \cdots \xi_{q-1} \begin{vmatrix} 1 & \cdots & \xi_1^{q-2} \\ \cdots & \cdots & \cdots \\ 1 & \cdots & \xi_{q-1}^{q-2} \end{vmatrix},$$

which is non zero, hence A is invertible and A^{-1} has its coefficients in $\mathbb{F}_q$. It follows that $a_i \in \mathbb{F}_q$ for $i = 0, \ldots, q-1$.

(2) Since $f(x) \equiv g(x) \pmod{x^q - x}$, let $h(x) \in \mathbb{F}_{q^r}[x]$ such that
$$f(x) = g(x) + (x^q - x)h(x).$$
We have $g(\mathbb{F}_q) = f(\mathbb{F}_q) \subseteq \mathbb{F}_q$. Since $\deg g \leq q - 1$, then we may apply (1) and conclude that $g(x) \in \mathbb{F}_q[x]$.

Exercise 7.2.
Let q be a prime power, H be a subgroup of $\mathbb{F}_q^\star$ of order d and $s = (q-1)/d$. Let $f(x)$ be a polynomial with coefficients in $\mathbb{F}_q$. Show that $f(H) \subset H$ if and only if there exist a non negative integer r and $g(x) \in \mathbb{F}_q[x]$ such that $f(x) \equiv x^r g(x)^s \pmod{x^d - 1}$.

Solution 7.2.

- **Sufficiency of the condition.**
Notice that $H = \{b^s, b \in \mathbb{F}_q\}$. Let $a \in H$. Since $a^r \in H$ and $g(a)^s \in H$ then $f(a) = a^r g(a)^s \in H$, thus $f(H) \subset H$.
- **Necessity of the condition.**
We recall Lagrange's interpolation theorem [Bourbaki (1950), Application: Formule d'interpollation de Lagrange, Chap. 4.2] or [Ayad (1997), Exercice 1.14]. Let K be a field, $a_1, \ldots, a_n$ be distinct elements of K and $b_1, \ldots, b_n$ be elements of K distinct or not. Then there exists one and only one polynomial $h(x)$ with coefficients in K of degree at most $n - 1$ such that $h(a_i) = b_i$ for $i = 1, \ldots, n$. This polynomial may be written explicitly. Moreover if no condition on the degree of $h(x)$, is prescribed, then any two polynomials $h_1(x)$ and $h_2(x)$ such that $h_1(a_i) = h_2(a_i) = b_i$, satisfy the condition
$$h_2(x) \equiv h_1(x) \left(\operatorname{mod} \prod_{i=1}^{n} (x - a_i) \right).$$
To get this result it is equivalent to prove that $h_2(x) \equiv h_1(x) \pmod{x - a_i}$ for $i = 1, \ldots, n$. But this last claim is true since $h_2(a_i) - h_1(a_i) = 0$ for $i = 1, \ldots, n$. This theorem of Lagrange in mind, write $f(x)$ in the form $f(x) = x^r f_1(x)$, where r is a non negative integer and $f_1(x)$ is a polynomial with coefficients in $\mathbb{F}_q$ such that $f_1(0) \neq 0$. For any $a \in H$, we have $f(a) = a^r f_1(a)$. since a^r and $f(a) \in H$, then $f_1(a) \in H$. According to the remark made above on H, there exists $b_a \in \mathbb{F}_q$ such that $f_1(a) = b_a^s$. Let $g(x) \in \mathbb{F}_q[x]$ such that $g(a) = b_a$ for any $a \in H$. Since, for any $a \in H$, $f(x)$ and $x^r g(x)^s$ take the same value on a, then by Lagrange's theorem, these polynomials are congruent modulo $\prod_{a \in H}(x - a)$ that is modulo $x^d - 1$.

Exercise 7.3.
Let q be a prime power and $n \geq 1$ be an integer. Let $f(X) \in \mathbb{F}_q[X]$ such that $\deg f < q^n$ and $f(X)$ permutes $\mathbb{F}_{q^n}$. Let σ be the permutation induced by $f(X)$ and let $g(X) \in \mathbb{F}_{q^n}[X]$ be the unique polynomial of degree at most $q^n - 1$ such that $g(x) = \sigma^{-1}(x)$ for any $x \in \mathbb{F}_{q^n}$. Show that $g(X) \in \mathbb{F}_q[X]$.

Solution 7.3.

- **First proof.** Let d be the order of σ, then for any $x \in \mathbb{F}_{q^n}$, we have $\sigma^d(x) = f^d(x) = x$ hence $\sigma^{-1}(x) = f^{d-1}(x)$ for any $x \in \mathbb{F}_{q^n}$. It follows that $g(X) \equiv f^{d-1}(X) \pmod{X^{q^n} - X}$ and the proof is complete.
- **Second proof.** Since, for any $x \in \mathbb{F}_{q^n}$ we have $f(g(x)) = x$, then there exists $h(X) \in \mathbb{F}_{q^n}[X]$ such that
$$f(g(x)) = X + (X^{q^n} - X)h(X).$$
For any $\mathbb{F}_q$-automorphism σ of $\mathbb{F}_{q^n}$, denote by $u^\sigma(X)$ the unique polynomial obtained from $u(x)$ by applying σ to the coefficients. Operating σ on the above identity leads to:
$$f^\sigma(g^\sigma(X)) = X + (X^{q^n} - X)h^\sigma(X),$$
hence
$$f(g^\sigma(X)) = X + (X^{q^n} - X)h^\sigma(X).$$
This shows that $g^\sigma(X)$, as a permutation of $\mathbb{F}_{q^n}$, coincides with σ^{-1}. Since $g(X)$ and $g^\sigma(X)$ have their degrees smaller than q^n, then $g^\sigma(X) = g(X)$. It follows that $g(X) \in \mathbb{F}_q[X]$.

Exercise 7.4.
Let n and m be positive integers, a and $b \in \mathbb{F}_q^\star$ and $f(X) = aX^n + bX^m$. Let $d = \gcd(n - m, q - 1)$. Suppose that $(-b/a)^{(q-1)/d} = 1$. Show that $f(X)$ does not permute $\mathbb{F}_q$.

Solution 7.4.
According to Bezout's Identity, there exist u and $v \in \mathbb{Z}$ such that
$$u(n - m) + v(q - 1) = d.$$
Let ξ be a generator of $\mathbb{F}_q^\star$, then ξ^d generates the unique subgroup of order $(q-1)/d$ of $\mathbb{F}_q^\star$ and $-b/a = \xi^{id}$ for some positive integer i. We deduce that
$$-b/a = \xi^{id} = \xi^{iu(n-m)} = \left(\xi^{iu}\right)^{n-m}.$$
This shows that $f(\xi^{iu}) = 0$ which means that the map induced by $f(X)$ is not injective.

Exercise 7.5.

Let q be a prime power. Recall that if $\sigma : \mathbb{F}_q \to \mathbb{F}_q$ is a map, then the unique polynomial $f(x)$ with coefficients in $\mathbb{F}_q$ and of degree at most $q-1$ such that $f(a) = \sigma(a)$ for any $a \in \mathbb{F}_q$ is given by

$$f(x) = \sum_{c \in \mathbb{F}_q} \sigma(c)(1 - (x-c)^{q-1}). \qquad \text{(Eq 1)}$$

(1) Let a and b be distinct elements of $\mathbb{F}_{||}$ and τ the transposition (ab). By using **(Eq 1)**, show that the unique polynomial $f(x) \in \mathbb{F}_q$ representing τ with $\deg f(x) \leq q-1$ is given by

$$f(x) = x + (a-b)(x-a)^{q-1} + (b-a)(x-b)^{q-1}.$$

Show that $\deg f(x) = q-2$.

(2) Let a, b and c be distinct elements of $\mathbb{F}_{||}$ and ρ the 3-cycle (abc). Show that the unique polynomial $g(x) \in \mathbb{F}_q$ representing ρ with $\deg g(x) \leq q-1$ is given by

$$g(x) = x + (a-b)(x-a)^{q-1} + (b-c)(x-b)^{q-1} + (c-a)(x-c)^{q-1}.$$

Show that $\deg g(x) = q-2$ if and only if -3 is not a square in $\mathbb{F}_q$.

Solution 7.5.

(1) Since $f(x)$ represents τ, then $f(x)-x$ represents $\tau - Id_{\mathbb{F}_q}$. Using **(Eq 1)**, we get

$$f(x) - x = (a-b)(x-a)^{q-1} + (b-a)(x-b)^{q-1},$$

hence

$$f(x) = x + (a-b)(x-a)^{q-1} + (b-a)(x-b)^{q-1}.$$

It is clear from this identity that the coefficient of x^{q-1} is zero. This could be also proved by arguing that $f(x)$ is a permutation polynomial and thus its degree is not a divisor of $q-1$. The coefficient α of x^{q-2} is given by

$$\alpha = -a(a-b)(q-1) + -b(b-a)(q-1),$$

that is $\alpha = (a-b)^2$, hence non zero. It follows that $\deg f(x) = q-2$.

(2) Here $g(x) - x$ represents $\rho - Id_{\mathbb{F}_q}$. By **(Eq 1)**, we obtain

$$g(x) - x = (a-b)(x-a)^{q-1} + (b-c)(x-b)^{q-1} + (c-a)(x-c)^{q-1},$$

hence

$$g(x) = x + (a-b)(x-a)^{q-1} + (b-c)(x-b)^{q-1} + (c-a)(x-c)^{q-1}.$$

As in **(1)**, the degree of $g(x)$ is not equal to $q-1$. The coefficient α of x^{q-2} is given by

$$\alpha = a(a-b) + b(b-c) + c(c-a).$$

We have $\alpha = 0$ if and only if a is a root of the polynomial $x^2 - (b+c)x + b^2 + c^2 - bc$. The discriminant of this polynomial is given by $D = -3(b-c)^2$. Thus $\alpha \neq 0$ if and only if -3 is not a square in $\mathbb{F}_q$.

Exercise 7.6.
Let K be a field of characteristic $p \geq 0$, $f(x)$ and $g(y)$ be non-constant polynomials with coefficients in K. If $p > 0$, suppose that $f(x) \notin K[x^p]$ or $g(x) \notin K[x^p]$.

(1) Show that $f(x) - g(y)$ is square free.
(2) Suppose that $K = \mathbb{F}_q$ and $f(x)$ is a permutation polynomial over $\mathbb{F}_q$. Show that $(f(x) - f(y))/(x-y)$ has no factor of the form $ax + by + c$ in $\mathbb{F}_q[x, y]$ with $a \neq 0$ or $b \neq 0$.

Solution 7.6.

(1) Suppose that $h^2(x,y) | f(x) - g(y)$, where $h(x,y)$ is a non-constant polynomial with coefficient in K, then $h(x,y)|f'(x)$ and $h(x,y)|g'(y)$. From the assumptions, one of the polynomials $f'(x)$ and $g'(y)$ is non zero. Without loss of generality, suppose that $f'(x) \neq 0$, then $h(x,y) \in K[x]$. It follows that $h(x,y)$ divides the content in $K[x]$ of $f(x) - g(y)$, which is equal to 1, hence a contradiction. We conclude that $f(x) - g(y)$ is square free.
(2) Suppose that $(ax+by+c) \mid (f(x)-f(y))/(x-y)$, with $a, b, c \in K$, $a \neq 0$ or $b \neq 0$, then by **(1)**, $ax + by + c \neq \lambda(x-y)$ for any $\lambda \in K$. Suppose that $b \neq 0$, then for any $x \in \mathbb{F}_q$, we have $f(x) - f((-c-ax)/b) = 0$, which implies that $f(x)$ is not a permutation polynomial.

Exercise 7.7.
Let $\sigma : \mathbb{F}_{2^n} \to \mathbb{F}_{2^n}$ be a map such that $\sigma(0) = \sigma(e) = 0$ for some $e \in \mathbb{F}_{2^n}^\star$ and $\sigma(i) \neq \sigma(j)$ for all $i, j \in \mathbb{F}_{2^n}^\star$ with $i \neq j$. Let $f(X) \in \mathbb{F}_{2^n}[X]$ be the unique polynomial of degree at most $2^n - 1$ such that $f(x) = \sigma(x)$ for any $x \in \mathbb{F}_{2^n}$. Show that $\deg f = 2^n - 1$.

Solution 7.7.
Since the map σ is not injective, then it is not surjective, hence there exists $\alpha \notin \sigma(\mathbb{F}_{2^n})$. Let

$$P(X) = f(X) + \alpha[(X+e)^{2^n-1} + 1],$$

then clearly $P(e) = \alpha$ and for any $x \neq e$, $P(x) = f(x)$, hence $P(X)$ is a permutation polynomial. Obviously, by the very definition of $P(X)$, $\deg P \leq 2^n - 1$. Since it is a permutation polynomial its degree must be at most $2^n - 2$. Since $\alpha \neq 0$, then the leading term of the polynomial $f(X)$ is αX^{2^n-1} and its degree is $2^n - 1$.

Second proof. We will use the following formula, which is Lagrange interpolation formula [Small (1991), Remarks, Chap. 2]

$$f(x) = \sum_{c \in \mathbb{F}_q} \sigma(c)(1 - (x - c)^{q-1}).$$

We obtain:

$$f(x) = \sum_{c \in \mathbb{F}_q, c \notin \{0,e\}} \sigma(c)(1 - (x - c)^{q-1}).$$

The coefficient λ of x^{q-1} is given by $\lambda = \sum_{c \notin \{0,e\}} \sigma(c)$. The assumptions shows that there exists $d \in \mathbb{F}_{2^n}^{\star}$ such that $d \neq \sigma(c)$ for any $c \in \mathbb{F}_{2^n}$. We may write λ in the form $\lambda = d_1 + d_2 + \cdots + d_{q-2}$, where the d_i are distinct and $d_i \notin \{0, d\}$. There exists i_0 such that $d_{i_0} = -d$. Since the opposite of any d_i for $i \neq i_0$ is equal to some d_j, then it follows that $\lambda = d_{i_0} \neq 0$, that is the coefficient x^{q-1} is nonzero.

Exercise 7.8.
Let $p \geq 3$ be a prime number and $\sigma = (01)$ be the transposition of $\{0, \ldots, p-1\}$. Let $f(X)$ be the unique polynomial with coefficients in $\mathbb{F}_p$ of degree at most $p - 1$ such that for any $x \in \mathbb{F}_p$, $f(x) = \sigma(x)$. Show that

$$f(X) = X + \prod_{k=2}^{p-1}(X - k).$$

Solution 7.8.
We have

$$f(x) - x = \sigma(x) - x = \begin{cases} 1 & \text{if } x = 0 \\ -1 & \text{if } x = 1 \\ 0 & \text{if } x \neq 0, 1 \end{cases}.$$

Since $\deg f \leq p - 1$, then

$$f(X) - X = (aX + b) \prod_{k=2}^{p-1}(X - k),$$

where $a, b \in \mathbb{F}_p$. We have

$$1 = f(0) = (-1)^{p-2}(p-1)!b \quad \text{and}$$
$$-1 = f(1) - 1 = (a+b)\prod_{k=2}^{p-1}(k-1) = -(a+b)(p-2)!,$$

hence $b = 1$ and $a = 0$.

Second proof. We will use the following formula, which is Lagrange interpolation formula [Small (1991), Remarks, Chap. 2]

$$f(X) - X = \sum_{c \in \mathbb{F}_p} (\sigma(c) - c)(1 - (X-c)^{p-1}).$$

We have

$$\sigma(c) - c = \begin{cases} 1 & \text{if } c = 0 \\ -1 & \text{if } c = 1 \\ 0 & \text{if } c \neq 0, 1 \end{cases},$$

hence

$$f(X) - X = (X-1)^{p-1} - X^{p-1}.$$

The degree of this polynomial is equal to $p-2$, and for any $a \in \{2, \dots, p-1\}$, we have $f(a) - a = 0$. It follows that $f(X) - X = \prod_{k=2}^{p-1}(X-k)$ and then

$$f(X) = X + \prod_{k=2}^{p-1}(X-k).$$

Exercise 7.9.
Let p be a prime number and $n \geq 2$ be an integer. Let $f : \mathbb{Z}/p^n\mathbb{Z} \to \mathbb{Z}/p^n\mathbb{Z}$ be the map such that $f(0) = 0$ and $f(x) = 1$ for $x \neq 0$. Show that there exists no polynomial $F(X) \in \mathbb{Z}/p^n\mathbb{Z}[X]$ such that $f(x) = F(x)$ for any $x \in \mathbb{Z}/p^n\mathbb{Z}$.

Solution 7.9.
Suppose the contrary. Since $f(0) = 0$, then $F(x) = a_m x^m + \cdots + a_1 x$, where $a_i \in \mathbb{Z}/p^n\mathbb{Z}$. Let $\hat{F}(x) \in \mathbb{Z}[x]$ be a lift of $F(x)$. We have $\hat{F}(p) \equiv 0 \pmod{p}$, which contradicts the assumption $F(p) = 1$.

Exercise 7.10.
Let $n \geq 2$ be an integer. Show that n is prime if and only if there exists a polynomial $f(x)$ with integral coefficients such that $f(0) \equiv 1 \pmod{n}$, $f(1) \equiv 0 \pmod{n}$ and $f(k) \equiv k \pmod{n}$ for $k = 2, \dots, n-1$, i.e. f induces the permutation (01) on the set $\mathbb{Z}/n\mathbb{Z}$.

Solution 7.10.
The necessity of the condition is well known. We prove the sufficiency. By contradiction, suppose that n is not prime and let p be a prime factor of n. Let $f(x) \in \mathbb{Z}[x]$ satisfying the prescribed conditions. Then $f(p) = a_0 + \cdots + a_m p^m \equiv 0 \pmod{p}$, hence $p \mid a_0 = f(0) \equiv 1 \pmod{p}$, which is a contradiction.

Exercise 7.11.
Let p be a prime number and let $f(x) \in \mathbb{Z}[x]$ be a polynomial permuting the elements of $\mathbb{Z}/p\mathbb{Z}$. Show that $f'(x)$ permutes the elements of $\mathbb{Z}/p\mathbb{Z}$ or there exists $a \in \mathbb{Z}$ such that $f(x) + a(x^p - x)$ permutes the elements of $\mathbb{Z}/p^2\mathbb{Z}$.

Solution 7.11.
Suppose that $f'(x)$ does not permute the elements of $\mathbb{Z}/p\mathbb{Z}$, then there exists $a \in \{0, \ldots, p-1\}$ such that for any $x \in \mathbb{Z}$ $f'(x) - a \not\equiv 0 \pmod{p}$. Let $g(x) = f(x) + a(x^p - x)$. It is clear that $g(x)$ induces the same function over $\mathbb{Z}/p\mathbb{Z}$ as $f(x)$, hence it permutes the elements of this ring. Notice that $g'(x) = f'(x) - a$. Let $b \in \mathbb{Z}$ and consider the equation $g(y) \equiv b \pmod{p^2}$. Let x_0 such that $g(x_0) \equiv b \pmod{p}$. We show that there exists a solution of the form $y_0 = x_0 + \lambda p$, where λ is an integer to be determined. We have

$$\begin{aligned} g(y_0) &\equiv g(x_0) + \lambda p g'(x_0) \pmod{p^2} \\ &\equiv b \pmod{p^2} \\ &\equiv b + \mu p + \lambda p g'(x_0) \pmod{p^2}. \end{aligned}$$

We deduce that $\mu + \lambda g'(x_0) \equiv 0 \pmod{p}$ and this equation determines λ and then y_0.

Exercise 7.12.
Let p be a prime number, e be a positive integer and $q = p^e$. Suppose that $q \geq 3$.

(1) Let $a \in \mathbb{F}_q^\star$ and

$$f_a(x) = -a^2\left[\left((x-a)^{q-2} + a^{-1}\right)^{q-2} - a\right]^{q-2}.$$

Show that the polynomial $f_a(x)$ induces the transposition $(0\,a)$ of $\mathbb{F}_q$.

(2) Deduce that any permutation of $\mathbb{F}_q$ is the composition of permutations induced by x^{q-2} and linear polynomials with coefficients in $\mathbb{F}_q$.

Solution 7.12.

(1) We have

$$f_a(0) = -a^2\left[\left((-a)^{q-2} + a^{-1}\right)^{q-2} - a\right]^{q-2}$$
$$= -a^2(-a)^{q-2} = -a^2(-a^{-1}) = a,$$

$$f_a(a) = -a^2\left[\left(a^{-1}\right)^{q-2} - a\right]^{q-2} = 0$$

and for $b \in \mathbb{F}_q$, $b \neq 0$, $b \neq a$,

$$\begin{aligned} f_a(b) &= -a^2\left[\left((b-a)^{q-2} + a^{-1}\right)^{q-2} - a\right]^{q-2} \\ &= -a^2\left[\left((b-a)^{-1} + a^{-1}\right)^{q-2} - a\right]^{q-2} \\ &= -a^2\left[\left(\frac{b}{a(b-a)}\right)^{q-2} - a\right]^{q-2} \\ &= -a^2\left[\frac{a(b-a)}{b} - a\right]^{q-2} \\ &= -a^2\left[\frac{-a^2}{b}\right]^{q-2} = b. \end{aligned}$$

This implies that $f_a(x)$ as a (polynomial) function of $\mathbb{F}_q$ into itself represents the transposition $(0\,a)$.

(2) Clearly $f_a(x)$ as a permutation is the composition of permutations induced by x^{q-2} and by permutations induced by linear polynomials. Since the transpositions of the form $(0\,a)$, when a runs in $\mathbb{F}_q^\star$ generates the set of permutations of $\mathbb{F}_q$, then the permutation induced by x^{q-2} together with those generated by linear polynomials generate the full symmetric group operating on $\mathbb{F}_q$.

Exercise 7.13.

Let $f(x)$ be a non constant polynomial with integral coefficients, p be a prime number and e be a positive integer. Let $a \in \mathbb{Z}$ such that $f(a) \equiv 0 \pmod{p^e}$.

(1) Show that the number of solutions b distinct modulo p^{e+1} of the equation $f(x) \equiv 0 \pmod{p^{e+1}}$ is equal to

(i) 0 if $f'(a) \equiv 0 \pmod{p}$ and $f(a) \not\equiv 0 \pmod{p^{e+1}}$.
(ii) 1 if $f'(a) \not\equiv 0 \pmod{p}$.
(iii) p if $f'(a) \equiv 0 \pmod{p}$ and $f(a) \equiv 0 \pmod{p^{e+1}}$.

(2) Let $n \geq 2$ be an integer. Deduce that $f(x)$ permutes the elements of $\mathbb{Z}/p^n\mathbb{Z}$ if and only if $f(x)$ permutes the elements of $\mathbb{Z}/p\mathbb{Z}$ and $f'(a) \not\equiv 0 \pmod{p}$ for any $a \in \mathbb{Z}$.

Solution 7.13.

(1) Let $b \in \mathbb{Z}$ such that $b = a + cp^e$ with $c \in \mathbb{Z}$, then the equation $f(x) \equiv 0 \pmod{p^{e+1}}$ is equivalent to $f(a)/p^e + cf'(a) \equiv 0 \pmod{p}$. If $f'(a) \not\equiv 0 \pmod{p}$, then this condition determines c in one and only one way. Therefore *(ii)* is proved. If $f'(a) \equiv 0 \pmod{p}$, then the same equation is equivalent to $f(a)/p^e \equiv 0 \pmod{p}$, i.e. $f(a) \equiv 0 \pmod{p^{e+1}}$. Thus **(i)** and **(iii)** are proved.

(2) Suppose that $f(x)$ permutes the elements of $\mathbb{Z}/p^n\mathbb{Z}$. Let $b \in \mathbb{Z}$. Consider the equation $f(x) \equiv b \pmod{p}$. Let $a \in \mathbb{Z}$ such that $f(a) \equiv b \pmod{p^n}$, then $f(a) \equiv b \pmod{p}$, hence the equation modulo p has a solution and $f(x)$ permutes the elements of $\mathbb{Z}/p\mathbb{Z}$. Let $a \in \mathbb{Z}$, $b = f(a)$ and $g(x) = f(x) - b$, then a is a solution of the equations $g(x) \equiv 0 \pmod{p}$ and $g(x) \equiv 0 \pmod{p^n}$. Applying **(1)** to $g(x)$, we conclude that we have the case **(ii)**, hence $f'(a) = g'(a) \not\equiv 0 \pmod{p}$.
We prove the converse. Let $b \in \mathbb{Z}$. Consider the equation $f(x) \equiv b \pmod{p^n}$. Let $a \in \mathbb{Z}$ such that $f(a) \equiv b \pmod{p}$. Applying **(ii)** to $g(x) = f(x) - b$, we obtain that there exists $a_2 \in \mathbb{Z}$ such that $a_2 \equiv a \pmod{p}$ and $f(a_2) \equiv b \pmod{p^2}$. By induction, we may prove that there exists $a_n \in \mathbb{Z}$ such that $f(a_n) \equiv b \pmod{p^n}$, hence f permutes $\mathbb{Z}/p^n\mathbb{Z}$.

Exercise 7.14.
Let q be a prime power and n be a positive integer. Let

$$G_n = \{f(x) \in \mathbb{F}_{q^n}[x], \deg f < q^n, f \text{ permutes } \mathbb{F}_{q^n}\}$$

and let

$$H_n = \{f(x) \in \mathbb{F}_q[x], \deg f < q^n,\ f \in G_n\}.$$

(1) Show that H_n is a subgroup of G_n.
(2) Let $\phi : \mathbb{F}_{q^n} \to \mathbb{F}_{q^n}$, be the Frobenius automorphism and let $f \in G_n$. Show that $f \in H_n$ if and only if $f \circ \phi = \phi \circ f$.
(3) Let d be a positive divisor of n and let

$$M_d = \{\alpha \in \mathbb{F}_{q^n}, \deg \alpha \text{ (over } \mathbb{F}_q) = d\}.$$

Let $\alpha_0 \in M_d$. Show that $M_d = \{f(\alpha_0),\ f \in H_n\}$.

Solution 7.14.

(1) $H_n \neq \emptyset$ since x belongs to this set. Obviously the composition of two elements of H_n belongs to H_n. According to **Exercise 7.3**, the symmetric of an element of H_n is an element of H_n. Therefore H_n is a subgroup of G_n.

(2) Suppose that $f \in H_n$, then

$$\phi(f(x)) = (f(x))^q = f(x^q) = f(\phi(x)).$$

Conversely let $f(x) = \sum a_i x^i \in G_n$. Suppose that $f(\phi(x)) = \phi(f(x))$, then $f(x^q) = f(x)^q$ for any $x \in \mathbb{F}_{q^n}$, i.e. $\sum a_i (x^q)^i = \sum a_i^q (x^q)^i$ for any $x \in \mathbb{F}_{q^n}$. Therefore $\sum a_i y^i = \sum a_i^q y^i$ for any $y \in \mathbb{F}_{q^n}$. Now the polynomials $f(y)$ and $g(y) = \sum a_i^q y^i$ both have their degrees smaller than q^n and are equal as functions over $\mathbb{F}_{q^n}$, hence they are equal as polynomials, which implies $f \in H_n$.

(3) Let $N_d = \{f(\alpha_0), f \in H_n\}$. We first show that $N_d \subset M_d$, that is $f(\alpha_0) \in M_d$ for any $f \in H_n$. For this purpose we prove that for any field F such that $\mathbb{F}_q \subset F \subset \mathbb{F}_{q^n}$, and any $f \in H_n$, we have $f(F) = F$. Let f and F be such objects and let $\gamma \in F$, then since $f(x) \in \mathbb{F}_q[x]$, $f(\gamma) \in F$. Therefore $f(F) \subset F$. Since F is finite and f is injective, then $f(F) = F$. We have $\mathbb{F}_{q^d} = M_d \cup_{\substack{e|d \\ e<d}} \mathbb{F}_{q^e}$, hence $f(\alpha_0) \in M_d$, thus $N_d \subset M_d$. We prove the reverse inclusion. Let $\beta \in M_d$, we show that there exists $f \in H_n$ such that $\beta = f(\alpha_0)$. If β is a conjugate of α_0 over $\mathbb{F}_q$, say $\beta = \phi^s(\alpha_0)$, then $f = \overline{\phi^s}$ is suitable, where $\overline{\phi^s}$ is the reduction of ϕ^s modulo $x^{q^n} - x$. Suppose that β and α_0 are not conjugate over $\mathbb{F}_q$. We define f as follows. First we define the images of the conjugates of α_0 and of β by putting them in the second line hereafter.

$$\alpha_0, \phi(\alpha_0), \ldots, \phi^{d-1}(\alpha_0), \beta, \phi(\beta), \ldots, \phi^{d-1}(\beta)$$

$$\beta, \phi(\beta), \ldots, \phi^{d-1}(\beta), \alpha_0, \phi(\alpha_0), \ldots, \phi^{d-1}(\alpha_0).$$

If x is not a conjugate of α_0 nor β, we set $f(x) = x$. Clearly f is a permutation of $\mathbb{F}_{q^n}$ and since $f \circ \phi = \phi \circ f$, then by **(2)**, $f \in H_n$.

Exercise 7.15.

Let q be a prime power and $a_1, \ldots, a_q$ be the distinct elements of $\mathbb{F}_q$. For any $b \in \mathbb{F}_q^\star$, let ϕ_b be the permutation of $\{1, \ldots, q\}$ such that for any $i \in$\{1,...,q\}, $\phi_b(i) = j$, where j is the unique index such that $ba_i = a_j$. For any $b \in \mathbb{F}_q^\star$ and any $F(x_1, \ldots, x_q) \in \mathbb{F}_q[x_1, \ldots, x_q]$, define the polynomial $F^b(x_1,\ldots,x_q)$ by $F^b(x_1,\ldots,x_q) = F(x_{\phi_b(1)},\ldots,x_{\phi_b(q)})$. If $F^b(x_1,\ldots,x_q) = F(x_1,\ldots,x_q)$, then F is said to be invariant under the action of b.

(1) Let $f(x)$ and $F(x_1,\ldots,x_q)$ be polynomials with coefficients in $\mathbb{F}_q$ of degrees d and D respectively. Let G be a subgroup of $\mathbb{F}_q^\star$ of order r such that F is invariant under the action of any $b \in G$. If $dD < r$, show that
$$F(f(a_1),\ldots,f(a_q)) = F(f(0),\ldots,f(0)).$$
(2) Let $g(x) \in \mathbb{F}_q[x]$, $d = \deg g$ and let $\sigma_1,\ldots,\sigma_q \in \mathbb{F}_q$ such that
$$\prod_{i=1}^{q}(x - g(a_i)) = x^q - \sigma_1 x^{q-1} + \cdots + (-1)^q\sigma_q.$$
Show that $\sigma_k = 0$ for $1 \leq k < (q-1)/d$.
(3) Let ξ be a root of $x^3 + x + 1$ in an algebraic closure of $\mathbb{F}_2$.
(a) Show that $\mathbb{F}_8 = \{a_i,\ i = 1,\ldots,8\}$, where $a_1 = 0$ and $a_i = \xi^{i-2}$ for $i = 2,\ldots,8$.
(b) Let $F(x_1,\ldots,x_8) = x_1\sum_{i=2}^{8} x_i$ and let $G = \mathbb{F}_8^\star$. Show that F is invariant under the action of any $b \in G$.
(c) Show that for any $f(x) \in \mathbb{F}_8[x]$, such that $1 \leq \deg f \leq 3$, we have
$$F(f(a_1),\ldots,f(a_8)) = f(0)^2.$$

Solution 7.15.

(1) Let
$$u(x) = F(f(a_1x),\ldots,f(a_qx)),$$
then $u(x) \in \mathbb{F}_q[x]$ and $\deg u \leq \deg f \deg F = dD < r$. For any $b \in G$, we have
$$\begin{aligned} u(b) &= F(f(a_1b),\ldots,f(a_qb)) \\ &= F(f(a_{\phi_b(1)}),\ldots,f(a_{\phi_b(q)}) \\ &= F^b(f(a_1),\ldots,f(a_q)) \\ &= F(f(a_1),\ldots,f(a_q)) \\ &= u(1). \end{aligned}$$
It follows that the polynomial $u(x) - u(1)$ has at least r distinct roots. Taking into account of the degree of $u(x) - u(1)$, we conclude that $u(x) - u(1) = 0$. Thus $u(x) = u(1)$. In particular, we have $u(0) = u(1)$, that is
$$F(f(0),\ldots,f(0)) = F(f(a_1),\ldots,f(a_q)).$$

(2) For any integer k such that $1 \le k < (q-1)/d$, we apply **(1)** with $f(x) = g(x)$, $F(x_1, \ldots, x_q) = \sum_{1 \le i_1 < \ldots < i_k \le q} x_{i_1} \cdots x_{i_k}$ and $G = \mathbb{F}_q^\star$.

Here

$$\deg f = d,\ \deg F = k,\ |G| = q-1 \text{ and } dk < |G|,$$

so that the assumptions in **(1)** are satisfied. We conclude that

$$\begin{aligned}\sigma_k &= \sum_{1 \le i_1 < \cdots < i_k \le q} g(a_{i_1}) \cdots g(a_{i_k}) \\ &= F(g(a_1), \ldots, g(a_q)) \\ &= F(g(0), \ldots, g(0)) \\ &= \binom{q}{k} g(0)^k.\end{aligned}$$

It is easy to see that $\binom{q}{k} = 0$ and the result follows.

(3)(a) Since x^3+x+1 has no root in $\mathbb{F}_2$, then this polynomial is irreducible over $\mathbb{F}_2$ and $\mathbb{F}_8 = \mathbb{F}_2(\xi)$. The order of ξ in $\mathbb{F}_8^\star$ must divide 7. Since $\xi \notin \mathbb{F}_2$, then this order is equal to 7. Thus the cyclic group $\mathbb{F}_8^\star$ of order 7 is generated by ξ and then $\mathbb{F}_8 = \{a_i,\ i = 1, \ldots, 8\}$ with $a_1 = 0$ and $a_i = \xi^{i-2}$ for $i = 2, \ldots, 8$.

(b) It is sufficient to prove that F is invariant under the action of ξ. We have

$$\begin{aligned}\xi a_1 &= 0 = a_1 \\ \xi a_2 \xi &= \xi = a_3 \\ \xi a_3 &= \xi^2 = a_4 \\ \xi a_4 &= \xi^3 = a_5 \\ \xi a_5 &= \xi^4 = a_6 \\ \xi a_6 &= \xi^5 = a_7 \\ \xi a_7 &= \xi^6 = a_8 \\ \xi a_8 &= 1 = a_2,\end{aligned}$$

so that ϕ_ξ is the cycle $(2\,3\,4\,5\,6\,7\,8)$.

(c) We have

$$F^\xi(x_1, \ldots, x_8) = x_1(x_3+x_4+x_5+x_6+x_7+x_8+x_2) = F(x_1, \ldots, x_8),$$

hence F is invariant under the action of the elements of G. If f satisfies the condition $1 \le \deg f \le 3$, then $\deg f \deg F = 2 \deg f \le 6 < |G| = 7$. Therefore, we may apply **(1)** and we obtain

$$F(f(a_1), \ldots, f(a_8)) = F(f(0), \ldots, f(0) = f(0)^2.$$

Exercise 7.16.
Let p be a prime number, r be a positive integer and $q = p^r$.

(1) Show that for $k = 0, \ldots, q-1$, $\binom{q-1}{k} \not\equiv 0 \pmod{p}$.
(2) Let $f(x) \in \mathbb{F}_q[x]$. Denote by $\mu_p(f)$, the smallest positive integer k if it exists, such that $\sum_{c \in \mathbb{F}_q} f(c)^k \neq 0$. If not set $\mu_p(f) = \infty$.

(a) If $\mu_p(f) < \infty$, show that $\mu_p(f) \leq q-1$.
(b) Suppose that $\mu_p(f) < \infty$. Let

$$F(x) = \sum_{k=1}^{q-1} (-1)^{q-k} \binom{q-1}{k} \left(\sum_{c \in \mathbb{F}_q} f(c)^k \right) x^{q-1-k} \in \mathbb{F}_q[x].$$

For any $a \in \mathbb{F}_q$, let $N_f(a) = |f^{-1}\{a\}|$. Show that, for any $a \in \mathbb{F}_q$, $F(a) = N_f(a)$ and that $\deg F(x) = q - 1 - \mu_p(f)$.
Deduce that $\mu_p(f) \leq |f(\mathbb{F}_q)| - 1$.
(c) Suppose that $\mu_p(f) < \infty$. Show that $\mu_p(f) \leq q - 1 - q/p$.

Solution 7.16.

(1) If $k = 0$ then $\binom{q-1}{k} = 1 \not\equiv 0 \pmod{p}$. Suppose that $k \geq 1$, then

$$\binom{q-1}{k} = \frac{(q-k)}{k} \frac{(q-(k-1))}{k-1} \cdots \frac{(q-2)}{2} \frac{(q-1)}{1}.$$

Suppose that $p^i \mid\mid k-j$ for some $0 \leq j \leq k-1$, then $i < r$ and $p^i \mid\mid q-(k-j)$. Therefore $p \nmid \binom{q-1}{k}$.
(2)(a) Let k be a positive integer such that $\sum_{c \in \mathbb{F}_q} f(c)^k \neq 0$. Divide k by q and obtain the equality $k = qt_1 + s_1$, with $0 \leq s \leq q-1$. We have

$$\sum_{c \in \mathbb{F}_q} f(c)^k = \sum_{c \in \mathbb{F}_q} f(c)^{t_1 + s_1} \neq 0.$$

We repeat the euclidean divisions by q and obtain the sequence (s_i, t_i) until the stage r where $s_r + t_r < q$. The integer $u = s_r + t_r < q$ satisfies the condition $\sum_{c \in \mathbb{F}_q} f(c)^u \neq 0$ and then $\mu_p(f) \leq q-1$.

(b) We have

$$\begin{aligned} N_f(a) &= \sum_{c\in\mathbb{F}_q} (1-(f(c)-a)^{q-1}) \\ &= -\sum_{c\in\mathbb{F}_q} (f(c)-a)^{q-1} \\ &= -\sum_{c\in\mathbb{F}_q}\sum_{k=0}^{q-1}\binom{q-1}{k} f(c)^k(-a)^{q-1-k} \\ &= \sum_{k=0}^{q-1}(-1)^{q-k}\binom{q-1}{k}\sum_{c\in\mathbb{F}_q} f(c)^k(a)^{q-1-k}. \end{aligned}$$

Since $\sum_{c\in\mathbb{F}_q} f(c)^k = 0$ for $k=0$, then $F(a) = N_f(a)$. From **(1)**, we know that $\binom{q-1}{k} \not\equiv 0 \pmod p$ for any $k = 0,\ldots,q-1$, hence $\deg F(x) = q-1-\mu_p(f)$. To prove the last statement in **(b)**, we use the fact that the number of roots of a non zero polynomial in a field is at most equal to the degree of the polynomial. Since $N_f(a) = 0$ for any $a \notin f(\mathbb{F}_q)$, then $q - |f(\mathbb{F}_q)| \le q-1-\mu_p(f)$ and $\mu_p(f) \le |f(\mathbb{F}_q)| - 1$.

(c) By **(2) (b)**, the polynomial function induced by F takes its values in $\mathbb{F}_p$. Hence, for any $a \in \mathbb{F}_q$, $F(a)^p - F(a) = 0$. Therefore $(x^q - x) \mid (F(x)^p - F(x))$. Since, for any $a \in \mathbb{F}_q$, $F(a) = N_f(a)$ then $F(x)$ is not the zero polynomial. We deduce that

$$q \le \deg(F(x)^p - F(x)) = p(q-1-\mu_p(f))$$

and then $\mu_p(f) \le q-1-q/p$.

Exercise 7.17.

Let q be a prime power, $f(x) = a_{q-2}x^{q-2} + \cdots + a_1x + a_0$ be a polynomial with coefficients in $\mathbb{F}_q$ of degree no greater than $q-2$. Let

$$A = \begin{pmatrix} a_0 & a_1 & \cdots & a_{q-2} \\ a_{q-2} & a_0 & \cdots & a_{q-3} \\ a_{q-3} & a_{q-2} & \cdots & a_{q-4} \\ \cdots & \cdots & \cdots & \cdots \\ a_1 & a_2 & \cdots & a_0 \end{pmatrix}$$

and let $P(x)$ be the characteristic polynomial of A.

(1) Show that $A = \sum_{i=0}^{q-2} a_i J^i$, where

$$J = \begin{pmatrix} 0 & 1 & 0 & \dots & 0 \\ 0 & 0 & 1 & \dots & 0 \\ \dots & \dots & \dots & \dots & \dots \\ 0 & 0 & 0 & \dots & 1 \\ 1 & 0 & 0 & \dots & 0 \end{pmatrix}.$$

(2) Show that, for any $\alpha \in \mathbb{F}_q^\star$,

$$A \begin{pmatrix} 1 \\ \alpha \\ \alpha^2 \\ \dots \\ \alpha^{q-2} \end{pmatrix} = f(\alpha) \begin{pmatrix} 1 \\ \alpha \\ \alpha^2 \\ \dots \\ \alpha^{q-2} \end{pmatrix}.$$

(3) Show that $f(x)$ permutes $\mathbb{F}_q$ if and only if $P(x) = (x - a_0)^{q-1} - 1$.

Solution 7.17.

(1) We may write A in the form $A = \sum_{i=0}^{q-2} a_i A_i$, where A_i is a $(q-1) \times (q-1)$ matrix with coefficients in $\mathbb{F}_q$, $A_0 = I$ and $A_1 = J$. Let $\{e_1, \dots, e_{q-1}\}$ be the canonical basis of $\mathbb{F}_q^{q-1}$, then viewing the linear map associated to J as a permutation of $\{e_1, \dots, e_{q-1}\}$, it is seen that this permutation is the cycle $(e_1\, e_{q-1}\, e_{q-2} \cdots e_2)$ and then $A_i = J^i$.

(2) For any $\alpha \in K^\star$, we have

$$J \begin{pmatrix} 1 \\ \alpha \\ \alpha^2 \\ \dots \\ \alpha^{q-2} \end{pmatrix} = \alpha \begin{pmatrix} 1 \\ \alpha \\ \alpha^2 \\ \dots \\ \alpha^{q-2} \end{pmatrix},$$

hence by **(1)**,

$$A \begin{pmatrix} 1 \\ \alpha \\ \alpha^2 \\ \dots \\ \alpha^{q-2} \end{pmatrix} = f(\alpha) \begin{pmatrix} 1 \\ \alpha \\ \alpha^2 \\ \dots \\ \alpha^{q-2} \end{pmatrix}.$$

(3) Suppose that $f(x)$ permutes $\mathbb{F}_q$. From **(2)**, we conclude that for any $\alpha \in K^\star$, $f(\alpha)$ is an eigenvalue of A. Since $f(\alpha_1) \neq f(\alpha_2)$ for $\alpha_1 \neq \alpha_2$, then

$$P(x) = \prod_{\alpha \in \mathbb{F}_q^\star} (x - f(\alpha)) = \prod_{\beta \in \mathbb{F}_q \setminus \{a_0\}} (x - \beta) = (x - a_0)^{q-1} - 1.$$

Conversely, suppose that $P(x) = (x - a_0)^{q-1} - 1$, then $P(x) = \prod_{\beta \in \mathbb{F}_q \setminus \{a_0\}} (x - \beta)$, hence all the eigenvalues of A have multiplicity 1 and belong to $\mathbb{F}_q$. Therefore any eigeinspace of A has dimension 1 over $\mathbb{F}_q$. Let α_1 and α_2 be distinct elements of $\mathbb{F}_q$ such that $f(\alpha_1)$ and $f(\alpha_2) \neq a_0$. From **(2)**, we conclude that $f(\alpha_1)$ (resp. $f(\alpha_2)$) is an eigenvalue of A and $(1, \alpha_1, \ldots, \alpha_1^{q-2})$ (resp. $(1, \alpha_2, \ldots, \alpha_2^{q-2})$) is a related eigenvector. Since $(1, \alpha_1, \ldots, \alpha_1^{q-2})$ and $(1, \alpha_2, \ldots, \alpha_2^{q-2})$ are linearly independent over $\mathbb{F}_q$, then $f(\alpha_1) \neq f(\alpha_2)$. We deduce that

$$|\{f(\alpha),\, \alpha \in \mathbb{F}_q,\, f(\alpha) \neq a_0\}| = q - 1$$

and then

$$\mathbb{F}_q = \{f(0)\} \cup \{f(\alpha),\, \alpha \in \mathbb{F}_q,\, f(\alpha) \neq a_0\}.$$

This implies that $f(x)$ is surjective, thus $f(x)$ permutes $\mathbb{F}_q$.

Exercise 7.18.

Let q be a prime power. For any polynomial $g(x)$ with coefficients in $\mathbb{F}_q$, denote by $\overline{g(x)}$ be the unique polynomial with coefficients in $\mathbb{F}_q$ such that $\deg \overline{g(x)} \leq q - 1$ and $g(x) \equiv \overline{g(x)} \pmod{x^q - x}$.

(1) Let $f(x)$ be a polynomial with coefficients in $\mathbb{F}_q$ of degree at most $q-1$. Show that the following conditions are equivalent.

 (i) $f(x)$ permutes $\mathbb{F}_q$.
 (ii) For any $c \in \mathbb{F}_q$, $\overline{(f(x) - c)^{q-1}}$ is monic of degree $q - 1$.
 (iii) For any $c \in \mathbb{F}_q$, $\overline{(f(x) - c)^{q-1}} \neq 1$.
 (iv) For any $c \in \mathbb{F}_q$, $\overline{(f(x) - f(c))^{q-1}} = (x - c)^{q-1}$.

 Hint. One may use the following result [Small (1991), Remarks, Chap. 2]. Let $\sigma : \mathbb{F}_q \to \mathbb{F}_q$ be a map. Then the unique polynomial $g(x)$ with coefficients in $\mathbb{F}_q$ such that $\deg g \leq q - 1$ and $\sigma(a) = g(a)$ for any $a \in \mathbb{F}_q$ is given by

$$g(x) = \sum_{a \in \mathbb{F}_q} \sigma(a)(1 - (x - a)^{q-1}). \qquad \text{(Eq 1)}$$

(2) Fix $c \in \mathbb{F}_q$ and consider the following propositions.

 (a) The equation $f(x) = c$ has one and only one solution in $\mathbb{F}_q$.
 (b) The polynomial $\overline{(f(x) - c)^{q-1}}$ is monic of degree $q - 1$.

 Show that **(a)** $\Rightarrow$ **(b)** but the converse does not hold.

Solution 7.18.

(1) • **(i)** $\Leftrightarrow$ **(ii)**. For any $c \in \mathbb{F}_q$, set

$$\overline{(f(x)-c)^{q-1}} = \sum_{k=0}^{q-1} b_k^{(c)} x^k.$$

Using the given hint, we obtain

$$\sum_{k=0}^{q-1} b_k^{(c)} x^k = \overline{(f(x)-c)^{q-1}} = \sum_{a \in \mathbb{F}_q} (f(a)-c)^{q-1}(1-(x-a)^{q-1}).$$

Hence

$$\begin{aligned} b_{q-1}^{(c)} &= -\sum_{a \in \mathbb{F}_q} (f(a)-c)^{q-1} \\ &= -|\{a \in \mathbb{F}_q, f(a) \neq c\}| \\ &= -(q - |\{a \in \mathbb{F}_q, f(a) = c\}|) \\ &= |\{a \in \mathbb{F}_q, f(a) = c\}| \\ &= N(c), \end{aligned}$$

where $N(c)$ denotes the number of solutions of the equation $f(x) = c$. We deduce that if $f(x)$ permutes $\mathbb{F}_q$ then $N(c) = 1$ for any $c \in \mathbb{F}_q$. Hence $b_{q-1}^{(c)} = 1$, which means that $\overline{(f(x)-c)^{q-1}}$ is monic of degree $q-1$. Conversely if **(ii)** holds for any $c \in \mathbb{F}_q$, then $N(c) \equiv 1 \pmod{p}$, hence $N(c) \neq 0$ for any $c \in \mathbb{F}_q$, which means that the polynomial map $f(x)$ is onto, thus $f(x)$ permutes $\mathbb{F}_q$.

- **(i)** $\Rightarrow$ **(iii)**. The proposition **(iii)** is equivalent to saying that the function $a \to (f(a)-c)^{q-1}$ is not the constant function $a \to 1$. Since the equation $f(x) = c$ has a solution $a \in \mathbb{F}_q$, then $(f(a)-c)^{q-1} = 0 \neq 1$.
- **(iii)** $\Rightarrow$ **(i)**. Let $c \in \mathbb{F}_q$, then $\overline{(f(x)-c)^{q-1}} \neq 1$, hence there exists $a \in \mathbb{F}_q$ such that $(f(a)-c)^{q-1} \neq 1$, thus $f(a)-c = 0$, which means that $f(x)$ permutes $\mathbb{F}_q$.
- **(i)** $\Rightarrow$ **(iv)**. We must show that the functions

$$\sigma_1 : a \to (a-c)^{q-1} \text{ and } \sigma_2 : a \to (f(a)-f(c))^{q-1}$$

coincide. If $a = c$, then $\sigma_1(a) = 0 = \sigma_2(a)$. If $a \neq c$, then $\sigma_1(a) = 1 = \sigma_2(a)$.
- **(iv)** $\Rightarrow$ **(i)**. Let a and $c \in \mathbb{F}_q$. Suppose that $f(a) = f(c)$, then

$$0 = (f(a)-f(c))^{q-1} = (a-c)^{q-1},$$

hence $a = c$, which implies that $f(x)$ is a permutation polynomial.

(2) • **(a)** ⇒ **(b)**. Use similar arguments as in the proof of **(i)** ⇔ **(ii)**.
- For the converse let $q = 4$, $K = \mathbb{F}_4 = \mathbb{F}_2(\alpha)$, where α is a root of x^2+x+1. Let $f(x) = x^3+\alpha^2x^2+\alpha x$, then $f(0) = f(1) = f(\alpha) = 0$ and $f(\alpha^2) = 1$, so that the equation $f(x) = 0$ has 3 solutions in $\mathbb{F}_4$. On the other hand, we have $\overline{f(x)^3} = f(x)$. It follows that **(b)** holds while **(a)** does not.

Exercise 7.19.
Let q be a prime power, G be the group of permutation polynomials $f(x) \in \mathbb{F}_q[x]$ of $\mathbb{F}_q$ such that $\deg f = 1$. Show that G is 2-transitive on $\mathbb{F}_q$ but not 3-transitive if $q > 3$.

Solution 7.19.
Let $\{a_1, a_2\}$ and $\{b_1, b_2\}$ be subsets of $\mathbb{F}_q$ of cardinality 2. Let $f(x) = \alpha x+\beta$ be an element of G. Then the conditions $f(a_1) = b_1$ and $f(a_2) = b_2$ are equivalent to the following system of linear equations $\alpha a_1 = b_1$, $\alpha a_2 = b_2$. The determinant of this system is equal to $a_1 - a_2$, hence non zero. It follows that it has a unique solution in $\mathbb{F}_q$, given by $\alpha = (b_1 - b_2)/(a_1 - a_2) \neq 0$ and $\beta = (a_1b_2 - a_2b_1)/(a_1 - a_2)$. Therefore G is 2-transitive.

Suppose that $q > 3$. Let $\alpha \in \mathbb{F}_q$ such that $\alpha \neq 0$ and $\alpha \neq 1$. Suppose that there exists $f(x) \in G$ such that $f(0) = 1$, $f(1) = 0$ and $f(\alpha) = \alpha$. Then $f(x)$ may be computed by application of the Lagrange interpolation formula [Small (1991), Remarks, Chap. 2]. We find

$$\begin{aligned} f(x) &= f(0)(x-1)(x-\alpha)/(-1)(-\alpha) \\ &\quad + f(1)x(x-\alpha)/(1)(1-\alpha) \\ &\quad + f(\alpha)x(x-1)/\alpha(\alpha-1) \\ &= \frac{2\alpha-1}{\alpha(\alpha-1)}x^2 + \frac{1-\alpha-\alpha^2}{\alpha(\alpha-1)}x + 1. \end{aligned}$$

If q is a power of 2, then the coefficient of x^2 is equal to $-1/\alpha(\alpha-1)$ hence non zero. If q is odd, we choose $\alpha \neq 1/2$ and we get the same conclusion. In any case $\deg f = 2$. Therefore G is not 3-transitive.

Remark. Here is an alternative proof of the fact that G is 2-transitive. Given $\{a_1, a_2\}$ and $\{b_1, b_2\}$ be subsets of $\mathbb{F}_q$ of cardinality 2, by Lagrange interpolation theorem [Bourbaki (1950), Application: Formule d'interpollation de Lagrange, Chap. 4.2] or [Ayad (1997), Exercice 1.14], there exists $f(x) \in \mathbb{F}_q[x]$ such that $\deg f \leq 1$, $f(a_1) = b_1$ and $f(a_2) = b_2$. Since $b_1 \neq b_2$, then $f(x)$ is not constant. Therefore $f(x) \in G$.

Chapter 8

Transcendental extensions, Linearly disjoint extensions, Luroth's theorem

Exercise 8.1.

(1) Let K be a field, $(u_n)_{n\in\mathbb{N}}$ be the sequence defined by $u_0 = 0$, $u_{2n} = u_n$ and $u_{2n+1} = 1 - u_n$. Show that the values of u_n are 0 or 1. This sequence is called the Thue-Morse sequence.

(2) For any non negative integer n, let $n = \sum_{i\geq 0} a_i 2^i$ be its 2-adic expansion and let $v_n = (\sum_{i\geq 0} a_i) \pmod 2$, where $\pmod 2$ means that v_n is the residue modulo 2 of $\sum_{i\geq 0} v_i$ belonging to $\{0, 1\}$. Show that $v_n = u_n$ for any $n \in \mathbb{N}$.

(3) Suppose that $K = \mathbb{F}_2$ and let $(u_n)_{n\in\mathbb{N}}$ be the sequence defined in **(1)**, let $f(x) = \sum_{n\geq 0} u_n x^n \in \mathbb{F}_2[[x]]$. Show that $f(x)$ is algebraic over $\mathbb{F}_2(x)$.

Solution 8.1.

(1) Easy by induction on n.

(2) We have $v_0 = u_0 = 0$. Let $n = a_0 + a_1 2 + \cdots + a_m 2^m$, then

$$2n = a_0 2 + a_1 2^2 + \cdots + a_m 2^{m+1} \quad \text{and}$$
$$2n + 1 = 1 + a_0 2 + a_1 2^2 + \cdots + a_m 2^{m+1},$$

hence

$$v_n = \sum_{i=0}^{m} a_i,$$
$$v_{2n} = \sum_{i=0}^{m} a_i = v_n \quad \text{and}$$
$$v_{2n+1} = 1 + \sum_{i=0}^{m} a_i = 1 - \sum_{i=0}^{m} a_i = 1 - v_n.$$

Therefore $v_n = u_n$ for any $n \geq 0$. The preceding computations are made in $\mathbb{F}_2 = \{0, 1\}$.

(3) We have

$$\begin{aligned} f(x) &= \sum_{n\geq 0} u_n x^n = \sum_{n\geq 0} u_{2n} x^{2n} + \sum_{n\geq 0} u_{2n+1} x^{2n+1} \\ &= \sum_{n\geq 0} u_n x^{2n} + \sum_{n\geq 0} (1+u_n) x^{2n+1} \\ &= \sum_{n\geq 0} u_n x^{2n} + x\sum_{n\geq 0} u_n x^{2n} + \sum_{n\geq 0} x^{2n+1} \\ &= (1+x)\sum_{n\geq 0} u_n x^{2n} + x\sum_{n\geq 0} x^{2n} \\ &= (1+x)f(x^2) + x\frac{1}{1+x^2} = (1+x)f(x)^2 + \frac{x}{(1+x)^2}. \end{aligned}$$

Hence

$$(1+x)^3 f(x)^2 + (1+x)^2 f(x) + x = 0.$$

This shows that $f(x)$ is algebraic over $\mathbb{F}_2(x)$.

Exercise 8.2.

Let K be a field, E and F be algebraic extensions of K contained in a field Ω. Let $\alpha_1, \ldots, \alpha_n \in \Omega$ be algebraically independent over K. Suppose that $E(\alpha_1, \ldots, \alpha_n) = F(\alpha_1, \ldots, \alpha_n)$. Show that $E = F$.

Solution 8.2.

Let K_0 be the algebraic closure of K in

$$L = E(\alpha_1, \ldots, \alpha_n) = F(\alpha_1, \ldots, \alpha_n).$$

Since E is algebraically closed in $E(\alpha_1, \ldots, \alpha_n)$ then $K_0 = E$. A similar argument shows that $K_0 = F$, hence $E = F$.

Second method. Since $\alpha_1, \ldots, \alpha_n$ are algebraically independent over E, then $E(\alpha_1, \ldots, \alpha_n)$ is isomorphic to a field of rational functions in n variables say $E(x_1, \ldots, x_n)$ over E and E is the set of constant rational functions. Similarly $F(\alpha_1, \ldots, \alpha_n)$ is isomorphic to a field of rational functions in n variables $F(x_1, \ldots, x_n)$ over F and F is the set of constant rational functions. Since these fields are equal, then $E = F$.

Third method. It is sufficient to prove that $E \subset F$. Let $\beta \in E$, then $\beta \in \bar{K} = \bar{F}$ and $\beta \in F(\alpha_1, \ldots, \alpha_n)$. Therefore $\beta \in \bar{K} \cap F(\alpha_1, \ldots, \alpha_n) = F$, since the fields which are the terms of the intersection are linearly disjoint over F [Lang (1965), Pro. 3, Chap. 10.5].

Exercise 8.3.
Let K be an algebraically closed field, E be an extension of K of finite type. Let $L \subset E$ be an algebraically closed field. Show that $L \subset K$.

Hint. One may use the fact that an intermediate field between a given field and an extension of finite type is of finite type over the base field (see [Clark (1993), Theorem 11.18] or [Schinzel (2000), Th. 2, Chap. 1.1]).

Solution 8.3.
Suppose, by contradiction, that L is not contained in K, then there exists $\alpha \in L$ such that α is transcendental over K. Let P be the prime subfield of K and $\bar{P}$ its algebraic closure in E, then $\bar{P} \subset K \cap L$. Let $R = \bar{P}(\alpha)$ and $\bar{R}$ be the algebraic closure of R in E (or in L) and let $M = K \cdot \bar{R}$. We may visualize these fields in the following diagram.

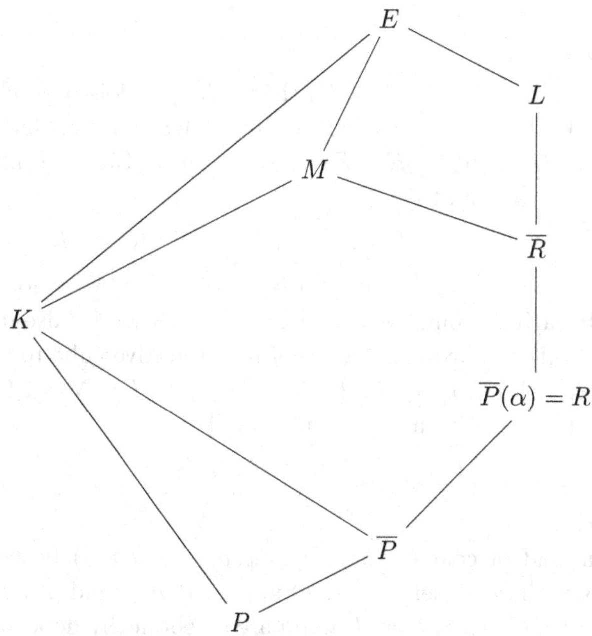

We have $R = \bar{P}(\alpha) = \bar{P}P(\alpha)$, hence

$$\bar{R} = \bar{P}\overline{P(\alpha)} = \overline{P(\alpha)} \text{ and } M = K \cdot \overline{P(\alpha)}.$$

Any element $\beta \in M$ has the form $\beta = (\sum ab)/(\sum cd)$ where $a, c \in K$ and $b, d \in \overline{P(\alpha)}$. Clearly a, b, c and d are algebraic over $K(\alpha)$, hence M is algebraic over $K(\alpha)$. Moreover since $K \subset M \subset E$ and E is of finite type over K, then so is M. It follows M is algebraic of finite degree over $K(\alpha)$. Let $q = [M : K(\alpha)]$. Let l be a prime number such that $l > q$ and let $f(x) = x^l - \alpha \in K(\alpha)[x]$. This polynomial is irreducible over $K(\alpha)$. Otherwise α would be an l-th power in $K(\alpha)$, implying that α would be algebraic over K contradicting our assumptions. Let μ be a root of $f(x)$, then $\mu \in R \subset M$. We deduce that

$$q = [M : K(\alpha)] \geq [K(\alpha)(\mu) : K(\alpha)] = l > q,$$

hence a contradiction.

Exercise 8.4.
Let K be a field, E and F be finite extensions of K. Suppose that F is separable over K and let N be the normal closure of F over K. If $E \cap N = K$, show that E and F are linearly disjoint over K. Show that the converse is false.

Solution 8.4.
Since N is a Galois extension of K, then EN is Galois over E, see [Lang (1965), Th. 4, Chap. 8.1]. We deduce that $\text{Gal}(EN, E) \simeq \text{Gal}(N, E \cap N)$. Since $E \cap N = K$, then $\text{Gal}(EN, E)$ is isomorphic to $\text{Gal}(N, K)$. We deduce $[EN : E] = [N : K]$ and then

$$[EN : K] = [EN : E][E : K] = [N : K][E : K]$$

which means that E and N are linearly disjoint over K. It follows that E and F are linearly disjoint over K. That the converse is false may be seen through the following example, where j is a primitive cube root of unity:

$K = \mathbb{Q}$, $E = \mathbb{Q}(j)$, $F = \mathbb{Q}(\sqrt[3]{2})$, $N = \mathbb{Q}(j, \sqrt[3]{2})$, $E \cap N = \mathbb{Q}(j) \neq \mathbb{Q}$.

Here E and F are linearly disjoint over $\mathbb{Q}$ since

$$[E \cdot F : \mathbb{Q}] = 6 = [E : \mathbb{Q}][F : \mathbb{Q}].$$

Exercise 8.5.
Let K be a field of characteristic 0, $f(x, y)$ and $g(x, y)$ be non-constant polynomials with coefficients in K. Suppose that f and g are absolutely irreducible over K. Show that f and g are algebraically dependent over K if and only if there exists $(a, b) \in K^2$ such that $a \neq 0$ and $g = af + b$.

Solution 8.5.
The sufficiency of the condition is obvious. We prove the necessity. Let $\phi(u, v) \in K[u, v]$ be the unique (up to multiplication by an element of K^*) irreducible polynomial such that $\phi(f, g) = 0$. Write ϕ in the form

$$\phi(u, v) = a_n(u)v^n + a_{n-1}(u)v^{n-1} + \cdots + a_0(u),$$

where $a_i(u) \in K[u]$ and $a_n(u)a_0(u) \neq 0$. Let $\alpha \in \overline{K(x)}$ such that $g(u, \alpha) = 0$, where $\overline{K(x)}$ denotes an algebraic closure of $K(x)$. Since $\phi(f, g) = 0$, then $\phi(f(x, \alpha), g(x, \alpha)) = 0$, hence $a_0(f(x, \alpha)) = 0$. Since $a_0(u) \neq 0$, then $f(x, \alpha)$ is algebraic over K. Therefore $f(x, \alpha) \in \bar{K} \cap K(x, \alpha)$. Since $g(x, y)$ is absolutely irreducible then $\bar{K} \cap K(x, \alpha) = K$ (see **Exercise 11.7 (2)**), thus $f(x, \alpha) \in K$. Let $\lambda \in K$ such that $f(x, \alpha) = \lambda$. Since $g(x, y)$ is the minimal polynomial of α over $K(x)$, then $g(x, y) \mid f(x, y) - \lambda$ in $K(x)[y]$.

Let $q(x, y) \in K(x)[y]$ such that

$$f(x, y) - \lambda = g(x, y)q(x, y).$$

Write $q(x, y)$ in the form $q(x, y) = \frac{A(x)}{D(x)} q_1(x, y)$, where $gcd(A(x), D(x)) = 1$ and $A(x)/D(x) = \text{cont}(q)$. We have

$$\text{cont}(f(x, y) - \lambda) = \frac{A(x)}{D(x)} \text{cont}(g) = \frac{A(x)}{D(x)}.$$

Since $\text{cont}(f(x, y) - \lambda) \in K[x]$, then $D(x) \mid A(x)$, that is $D(x) \in K^*$. Therefore $g(x, y) \mid f(x, y) - \lambda$ for some $\lambda \in K$. Similarly, we may prove the existence of $\mu \in K$ such that $f(x, y) \mid g(x, y) - \mu$. We deduce that $\deg f = \deg g$. Let $a \in K^*$ such that

$$g(x, y) - \mu = af(x, y),$$

then $g(x, y) = af(x, y) + \mu$.

Exercise 8.6.

Let K be a field, $u(s)$ and $v(t)$ be polynomials with coefficients in K, not both constant. Let $\phi(s, t)$ be an irreducible polynomial over K of the form $\phi(s, t) = as^m + \sum_{\substack{i<m \\ j<n}} a_{ij} s^i t^j + bt^n$ with a, $b \neq 0$ and let S and T such that $\phi(S, T) = 0$.

(1) Show that, up to multiplication by a constant in K^*, there exists an irreducible polynomial $P(x, y) \in K[x, y]$ such that $P(u(S), v(T)) = 0$. Show that if $F(x, y) \in K[x, y]$ satisfies the condition $F(u(S), v(T)) = 0$, then $P(x, y)$ divides $F(x, y)$ in $K[x, y]$.

(2) Let I be the ideal of $K[s, t, x, y]$ generated by $u(s) - x, v(t) - y, \phi(x, y)$. Show that $I \cap K[x, y] = (P(x, y))$.

(3)(a) Show that there exist λ, $\mu \in K^*$ and an irreducible polynomial $G(t, x) \in K[t, x]$ such that

$$\text{Res}_s(\phi(s, t), u(s) - x) = \lambda G(t, x)^{k_1} \quad \text{(Eq 1)}$$

and

$$\text{Res}_t(G(t, x), v(t) - y) = \mu P(x, y)^{k_2}, \quad \text{(Eq 2)}$$

where k_1 and k_2 are positive integers. Show that $G(x, y)$ and $P(x, y)$, viewed as elements of $K(x)[y]$, (resp. $K(y)[x]$), have their leading coefficients in K.

(b) Show that $k_1 k_2 = [K(S, T) : K(u(S), v(T)]$.

(c) Show that

$$\deg u \deg_t \phi(s, t) \deg_x P(x, y) = \deg v \deg_s \phi(s, t) \deg_y P(x, y) \tag{Eq 3}$$

and

$$k_1 k_2 \mid \gcd(\deg u \deg_t \phi(s, t), \deg v \deg_s \phi(s, t)). \tag{Eq 4}$$

(d) Show that $\phi(x, y) \mid P(u(x), v(y)$ in $K[x, y]$.

(4) Let $a(z)$ and $b(z)$ be polynomials with coefficients in K, not both constant. Show that, up to multiplication by a constant in $K^\star$, there exists one and only one irreducible polynomial $P(x, y) \in K[x, y]$ such that $P(a(z), b(z)) = 0$. Show that $P(x, y)$ is given by

$$\text{Res}_z(a(z) - x, b(z) - y) = \mu P(x, y)^k, \tag{Eq 5}$$

where $\mu \in K^\star$ and $k = [K(z) : K(a(z), b(z))]$.

Solution 8.6.

(1) Consider the chain of fields

$$K \subset K(u(S), v(T)) \subset K(S, T).$$

Since S and T are algebraically dependent over K, then so are $u(S)$ and $v(T)$. Hence, there exists an irreducible polynomial $P(x, y) \in K[x, y]$ such that $P(u(S), v(T)) = 0$. Let $C(x)$ be the leading coefficient of P, as a polynomial in y, then $P(x, y)/C(x)$ is the minimal polynomial of $v(T)$ over $K(u(S))$. We deduce that if a polynomial $F(x, y)$ with coefficients in K satisfies the condition $F(u(S), v(T)) = 0$, then $P(x, y)/C(x)$ divides $F(x, y)$ in $K(x)[y]$, thus $P(x, y)$ divides $F(x, y)$ in $K(x)[y]$. Set $F(x, y) = P(x, y)Q(x, y)$, where $Q(x, y) \in K(x)[y]$. We have

$$\text{cont}(F) = \text{cont}(P)\,\text{cont}(Q) = \text{cont}(Q),$$

hence $\text{cont}(Q) \in K[x]$ and then $Q(x, y) \in K[x, y]$. This implies that $P(x, y) \mid F(x, y)$ in $K[x, y]$

Remark. We may prove this property of divisibility in the following way. Let $\theta : K[x, y] \to K[S, T]$ be the unique K-morphism of algebras such that $\theta(x) = u(S)$ and $\theta(y) = v(T)$, then

$$K[x, y]/\operatorname{Ker}\psi \simeq \operatorname{Im}\theta = K[u(S), v(T)].$$

Therefore $\operatorname{Ker}\theta$ is a non zero prime ideal of $K[x,y]$ containing $P(x,y)$. We prove that $\operatorname{Ker}\theta = P(x,y)K[x,y]$. Since the image of θ is not a field, then $\operatorname{Ker}\theta$ is not maximal. Let $\mathcal{M}$ be a maximal ideal of $K[x,y]$ containing $\operatorname{Ker}\theta$. Consider the chain of prime ideals

$$\{0\} \subset P(x,y)K[x,y] \subset \operatorname{Ker}\theta \subset \mathcal{M}.$$

Since the Krull dimension of $K[x,y]$ is equal to 2, then $\operatorname{Ker}\theta = (P)$. Hence $P(x,y) \mid F(x,y)$ for any $F(x,y) \in K[x,y]$ satisfying the condition $F(u(S), v(T)) = 0$.

(2) Let $\alpha : K[s,t,x,y] \to K[S,T]$ be the unique K-morphism of algebras such that

$$\alpha(s) = S,\ \alpha(t) = T,\ \alpha(x) = u(S) \text{ and } \alpha(y) = v(T).$$

Then $\operatorname{Ker}\alpha$ is a prime ideal of $K[s,t,x,y]$. We first show that $\operatorname{Ker}\alpha = I$. Clearly $I \subset \operatorname{Ker}\alpha$. For the reverse inclusion, let $f(s,t,x,y)$ be an element of $\operatorname{Ker}\alpha$. By successive Euclidean divisions, we obtain

$$f = (x - u(s))q_1 + r_1, r_1 = (y - v(t))q_2 + r_2 \quad \text{and} \quad r_2 = \phi(s,t)q_3 + r_3,$$

where $q_1 \in K[s,t,x,y]$, $r_1 \in K[s,t,y]$, $q_2 \in K[s,t,y]$, $r_2 \in K[s,t]$, $q_3, r_3 \in K[s,t]$ and $\deg_s r_3 < \deg_s \phi$. Since $r_1, r_2, r_3 \in \operatorname{Ker}\alpha$, we have $r_3(S,T) = 0$, hence $\phi(s,t)$ divides $r_3(s,t)$ and then $r_3 = 0$. We deduce successively that $r_2 \in I$, $r_1 \in I$ and $f \in I$. Thus $\operatorname{Ker}\alpha \subset I$. We now have

$$I \cap K[x,y] = \operatorname{Ker}\alpha \cap K[x,y] = (P(x,y)).$$

(3)(a) Let $l = \deg u$ and $\beta_1, \ldots, \beta_m$ be the roots of $\phi(s,t)$ in $\overline{K(t)}$. Then

$$\begin{aligned}
\operatorname{Res}_s(\phi(s,t), u(s) - x) &= a^l \prod_{i=1}^{m} (u(\beta_i) - x) \\
&= (-1)^m a^l \prod_{i=1}^{m} (x - u(\beta_i)) \\
&= (-1)^m a^l \operatorname{Char}(u(\beta_1), K(t), x) \\
&= (-1)^m a^l \operatorname{Irr}(u(\beta_1), K(t), x)^{k_1},
\end{aligned}$$

where $\operatorname{Char}(u(\beta_1), K(t), x)$ is the characteristic polynomial of $u(\beta_1)$ in the extension $K(t,\beta_1)/K(t)$ and

$$k_1 = [K(t,\beta_1) : K(t, u(\beta_1))].$$

Since β_1 is integral over $K[t]$, then so is $u(\beta_1)$, thus $\operatorname{Irr}(u(\beta_1), K(t), x) \in K[t,x]$. If we set $G(t,x) = \operatorname{Irr}(u(\beta_1), K(t), x)$

and $\lambda = (-1)^m a^l$, then formula **(Eq 1)** is proved and it is seen that the leading term of $G(t,x)$ viewed as a polynomial in x does not depend on t.
We prove that $G(t,x)$ considered as a polynomial in t has a leading term independent from x. Let u_l be the leading coefficient of $u(s)$ and $\gamma_1,\dots,\gamma_l$ be the roots of $u(s)-x$ in $\overline{K(x)}$. Then

$$\begin{aligned}\mathrm{Res}_s(\phi(s,t),u(s)-x) &= (-1)^{ml}\prod_{i=1}^{l}(\phi(\gamma_i,t) \\ &= (-1)^{ml}u_l^m\prod_{i=1}^{l}(bt^n+b_{n-1}(\gamma_i)t^{n-1}+\dots+b_0(\gamma_i) \\ &= (-1)^{ml}u_l^m t^{nl}+S_{nl},\end{aligned}$$

where S_{nl} is the sum of the terms of degree in t less than nl. This proves that the leading term of $G(t,x)$, as a polynomial in t does not depend on x. Using similar arguments as in the computation of the preceding resultant, we conclude that there exist a positive integer k_2, a non zero constant $\mu \in K$ and an irreducible polynomial $H(x,y)\in K[x,y]$ such that

$$\mathrm{Res}_t(G(t,x),v(t)-y) = \mu H(x,y)^{k_2}.$$

Moreover $k_2 = [K(x,\delta_1) : K(x)]$, where δ_1 satisfies the condition $G(\delta_1,x)=0$. To obtain **(Eq 2)**, we must verify that $H(x,y) = \gamma P(x,y)$ for some $\gamma\in K^\star$, that is $H(u(S),v(T))=0$.
Since the polynomials in s, $\phi(s,T)$ and $u(s)-u(S)$ have the common root $s=S$, then

$$\mathrm{Res}_s(\phi(s,T),u(s)-u(S) = \lambda G(t,u(S))^{k_1} = 0,$$

hence $G(T,u(S))=0$. Because the polynomials $G(t,u(s))$ and $v(t)-v(T)$ have the common root $t=T$, then

$$\mathrm{Res}_t(G(t,u(S)),v(t)-v(t)) = \mu H(u(S),v(T))^{k_2} = 0,$$

hence $H(u(S),v(T))=0$. Thus $H(x,y)=\gamma P(x,y)$. Using similar arguments than those used above for $G(x,y)$, we conclude that $P(x,y)$, viewed as an element of $K(x)[y]$, (resp. $K(y)[x]$), has its leading coefficient in K.

(b) Set $t=T$ in **(Eq 1)** and obtain

$$\mathrm{Res}_s(\phi(s,T),u(s)-x) = \lambda G(T,x)^{k_1}.$$

This shows that $G(T,x) = \mathrm{Irr}(u(S), K(T), x)$, hence

$$\begin{aligned} k_1 &= \deg_s \phi(s,t)/\deg_x G(t,x) \\ &= [K(S,T) : K(T)]/[K(T,u(S)) : K(T)] \\ &= [K(T,S) : K(T,u(S))]. \end{aligned}$$

Set $x = u(S)$ in **(Eq 2)** and obtain

$$\mathrm{Res}_t(G(t,u(S)), v(t) - y) = \mu P(u(S), y)^{k_2},$$

hence

$$P(u(S), y) = \mathrm{Irr}(v(T), K(u(S), y))$$

and then

$$k_2 = [K(u(S),T) : K(u(S), v(T))].$$

It follows that

$$k_1 k_2 = [K(S,T) : K(u(S), v(T))].$$

(c) From the diagram, hereafter, it is easy to deduce **(Eq 3)** and **(Eq 4)**.

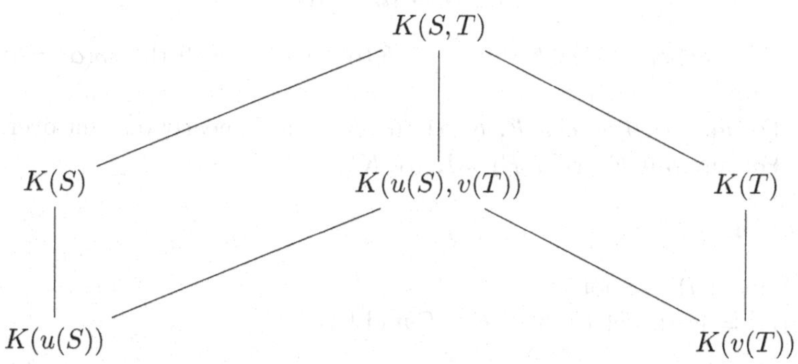

(d) We have

$$P(u(S), v(T) = 0 = \phi(S,T).$$

By **(1)**, we conclude that $\phi(x,y) \mid P(u(x), v(y)$ in $K[x,y]$.

(4) Let $\phi(s,t) = s - t$. Here **(Eq 1)** reads

$$\mathrm{Res}_s(\phi(s,t), a(s) - x) = \mathrm{Res}_s(s - t, a(s) - x) = a(t) - x$$

so that $G(t,x) = a(t) - x$. According to **(2)**, we have

$$\mathrm{Res}_t(G(t,x), b(t) - y) = \mu P(x,y)^{k_2},$$

where $P(x, y)$ is the unique (up to multiplication by a constant in $K^\star$) irreducible polynomial with coefficients in K such that $P(a(z), b(z)) = 0$. Since $k_1 = 1$, then by **(3)(a)**,

$$k = k_2 = k_1 k_2 = [K(z) : K(a(z), b(z))].$$

Remark. This exercise shows that for any point $(s, t) \in \overline{K}^2$ lying on the curve $\phi(x, y) = 0$, the point $(u(s), v(t))$ lyes on the curve $P(x, y) = 0$. Is this last curve completely recovered by the points $(u(s), v(t))$? In **(3)(d)** it is proved that $\phi(x, y) \mid P(u(x), v(y)$ in $K[x, y]$. May it happen that the equality $\phi(x, y) = aP(u(x), v(y)$ holds with $a \in K^\star$.

Exercise 8.7.
Let K be a field, Ω an algebraic closure of K, E and F be extensions of K contained in Ω. Show that the following assertions are equivalent.

(i) E and F are linearly disjoint over K.
(ii) For any $\alpha \in E$, $K(\alpha)$ and F are linearly disjoint over K.
(iii) For any $(\alpha, \beta) \in E \times F$ and any K-embeddings

$$\sigma : K(\alpha) \to \Omega,\ \tau : K(\beta) \to \Omega,$$

there exists a K-embedding $\rho : K(\alpha, \beta) \to \Omega$ such that $\rho(\alpha) = \sigma(\alpha)$ and $\rho(\beta) = \tau(\beta)$.
(iv) For any $(\alpha, \beta) \in E \times F$, $K(\alpha)$ and $K(\beta)$ are linearly disjoint over K.
(v) For any $\alpha \in E$, $\mathrm{Irr}(\alpha, F) = \mathrm{Irr}(\alpha, K)$.

Solution 8.7.

- *(i)* $\Rightarrow$ *(ii)*. Obvious.
- *(ii)* $\Rightarrow$ *(iii)*. Let $(\alpha, \beta) \in E \times F$ and let

$$\sigma : K(\alpha) \to \Omega \quad \text{and} \quad \tau : K(\beta) \to \Omega$$

be K-embeddings. Let $n = [K(\beta) : K]$, then $1, \beta, \ldots, \beta^{n-1}$ are linearly independent over K, hence linearly independent over $K(\alpha)$. Since $\mathrm{Irr}(\beta, K(\alpha)) \mid \mathrm{Irr}(\beta, K)$, then $\mathrm{Irr}(\beta, K(\alpha)) = \mathrm{Irr}(\beta, K)$ and then $[K(\alpha, \beta) : K(\alpha)] = n$. Since $\tau(\beta)$ is a root of $\mathrm{Irr}(\beta, K)$ then $\tau\beta)$ is a conjugate of β over $K(\alpha)$. Therefore there exists an extension ρ of σ to $K(\alpha, \beta)$ such that $\rho(\beta) = \tau(\beta)$.
- *(iii)* $\Rightarrow$ *(iv)*. We show that

$$[K(\alpha, \beta) : K(\alpha)] = [K(\beta) : K].$$

Since $K(\beta)$ (resp. $K(\alpha,\beta)$) is separable over K (resp. $K(\alpha)$) it is equivalent to prove that

$$[K(\alpha,\beta):K(\alpha)]_s = [K(\beta):K]_s.$$

Let $\tau : K(\beta) \to \Omega$ be any K embedding and let $i : K(\alpha) \to \Omega$ be the canonical injection then by (iii) there exists $\rho : K(\alpha,\beta) \to \Omega$ such that $\rho(\alpha) = i(\alpha) = \alpha$ and $\rho(\beta) = \tau(\beta)$.
Therefore i has at least $[K(\beta):K]$ extensions to $K(\alpha,\beta)$. Since this number of extensions is smaller than $[K(\beta):K]$, we obtain

$$[K(\alpha,\beta):K(\alpha)]_s = [K(\beta):K] = [K(\beta):K]_s.$$

- $(iv) \Rightarrow (v)$. Let $\alpha \in E$, $n = [K(\alpha):K]$ and let $\{\beta_1,\ldots,\beta_s\}$ be the set of coefficients of $\mathrm{Irr}(\alpha,F)$. By the primitive element theorem, let $\beta \in F$ such that $K(\beta_1,\ldots,\beta_s) = K(\beta)$, then $\mathrm{Irr}(\alpha,F) = \mathrm{Irr}(\alpha,K(\beta))$. By (iv), $K(\alpha)$ and $K(\beta)$ are linearly disjoint over K, hence $1,\alpha,\ldots,\alpha^{n-1}$ are linearly independent over $K(\beta)$. Therefore

$$[K(\beta,\alpha):K(\beta)] \geq [K(\alpha):K].$$

Since the reverse inequality is obvious, then

$$[K(\beta,\alpha):K(\beta)] = [K(\alpha):K].$$

We deduce that

$$\mathrm{Irr}(\alpha,F) = \mathrm{Irr}(\alpha,K(\beta)) = \mathrm{Irr}(\alpha,K).$$

- $(v) \Rightarrow (i)$. Let $\alpha_1,\ldots,\alpha_m$ be elements of E linearly independent over K. We show that they are linearly independent over F. Let $\alpha \in E$ such that $K(\alpha) = K(\alpha_1,\ldots,\alpha_m)$. Let $n = [K(\alpha):K]$, then $n \geq m$ and by **(v)**, $1,\alpha,\ldots,\alpha^{n-1}$ are linearly independent over F and $[F(\alpha):F] = n$. Complete the family $\alpha_1,\ldots,\alpha_m$ by elements $\alpha_{m+1},\ldots,\alpha_n \in K(\alpha)$ and obtain a basis of $K(\alpha)$ over K. We show that $\{\alpha_1,\ldots,\alpha_m,\alpha_{m+1},\ldots,\alpha_n\}$ generates $F(\alpha)$ over F. Let $\beta \in F$. Then $\beta = \sum_{i=0}^{n-1} f_i\alpha^i$, where $f_i \in F$ for $i = 0,\ldots,n-1$.
For any $i \in \{0,\ldots,n-1\}$ we have $\alpha^i = \sum_{j=1}^{n} a_{ij}\alpha_j$, hence

$$\beta = \sum_{i=0}^{n-1} f_i \left(\sum_{j=1}^{n} a_{ij}\alpha_j\right) = \sum_{j=1}^{n}\left(\sum_{i=0}^{n-1} f_i a_{ij}\right)\alpha_j,$$

therefore our claim on the generating set of $F(\alpha)$ over F is proved. Since $[F(\alpha : F] = n$, then $\alpha_1,\ldots,\alpha_n$ are linearly independent over F. We deduce that $\alpha_1,\ldots,\alpha_m$ are linearly independent over F.

Exercise 8.8.
Let K be a field, E and F be algebraic extensions of K contained in some field Ω. Recall that if we define a multiplication in $E \bigotimes_K F$ by setting

$$\left(x_1 \bigotimes y_1\right) \cdot \left(x_2 \bigotimes y_2\right) = x_1x_2 \bigotimes y_1y_2,$$

then $E \bigotimes_K F$ is a K-algebra.

(1) Let $\Phi : E \bigotimes_K F \to E \cdot F$ be the unique K-linear map such that $\Phi(x \bigotimes y) = xy$. Show that Φ is a surjective morphism of K-algebras. Deduce that $E \bigotimes_K F/Ker\Phi$ is a field isomorphic to $E \cdot F$.
(2) Let $(e_i)_{i \in I}$ (resp. $(f_j)_{j \in J}$) be a basis of E (resp. F) over K. Let $V = \{(a_{ij}) \in K^{(I \times J)}, \sum a_{ij} e_i f_j = 0\}$. Show that V is a vector space over K and $V \simeq \operatorname{Ker} \Phi$.
(3) Show that E and F are linearly disjoint over K if and only if Φ is injective.
(4) Suppose that E/K and F/K are finite of degree m and n respectively. Let $\delta = \operatorname{lcm}(m, n)$. Show that

$$[E \cdot F : K] = \lambda\delta, \ \operatorname{Dim}_K \operatorname{Ker} \Phi = \mu\delta,$$

where λ, μ are non negative integers, $\lambda \geq 1$ and $\lambda + \mu = \frac{mn}{\delta}$.
(5) Compute $\operatorname{Ker} \Phi$ when $K = \mathbb{Q}$, $E = \mathbb{Q}(\sqrt[3]{2})$ and $F = \mathbb{Q}(j\sqrt[3]{2})$, where j is a primitive cube root of unity.
(6) Let $g(x)$ and $h(x)$ be non constant polynomials with coefficients in K of degree m and n respectively. Let $\alpha_1, \ldots, \alpha_m$ (resp. $\beta_1, \ldots, \beta_n$) be the roots of $g(x)$ (resp. $h(x)$) in an algebraic closure of K. Define the polynomial $(g \bigotimes h)(x)$ by

$$\left(g \bigotimes h\right)(x) = a_m^n b_n^m \prod_{j=1}^{n} \prod_{i=1}^{m} (x - \alpha_i\beta_j),$$

where a_m (resp, b_n) is the leading coefficient of $g(x)$ (resp. $h(x)$). Show that $(g \bigotimes h)(x) \in K[x]$.
(7) Show that $x^4 + 1$ may be represented in the form $x^4 + 1 = (g \bigotimes h)(x)$, where $g(x) \in \mathbb{Q}(\sqrt{2})[x]$ and $h(x) \in \mathbb{Q}(i)[x]$.
(8) Let $f(x)$ be a non constant polynomial with coefficients in K and let γ be a root of $f(x)$ in an algebraic closure of K. Suppose that $f(x)$ is irreducible over K and $f(x) = (g \bigotimes h)(x)$, where $g(x)$ and $h(x) \in K[x]$. Show that $g(x)$ and $h(x)$ are irreducible over K. Show that there exist two subfields E and F of $K(\gamma)$, linearly disjoint over K such that $K(\gamma) = E \cdot F$.

Solution 8.8.

(1) To prove that Φ is a morphism of K-algebras, it is sufficient to show that $\Phi((x_1 \bigotimes y_1)(x_2 \bigotimes y_2)) = \Phi(x_1 \bigotimes y_1) \cdot \Phi(x_2 \bigotimes y_2)$.
We have

$$\begin{aligned}\Phi\left(\left(x_1 \bigotimes y_1\right)\left(x_2 \bigotimes y_2\right)\right) &= \Phi\left(x_1 \cdot x_2 \bigotimes x_2 \cdot y_2\right) \\ &= x_1 x_2 y_1 y_2 = x_1 x_2 \cdot y_1 y_2 \\ &= \Phi\left(x_1 \bigotimes y_1\right) \cdot \Phi\left(x_2 \bigotimes y_2\right).\end{aligned}$$

We show that Φ is surjective. Since E/K and F/K are algebraic, then $E.F = \{e.f,\ e \in E,\ f \in F\}$. Let $\alpha = \sum_{\substack{e \in E \\ f \in F}} e \cdot f$ be an element of $E \cdot F$, where the sum is finite, then $\alpha = \sum \Phi(e \bigotimes f) = \Phi(\sum e \bigotimes f)$, hence $\alpha \in Im\Phi$. Therefore Φ is surjective. We deduce that

$$\left(E \bigotimes_K F\right) / \mathrm{Ker}\, \Phi \simeq \mathrm{Im}\, \Phi = E \cdot F$$

and then $E \bigotimes_K F/\mathrm{Ker}\, \Phi$ is a field isomorphic to $E \cdot F$.

(2) Since the elements $e \bigotimes f$ for $e \in E$ and $f \in F$ generate $E \bigotimes_K F$, then the family $(e_i \bigotimes f_j)_{(i,j) \in I \times J}$ is also a generating set of $E \bigotimes_K F$. Let $\mu = \sum_{(i,j)} a_{ij} e_i \bigotimes f_j$ be an element of $E \bigotimes_K F$, then

$$\begin{aligned}\mu \in \mathrm{Ker}\, \Phi &\Leftrightarrow \sum_{(i,j)} a_{ij} \Phi\left(e_i \bigotimes f_j\right) = 0 \\ &\Leftrightarrow \sum_{(i,j)} a_{ij} e_i f_j = 0 \\ &\Leftrightarrow (a_{ij})_{(i,j)} \in V,\end{aligned}$$

hence V is a vector space over K isomorphic to $\mathrm{Ker}\, \Phi$.

(3) We have

$$\begin{aligned}\Phi \text{ injective} &\Leftrightarrow \mathrm{Ker}\, \Phi = \{0\} \\ &\Leftrightarrow V = \{0\} \Leftrightarrow \{e_i f_j\} \text{ is a basis of } E \cdot F \text{ over } K \\ &\Leftrightarrow E \text{ and } F \text{ are linearly disjoint over } K.\end{aligned}$$

(4) We have

$$mn = \mathrm{Dim}_K\, E \cdot \mathrm{Dim}_K\, F = \mathrm{Dim}_K\, E \bigotimes_K F.$$

By the isomorphism established in **(1)** we deduce that

$$\mathrm{Dim}_K\, E \bigotimes_K F - \mathrm{Dim}_K\, \mathrm{Ker}\, \Phi = \mathrm{Dim}_K\, E \cdot F,$$

hence

$$mn = \mathrm{Dim}_K \operatorname{Ker}\Phi + \mathrm{Dim}_K E \cdot F. \qquad \text{(Eq 1)}$$

Obviously $[E \cdot F : K]$ is a common multiple of m and n. Therefore $[E \cdot F : K] = \lambda\delta$, where λ is a positive integer. From **(Eq 1)**, it follows immediately that δ divides $\mathrm{Dim}_K \operatorname{Ker}\Phi$, hence $\mathrm{Dim}_K \operatorname{Ker}\Phi = \mu\delta$, where μ is a non negative integer. From **(Eq 1)** again, we get $\lambda + \mu = \frac{mn}{\delta}$.

(5) Let $\alpha = \sqrt[3]{2}$ and $\beta = j\sqrt[3]{2}$. Set $x_i = \alpha^i$ for $i = 0, 1, 2$ and $y_k = \beta^k$ for $k = 0, 1, 2$, then $(x_i \bigotimes y_k)$ for $(i, k) \in \{0, 1, 2\}^2$ is a basis of $E \bigotimes F$ over K. Let $z = \sum a_{ik} x_i \bigotimes y_k$ be an element of $E \bigotimes_K F$, then

$$\begin{aligned} z \in \operatorname{Ker}\Phi &\Leftrightarrow \sum_{(i,k)\in\{0,1,2\}^2} a_{ik}\alpha^i\beta^k = 0 \\ &\Leftrightarrow a_{00} + a_{01}j\sqrt[3]{2} + a_{02}j^2\sqrt[3]{4} + a_{10}\sqrt[3]{2} + a_{11}j\sqrt[3]{4} \\ &\qquad + 2a_{12}j^2 + a_{20}\sqrt[3]{4} + 2a_{21}j + 2a_{22}j^2\sqrt[3]{2} = 0. \end{aligned}$$

Since $j^2 = -1 - j$ and since $1, \alpha, \alpha^2, j, j\alpha, j\alpha^2$ are linearly independent over K, we obtain

$$\begin{aligned} z \in \operatorname{Ker}\Phi &\Leftrightarrow a_{00} - 2a_{12} + (2a_{21} - 2a_{12})j + (a_{10} - 2a_{22})\sqrt[3]{2} \\ &\qquad + (a_{20} - a_{02})\sqrt[3]{4} + (a_{01} - 2a_{22})j\sqrt[3]{2} + (a_{11} - a_{02})j\sqrt[3]{4} = 0 \\ &\Leftrightarrow a_{00} - 2a_{12} = 0,\ a_{21} - a_{12} = 0,\ a_{10} - 2a_{22} = 0, \\ &\qquad a_{20} - a_{02} = 0,\ a_{01} - 2a_{22} = 0,\ a_{11} - a_{02} = 0 \\ &\Leftrightarrow a_{00} = 2a_{12},\ a_{21} = a_{12},\ a_{10} = a_{01} = 2a_{22},\ a_{20} = a_{11} = a_02 \\ &\Leftrightarrow z = a_{12}\left(2\left(1 \bigotimes 1\right) + \sqrt[3]{2} \bigotimes j^2\sqrt[3]{4} + \sqrt[3]{4} \bigotimes j\sqrt[3]{2}\right) \\ &\qquad + a_{22}\left(2\left(1 \bigotimes j\sqrt[3]{2}\right) + 2\left(\sqrt[3]{2} \bigotimes 1\right) + \sqrt[3]{4} \bigotimes j^2\sqrt[3]{4}\right) \\ &\qquad + a_{20}\left(1 \bigotimes j^2\sqrt[3]{4} + \sqrt[3]{2} \bigotimes j\sqrt[3]{2} + \sqrt[3]{4} \bigotimes 1\right). \end{aligned}$$

It is obvious that the elements of

$$\begin{aligned} \mathcal{B} = \Big\{ & 2\left(1 \bigotimes 1\right) + \sqrt[3]{2} \bigotimes j^2\sqrt[3]{4} + \sqrt[3]{4} \bigotimes j\sqrt[3]{2}, \\ & 2\left(1 \bigotimes j\sqrt[3]{2}\right) + 2\left(\sqrt[3]{2} \bigotimes 1\right) + \sqrt[3]{4} \bigotimes j^2\sqrt[3]{4}, \\ & 1 \bigotimes j^2\sqrt[3]{4} + \sqrt[3]{2} \bigotimes j\sqrt[3]{2} + \sqrt[3]{4} \bigotimes 1 \Big\} \end{aligned}$$

belong to $\operatorname{Ker}\Phi$ and are linearly independent over $\mathbb{Q}$. Therefore $\operatorname{Ker}\Phi$ is a sub algebra of $E \bigotimes_K F$ of dimension 3 over $\mathbb{Q}$ whose basis is $\mathcal{B}$.

(6) We have

$$\begin{aligned}
g \bigotimes h(x) &= a_m^n b_n^m \prod_{j=1}^{n} \prod_{i=1}^{m} (x - \alpha_i \beta_j) \\
&= b_n^m \prod_{j=1}^{n} (a_m \prod_{i=1}^{m} (x - \alpha_i \beta_j)) \\
&= b_n^m \prod_{j=1}^{n} \mathrm{Res}_y(g(y), x - y\beta_j) \\
&= b_n^m \mathrm{Res}_y(g(y), \prod_{j=1}^{n} (x - y\beta_j)) \\
&= b_n^m \mathrm{Res}_y(g(y), \frac{1}{b_n} \mathrm{Res}_z(h(z), x - yz) \\
&= \mathrm{Res}_y(g(y), Res_z(h(z), x - yz).
\end{aligned}$$

This implies that $g \bigotimes h(x) \in K[x]$.

(7) The roots of $x^4 + 1$ in $\mathbb{C}$ are given by

$$\alpha = \frac{\sqrt{2}}{2}(1 + i), \ -\alpha, \ \bar{\alpha} = \frac{\sqrt{2}}{2}(1 - i) \text{ and } -\bar{\alpha},$$

hence $x^4 + 1 = g \bigotimes h(x)$ where

$$g(x) = x^2 - \frac{1}{2} \text{ and } h(x) = x^2 - 2x + 2.$$

(8) Since $f(x) = g \bigotimes h(x)$, there exist a root α (resp. β) of $g(x)$ (resp. $h(x)$) such that $\gamma = \alpha \cdot \beta$. The following diagram describes the situation.

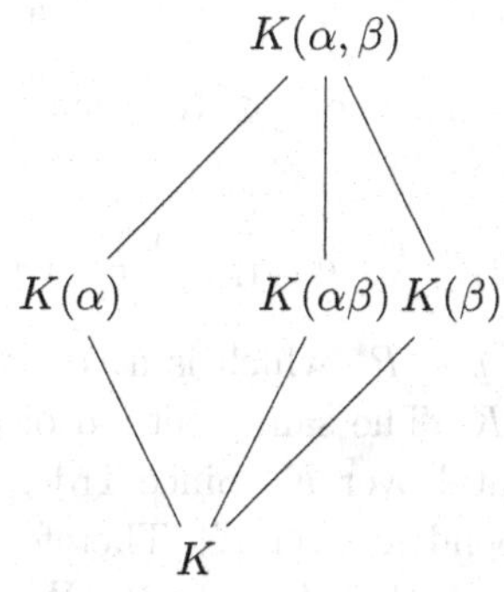

Since $f(x)$ is irreducible over K, then

$$[K(\alpha\beta) : K] = \deg f = \deg g \deg h,$$

hence $K(\alpha,\beta) = K(\alpha\beta)$. We deduce that $[K(\alpha,\beta):K] = \deg g \deg h$ and since

$$[K(\alpha : K)] \leq \deg g, \ [K(\beta):K] \leq \deg h,$$

then $[K(\alpha):K] = \deg g$ and $[K(\beta):K] = \deg h$, which implies that g and h are irreducible over K. Now since

$$[K(\alpha,\beta):K] = [K(\alpha):K][K(\beta):K],$$

then $K(\alpha)$ and $K(\beta)$ are linearly disjoint over K.

Exercise 8.9.
Let K be a field, R be an integral domain containing K and F be its fraction field. Suppose that $\operatorname{Trdeg} F/K \leq 1$.

(1) Show that $\operatorname{Dim} R \leq 1$.
(2) Give an example where $\operatorname{Trdeg} F/K = 1$ and $\operatorname{Dim} R = 0$.

Solution 8.9.

(1) If $\operatorname{Trdeg} F/K = 0$, then F is algebraic over K. Since $K \subset R \subset F$, then R itself is a field. Therefore $R = F$ and then $\operatorname{Dim} R = \operatorname{Dim} F = 0$. Suppose now that $\operatorname{Trdeg} F/K = 1$ and that $\operatorname{Dim} R \geq 2$. There exist two prime ideals of R, $\mathcal{P}$ and $\mathcal{Q}$ such that $\{0\} \subsetneq \mathcal{P} \subsetneq \mathcal{Q}$. Choose elements $p \in \mathcal{P}$ and $q \in \mathcal{Q}$ such that $p \neq 0$ and $q \notin \mathcal{P}$. If p is algebraic over K, then there exist $a_0, \ldots, a_{n-1} \in K$ such that

$$p^n + a_{n-1}p^{n-1} + \cdots + a_0 = 0$$

and $a_0 \neq 0$. Moreover since $p \notin K$, then $n \geq 2$. We deduce that $pr = -1$, where

$$r = (p^{n-1} + a_{n-1}p^{n-2} + \cdots + a_1)a_0^{-1}$$

belongs to R, hence $p \in R^*$ which is a contradiction. Therefore p is transcendental over K. The same method of proof, used for p, shows that q is transcendental over K. Since $\operatorname{Trdeg} F/K = 1$, then p and q are algebraically dependent over K. Therefore, there exists $f(x,y) \in K[x,y]$, irreducible such that $f(p,q) = 0$. We may write this equation in the form:

$$f_m(q)p^m + f_{m-1}(q)p^{m-1} + \cdots + f_1(q)p + f_0(q) = 0,$$

where m is a positive integer, $f_i(q) \in K[q]$ for $i = 0, \ldots, m$ and $f_0(q) \neq 0$. Set $f_0(q) = a_n q^n + \cdots + a_0$, where n is a nonnegative integer and $a_0, \ldots, a_n \in K$, with $a_n \neq 0$. Since

$$a_0 = -\sum_{i=1}^{m} f_i(q)p^i - \sum_{j=1}^{n} a_j q^j \in \mathcal{Q},$$

we conclude that $a_0 = 0$ and then

$$\sum_{i=1}^{m} f_i(q)p^i = -\sum_{j=1}^{n} a_j q_j.$$

We have established an identity of the form $p' = \sum_{j=1}^{n} b_j q^j$ where $p' \in \mathcal{P}$ and $b_j \in K$ for $j = 1, \ldots, n$. We may suppose that for these kind of identities, n is minimal. We have $q(\sum_{j=1}^{n} a_j q^{j-1}) \in \mathcal{P}$. Since $q \notin \mathcal{P}$ and $\mathcal{P}$ is a prime ideal, then $\sum_{i=j}^{n} a_j q^{j-1} = p'' \in \mathcal{P}$. Therefore $\sum_{j=0}^{n-1} a_{j+1} q^j = p''$ and we have found a smaller relation which is a contradiction. We conclude that $\operatorname{Dim} R \leq 1$.

(2) Let K be a field and F be a transcendental extension of K of transcendence degree equal to 1 and let $R = F$, then $\operatorname{Trdeg} F/K = 1$ and $\operatorname{Dim} R = 0$.

Exercise 8.10.

Let K be a field and x be an indeterminate over K.

(1) Let F be a field such that $K \subsetneq F \subset K(x)$. Show that the following assertions are equivalent.

 (i) There exists $P(x) \in K[x]$ such that $F = K(P)$.
 (ii) F contains a non-constant polynomial with coefficients in K.

(2) Let $f(x) \in K[x]$. Show that the following propositions are equivalent.

 (a) There exist $g(x), h(x) \in K[x]$ such that $\deg g \geq 2$, $\deg h \geq 2$ and $f(x) = g(h(x))$.
 (b) There exists a field F such that $K \subsetneq F \subsetneq K(x)$.

(3) Let $f(y) \in K[y]$ and $g(z) \in K[z]$ be two non constant polynomials. Suppose that $f(y)$ (resp. $g(z)$) is not the composition of two polynomials of degree ≥ 2 in $K[y]$ (resp. $K[z]$). Suppose that $f(y) - g(z)$ is reducible in $K[y, z]$ and has a factorization of the form

$$f(y) - g(z) = A(y, z)B(y, z).$$

 (u) Show that $\deg_y A$, $\deg_z A$, $\deg_y B$ and $\deg_z B$ are non zero.

(v) Let $m = \deg f$, $n = \deg g$, $\alpha_1, \ldots, \alpha_m$ (resp. $\beta_1, \ldots, \beta_n$) be the roots of $f(y) - x$ (resp. $g(z) - x$) in an algebraic closure of $K(x)$. Show that there exists $j \in \{1, \ldots, n\}$ such that $A(\alpha_1, \beta_j) = 0$. Show that $K(\beta_j)$ and $K(\alpha_1, \ldots, \alpha_m)$ are not linearly disjoint over $K(x)$.

(w) Deduce that $K(\alpha_1, \ldots, \alpha_m) = K(\beta_1, \ldots, \beta_n)$.

Solution 8.10.

(1) • *(i)* $\Rightarrow$ *(ii)*. Since $K \subsetneq F$ and $F = K(P)$, then P is non constant. Therefore F contains a non constant polynomial.

• *(ii)* $\Rightarrow$ *(i)*. Suppose that F contains a non constant polynomial $R(x) \in K[x]$. Luroth's Theorem [Schinzel (2000), Th. 2, Chap. 1.1] implies that $F = K(\alpha)$, where $\alpha = \frac{u(x)}{v(x)} \in K(x)$ with $\gcd(u(x), v(x)) = 1$. We show that $\alpha(x) \in K[x]$ or there exists $c \in K$ such that $\frac{1}{\alpha(x)-c} \in K[x]$. We choose $P(x) = \alpha(x)$ in the first case and $P(x) = \frac{1}{\alpha(x)-c}$ in the second case and we are done. Since $R(x) \in F = K(\alpha)$, there exists $\theta(x) \in K(x)$ such that $R(x) = \theta(\alpha)$. Set $\theta = A(x)/B(x)$ with

$$\begin{aligned} A(x) &= a_0x^n + a_1x^{n-1} + \cdots + a_n, \\ B(x)) &= b_0x^m + b_1x^{m-1} + \cdots + b_m, \end{aligned}$$

$a_0b_0 \neq 0$ and $\gcd(A(x), B(x)) = 1$. We distinguish three cases:

⋆ $\deg u > \deg v$.
We have

$$R(x) = \theta(\alpha) = \frac{(a_0u^n + a_1u^{n-1}v + \cdots + a_nv^n)v^m}{(b_0u^m + b_1u^{m-1}v + \cdots + b_mv^m)v^n},$$

hence

$$\begin{aligned} \deg R &= n \deg u + mdegv - m \deg u - n \deg v \\ &= (n-m)(\deg u - \deg v) > 0, \end{aligned}$$

where the degree of a rational function over K is the difference between the degree of the numerator and the degree of the denominator. We deduce that $n > m$ and

$$R(x) = \frac{a_0u^n + a_1u^{n-1}v + \cdots + a_nv^n}{(b_0u^m + b_1u^{m-1}v + \cdots + b_mv^m)v^{n-m}}.$$

It follows that any prime factor of $v(x)$ divides $u(x)$, thus $\alpha(x) \in K[x]$.

⋆ $\deg u < \deg v$.
Let $\mu(x) = \theta(\frac{1}{x})$. Then

$$R(x) = \theta(\alpha(x)) = \mu\left(\frac{1}{\alpha(x)}\right) = \mu\left(\frac{u(x)}{v(x)}\right).$$

Therefore we are moved to the preceding case. We conclude that $\frac{v(x)}{u(x)} = \frac{1}{\alpha(x)} \in K[x]$.

⋆ $\deg u = \deg v$.
Let c be the quotient of the leading coefficient of u by that of v. We have

$$\deg(u - cv) < \deg(u) = \deg v$$

and

$$R(x) = \gamma(\alpha - c) = \gamma\left(\frac{u - cv}{v}\right),$$

where $\gamma(x) = \theta(x + c)$. In this case, we are moved to the second case. We conclude that

$$\frac{v(x)}{u(x) - cv(x)} = \frac{1}{\alpha(x) - c} \in K[x].$$

(2) • $(a) \Rightarrow (b)$. Since $f(x) = g(h(x))$, then $K(f) \subset K(h) \subset K(x)$. Since

$$[K(x) : K(f)] = \deg f, \; [K(x) : K(h)] = \deg h$$

and $\deg h < \deg f$, then $F = K(h)$ is a strict intermediate field between $K(f)$ and $K(x)$.

• $(b) \Rightarrow (a)$. Let F be a field such that $K(f) \subsetneq F \subsetneq K(x)$. Since F contains the non-constant polynomial $f(x) \in K[x]$, then there exists $h(x) \in K[x]$, non-constant, such that $F = K(h)$. We have $f(x) = g(h(x))$, where $g(x) = \frac{u(x)}{v(x)} \in K(x)$ and $\gcd(u(x), v(x)) = 1$, hence $f(x) = \frac{u(h(x))}{v(h(x))}$. Since $f(x)$ is a polynomial then $v(h(x)) \mid u(h(x))$. By Bezout's identity there exist $A(x), B(x) \in K[x]$ such that $A(x)u(x) + B(x)v(x) = 1$. We deduce that

$$A(h(x))u(h(x)) + B(h(x))V(h(x)) = 1,$$

hence $v(h(x))$ is a constant. Therefore $v(x)$ is constant (recall that $h(x)$ is not constant). It follows that $g(x) \in K[x]$ and $f(x) = g(h(x))$. The assertion on the degrees of g and h is easy to prove and will be omitted.

(3)(u) Obvious.

(v) For any $j \in \{1, \ldots, n\}$ we have

$$0 = f(\alpha_1) - g(\beta_j) = A(\alpha_1, \beta_j)B(\alpha_1, \beta_j).$$

From **(1)**, it follows that there exists $j \in \{1, \ldots, n\}$ such that $A(\alpha_1, \beta_j) = 0$. We deduce that

$$[K(\alpha, \beta_j) : K(\alpha_1)] < n = [K(\beta_j) : K(x)],$$

hence $K(\beta_j)$ and $K(\alpha_1, \ldots, \alpha_m)$ are not linearly disjoint over $K(x)$.

(w) Since $K(\alpha_1, \ldots, \alpha_m)$ is a Galois extension of $K(x)$, by [Lang (1965), Th. 4, Chap. 8.1], it is linearly disjoint over $K(x)$ from any extension F of $K(x)$ such that

$$F \cap K(\alpha_1, \ldots, \alpha_m) = K(x).$$

We deduce that

$$K(x) \subsetneq K(\alpha_1, \ldots, \alpha_m) \cap K(\beta_j).$$

We now have

$$K(x) \subsetneq K(\alpha_1, \ldots, \alpha_m) \cap K(\beta_j) \subset K(\beta_j),$$

$x = g(\beta_j)$ and $g(x)$ is indecomposable over K, then by **(1)**, there is no strict intermediate field between $K(x)$ and $K(\beta_j)$. Therefore

$$K(\alpha_1, \ldots, \alpha_m) \cap K(\beta_j) = K(\beta_j).$$

This implies $K(\beta_j) \subset K(\alpha_1, \ldots, \alpha_m)$. By conjugating, we conclude that $K(\beta_1, \ldots, \beta_n) \subset K(\alpha_1, \ldots, \alpha_m)$. By symmetry, we obtain the reverse inclusion. Therefore $f(y) - x$ and $g(z) - x$ have the same splitting field over $K(x)$.

Exercise 8.11.
Let K be a field, $f(x)$ and $g(x)$ be non constant polynomials with coefficients in K.

(1) Show that $K(f)$ and $K(g)$ are not linearly disjoint over K.
(2) Suppose that $K = \mathbb{C}$.
   (a) Show that $\mathbb{C}(f) \cap \mathbb{C}(g) = \mathbb{C}$ if and only if $\mathbb{C}[f] \cap \mathbb{C}[g] = \mathbb{C}$.
   (b) Show that $\mathbb{C}(x^2+x) \cap \mathbb{C}(x^2) = \mathbb{C}$ and $\mathbb{C}(x^n) \cap \mathbb{C}(x^m) = \mathbb{C}(x^k)$, where m and n are given positive integers and k is their least common multiple.
(3) Give an example of two fields E and F contained in some field Ω such that there exists $\alpha \in \Omega$, algebraic over E and over F but transcendental over $E \cap F$.

Solution 8.11.

(1) Since $f(x)$ and $g(x)$ are elements of $K(x)$ and $\text{Trdeg}\, K(x)/K = 1$, then $f(x)$ and $g(x)$ are algebraically dependent over K. It follows that there exist $a_0(g), \ldots, a_m(g) \in K[g]$, with $a_m(g) \neq 0$ such that

$$a_0(g) + a_1(g)f + \cdots + a_m(g)f^m = 0.$$

Therefore $1, f, \ldots, f^m$ are linearly dependent over $K(g)$. Since f is transcendental over K, then these elements are linearly independent over K, hence $K(f)$ and $K(g)$ are not linearly disjoint over K.

(2)(a) The necessity of the condition is obvious. We prove its sufficiency. Suppose by contradiction that there exists

$$h(x) \in (\mathbb{C}(f) \cap \mathbb{C}(g)) \setminus \mathbb{C}.$$

Write $h(x)$ in the forms

$$h(x) = A(f(x))/B((f(x)) = C(g(x))/D(g(x)),$$

where $A(x), B(x), C(x)$ and $D(x)$ are polynomials with complex coefficients such that

$$\gcd(A(x), B(x)) = \gcd(C(x), D(x)) = 1.$$

We may suppose that $h(x)$ is chosen in order to satisfy the condition $\deg B + \deg D$ is minimal. Then

$$A(f(x))D(g(x)) = B((f(x))C(g(x)).$$

Since $\gcd(A(f(x)), B(f(x))) = \gcd(C(g(x)), Dg((x))) = 1$, then $A(f(x)) \mid C(g(x))$ and $C(g(x)) \mid A(f(x))$, hence there exists $\lambda \in \mathbb{C}$ such that $C(g(x)) = \lambda A(f(x))$. It follows that $D(g(x)) = \lambda B(f(x))$. By differentiating the preceding identities, we get

$$C'(g(x))g'(x) = \lambda A'(f(x))f'(x).$$

It follows that

$$D'(g(x))g'(x) = \lambda B'(f(x))f'(x).$$

If $B'(f(x)) = 0$, then $B'(x) = 0$ and $D'(g(x)) = D'(x) = 0$, so that $h(x) \in (\mathbb{C}[f] \cap \mathbb{C}[g]) \setminus \mathbb{C}$, which is a contradiction. We now have

$$A'(f(x))/B'((f(x)) = C'(g(x))/D'(g(x)),$$

which contradicts our assumption on the minimality of $\deg B + \deg D$.

(b) To show that $\mathbb{C}(x^2+x)\cap\mathbb{C}(x^2)=\mathbb{C}$, it is sufficient according to **(a)**, to prove that $\mathbb{C}[x^2+x]\cap\mathbb{C}[x^2]=\mathbb{C}$. Let $h(x)$ be an element of $\mathbb{C}[x^2+x]\cap\mathbb{C}[x^2]$, then

$$\begin{aligned} h(x) &= a_n(x^2+x)^n + a_{n-1}(x^2+x)^{n-1}+\cdots+a_0 \\ &= b_n(x^2)^n + b_{n-1}(x^2)^{n-1}+\cdots+b_0, \end{aligned}$$

where $a_0,\dots,a_n$ and $b_0,\dots,b_n$ are complex numbers. Suppose that $n\ge 1$ and $a_nb_n\neq 0$. Equating the leading coefficients (resp. the coefficients of x^{2n-1}) in these two expressions of $h(x)$, we obtain: $a_n=b_n$ and $na_n=0$, which is a contradiction. Thus $n=0$ and $h(x)\in\mathbb{C}$. We now consider the other intersection. Obviously $\mathbb{C}(x^k)\subseteq\mathbb{C}(x^n)\cap\mathbb{C}(x^m)$. We prove the reverse inclusion. Let $h(x)$ be an element of the intersection then

$$h(x)=A(x^m)/B((x^m)=C(x^n)/D(x^n),$$

where $A(x),B(x),C(x)$ and $D(x)$ are polynomials with complex coefficients such that $\gcd(A(x),B(x))=\gcd(C(x),D(x))=1$. As in **(a)**, we obtain

$$C(x^n)=\lambda A(x^m) \text{ and } D(x^n)=\lambda B(x^m).$$

Let $U(x)=\sum a_ix^i$ be the polynomial appearing on the left (and on the right) side of the first identity, then if $a_i\neq 0$, we have $i\not\equiv 0 \pmod m$ and $i\not\equiv 0 \pmod n$. It follows that $i\not\equiv 0 \pmod k$, whenever $a_i\neq 0$, that is $U(x)\in\mathbb{C}[x^k]$. We may obtain the same conclusion for $V(x)=D(x^n))=\lambda B(x^m)$. We conclude that $h(x)\in\mathbb{C}[x^k]$.

(3) Consider $E=\mathbb{C}(x^2)$, $F=\mathbb{C}(x^2+x)$, $\Omega=E=\mathbb{C}(x)$ and $\alpha=x$.

Exercise 8.12.
Let K be a field, $f(x,y)$, $g(x,y)$ and $h(x,y)$ be polynomials with coefficients in K such that $f(x,y)$ is not constant.

(1) If f is irreducible over K, show that there exist $R(x,y)$ and $q(x,y)\in K[x,y]$ such that R is irreducible and $R(g,h)=f(x,y)q(x,y)$.
(2) If f is irreducible over K, let

$$I=\{A(x,y)\in K[x,y], \text{ such that } A(g,h)\equiv 0 \pmod{f(x,y)}\}.$$

Show that I is a principal prime ideal of $K[x,y]$ generated by $R(x,y)$. Deduce that $K[g,h]\cap fK[x,y]\neq(0)$ if and only if g and h are algebraically independent over K.

(3) If f is arbitrary (irreducible or not), show that there exists $S(x,y) \in K[x,y] \setminus K$ such that $S(g,h) \equiv 0 \pmod{f(x,y)}$.

Solution 8.12.

(1) Since f is non constant, then $\deg_x f \geq 1$ or $\deg_y f \geq 1$. Suppose that $\deg_y f \geq 1$, the proof is similar if $\deg_x f \geq 1$ and let α be a root of $f(x,y)$ in an algebraic closure of $K(x)$. Let $E = K(x,\alpha)$, then E is a transcendental extension of K and $\operatorname{Trdeg} E/K(x) = 1$. Therefore there exists $R(u,v) \in K[u,v]$ such that $R(g(x,\alpha), h(x,\alpha)) = 0$ and we may suppose that R is irreducible over K. It follows that there exists $q(x,y) \in K(x)[y]$ such that

$$R(g(x,y), h(x,y)) = f(x,y)q(x,y).$$

Using the content in $K(x)$, we obtain

$$\operatorname{cont}(R(g(x,y), h(x,y))) = \operatorname{cont}(f(x,y)) \operatorname{cont}(q(x,y)) = \operatorname{cont}(q(x,y)).$$

We conclude that $\operatorname{cont}(q(x,y)) \in K[x]$, which implies $q(x,y) \in K[x,y]$.

(2) Let $\psi : K[x,y] \to K[x,y]/fK[x,y]$ be the morphism of rings such that $\psi(A(x,y)) = \overline{A(g,h)}$, then $\operatorname{Ker} \psi = I$ so that I is a prime ideal of $K[x,y]$ and we have $R(x,y)K[x,y] \subseteq I$. Since the image of ψ is not a field, then $\operatorname{Ker} \psi$ is not maximal. Let $\mathcal{M}$ be a maximal ideal of $K[x,y]$ containing $\operatorname{Ker} \psi$. Consider the chain of prime ideals

$$\{0\} \subset R(x,y)K[x,y \subset \operatorname{Ker} \psi \subset \mathcal{M}.$$

Since the Krull dimension of $K[x,y]$ is equal to 2, then

$$R(x,y)K[x,y] = \operatorname{Ker} \psi = I.$$

Suppose that $g(x,y)$ and $h(x,y)$ are algebraically dependent over K and let $A(u,v)$ be an irreducible polynomial with coefficients in K such that $A(g,h) = 0$. Since $A \in I$, then up to a multiplicative constant, the polynomials $A(x,y)$ and $R(x,y)$ are equal. Let $u(x,y) \in K[g,h] \cap fK[x,y]$, then $u(x,y) = v(g,h) \equiv 0 \pmod{f}$, hence $v(x,y) \in I$ and then $Ax,y) \mid v(x,y)$. Since $A(g,h) = 0$, then $u(x,y) = v(g,h) = 0$. Suppose that $K[g,h] \cap fK[x,y] = (0)$. Since $R(g,h)$ belongs to this intersection, then $R(g,h) = 0$, which implies g and h are algebraically dependent over K.

(3) Let

$$f(x,y) = f_1(x,y) \cdots f_r(x,y)$$

be the factorization of f into irreducible factors over K, distinct or not. For each $i \in \{1, \ldots, r\}$ let $R_i(x, y)$ be the polynomial defined in **(1)** and satisfying the condition $R_i(g, h) \equiv 0 \pmod{f_i(x, y)}$, then $\prod_{i=1}^{r} R_i(g, h) \equiv 0 \pmod{f(x, y)}$, so that $S(x, y) = \prod_{i=1}^{r} R_i(x, y)$ satisfies the required conditions.

Exercise 8.13.

Let m, n be coprime positive integers and q be a prime power. Let $\alpha \in \mathbb{F}_{q^n}$, $\beta \in \mathbb{F}_{q^m}$ and $\gamma = \alpha\beta$. Show that γ generates a normal basis of $\mathbb{F}_{q^{mn}}$ if and only if α (resp. β) generates a normal basis of $\mathbb{F}_{q^n}$ (resp. $\mathbb{F}_{q^m}$) over $\mathbb{F}_q$.

Solution 8.13.

- **Necessity of the conditions.** We prove that β generates a normal basis of $\mathbb{F}_{q^m}$. The same proof works for α and will be omitted. Since $\mathbb{F}_{q^m}$ and $\mathbb{F}_{q^n}$ are linearly disjoint over $\mathbb{F}_q$, we have
$$\mathrm{Tr}_{\mathbb{F}_{q^{mn}}/\mathbb{F}_{q^m}}(\gamma) = \mathrm{Tr}_{\mathbb{F}_{q^{mn}}/\mathbb{F}_{q^m}}(\alpha\beta) = \beta\,\mathrm{Tr}_{\mathbb{F}_{q^{mn}}/\mathbb{F}_{q^m}}(\alpha) = \beta\,\mathrm{Tr}_{\mathbb{F}_{q^n}/\mathbb{F}_q}(\alpha).$$
Since γ generates a normal basis of $\mathbb{F}_{q^{mn}}$, then the left side of the above identity is non zero. Therefore $\mathrm{Tr}_{\mathbb{F}_{q^n}/\mathbb{F}_q}(\alpha) \neq 0$. It follows that β generates a normal basis of $\mathbb{F}_{q^m}$ over $\mathbb{F}_q$ if and only if $\beta\,\mathrm{Tr}_{\mathbb{F}_{q^n}/\mathbb{F}_q}(\alpha)$ does, that is $\mathrm{Tr}_{\mathbb{F}_{q^{mn}}/\mathbb{F}_{q^m}}(\gamma)$ does. This last claim is true by **Exercise 6.19**. So β generates a normal basis of $\mathbb{F}_{q^m}$ over $\mathbb{F}_q$.
- **Sufficiency of the conditions.** Consider the following diagram of fields.

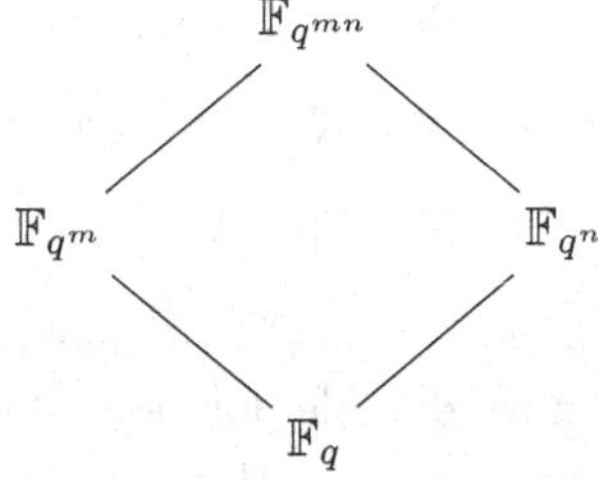

Since $\gcd(m, n) = 1$, then the fields $\mathbb{F}_{q^m}$ and $\mathbb{F}_{q^n}$ are linearly disjoint over $\mathbb{F}_q$ and
$$\mathrm{Gal}(\mathbb{F}_{q^{mn}}, \mathbb{F}_q) \simeq \mathrm{Gal}(\mathbb{F}_{q^m}, \mathbb{F}_q) \times \mathrm{Gal}(\mathbb{F}_{q^n}, \mathbb{F}_q).$$
The list of conjugates of γ is given by $\alpha^{q^i}\beta^{q^j}$ for $i = 0, \ldots, n-1$ and $j = 0, \ldots, m-1$. Now $\{\alpha, \ldots, \alpha^{q^{n-1}}\}$ is a basis of $\mathbb{F}_{q^n}$ over $\mathbb{F}_q$ and $\{\beta, \ldots, \beta^{q^{m-1}}\}$ is a basis of $\mathbb{F}_{q^{mn}}$ over $\mathbb{F}_{q^n}$, hence
$$\{\alpha^{q^i}\beta^{q^j}, \text{ for } i = 0, \ldots, n-1 \text{ and } j = 0, \ldots, m-1\}$$
is a basis of $\mathbb{F}_{q^{mn}}$ over $\mathbb{F}_q$.

Exercise 8.14.
Let q be a prime power, $\{\alpha_1, \ldots, \alpha_n\}$ be a normal basis of $\mathbb{F}_{q^n}$ over $\mathbb{F}_q$ and $\{\beta_1, \ldots, \beta_n\}$ be its dual basis. Show that this second basis is a normal basis of $\mathbb{F}_{q^n}$ over $\mathbb{F}_q$.

One may use the following matrices

$$A = \begin{pmatrix} \alpha_1 & \alpha_1^q & \ldots & \alpha_1^{q^{n-1}} \\ \alpha_2 & \alpha_2^q & \ldots & \alpha_2^{q^{n-1}} \\ \ldots & \ldots & \ldots & \ldots \\ \alpha_n & \alpha_n^q & \ldots & \alpha_n^{q^{n-1}} \end{pmatrix} \text{ and } B = \begin{pmatrix} \beta_1 & \beta_2 & \ldots & \beta_n \\ \beta_1^q & \beta_2^q & \ldots & \beta_n^q \\ \ldots & \ldots & \ldots & \ldots \\ \beta_1^{q^{n-1}} & \beta_2^{q^{n-1}} & \ldots & \beta_n^{q^{n-1}} \end{pmatrix}.$$

Solution 8.14.
For any matrix C, we denote by C^t its transpose. We may suppose that $\alpha_1, \ldots, \alpha_n$ are labeled such that $\alpha_i = \alpha_1^{q^{i-1}}$. Our assumption implies that $AB = I_n$, hence $BA = I_n$. Since A is symmetric, we deduce that

$$(AB)^t = B^t A^t = B^t A = I_n.$$

Therefore $B^t A = I_n = BA$, which implies $B^t = B$ and then $\beta_i = \beta^{q^{i-1}}$. This means that $\{\beta_1, \ldots, \beta_n\}$ is a normal basis of $\mathbb{F}_{q^n}$ over $\mathbb{F}_q$.

Exercise 8.15.
Let q be a prime power, m and n be coprime positive integers, $\alpha_1, \ldots, \alpha_m$ be elements of $\mathbb{F}_{q^m}$. Show that they form a basis of $\mathbb{F}_{q^m}$ over $\mathbb{F}_q$ if and only if they form a basis of $\mathbb{F}_{q^{mn}}$ over $\mathbb{F}_{q^n}$.

Solution 8.15.
Consider the following diagram of fields:

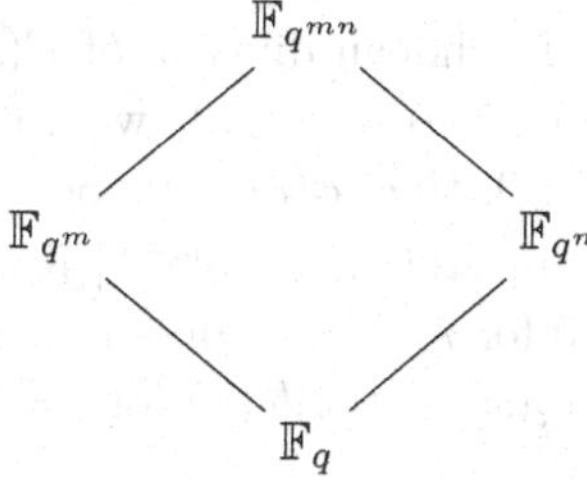

Since $\gcd(m, n) = 1$, then the fields $\mathbb{F}_{q^m}$ and $\mathbb{F}_{q^n}$ are linearly disjoint over $\mathbb{F}_q$. Therefore $\alpha_1, \ldots, \alpha_m$ are linearly independent over $\mathbb{F}_q$ if and only if they are linearly independent over $\mathbb{F}_{q^n}$. To complete the proof, we observe that $[\mathbb{F}_{q^{mn}} : \mathbb{F}_{q^n}] = [\mathbb{F}_{q^m} : \mathbb{F}_q] = m$.

Exercise 8.16.
Let K be field, E and F linearly disjoint extensions of K. Show that $E \cap F = K$. Show that the converse is false.

Solution 8.16.
Suppose that $E \cap F \neq K$ and let $\alpha \in E \cap F \setminus K$, then $1, \alpha$ are linearly independent over K, but linearly dependent over F, a contradiction. Therefore $E \cap F = K$.

For the converse, let $E = \mathbb{Q}(2^{1/3})$ and $F = \mathbb{Q}(j2^{1/3})$, where j is a primitive cube root of unity, then $E \cap F = \mathbb{Q}$ and E, F are not linearly disjoint over $\mathbb{Q}$.

Exercise 8.17.
Let K be a field, $f(x)$ and $g(y)$ be irreducible polynomials over K. Let α (resp. β) be a root of $f(x)$ (rep. $g(y)$ in an algebraic closure of K. Show that the ideal generated by $f(x)$ and $g(y)$ in $K[x, y]$ is maximal if and only $K(\alpha)$ and $K(\beta)$ are linearly disjoint over K.

Solution 8.17.
We first show that

$$K[x, y]/(f(x), g(y)) \simeq K(\alpha)[y]/(g(y)).$$

Let

$$\phi : K[x, y] \to K(\alpha)[y]/(g(y))$$

be the unique K-morphism of rings such that

$$\phi(x) = \alpha \text{ and } \phi(y) = y + (g(y)).$$

Obviously ϕ is surjective. Let $m = \deg g$ and let $F(x, y) \in K[x, y]$ such that $\phi(F) = 0$. Performing an Euclidean division of $F(x, y)$ by $g(y)$ in $K[x][y]$, we obtain $F(x, y) = g(y)q(x, y) + r(x, y)$, where q and $r \in K[x, y]$ and $\deg_y r < m$. Since $\phi(F) = 0$, then $\phi(r) = 0$. Set $r(x, y) = \sum_{i=1}^{m-1} a_i(x)y^i$, then $\sum_{i=1}^{m-1} a_i(\alpha)\bar{y}^i = 0$. Since $\{1, \bar{y}, \ldots, \bar{y}^{m-1}\}$ is a basis of $K(\alpha)[y]/(g(y))$ over $K(\alpha)$, then $a_i(\alpha) = 0$ for $i = 0, \ldots, m-1$. It follows that $f(x) \mid a_i(x)$ for $i = 0, \ldots, m-1$. Set $a_i(x) = f(x)b_i(x)$ for $i = 0, \ldots, m-1$, then

$$F(x, y) = g(y)q(x, y) + f(x) \sum_{i=1}^{m-1} a_i(x)y^i.$$

Thus F belongs to the ideal generated by $f(x)$ and $g(y)$, that is $\operatorname{Ker} \phi$ is contained in this ideal. The proof of the reverse inclusion is obvious, so that

$\operatorname{Ker}\phi = (f(x), g(y))$. The first isomorphism theorem implies the mentioned isomorphism. We now have $(f(x), g(y))$ is maximal in $K[x, y]$ if and only if $K(\alpha)[y]/(g(y))$ is a field, thus if and only if $g(y)$ is irreducible over $K(\alpha)$. Therefore if and only $K(\alpha)$ and $K(\beta)$ are linearly disjoint over K.

Exercise 8.18.

Let K be a field, $f(x, y)$ and $g(u, z)$ be irreducible polynomials with coefficients in K such that $\deg_y f = d_1$ and $\deg_z g = d_2$. Let $\alpha \in \overline{K(x)}$ and $\beta \in \overline{K(u)}$ such that $f(x, \alpha) = 0$ and $g(u, \beta) = 0$. Suppose that f is absolutely irreducible.

(1) Show that $[K(x, u, \beta) : K(x, u)] = d_2$ and $[K(x, u, \beta, \alpha) : K(x, u, \beta)] = d_1$. Deduce that $K(x, u, \beta)$ and $K(x, u, \alpha)$ are linearly disjoint over $K(x, u)$.

(2) By considering the example $K = \mathbb{Q}$, $f(x, y) = y^4 - 2x^2$ and $g(u, z) = z^2 - 2u^2$, show that the conclusion of **(1)** does not hold if we remove the condition of absolute irreducibility.

Solution 8.18.

(1) Consider the following diagram of fields:

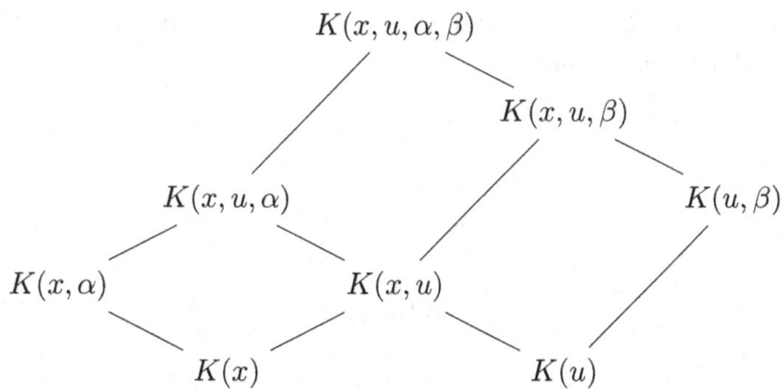

We first prove that $[K(x, u, \beta) : K(x, u)] = d_2$. We have $K(u, x)/K(u)$ is purely transcendental while $K(u, \beta)/K(u)$ is algebraic, hence $K(u, x)$ and $K(u, \beta)$ are linearly disjoint over $K(u)$ [Lang (1965), Pro. 3, Chap. 10.5]. It follows that

$$[K(x, u, \beta) : K(x, u)] = [K(u, \beta) : K(u)] = d_2.$$

We now prove that $[K(x, u, \beta, \alpha) : K(x, u, \beta)] = d_1$. It is sufficient to prove that $f(x, y)$ is irreducible over $K(u, \beta) = K(u)[\beta]$. Suppose that

$f(x, y = f_1(x,y)f_2(x,y)$, where f_1 and $f_2 \in K(u)[\beta][x,y]$ with $\deg_y f_1$ and $\deg_y f_2 \geq 1$. Set

$$f_1(x,y) = \sum a_{jk}x^j y^k \quad \text{and} \quad f_2(x,y) = \sum b_{jk}x^j y^k,$$

where a_{jk} and $b_{jk} \in K(u)[\beta]$. We may write a_{jk} and b_{jk} in the form

$$a_{jk} = \left(\sum A^h_{jk}(u)\beta^h\right)/D(u) \quad \text{and} \quad b_{jk} = \left(\sum B^h_{jk}(u)\beta^h\right)/D(u).$$

Choose $u_0 \in \bar{K}$ such that $D(u_0) \neq 0$ and such that, when the polynomial $g(u,z)$ is expressed through the decreasing powers of z, then its leading coefficient does not vanish at $u = u_0$. Choose $z_0 \in \bar{K}$ such that $g(u_0, z_0) \neq 0$. If we substitute u_0 for u and z_0 for z, we obtain the factorization $f(x,y) = \overline{f}_1(x,y)\overline{f}_2(x,y)$, where $\overline{f}_1$ and $\overline{f}_2$ are the polynomials obtained by the substitution from f_1 and f_2 respectively. This contradicts the absolute irreducibility of $f(x,y)$, hence $[K(u,x,\beta,\alpha) : K(u,x,\beta)] = d_1$.
It follows that

$$\begin{aligned}[K(u,x,\beta,\alpha) : K(u,x)] &= d_1 d_2 \\ &= [K(u,x,\alpha) : K(u,x)][K(u,x,\beta) : K(u,x)]\end{aligned}$$

which means that $K(u,x,\alpha)$ and $K(u,x,\beta)$ are linearly disjoint over $K(u,x)$.

(2) Here we have $\beta = \pm u\sqrt{2}$ and $\alpha^2 = \pm x\sqrt{2}$, $d_1 = 4$ and $d_2 = 2$. We have the following diagram:

$$\begin{array}{c} K(x,u,\alpha,\beta) \\ | \\ K(u,x,\sqrt{2}) \\ | \\ K(u,x) \end{array}$$

hence

$$8 = d_1 d_2 \neq [K(u,x,\beta,\alpha) : K(u,x)] = 4.$$

Exercise 8.19.

Let K be a field, A be a graded commutative K-algebra without divisors of 0, L be its fraction field and d be a non negative integer. A rational function $f = g/h \in L$ is said to be homogeneous of degree d if g and h are homogeneous and $\deg g - \deg h = d$. Let

$$L_0 = \{f \in L, f \text{ homogeneous}, \deg f = 0\} \cup \{0\}.$$

(1) Show that L_0 is a field and $K \subset L_0 \subset L$.
(2) Let $g \in L$, homogeneous, of degree $e \geq 1$. Show that g is transcendental over L_0. If moreover g is chosen to satisfy the condition that its degree $e \geq 1$ is minimal, show that $L = L_0(g)$.
(3) Suppose that there exist $f_1, \ldots, f_r \in L$ be homogeneous such that $L = K(f_1, \ldots, f_r)$. Show that there exist $g_1, \ldots, g_{r-1} \in L_0$ such that $L_0 = K(g_1, \ldots, g_{r-1})$. Prove that $\gcd(\deg f_1, \ldots, \deg f_r) = e$, where e is defined in **(2)**.
Hint. One may use **Exercise 1.5** and **Exercise 10.26**.
(4) Let $x_1, \ldots, x_n$ be a set of algebraically independent variables over K. Suppose that $A = K[x_1, \ldots, x_n]$. Determine in this case L_0.

Solution 8.19.

(1) Clear.
(2) Set $g = g_1/g_2$ where $g_1, g_2 \in A$, homogeneous with $\deg g_1 > \deg g_2$. Suppose that g is algebraic over L_0, then g satisfies an equation of the form

$$(a_n/b_n)g^n + \cdots + (a_k/b_k)g^k + \cdots + (a_0/b_0) = 0$$

with $n \geq 1$ and $\deg a_i = \deg b_i$. Let $D = \text{lcm}(b_0, \ldots, b_n)$. Replacing g by its value and multiplying by D, this equation becomes

$$(Da_n/b_n)g_1^n + \cdots + (Da_k/b_k)g_1^k g_2^{n-k} + \cdots + (Da_0/b_0)g_2^n = 0.$$

We have $\deg(Da_n/b_n)g_1 = \deg D + n \deg g_1$ and for any $k < n$,

$$\begin{aligned}\deg(Da_k/b_k)g_1^k g_2^{n-k} &= \deg D + k \deg g_1 + (n-k)\deg g_2 \\ &< \deg D + k \deg g_1 + (n-k)\deg g_1 \\ &= \deg D + n \deg g_1,\end{aligned}$$

hence a contradiction. We conclude that g is transcendental over L_0. Suppose now that e is minimal. To prove that $L \subset L_0(g)$ it is sufficient to show that any $f \in A$, homogeneous, belongs to $L_0(g)$. The inclusion in the opposite direction is obvious and will be omitted. Let $f \in A$ homogeneous, then $\deg f = eq + r$, where q and r are integers and $0 \leq r < e$. We have $\deg(f/g^q) = \deg f - qe = r$ and f/g^q is homogeneous. Since e is minimal, then $r = 0$, hence $f/g^q \in L_0$ and $f \in L_0(g)$.
(3) Let $d_i = \deg f_i$ for $i = 1, \ldots, r$ and $d = \gcd(d_1, \ldots, d_r)$. By Bezout's identity, there exist integers $u_1, \ldots, u_r$ such that $u_1 d_1 + \cdots + u_r d_r = 1$. The integers $u_1, \ldots, u_r$ are coprime (see **Exercise 1.5**), hence by **Exercise 10.26**, there exists a matrix with integral coefficients $M =$

$(c_i^j) \in \mathbb{Z}^{r\times r}$ such that $c_1^j = u_j$ and $\det M = 1$. Set $M^{-1} = (b_i^j)$. Let $\phi_i = \prod_{j=1}^r f_j^{c_i^j}$ for $i = 1, \dots, r$, which means

$$(\phi_1, \dots, \phi_r) = (f_1, \dots, f_r)^M.$$

We deduce that

$$(f_1, \dots, f_r) = (\phi_1, \dots, \phi_r)^{M^{-1}},$$

hence $f_i = \prod_{j=1}^r \phi_i^{b_i^j}$. This shows that $L = K(\phi_1, \dots, \phi_r)$. We have

$$\deg \phi_1 = \sum c_i^j d_j = d \quad \text{and} \quad \deg \phi_i = \sum c_i^j d_j \equiv 0 \pmod d$$

for $i \geq 2$. For $i = 1, \dots, r-1$, let $g_i = \phi_{i+1}/\phi_1^{\deg \phi_{i+1}/d}$. Then $\deg g_i = 0$, hence $g_i \in L_0$. We have proved that $L = K(\phi_1, \dots, \phi_r)$. It is easy to see that $L = K(g_1, \dots, g_{r-1}, \phi_1)$. Consider the chain of fields:

$$K(g_1, \dots, g_{r-1}) \subset L_0 \subset K(g_1, \dots, g_{r-1}, \phi_1) = L.$$

By **(2)**, L/L_0 is a purely transcendental extension, hence ϕ_1 is transcendental over L_0, thus also over $K(g_1, \dots, g_{r-1})$. It follows that $L/K(g_1, \dots, g_{r-1})$ is purely transcendental. Since L is algebraic over any field F such that

$$K(g_1, \dots, g_{r-1}) \subsetneq F \subset L,$$

then

$$L_0 = K(g_1, \dots, g_{r-1}), \; L = L_0(\phi_1)$$

and that the degree of ϕ_1 is minimal. Hence

$$e = \deg \phi_1 = d = \gcd(\deg f_1, \dots, \deg f_r).$$

(4) In this case $L = K(x_1, \dots, x_n)$. Let $\alpha \in L_0$. We may write α in the form

$$\alpha = \frac{\sum_{i_1+\cdots+i_n=d} a_{i_1\cdots i_n} x_1^{i_1} \cdots x_n^{i_n}}{\sum_{j_1+\cdots+j_n=d} b_{j_1\cdots j_n} x_1^{j_1} \cdots x_n^{j_n}}.$$

Dividing numerator and denominator of this fraction by x_n^d, we get

$$\alpha = \frac{\sum_{i_1+\cdots+i_n=d} a_{i_1\cdots i_n} (x_1/x_n)^{i_1} \cdots (x_{n-1}/x_n)^{i_{n-1}}}{\sum_{j_1+\cdots+j_n=d} b_{j_1\cdots j_n} (x_1/x_n)^{j_1} \cdots (x_{n-1}/x_n)^{j_{n-1}}},$$

hence $\alpha \in K(x_1/x_n, \dots, x_{n-1}/x_n)$, thus $L_0 \subset K(x_1/x_n, \dots, x_{n-1}/x_n)$. The reverse inclusion is obvious. We claim and we omit the prof that $L = K(x_1/x_n, \dots, x_{n-1}/x_n, x_n)$.

Exercise 8.20.
Let K be a field of characteristic 0, $x_1, \ldots, x_n$ be algebraically independent variables over K. Denote by $\overrightarrow{x}$ the n-tuple $(x_1, \ldots, x_n)$. Let Ω be an algebraic closure of $K(x_1, \ldots, x_n)$.

(1) Let $f_1(\overrightarrow{x}), \ldots, f_n(\overrightarrow{x})$ be elements of Ω and let J be the matrix, called the Jacobian matrix of $f_1(\overrightarrow{x}), \ldots, f_n(\overrightarrow{x})$, and given by $J = \begin{pmatrix} \frac{\partial f_1}{\partial x_1} & \cdots & \frac{\partial f_n}{\partial x_1} \\ \vdots & \ddots & \vdots \\ \frac{\partial f_1}{\partial x_n} & \cdots & \frac{\partial f_n}{\partial x_n} \end{pmatrix}$. Show that the following conditions are equivalent.

   (i) The elements $f_1(\overrightarrow{x}), \ldots, f_n(\overrightarrow{x})$ are algebraically independent over K.
   (ii) The matrix J is non singular.

   Show that the condition on the characteristic of K cannot be omitted.

(2) Deduce that if $g_1(\overrightarrow{x}), \ldots, g_n(\overrightarrow{x})$ are rational functions with coefficients in K such that $g_i(\overrightarrow{x}) \in K(x_1, \ldots, x_i)$ and $\deg_{x_i} g_i \geq 1$ for $i = 1, \ldots, n$, then $g_1(\overrightarrow{x}), \ldots, g_n(\overrightarrow{x})$ are algebraically independent over K. Here $\deg_{x_i} g_i$ denotes the maximum among the degrees in x_i of the numerator and the denominator of $g_i(\overrightarrow{x})$.

(3) Let r be a positive integer such that $r \leq n$ and $h_1(\overrightarrow{x}), \ldots, h_r(\overrightarrow{x})$ be elements of Ω. Let $J_r = \begin{pmatrix} \frac{\partial h_1}{\partial x_1} & \cdots & \frac{\partial h_r}{\partial x_1} \\ \vdots & \ddots & \vdots \\ \frac{\partial h_1}{\partial x_n} & \cdots & \frac{\partial h_r}{\partial x_n} \end{pmatrix}$. Show that the following assertions are equivalent.

   (i) The elements $h_1(\overrightarrow{x}), \ldots, h_r(\overrightarrow{x})$ are algebraically independent over K.
   (ii) The rank of the matrix J_r is equal to r.

(4) Let r, $h_1(\overrightarrow{x}), \ldots, h_r(\overrightarrow{x})$ and J_r as in **(3)**. Show that

$$\operatorname{Rank}(J_r) = \operatorname{Trdeg}(K(h_1(\overrightarrow{x}), \ldots, h_r(\overrightarrow{x}))/K).$$

Solution 8.20.

(1) • *(ii)* $\Rightarrow$ *(i)*. Suppose that $f_1(\overrightarrow{x}), \ldots, f_n(\overrightarrow{x})$ are algebraically dependent over K and let $\phi(y_1, \ldots, y_n) \in K[y_1, \ldots, y_n]$, irreducible such that $\phi(f_1, \ldots, f_n) = 0$. Then differentiating this identity relatively

to x_k, for any $k \in \{1, \ldots, n\}$, we get

$$\sum_{i=1}^{n} \frac{\partial \phi}{\partial y_i}(f_1, \ldots, f_n).\frac{\partial f_i}{\partial x_k} = 0.$$

This shows that

$$\frac{\partial \phi}{\partial y_1}(f_1, \ldots, f_n), \ldots, \frac{\partial \phi}{\partial y_n}(f_1, \ldots, f_n)$$

is a solution of a homogeneous system of linear equations, where the coefficients appearing in the i-th equation are the ones in the i-th row of J. Since the characteristic of K is 0, then

$$\deg_{y_i} \frac{\partial \phi}{\partial y_i}(y_1, \ldots, y_n) = \deg_{y_i} \phi(y_1, \ldots, y_n) - 1.$$

If $\frac{\partial \phi}{\partial y_i}(f_1, \ldots, f_n) = 0$ for some $i \in \{1, \ldots, n\}$, then

$$\phi(y_1, \ldots, y_n) \mid \frac{\partial \phi}{\partial y_i}(y_1, \ldots, y_n),$$

thus $\frac{\partial \phi}{\partial y_i}(y_1, \ldots, y_n) = 0$ which implies that ϕ does not depend on y_i. It follows that there exists $i \in \{1, \ldots, n\}$ such that $\frac{\partial \phi}{\partial y_i}(f_1, \ldots, f_n) \neq 0$, hence the mentioned solution of the homogeneous system is non trivial. We conclude that the determinant of the system, that is $Det(J)$, is zero.

- *(i)* $\Rightarrow$ *(ii)*. Let $E = K(x_1, \ldots, x_n, f_1, \ldots, f_n)$. Since $f_1, \ldots, f_n$ are algebraic over $K(x_1, \ldots, x_n)$, then $\mathrm{Trdeg}(E/k) = n$. The assumptions show that $\{f_1, \ldots, f_n\}$ is a transcendence basis of E over K. It follows that for any $i \in \{1, \ldots, n\}$, $x_i, f_1, \ldots, f_n$ are algebraically dependent over K. Hence there exist

$$\phi_i(y_0, y_1, \ldots, y_n) \in K[y_0, y_1, \ldots, y_n]$$

such that

$$\phi_i(x_i, f_1, \ldots, f_n) = 0, \quad i = 1, \ldots, n.$$

Moreover for any $i \in \{1, \ldots, n\}$, we have $\deg_{y_0} \phi_i \geq 1$. Differentiate each of the preceding equations relatively to x_k for $k = 1, \ldots, n$ and obtain

$$\delta_{ik} \frac{\partial \phi_i}{\partial y_0}(x_i, f_1, \ldots, f_n) + \sum_{j=1}^{n} \frac{\partial f_j}{\partial x_k} \frac{\partial \phi_i}{\partial y_j}(x_i, f_1, \ldots, f_n) = 0,$$

where $\delta_{ik} = 1$ if $i = k$ and 0 otherwise. We may write these equations in the form

$$\sum_{j=1}^{n} \frac{\partial f_j}{\partial x_k} \left(\frac{-\frac{\partial \phi_i}{\partial y_j}(x_i, f_1, \ldots, f_n)}{\frac{\partial \phi_i}{\partial y_0}(x_i, f_1, \ldots, f_n)} \right) = \delta_{ik},\ i = 1, \ldots, n,\ k = 1, \ldots, n.$$

This set of equations shows that there exists an $n \times n$ matrix H such that $JH = I_n$, where I_n is the identity matrix. This implies that J is non singular.

Here we look at the condition on the characteristic of K. Let K be any field of characteristic $p > 0$, $n = 2$, $f_1(x_1, x_2) = x_1^p$ and $f_2(x_1, x_2)$ be any polynomial with coefficients in K such that $\deg_{x_2} f_2 \geq 1$, then while these polynomials are algebraically independent over K, their Jacobian matrix is 0. Therefore the equivalence of **(i)** with **(ii)** cannot be generalized for arbitrary characteristic.

(2) Here the Jacobian matrix of $g_1, \ldots, g_n$ is triangular and has no zero on its main diagonal, hence $Det(J) \neq 0$ and then by **(1)**, $g_1, \ldots, g_n$ are algebraically independent over K.

(3) • *(i)* ⇒ *(ii)*. Since $\text{Trdeg}(\Omega/K) = n$ and $h_1, \ldots, h_r$ are algebraically independent over K, then we may complete this set of algebraic function by say $h_{r+1}, \ldots, h_n$ and obtain a transcendence basis of Ω over K. The rank of the Jacobian matrix $\left(\frac{\partial h_i}{\partial x_j}\right)_{\substack{1 \leq i \leq n \\ 1 \leq j \leq n}}$ of $h_1, \ldots, h_n$ is equal to n, hence the extracted matrix formed by the r first columns has rank r. Therefore $\text{Rank}(J_r) = r$.

- *(ii)* ⇒ *(i)*. We may suppose that the matrix $H = \left(\frac{\partial h_i}{\partial x_j}\right)_{\substack{1 \leq i \leq r \\ 1 \leq j \leq r}}$ extracted from J_r is non singular. Let $h_i = x_i$ for $i = r+1, \ldots, n$ and let J be the Jacobian matrix of $h_1, \ldots, h_n$, then it is seen that $Det(J) = Det(H) \neq 0$. Thus, by **(1)**, $h_1, \ldots, h_n$ are algebraically independent over K. It follows that $h_1, \ldots, h_r$ are algebraically independent over K.

(4) Let $s = \text{Rank}(J_r)$ and $t = \text{Trdeg}(K(h_1(\overrightarrow{x}), \ldots, h_r(\overrightarrow{x}))/K)$. We may suppose that the s first column of J_r are linearly independent. The matrix extracted from J_r and constituted by these $s$column has rank s, hence by **(3)**, $h_1, \ldots, h_s$ are algebraically independent over K. Thus $s \leq t$. On the other hand, since $t \leq r$, we may suppose that $h_1, \ldots, h_t$ are algebraically independent over K. The Jacobian matrix H of $h_1, \ldots, h_t$ is then of rank t, by **(3)**. Since H is extracted from J_r, then $s = \text{Rank}(J_r) \geq t$. We conclude that $s = t$.

Exercise 8.21.
Let K be a field, $f(x) = \frac{a(x)}{b(x)}$ and $g(x) = \frac{c(y)}{d(y)}$ be non constant rational functions with coefficients in K such that $\gcd(a(x), b(x)) = \gcd(c(x), d(x)) = 1$.

(1) Show that the following propositions are equivalent.

(i) There exists $h(x) \in K(x) \backslash K$ such that $f = h \circ g$.
(ii) The polynomial $d(x)d(y)(g(x) - g(y))$ divides $b(x)b(y)(f(x) - f(y))$ in $K[x, y]$.

(2) Let $J(x, y) \in K[x, y]$. Show that if there exist $u(x), v(x) \in K[x]$ such that $J(x, y) = u(x)v(y) - v(x)u(y)$, then

(a) $J(y, x) = -J(x, y)$
(b) $J(w, x)J(y, z) + J(w, y)J(z, x) + J(w, z)J(x, y) = 0$.

Show that the converse holds if K is infinite.

(3) Let $f(x) = (x^4 - x^3 - 8x - 1)/(2x^4 + x^3 - 16x + 1) \in \mathbb{C}(x)$. Decompose $f(x)$ in the form $f(x) = h(g(x))$, where $g(x)$ and $h(x) \in \mathbb{C}(x)$ with $\deg g = \deg h = 2$.

(4) Deduce that there exist exactly three fields F_i for $i = 1, 2, 3$ such that $\mathbb{C}(f) \subsetneq F_i \subsetneq \mathbb{C}(x)$. Moreover they are given by $F_i = k(g_i)$, where

$$\begin{aligned} g_1(x) &= (x^2 - 2x)/(2x^2 - 3x + 1), \\ g_2(x) &= (x^2 - 2jx)/(2x^2 - 3jx + j^2) \quad \text{and} \\ g_3(x) &= (x^2 - 2j^2x)/(2x^2 - 3j^2x + j). \end{aligned}$$

(5) Show that $\mathbb{C}(g_i, g_j) = \mathbb{C}(x)$ for $i, j \in \{1, 2, 3\}, i \neq j$. Express x as a rational function in g_1, g_2 with complex coefficients.

Solution 8.21.

(1) • **(i) ⇒ (ii)**. We have

$$\begin{aligned} d(x)d(y)(g(x) - g(y)) &= c(x)d(y) - c(y)d(x) \quad \text{and} \\ b(x)b(y)(f(x) - f(y)) &= a(x)b(y) - a(y)b(x). \end{aligned}$$

The assumption **(i)** implies that $K(f) \subset K(g) \subset K(x)$. Moreover x is a root of $a(z) - b(z)f \in K(f)[z]$ and of $c(z) - d(z)g \in K(g)[z]$. This last polynomial is irreducible in $K[g, z]$, hence also in $K(g)[z]$, then it is equal to the minimal polynomial of x over $K(g)$ up to a multiplication by some element of the form $\lambda - \mu g \in K[g]$. Since x is also a root of

$$a(z) - b(z)f = a(z) - b(z)h(g),$$

then $c(z)-d(z)g$ divides $a(z)-h(g)b(z)$ in $K(g)[z]$, hence $c(z)-d(z)g$ divides $a(z)-h(g)b(z)$ in $K[g,z]$. Set

$$a(z)-h(g)b(z)=(c(z)-gd(x))q(g,z),$$

where $q(g,z)\in K[g][z]$. We deduce that

$$a(z)-\frac{a(x)}{b(x)}b(z)=\left(c(z)-\frac{c(x)}{d(x)}d(z)\right)q\left(\frac{c(x)}{d(x)},z\right).$$

Multiply by $b(x)d(x)^{e+1}$ and obtain:

$$d(x)^{e+1}(a(z)b(x)-a(x)b(z))=(c(z)d(x)-c(x)d(z))d(x)^{e}q\left(\frac{c(x)}{d(x)},z\right).$$

- **(ii) $\Rightarrow$ (i).** Since $c(x)d(y)-c(y)d(x)$ divides $a(x)b(y)-a(y)b(x)$ in $K[x,y]$, then $c(y)-\frac{c(x)}{d(x)}d(y)$ divides $a(y)-\frac{a(x)}{b(x)}b(y)$ in $K(x)[y]$, that is $c(y)-g(x)d(y)$ divides $a(y)-f(x)b(y)$ in $K(x)[y]$. Performing Euclidean divisions in $K(g(x))[y]$, we obtain

$$a(y)=(c(y)-g(x)d(y))q_1(g,y)+r_1(g,y)\quad\text{and}$$
$$b(y)=(c(y)-g(x)d(y))q_2(g,y)+r_2(g,y),$$

where $q_1,q_2,r_1,r_2\in K(g(x))[y]$ and $\deg r_1<\deg g$. $\deg_y r_2<\deg g$. Moreover $r_1\neq 0$ and $r_2\neq 0$. We deduce that

$$a(y)-f(x)b(y)=(c(y)-g(x)d(y))(q_1-f(x)q_2)+r_1-f(x)r_2.$$

Our assumption implies $r_1-f(x)r_2=0$. Let $p(g)$ (resp. $q(g)$) be the leading coefficient of r_1 (resp. r_2) as a polynomial in y with coefficients in $K(g)$, then $f(x)=\frac{p(g)}{q(g)}=h(g)$ with $h(x)=\frac{p(x)}{q(x)}$.

(2) Let $u(x),v(x)\in K[x]$ such that

$$J(x,y)=u(x)v(y)-v(x)u(y),$$

then clearly **(a)** holds. The identity **(b)** is a particular case of the following one:

$$(AB-CD)(FG-HE)+(AE-FD)(HB-CG)+(AG-HD)(CE-FB)=0$$

after setting

$$A=u(w),\ B=v(x),\ C=u(x),\ D=v(w),$$
$$E=v(y),\ F=u(y),\ G=v(z),\ \text{and}\ H=u(z).$$

This last identity may be verified by writing its left side in the form $A\gamma+D\mu$. It is readily seen that $\gamma=\mu=0$.

We now prove the converse. Suppose that $J(x,y)$ satisfies **(a)** and **(b)**. Let $w=\alpha$ and $z=\beta$ in **(b)**, where α and β are elements of K to be determined. Then

$$J(\alpha,x)J(y,\beta)+J(\alpha,y)J(\beta,x)+J(\alpha,\beta)J(x,y)=0.$$

Since K is infinite and $J\neq 0$, then there exists $(\alpha,\beta)\in K^2$ such that $J(\alpha,\beta)\neq 0$.
Therefore, we get:

$$J(x,y)=-J(\alpha,x)\frac{J(y,\beta)}{J(\alpha,\beta)}-J(\alpha,y)\frac{J(\beta,x)}{J(\alpha,\beta)}.$$

Using the identity **(a)**, we obtain

$$J(x,y)=J(x,\alpha)\frac{J(y,\beta)}{J(\alpha,\beta)}-J(y,\alpha)\frac{J(x,\beta)}{J(\alpha,\beta)},$$

so that

$$J(x,y)=u(x)v(y)-u(y)v(x)$$

where $u(x)=J(x,\alpha)$ and $v(x)=\frac{J(x,\beta)}{J(\alpha,\beta)}$. The assumption $J(\alpha,\beta)\neq 0$, made on (α,β) implies that $J(x,\beta)\neq 0$, hence $v(x)\neq 0$.

(3) We begin with a claim. Let $J(x,y)\in K[x,y]$ such that $J(y,x)=-J(x,y)$, then $J(x,y)=(x-y)H(x,y)$, where $H(x,y)\in K[x,y]$ is symmetric. For the proof, let

$$J(x,y)=(x-y)H(x,y)+R(x)$$

be the Euclidean division of J by $x-y$ in $K[x][y]$, then $R\in K[x]$. We have $0=J(x,x)=R(x)$. We deduce that $H(x,y)$ is symmetric.
Let $g(x)\in K(x)$, $\deg g=2$ such that $f(x)=h(g(x))$ for some $h(x)\in K(x)$. Let $a(x),b(x)\in K[x]$ such that $\gcd(a(x),b(x))=1$ and $g(x)=a(x)/b(x)$. By **(1)** the polynomial

$$a(x)b(y)-a(y)b(x) \text{ divides } u(x)v(y)-u(y)v(x),$$

where $u(x)$ (resp. $v(x)$) is the numerator (resp. denominator) of $f(x)$. Easy computations show that

$$F(x,y):=\frac{u(x)v(y)-u(y)v(x)}{x-y}=3x^3y^3+3(x+y)(x^2+y^2+8xy)-24.$$

Let $G(x,y)=\frac{a(x)b(y)-a(y)b(x)}{x-y}$. Write this polynomial in the form of a sum of its homogeneous components. Its leading homogeneous component say $G^+(x,y)$ divides $3x^3y^3$ and is symmetric. Suppose first that

$v(x) \in \mathbb{C}^*$, that is $g(x) \in \mathbb{C}[x]$, then we may express the numerator and the denominator of $f(x)$ in the form:

$$x^4 - x^3 - 8x - 1 = a(g(x)) \quad \text{and}$$
$$2x^4 + x^3 - 16x + 1 = b(g(x)),$$

where $a(x), b(x) \in \mathbb{C}[x]$ and $\gcd(a(x), b(x)) = 1$. Differentiating these equations, we obtain

$$4x^3 - 3x^2 - 8 = a'(g(x))g'(x) \quad \text{and}$$
$$8x^3 + 3x^2 - 16 = b'(g(x))g'(x).$$

It follows that $g'(x)$ divides the polynomials $4x^3 - 3x^2 - 8$ and $8x^3 + 3x^2 - 16$. But is easy to see that these two polynomials have no common root, hence they are relatively prime which implies $g'(x) \in \mathbb{C}$ and then $\deg g = 1$, which contradicts our assumption. We conclude that $g(x)$ is not polynomial. Now $g(x) = \frac{u(x)}{v(x)}$ with u, v non constant and $Max(\deg u, \deg v) = 2$. In this case it is seen that $\deg G^+(x, y) \leq 2$. Since G^+ is symmetric and divides $3x^3y^3$, then $G^+(x, y) = axy$, where $a \in \mathbb{C}^\star$. Moreover, replacing $u(x)$ by $u(x)/a$, if necessary, we may suppose that $G^+(x, y) = xy$. Let $H(x, y) \in \mathbb{C}[x, y]$ such that $F(x, y) = G(x, y)H(x, y)$. Set

$$G(x, y) = G_2 + G_1 + G_0 \quad \text{and}$$
$$H(x, y) = H_4 + H_3 + H_2 + H_1 + H_0,$$

where G_2, G_1, G_0 (resp. H_4, H_3, H_2, H_1, H_0) represent the homogeneous components of G (resp. H). Here the indices of the components are equal to their degrees. By identification in the identity $F = G \cdot H$, we obtain the following equations

$$G_2H_4 = 3x^3y^3 \qquad \text{(Eq 1)}$$
$$G_2H_3 + G_1H_4 = 0 \qquad \text{(Eq 2)}$$
$$G_2H_2 + G_1H_3 + G_0H_4 = 0 \qquad \text{(Eq 3)}$$
$$G_2H_1 + G_1H_2 + G_0H_3 = 3(x + y)(x^2 + y^2 + 8xy) \qquad \text{(Eq 4)}$$
$$G_2H_0 + G_1H_1 + G_0H_2 = 0 \qquad \text{(Eq 5)}$$
$$G_1H_0 + G_0H_1 = 0 \qquad \text{(Eq 6)}$$
$$G_0H_0 = -24. \qquad \text{(Eq 7)}$$

Since $G_2 = xy$, then by **(Eq 1)**, $H_4 = 3x^2y^2$. Replacing in **(Eq 2)**, G_2 and H_4 by their values, we obtain: $H_3 = -3xyG_1$. Since G_1 is homogeneous of degree 1 and symmetric, then $G_1 = a(x+y)$ with $a \in \mathbb{C}^\star$, hence $H_3 = -3axy(x+y)$. Set $G_0 = b$ with $b \in \mathbb{C}$, then replacing in **(Eq 3)**, G_1, G_2, H_3, H_4 and G_0 by their values, we get: $H_2 = 3a^2(x+y)^2 - 3bxy$. From **(Eq 6)**, we deduce that $H_1 = \frac{24a}{b^2}(x+y)$. We put in **(Eq 4)** the values of G_2, H_1, G_1, H_2, G_0 and H_3 and we obtain

$$3a^3(x^2+y^2) + (\frac{24a}{b^2} + 6a^3 - 6ab)xy = 3(x^2 + 8xy + y^2),$$

hence $a^3 = 1$ and $\frac{a(4-b^3+a^2b^2)}{b^2} = 4$. Therefore a and b are determined by the following equations $a^3 = 1$ and $ab^3 + 3b^2 - 4a = 0$. Now a is a cube root of unity and b is a root of the polynomial

$$c(x) = ax^3 + 3x - 4a = a(x-a^2)(x^2 + 4a^2x + 4a) = a(x-a^2)(x+2a^2)^2.$$

We deduce that $b = a^2$ or $b = -2a^2$. We examine each of these values of b.

- **Case** $b = a^2$.
  Recall the preceding computations of the H_i, G_j. We have:

$$\begin{aligned} &G_0 = a^2, \quad G_1 = a(x+y), \quad G_2 = xy, \quad H_0 = -24a, \\ &H_1 = 24(x+y), \quad H_2 = 3a^2(x^2+y^2+xy), \\ &H_3 = -3axy(x+y) \quad \text{and} \quad H_4 = 3x^2y^2. \end{aligned}$$

  Replace them in **(Eq 5)**, and obtain $x^2 + y^2 + 2xy = 0$, which is a contradiction. Therefore, this value of b must be rejected.
- **Case** $b = -2a^2$.
  Here we have

$$G_0 = -2a^2, \quad G_1 = a(x+y) \quad \text{and} \quad G_2 = xy.$$

  It follows that the polynomial $J(x,y) = a(x)b(y) - a(y)b(x)$ is given by:

$$\begin{aligned} J(x,y) = (x-y)G(x,y) &= (x-y)(-2a^2 + a(x+y) + xy) \\ &= x^2y - xy^2 + ax^2 - ay^2 - 2a^2x + 2a^2y. \end{aligned}$$

  We verify if the conditions **(a)** and **(b)** of **(2)** are satisfied.

Obviously **(a)** is fulfilled. We have

$$\begin{aligned}
&J(w,x)J(y,z)+J(w,y)J(z,x)+J(w,z)J(x,y)\\
&=(w^2x-wx^2+aw^2-ax^2-2a^2w+2a^2x)\\
&\quad.(y^2z-zy^2+ay^2-az^2-2a^2y+2a^2z)\\
&\quad+(w^2y-wy^2+aw^2-ay^2-2a^2w+2a^2y)\\
&\quad.(z^2x-xz^2+az^2-ax^2-2a^2z+2a^2x)\\
&\quad+(w^2z-z^2w+aw^2-az^2-2a^2w+2a^2z)\\
&\quad.(x^2y-xy^2+ax^2-ay^2-2a^2x+2a^2y)=0.
\end{aligned}$$

So that condition **(b)** is satisfied. Since $\mathbb{C}$ is infinite, we may choose $(\alpha,\beta)\in\mathbb{C}^2$ such that $J(\alpha,\beta)\neq 0$. For example $\alpha=0$ and $\beta=a$ works. We have $J(0,a)=1$ and then

$$a(x)=G(x,0)=ax^2-2a^2x \text{ and } b(x)=G(x,a)=2ax^2-3a^2x+1.$$

Therefore

$$g(x)=\frac{ax^2-2a^2x}{2ax^2-3ax+1}=\frac{x^2-2ax}{2x^2-3ax+a^2}.$$

We now compute the rational function $h(x)\in\mathbb{C}(x)$ such that $f(x)=h(g(x))$. Set $h(x)=\frac{c(x)}{d(x)}$, where $c(x),d(x)\in\mathbb{C}[x]$, $\gcd(c(x),d(x))=1$ and $Max(\deg c,\deg d)=2$. Suppose first that $\deg c=2$ and $\deg d=1$, then

$$\frac{b^2(x)c\left(\frac{a(x)}{b(x)}\right)}{b^2(x)d\left(\frac{a(x)}{b(x)}\right)}=f(x)=\frac{x^4-x^3-8x-1}{2x^4+x^3-16x+1},$$

hence $b(x)|2x^4+x^3-16x+1$, that is $2x^2-3ax+a^2|2x^4+x^3-16x+1$. This is a contradiction since $b(a)=0$ and $v(a)\neq 0$. Similarly we eliminate the case $\deg c=1$ and $\deg d=2$. It remains to consider the case $\deg c=\deg d=2$. We may suppose that $c(x)$ is monic. Set

$$c(x)=x^2+\lambda x+\mu \text{ and } d(x)=\alpha x^2+\beta x+\gamma.$$

Using the identity $h(g(x))=f(x)$, we obtain

$$\begin{aligned}
&\frac{(x^2-2ax)^2+\lambda(x^2-2ax)(2x^2-3ax+a^2)+\mu(2x^2-3ax+a^2)^2}{\alpha(x^2-ax)^2+\beta(x^2-ax)(2x^2-3ax+a^2)+\gamma(2x^2-3ax+a^2)^2}\\
&=\frac{x^4-x^3-8x-1}{2x^4+x^3-16x+1}.
\end{aligned}$$

This identity may be written in the form

$$A(x)/B(x) = \frac{x^4 - x^3 - 8x - 1}{2x^4 + x^3 - 16x + 1},$$

where

$$\begin{aligned} A(x) &= (1 + 2\lambda + 4\mu)x^4 - a(4 + 7\lambda + 12\mu)x^3 \\ &\quad + a^2(4 + 7\lambda + 13\mu)x^2 - (2\lambda + 6\mu)x + a\mu \end{aligned}$$

and

$$\begin{aligned} B(x) &= (\alpha + 2\beta + 4\gamma)x^4 - a(4\alpha + 7\beta + 12\gamma)x^3 \\ &\quad + a^2(4\alpha + 7\beta + 13\gamma)x^2 - (2\beta + 6\gamma)x + a\gamma \end{aligned}$$

It follows that:

$$\begin{aligned} -a(4 + 7\lambda + 12\mu)/(1 + 2\lambda + 4\mu) &= -1 \\ a^2(4 + 7\lambda + 12\mu)/(1 + 2\lambda + 4\mu) &= 0 \\ -(4 + 7\lambda + 12\mu)/(1 + 2\lambda + 4\mu) &= -8 \\ a\mu/(1 + 2\lambda + 4\mu) &= -1 \end{aligned}$$

and

$$\begin{aligned} (4\alpha + 7\beta + 12\gamma)/(1 + 2\lambda + 4\mu) &= 2 \\ -a(4\alpha + 7\beta + 12\gamma)/(1 + 2\lambda + 4\mu) &= 1 \\ a^2(4\alpha + 7\beta + 12\gamma)/(1 + 2\lambda + 4\mu) &= 0 \\ -(2\beta + 6\gamma)/(1 + 2\lambda + 4\mu) &= -16 \\ \alpha\gamma/(1 + 2\lambda + 4\mu) &= 1. \end{aligned}$$

Solving the first four equations, we obtain $\lambda = -\frac{3+4a}{2+7a}$ and $\mu = \frac{1}{2+7a}$. We deduce that $1 + 2\lambda + 4\mu = -a/(2 + 7a)$. The five last equations lead to

$\alpha = (-2+14a)/(2+7a)$, $\beta = (3-8a)/(2+7a)$ and $\gamma = -1/(2+7a)$.

We conclude that

$$h(x) = \frac{(2 + 7a)x^2 - (3 + 4a)x + 1}{(-2 + 14a)x^2 + (3 - 8a)x - 1}.$$

Conversely, it is easy to verify that this $h(x)$ satisfies the identity $g(h(x)) = f(x)$. Indeed there are three decompositions of $f(x)$ in the form $f(x) = h(g(x))$, where $\deg g = \deg h = 2$. Each one corresponds to the values of $g(x)$ and $h(x)$ computed above and to the choice $a \in \{1, j, j^2\}$, where j is a primitive cube root of unity in $\mathbb{C}$.

(4) Let F be a field such that $\mathbb{C}(f) \subset\neq F \subset\neq \mathbb{C}(x)$. By Luroth's Theorem [Schinzel (2000), Th. 2, Chap. 1.1], there exists $g(x) \in \mathbb{C}(x)$ such that $F = \mathbb{C}(g)$. Moreover since $[\mathbb{C}(x) : \mathbb{C}(f)] = d = 2$, then $[\mathbb{C}(x) : F] = 2$, then $\deg g = 2$. Since $f \in F$, then $f = h(g)$ where $h(x) \in \mathbb{C}(x)$ and $\deg h = 2$. It follows by **(3)** that $g(x) = \frac{x^2-2ax}{2x^2-3ax+a^2}$, where a is a cube root of unity in $\mathbb{C}$. Fix a primitive cube root of unity j and let F_1, F_2, F_3 be the fields corresponding to $a = 1$ or $a = j$ or $a = j^2$ respectively. Denote by $g_1(x), g_2(x)$ and $g_3(x)$ the corresponding rational functions. We show that the fields F_1, F_2, F_3 are pairwise distinct. Suppose that $F_2 = F_1$, then there exist $p, q, r, s \in \mathbb{C}$ such that

$$\frac{x^2 - 2jx}{2x^2 - 3jx + j^2} = \frac{p\frac{x^2-2x}{2x^2-3x+1} + q}{r\frac{x^2-2x}{2x^2-3x+1} + s}$$

and $ps - rq \neq 0$. Hence

$$\frac{(p + 2q)x^2 - (2p + 3q)x + q}{(r + 2s)x^2 - (2r + 3s)x + s} = \frac{x^2 - 2jx}{2x^2 - 3jx + j^2}.$$

We conclude immediately that $q = 0$, and then

$$\frac{x^2 - 2x}{\frac{r+2s}{p}x^2 - \frac{2r+3s}{p}x + \frac{s}{p}} = \frac{x^2 - 2jx}{2x^2 - 3jx + j^2},$$

which is a contradiction. Similarly we may prove that $F_1 \neq F_3$ and $F_2 \neq F_3$.

(5) The assertion on the composition of the fields F_i is obvious. We write x in the form: $x = \phi(g_1, g_2)/(\psi(g_1, g_2)$, with

$$g_1(x) = (x^2 - 2x)/(2x^2 - 3x + 1) \quad \text{and}$$
$$g_2(x) = (x^2 - 2jx)/(2x^2 - 3jx + j^2).$$

From the first equation, we deduce that

$$x^2 = (2x + g_1(1 - 3x))/(1 - 2g_1).$$

Replace this value of x^2 in the second equation and obtain:

$$g_2 = \frac{x(2 - 2j + g_1(4j - 3)) + g_1}{x(4 - 3j + g_1(6j - 6)) + j^2 + g_1(2 - 2j^2)}.$$

Therefore

$$x = \frac{j^2 - g_1 + (2 - 2j^2)g_1g_2}{2 - 2j + (4j - 3)g_1 + (3j - 4)g_2 + (6 - 6j)g_1g_2}.$$

Exercise 8.22.

Let K be a field, Ω be a algebraic closure of K, $u(t)$ and $v(t)$ be non constant polynomials with coefficients in K.

(1) Show that there exists $P(x, y) \in K[x, y]$ irreducible such that $P(u(t), v(t)) = 0$. Show that, for any $f(x, y) \in K[x, y]$, if $f(u(t), v(t)) = 0$ then $P(x, y) \mid f(x, y)$.

(2)(a) Show that there exist $u_1(t)$ and $v_1(t) \in K[t]$ such that $P(u_1(t), v_1(t)) = 0$, $\deg_x P = \deg v_1$ and $\deg_y P = \deg u_1$.

(b) Considered as a polynomial in x (resp. y) with coefficients in $K[y]$ (resp. $K[x]$), show that the leading coefficients of $P(x, y)$ is in K^*.

(3) Show that $P(x, y)$ is irreducible in $\Omega[x, y]$.

(4) Show that $P(x, y)$ is irreducible in $\Omega((\frac{1}{x}))[y]$ and in $\Omega((\frac{1}{y}))[x]$.

Solution 8.22.

(1) Consider the chain of fields $K(u) \subset K(u, v) \subset K(t) = K(u, t)$. Since t is algebraic over $K(u)$, then v is algebraic over $K(u)$. Let

$$F(y) = y^m + \frac{a_{m-1}(u)}{D(u)} y^{m-1} + \cdots + \frac{a_0(u)}{D(u)}$$

be the minimal polynomial of v over $K(u)$, where $D(x)$, $a_i(x) \in K[x]$ and $\gcd(D(x), a_0(x), \ldots, a_{m-1}(x)) = 1$. Let

$$P(x, y) = D(x)y^m + a_{m-1}(x)y^{m-1} + \cdots + a_0(x),$$

then $P(x, y)$ is irreducible over K and $P(u(t), v(t)) = 0$. Let $f(x, y) \in K[x, y]$ such that $f(u(t), v(t)) = 0$. Let $q(x, y), r(x, y) \in K(x)[y]$ such that $f = Pq + r$ and $\deg_y r < \deg_y P$. We have $r(u(t), v(t)) = 0$, hence $P(x, y)$ divides $r(x, y)$ in $K(x)[y]$. It follows that $r(x, y) = 0$ and then $f = Pq$. We compare the contents in $K(x)$ of the polynomials appearing in the two sides of this identity. We obtain

$$\operatorname{cont}(f) = \operatorname{cont}(P) \operatorname{cont}(q) = \operatorname{cont}(q).$$

It follows that $\operatorname{cont}(q) \in K[x]$. Therefore $q \in K[x, y]$ and then $P(x, y) \mid f(x, y)$ in $K[x, y]$.

Remark. We may prove this property of divisibility in the following way. Let $\psi : K[x, y] \to K[t]$ be the unique K-morphism of algebras such that $\psi(x) = u(t)$ and $\psi(y) = v(t)$, then

$$K[x, y]/\operatorname{Ker}\psi \simeq \operatorname{Im}\psi \subset K[t].$$

Therefore $\operatorname{Ker}\psi$ is a non zero prime ideal of $K[x, y]$ containing $P(x, y)$. We prove that $\operatorname{Ker}\psi = P(x, y)K[x, y]$. Since the image of ψ is not a field, then $\operatorname{Ker}\psi$ is not maximal. Let $\mathcal{M}$ be a maximal ideal of $K[x, y]$ containing $\operatorname{Ker}\psi$. Consider the chain of prime ideals

$$\{0\} \subset P(x, y)K[x, y \subset \operatorname{Ker}\psi \subset \mathcal{M}.$$

Since the Krull dimension of $K[x,y]$ is equal to 2, then $\operatorname{Ker}\psi = (P)$. Hence $P(x,y) \mid f(x,y)$ for any $f(x,y) \in K[x,y]$ satisfying the condition $f(u(t),v(t)) = 0$.

(2)(a) Luroth's Theorem [Schinzel (2000), Th. 2, Chap. 1.1], shows that there exists $\phi(t) \in K(t)$ such that $K(u,v) = K(\phi)$. Moreover **Exercise 8.10** implies that we may choose $\phi(t) \in K[t]$. This choice being made, let $u_1(x) = A(x)/B(x) \in K(x)$ such that $\gcd(A,B) = 1$ and

$$u(t) = u_1(\phi(t)) = A(\phi(t))/B(\phi(t)).$$

Bezout's Theorem implies that there exist $\lambda(x)$ and $\mu(x) \in K[x]$ such that $\lambda(x)A(x) + \mu(x)B(x) = 1$. We deduce that

$$\lambda(\phi(t))A(\phi(t)) + \mu(\phi(t))B(\phi(t)) = 1.$$

Since $B(\phi(t)) \mid A(\phi(t))$, then $B(\phi(t)) \in K^*$. Therefore $B(t) \in K^*$, ie $u_1(x) \in K[x]$. Similarly if we set $v(t) = v_1(\phi(t))$, then $v_1(x) \in K[x]$. We have

$$0 = P(u(t),v(t)) = P(u_1(\phi(t)), v_1(\phi(t))),$$

hence $P(u_1(t), v_1(t)) = 0$. Since

$$K(u_1(\phi(t)), v_1(\phi(t))) = K(\phi(t)),$$

then $K(u_1(z), v_1(z)) = K(z)$. Let $A(x)$ (resp. $B(y)$) be the leading coefficient of $P(x,y)$ considered as a polynomial in y (resp. in x), then $\frac{P(u_1(z),y)}{A(u_1(z))}$ is the minimal polynomial of $v_1(z)$ over $K(u_1(z))]$, hence

$$\deg_y P = [K(u_1,v_1) : K(u_1)] = [K(z) : K(u_1(z))] = \deg u_1.$$

A similar reasoning leads to $\deg_x P(x,y) = \deg v_1$.

(b) Clearly z is integral over $K[u_1(z)]$. Therefore $v_1(z)$ in integral over $K[u_1(z)]$ and then the leading coefficient of $P(x,y)$ as a polynomial in x is in K^*. Therefore we may suppose that $P(x,y)$ is monic relatively to y.

(3) We apply the results proved in **(1)**, **(2)(a)** and **(2)(b)** for $u(t), v(t)$ when K is replaced by Ω. We obtain the polynomials $\overline{P(x,y)} \in \Omega[x,y]$, $\overline{\phi(t)}$, $\overline{u_1(t)}, \overline{v}_1(t) \in \Omega[t]$ such that $\overline{P(x,y)}$ is irreducible, monic in x with moreover the following properties

$$\bar{P}(u(t),v(t)) = \bar{P}(\bar{u}_1(t), \bar{v}_1(t)) = 0,$$

$$\deg_x \bar{P} = \deg \bar{v}_1,\ \deg_y \bar{P}(x,y) = \deg \bar{u}_1 \text{ and } \Omega(u(t),v(t)) = \Omega(\bar{\phi}(t)).$$

By **(1)** we conclude that $\bar{P}(x,y) \mid P(x,y)$ in $\Omega[x,y]$. From

$$K(u(t),v(t)) = K(\phi(t))$$

we deduce that $\Omega(u(t),v(t)) = \Omega(\phi(t))$, hence we may suppose that $\bar{\phi}(t) = \phi(t)$ and then $\overline{u_1(t)} = u_1(t)$ and $\overline{v_1(t)} = v_1(t)$.
Now since

$$\deg_y \bar{P}(x,y) = \deg u_1(t) = \deg_y P(x,y),$$

$\bar{P}(x,y)|P(x,y)$ and these two polynomials are monic in y, then $\bar{P}(x,y) = P(x,y)$. We conclude that $P(x,y)$ is irreducible over Ω.

(4) Let $n = \deg u_1(t) = \deg_y P(x,y)$. Set

$$P(x,y) = y^n + a_{n-1}(x)y^{n-1} + \cdots + a_0(x),$$

where $a_i(x) \in K[x]$ for $i = 1,\ldots,n-1$. Let $\hat{\Omega}$ be an algebraic closure of $\Omega((x))$ and let $\beta : \omega(u_1(t)) \to \Omega((x))$ be the unique Ω-embedding such that $\beta(u_1) = \frac{1}{x}$. Since v_1 is algebraic over $K(u_1)$, there exists an extension $\hat{\beta} : \Omega(u_1,v_1) = \Omega(t) \to \hat{\Omega}$. Since $P(u_1,v_1) = 0$, then $\hat{\beta}(P(u_1,v_1)) = 0$, hence $P(\frac{1}{x},\hat{\beta}(v_1)) = 0$. Therefore $\hat{\beta}(v_1)$ is a root of the monic polynomial $P(\frac{1}{x},y) \in \Omega((x))[y]$ of degree n. Since

$$\Omega(u_1)(v_1) = \Omega(t) = \Omega(u_1)\left(\frac{1}{t}\right),$$

then

$$\Omega((x))(\hat{\beta}(v_1)) = \Omega((x))\left(\frac{1}{\hat{\beta}(t)}\right),$$

hence

$$[\Omega((x))(\hat{\beta}(v_1)) : \Omega((x))] = \left[\Omega((x))\left(\frac{1}{\hat{\beta}(t)}\right) : \Omega((x))\right]. \qquad \text{(Eq 1)}$$

Set

$$u_1(t) = a_0t^n + a_1t^{n-1} + \cdots + a_n,$$

where $a_0,\ldots,a_n \in K$ and $a_0 \neq 0$, then

$$t^n + \frac{a_1}{a_0}t^{n-1} + \cdots + \frac{a_n}{a_0} - \frac{u_1}{a_0} = 0.$$

Set $c = \frac{1}{a_0}$ and $c_i = \frac{a_i}{a_0}$ for $i \in \{1,\ldots,n\}$, then

$$t^n + c_1t^{n-1} + \cdots + c_n - cu_1 = 0.$$

Divide this equation by $u_1 t^n$ and obtain

$$(c_n u_1^{-1} - c)\left(\frac{1}{t}\right)^n + c_{n-1}u_1^{-1}\left(\frac{1}{t}\right)^{n-1} + \cdots + c_1 u_1^{-1}\frac{1}{t} + u_1^{-1} = 0.$$

The element $w = c_n u_1^{-1} - c$ is non zero and we obtain:

$$\left(\frac{1}{t}\right)^n + c_{n-1}u_1^{-1}w^{-1}\left(\frac{1}{t}\right)^{n-1} + \cdots + c_1 u_1^{-1}w^{-1}\frac{1}{t} + u_1^{-1}w^{-1} = 0.$$

We deduce that

$$\left(\hat{\beta}\left(\frac{1}{t}\right)\right)^n + c_{n-1}x\hat{\beta}(w)^{-1}\left(\hat{\beta}\left(\frac{1}{t}\right)\right)^{n-1} + \cdots + c_1 x\hat{\beta}(w)^{-1}\hat{\beta}\left(\frac{1}{t}\right) + x\hat{\beta}(w)^{-1} = 0.$$

We have

$$\hat{\beta}(w) = \hat{\beta}(c_n u^{-1} - c) = c_n x - c,$$

hence $\hat{\beta}(w)$ is a unit of $\Omega[[x]]$ and then $\hat{\beta}(w)^{-1} \in \Omega[[x]]$ is also a unit. Now $\hat{\beta}(\frac{1}{t})$ is a root of the polynomial $h(x,y) \in \Omega((x))[y]$,

$$h(x,y) = y^n + c_{n-1}x\hat{\beta}(w)^{-1}y^{n-1} + \cdots + c_1 x\hat{\beta}(w)^{-1} + x\hat{\beta}(w)^{-1}.$$

By application of Eisenstein's irreducibility criterion we conclude that $h(x,y)$ is irreducible over $\Omega((x))$. It follows that

$$\left[\Omega(x)\left(\hat{\beta}\left(\frac{1}{t}\right)\right) : \Omega((x))\right] = \deg_y h = n$$

and then, by **(Eq 1)**,

$$[\Omega((x))(\hat{\beta}(v_1)) : \Omega((x))] = n = \deg_y P\left(\frac{1}{x}, y\right).$$

We conclude that $P(\frac{1}{x}, y)$ is irreducible over $\Omega((x))$.

Chapter 9

Multivariate polynomials

Exercise 9.1.
Let $f(x,y)$ and $g(x,y)$ be polynomials with integral coefficients. Suppose that $\mathbb{C}[f,g]=\mathbb{C}[x,y]$ and $\frac{\delta f}{\delta x}(0,0)\frac{\delta g}{\delta y}(0,0)-\frac{\delta f}{\delta y}(0,0)\frac{\delta g}{\delta x}(0,0)=\pm 1$.

(1) Show that $\mathbb{Z}[f,g]=\mathbb{Z}[x,y]$.
(2) Show that the condition on the partial derivatives of of $f(x)$ and $g(x)$ cannot be omitted.

Solution 9.1.

(1) Let $f_1(x,y)=f(x,y)-f(0,0)$ and $g_1(x,y)=g(x,y)-g(0,0)$, then

$$f_1(x,y),g_1(x,y)\in\mathbb{Z}[x,y],$$

$$\mathbb{C}[f_1,g_1]=\mathbb{C}[f,g]=\mathbb{C}[x,y]$$

and

$$\frac{\delta f_1}{\delta x}(0,0)\frac{\delta g_1}{\delta y}(0,0)-\frac{\delta f_1}{\delta y}(0,0)\frac{\delta g_1}{\delta x}(0,0)=\pm 1.$$

Clearly if $\mathbb{Z}[f_1,g_1]=\mathbb{Z}[x,y]$, then $\mathbb{Z}[f,g]=\mathbb{Z}[x,y]$, hence we may suppose that $f(0,0)=0$ and $g(0,0)=0$. Set

$$f(x,y)=f_{10}x+f_{01}y+\text{ higher degree terms}\quad\text{and}$$
$$g(x,y)=g_{10}x+g_{01}y+\text{ higher degree terms},$$

where $f_{10}g_{01}-f_{01}g_{10}=\pm 1$. Making the following change of variables,

$$x'=f_{10}x+f_{01}y,\quad y'=g_{10}x+g_{01}y,$$

allows us to write f and g in the form: $f=x'+$ higher degree terms and $y=y'+$ higher degree terms, hence we may suppose that

$$f=x+\text{ higher degree terms}$$

and

$$g = y + \text{ higher degree terms.}$$

Since $\mathbb{C}[f,g] = \mathbb{C}[x,y]$, then $x = \sum_{(i,j)} c_{ij} f^i g^j$, where $c_{ij} \in \mathbb{C}$ for any (i,j). We are going to prove that $c_{ij} \in \mathbb{Z}$, that is $x \in \mathbb{Z}[f,g]$. It will appear from the proof that the same reasoning works for y.
Define the following order relation in $\mathbb{Z}^2$. We say that $(i,j) > (i',j')$ if $i+j > i'+j'$ or $i+j = i'+j'$ and $i > i'$. It is a total order in $\mathbb{Z}^2$. We prove by induction that $c_{ij} \in \mathbb{Z}$. Since $f(0,0) = g(0,0)$, then $c_{00} = 0$. Let $(i,j) \in \mathbb{Z}^2$ such that $(i,j) > (0,0)$ and suppose that $c_{ij} \in \mathbb{Z}$ for any (i',j') satisfying the condition $(i',j') < (i,j)$. The polynomial $x - \sum_{(i',j')<(i,j)} c_{i'j'} f^{i'} g^{j'}$ has integral coefficients, hence $\sum_{(i',j')\geq(i,j)} c_{i'j'} f^{i'} g^{j'} \in \mathbb{Z}[x,y]$. This last polynomial has the form:

$$\begin{aligned} c_{ij} f^i g^j &+ c_{i+1j-1} f^{i+1} g^{j-1} + c_{i+2j-2} f^{i+2} g^{j-2} \\ &+ \cdots + c_{0i+j+1} y^{i+j+1} + c_{1i+j} f g^{i+j} + \cdots . \end{aligned}$$

Note that the monomial $x^i y^j$ appears only once in the above summation and that its coefficient is equal to c_{ij}. In particular $c_{ij} \in \mathbb{Z}$ and thus concludes the proof.

(2) Let $f(x,y) = 2x$ and $g(x,y) = y$ then $x = \frac{1}{2} f$ and $y = g$, hence $\mathbb{C}[f,g] = \mathbb{C}[x,y]$. We prove by two methods that $\mathbb{Z}[f,g] \neq \mathbb{Z}[x,y]$. Clearly

$$\mathbb{Z}[f,g] \subseteq y\mathbb{Z}[x] \bigoplus 2x\mathbb{Z}[x] \bigoplus \mathbb{Z}$$

and x does not belong to the right hand side.
Second method. Suppose that $x \in \mathbb{Z}[f,g]$ then

$$\frac{1}{2} f = a_n(g) f^n + a_{n-1}(g) f^{n-1} + \cdots + a_0(g).$$

Since f and g are transcendental over $\mathbb{C}$ and algebraically independent, we may differentiate relatively to g and obtain:

$$0 = a_n'(g) f^n + a_{n-1}'(g) f^{n-1} + \cdots + a_0'(g),$$

hence

$$a_n'(g) = a_{n-1}(g) = \cdots = a_0'(g) = 0.$$

It follows that $a_n(g), a_{n-1}(g), \ldots, a_0(g) \in \mathbb{Z}$ and then $n = 1$, $\frac{1}{2} f = a_n f$, hence a contradiction.

Exercise 9.2.
Let D be an integral domain and $\{x_i\}_{i\in I}$ be a family of indeterminates over D. Let A be a ring such that $D \subset A \subset D[\{x_i\}_{i\in I}]$ and A satisfies the descending chain condition on prime ideals. Show that $A = D$ or there exist $i_1, \ldots, i_m \in I$ and a morphism of algebras $\phi : D[\{x_i\}_{i\in I}] \to D[x_{i_1}, \ldots, x_{i_m}]$ such that its restriction to A is one to one.

Solution 9.2.
Let $R = D[\{x_i\}_{i\in I}]$ and let M_0 be the ideal of R generated by the x_i for $i \in I$. Then M_0 is prime, since a polynomial in n variables with coefficients in D belongs to M_0 if and only if it vanishes for the n-tuple $(0, \ldots, 0)$. Let $P_0 = M_0 \cap A$, then P_0 is a prime ideal of A. Suppose that $A \neq D$ and let $g(x_{i_1}, \ldots, x_{i_n}) \in A\backslash D$. Clearly

$$g(x_{i_1}, \ldots, x_{i_n}) - g(0, \ldots, 0) \in A\backslash D.$$

Any monomial of this polynomial has the form $dx_{i_1}^{e_1} \cdots x_{i_n}^{e_n}$ where $d \in D$, $e_1, \ldots, e_n$ are non negative integers and $e_1 + \cdots + e_n \geq 1$. Therefore one of these exponents say e_1 is positive.

It follows that this monomial has the form $x_{i_1}(dx_{i_1}^{e_1-1} \cdots x_{i_n}^{e_n})$, hence it is an element of M_0. We deduce that $g(x_{i_1}, \ldots, x_{i_n}) - g(0, \ldots, 0) \in M_0$ and then $P_0 \neq \{0\}$. Let $f_0 \in P_0\backslash\{0\}$, then f_0 is a polynomial in a finite number variables, say $x_{i_1}, ..., x_{i_{t_0}}$. Let M_1 be the ideal of R generated by $\{x_i\}_{i\in I_1}$, where $I_1 = I\backslash\{i_1, \ldots, i_{t_0}\}$, then M_1 is prime and $M_1 \cap A$ is a prime ideal of A. Set $P_1 = M_1 \cap A$, then $P_0 \supset P_1$. If $P_1 \neq \{0\}$ select $f_1 \in P_1\backslash\{0\}$. This polynomial depends on finite number of variables say $x_{i_{t_0+1}}, \ldots, x_{i_{t_0+t_1}}$.

Let $I_2 = I_1/\{i_{t_0+1}, \ldots, i_{t_0+t_1}\}$. We define similarly M_2 and P_2. Continuing in this way, we construct a sequence of prime ideals of A,

$$P_0 \supset P_1 \supset \cdots \supset P_k \supset \cdots,$$

which must be finite by hypothesis. We conclude that there exists $k \geq 1$ such that $P_k = \{0\}$. It follows that there exists a finite number of indeterminates $x_{i_1}, \ldots, x_{i_s}$ such that the ideal M of R generated by the variables x_j for $j \in I\backslash\{i_1, \ldots, i_s\}$ satisfies the condition $M \cap A = \{0\}$. Let $\phi : R \to D[x_{i_1}, \ldots, x_{i_s}]$ be the unique morphism of algebras such $\phi(x_i) = x_i$ if $i \in \{i_1, \ldots, i_s\}$ and $\phi(x_i) = 0$ if $i \notin \{i_1, \ldots, i_s\}$. Clearly $\operatorname{Ker}\phi = M$. We deduce that $\operatorname{Ker}\phi_{|A} = M \cap A = \{0\}$. Therefore $\phi_{|A}$ is one to one.

Exercise 9.3.
Let K be a field, $n \geq 2$ be an integer and $f(x_1, \ldots, x_n)$ be a non constant polynomial with coefficients in K of degree d. Let

$$y_1 = x_1, y_2 = x_2 + c_2x_1, \ldots, y_n = x_n + c_nx_1$$

and let

$$g(y_1, \ldots, y_n) = f(y_1, y_2 - c_2 y_1, \ldots, y_n - c_n y_1).$$

Suppose that $|K| > d$. Show that there exist $c_2, \ldots, c_n \in K$ such that

$$g(y_1, \ldots, y_n) = a y_1^d + A_{d-1}(y_2, \ldots, y_n) y_1^{d-1} + \cdots + A_0(y_2, \ldots, y_n),$$

where $a \in K^\star$ and the polynomials A_i have their coefficients in K.

Solution 9.3.

Write $f(x_1, \ldots, x_n)$ in the form

$$f(x_1, \ldots, x_n) = f_d(x_1, \ldots, x_n) + f_{d-1}(x_1, \ldots, x_n) + \cdots + f_0(x_1, \ldots, x_n),$$

where $f_j(x_1, \ldots, x_n)$ is 0 or homogeneous of degree j and $f_d(x_1, \ldots, x_n) \neq 0$. Then

$$\begin{aligned} g(y_1, \ldots, y_n) = {} & f_d(y_1, y_2 - c_2 y_1, \ldots, y_n - c_n y_1) \\ & + f_{d-1}(y_1, y_2 - c_2 y_1, \ldots, y_n - c_n y_1) \\ & + \cdots + f_0(y_1, y_2 - c_2 y_1, \ldots, y_n - c_n y_1). \end{aligned}$$

It is clear that $\deg_{y_1} f_j(y_1, y_2 - c_2 y_1, \ldots, y_n - c_n y_1) < d$ for $j = 0, \ldots, d-1$. Set

$$f_d(x_1, \ldots, x_n) = \sum_{i_1 + \cdots + i_n = d} a_{i_1 \cdots i_n} x_1^{i_1} \cdots x_n^{i_n},$$

then

$$g_d(y_1, \ldots, y_n) = \sum_{i_1 + \cdots + i_n = d} a_{i_1 \cdots i_n} y_1^{i_1} (y_2 - c_2 y_1)^{i_2} \cdots (y_n - c_n y_1)^{i_n}.$$

From this we conclude that the coefficient, say a, of y_1^d belongs to K and is given by

$$a = \sum_{i_1 + \cdots + i_n = d} a_{i_1 \cdots i_n} (-c_2)^{i_2} \cdots (-c_n)^{i_n} = f_d(1, -c_2, \ldots, -c_n).$$

The polynomial $f_d(1, y_2, \ldots, y_n)$ is non zero. Otherwise the non zero homogeneous polynomial $f_d(y_1, \ldots, y_n)$ would be divisible by $y_1 - 1$, which is a contradiction. (A divisor of an non zero homogeneous polynomial is itself homogeneous!) The degree d' of the polynomial in $n-1$ variables, $f_d(1, y_2, \ldots, y_n)$ satisfies the condition $d' \leq d < |K|$, hence there exists $(-c_2, \ldots, -c_n) \in K^{n-1}$ such that $f_d(1, -c_2, \ldots, -c_n) \neq 0$, that is $a \neq 0$.

Exercise 9.4.

Let m and n be positive integers such that $1 \leq m \leq n$. Let K be a field and $x_1, \ldots, x_n$ be algebraically independent variables over K. Denote by $\vec{x}$ the n-tuple $(x_1, \ldots, x_n)$. For any $i = 1, \ldots, m$, let $g_i(x_i)$ be a non constant polynomial with coefficients in K.

(1) Show that any $f(\vec{x}) \in K[\vec{x}]$ may be written in one and only one way in the form

$$f(\vec{x}) = \sum a_{j_1\cdots j_m}(\vec{x}) g_1(x_1)^{j_1} \cdots g_m(x_m)^{j_m},$$

where the sum is finite, $a_{j_1\cdots j_m}(\vec{x}) \in K[\vec{x}]$, $\deg_{x_i} a_{j_1\cdots j_m}(\vec{x}) < \deg g_i$ for $i = 1, \ldots, m$ and $\deg_{x_i} a_{j_1\cdots j_m}(\vec{x}) < \deg_{x_i} f$ for $m < i \leq n$.

(2) Suppose that for any $i = 1, \ldots, m$, $\deg g_i \geq 2$. Show that for any $i = 1, \ldots, m$, $x_i \notin K[g_1(x_1), \ldots, g_m(x_m)]$.

Solution 9.4.

(1) We proceed by induction on m. Suppose first that $m = 1$. The proof follows the one given in [Lang (1965), Th. 9, Chap. 5.5]. In this theorem, the involved polynomials are polynomials in one variable with coefficients in a field K, which is not the case in this exercise.
The Euclidean division in $K[x_2, \ldots, x_n][x_1]$ of $f(\vec{x})$ by $g_1(x_1)$ leads to

$$f(\vec{x}) = g_1(x_1)q(\vec{x}) + b_0(\vec{x}),$$

where $q(\vec{x})$ and $b_0(\vec{x}) \in K[\vec{x}]$ and $\deg_{x_1} b_0 < \deg g_1$. We show that $\deg_{x_i} b_0 \leq \deg_{x_i} f$ for $i = 2, \ldots, n$. We fix i and we distinguish three cases.

- $\deg_{x_i} q = \deg_{x_i} b_0$.
  Let $B(x_1, \ldots, \hat{x}_i, \ldots, x_n)$ and $C(x_1, \ldots, \hat{x}_i, \ldots, x_n)$ be the leading coefficients of $q(\vec{x})$ and $b_0(\vec{x})$ respectively, when these polynomials are written as decreasing powers of x_i. Here the hated variable x_i in the given polynomials means that this variable is not present in the polynomials. If $Bg_1 + C = 0$, then $g_1 \mid C$, hence $\deg_{x_1} b_0 \geq \deg_{x_1} C \geq \deg g_1$, which is a contradiction. Therefore, $Bg_1 + C \neq 0$ and then $\deg_{x_i} q = \deg_{x_i} b_0 = \deg_{x_i} f$.
- $\deg_{x_i} q < \deg_{x_i} b_0$.
  In this case, we have $\deg_{x_i} b_0 = \deg_{x_i} f$.
- $\deg_{x_i} q > \deg_{x_i} b_0$.
  Here we have $\deg_{x_i} b_0 < \deg_{x_i} q = \deg_{x_i} f$.

In any case, we have proved that $\deg_{x_i} b_0 \leq \deg_{x_i} f$.
By induction, suppose that

$$q(\vec{x}) = b_1(\vec{x}) + b_2(\vec{x})g_1(x_1) + \cdots + b_d(\vec{x})g_1(x_1)^d,$$

where $\deg_{x_1} b_i < \deg g_1$ and $\deg_{x_i} b_i \leq \deg_{x_i} q$ for $i \geq 2$, then

$$f(\vec{x}) = b_0(\vec{x}) + b_1(\vec{x})g_1(x_1) + \cdots + b_{d+1}(\vec{x})g_1(x_1)^{d+1}. \qquad \text{(Eq 1)}$$

Here the conditions on the degrees of the b_i, relatively to x_1 and relatively to x_i are satisfied. Since $b_i(\vec{x})$ is the remainder of the Euclidean division of

$$\left(f(\vec{x}) - b_0(\vec{x}) - b_1(\vec{x})g_1(x_1) - \cdots - b_{i-1}(\vec{x})g_1(x_1)^{i-1}\right)/g_1(x_1)^{i-1}$$

by $g_1(x_1)$, then this representation of $f(\vec{x})$ is unique.

Suppose that the representation of any polynomial, with the conditions on the degrees, exists and is unique for any system of $m-1$ polynomials. Write each polynomial $b_i(\vec{x})$ appearing above as a polynomial in $g_2(\vec{x}), \ldots, g_m(\vec{x})$ with coefficients in $K[(\vec{x})]$ and with the degrees conditions, that is

$$b_i(\vec{x}) = \sum c^{(i)}_{h_1\cdots h_m}(\vec{x})g_2(x_2)^{h_2(i)} \cdots g_m(\vec{x})^{h_m(i)},$$

with $\deg_{x_j} c^{(i)}_{h_1\cdots h_m}(\vec{x}) < \deg g_j$ for $j = 2, \ldots, m$ and $\deg_{x_j} c^{(i)}_{h_1\cdots h_m}(\vec{x}) \leq \deg_{x_j} b_i(\vec{x})$. Then

$$f(\vec{x}) = \sum c^{(i)}_{h_1\cdots h_m}(\vec{x})g_1(x_1)^i g_2(x_2)^{h_2(i)} \cdots g_m(\vec{x})^{h_m(i)},$$

so that the representation of $f(\vec{x})$ exists. We prove the uniqueness of the representation. Suppose that beside the representation

$$f(\vec{x}) = \sum a_{j_1\cdots j_m}(\vec{x})g_1(x_1)^{j_1} \cdots g_m(x_m)^{j_m},$$

we have the second one

$$f(\vec{x}) = \sum b_{i_1\cdots i_m}(\vec{x})g_1(x_1)^{i_1} \cdots g_m(x_m)^{i_m}.$$

Rewriting each of these expressions as sums of linear combinations of ascending powers of $g_1(x_1)$, we conclude by the uniqueness of the representation in the unique polynomial $g_1(x_1)$, that for fixed k,

$$\sum_{j_1=k} a_{j_1\cdots j_m}(\vec{x})g_1(x_1)^{j_1} \cdots g_m(x_m)^{j_m}$$

$$= \sum_{j_1=k} b_{i_1\cdots i_m}(\vec{x})g_1(x_1)^{i_1} \cdots g_m(x_m)^{i_m}.$$

Here we have two representations of the same polynomial in the system $\{g_2(x_1), \ldots, g_m(x_m)\}$. By the inductive hypothesis we conclude that the coefficients in the representations are equal.

(2) The unique representation of x_i in the system $\{g_1, \ldots, g_m\}$ is given by

$$x_i = \sum a_{j_1\cdots j_m}(\vec{x})g_1(x_1)^{j_1} \cdots g_m(x_m)^{j_m},$$

where $a_{j_1\cdots j_m}(\vec{x}) = 0$ if $(j_1, \ldots, j_m) \neq (0, \ldots, 0)$ and $a_{j_1\cdots j_m}(\vec{x}) = x_i$ if $(j_1, \ldots, j_m) = (0, \ldots, 0)$. Therefore x_i cannot be expressed in the form $x_i = u(g_1(x_1), \ldots, g_m(x_m))$, where $u(y_1, \ldots, y_m)$ is a polynomial with coefficients in K because such an expression is a representation in the system $\{g_1, \ldots, g_m\}$.

Exercise 9.5.

(1) Let $f(x)$ and $h(x)$ be polynomials with integral coefficients of degree m and n respectively and let p be a prime number. Suppose that $h(x)$ is monic, irreducible over $\mathbb{F}_p$, m and $n \geq 2$, $\gcd(m, n) \neq 1$ and p does not divide the leading coefficient of $f(x)$. Let $g(x) = h(x)/p$. Show that $f(\mathbb{Q}) \cap g(\mathbb{Q}) = \emptyset$.

(2) Given a polynomial $F(x)$ with rational coefficients of degree $m \geq 2$ and a positive integer n such that $\gcd(m, n) \neq 1$, show that there exists a polynomial $G(x)$ of degree n for which the equation $F(x) = G(y)$ has no solution $(x, y) \in \mathbb{Q}^2$.

Solution 9.5.

(1) Suppose that $f(\mathbb{Q}) \cap g(\mathbb{Q}) \neq \emptyset$ and let $(a, b) \in \mathbb{Q}^2$ such that $pf(a) = h(b)$.

- Case $\nu_p(b) \geq 0$.
  Since $h(x)$ is irreducible modulo p, then $\nu_p(h(b)) = 0$. From the above equation, we conclude that $\nu_p(f(a)) = -1$. It follows that $\nu_p(a) < 0$ and then $\nu_p(f(a)) = m\nu_p(a) = -1$, which is a contradiction since $m \geq 2$.
- Case $\nu_p(b) < 0$.
  In this case, we have $\nu_p(h(b)) < 0$, and by the above equation satisfied by (a, b), we get $\nu_p(f(a)) < 0$ and then $\nu_p(a) < 0$. We now have

  $$\nu_p(f(a)) = m\nu_p(a) \quad \text{and} \quad \nu_p(h(b)) = n\nu_p(b).$$

  The equation relating a and b implies that $1 + m\nu_p(a) = n\nu_p(b)$, which in turn shows that $\gcd(m, n) = 1$, which contradicts our assumptions. Thus the intersection is empty.

(2) We may write $F(x)$ in the form $F(x) = \frac{a}{d} f(x)$, where a and d are relatively prime integers and $f(x)$ is a polynomial with integral coefficients. Let p be prime number which is not a divisor of the leading coefficient of $f(x)$. Since $m = \deg f \geq 2$, we may find a polynomial $h(x)$ of degree n such that according to (1), the equation $f(x) = h(x)/p$ has no rational solution. Put $G(x) = ah(x)/p$, then $\deg G = n$ and the equation $F(x) = G(y)$ has no rational solution.

Exercise 9.6.
Let K be a field, $n \geq 2$ be an integer and $g, h \in K[x_1, \dots, x_n]$ be non constant. Suppose that the leading homogeneous form of $g(x)h(x)$ is square free. Show that

$$gK[x_1, \dots, x_n] + hK[x_1, \dots, x_n] \neq K[x_1, \dots, x_n].$$

Hint. Use Exercise 3 of this chapter.

Solution 9.6.

- We may suppose that K is infinite. For if K were finite, let $K_1 = K(t)$, where t is a new variable, then if

$$gK_1[x_1, \dots, x_n] + hK_1[x_1, \dots, x_n] \neq K_1[x_1, \dots, x_n],$$

we conclude that

$$1 \notin gK_1[x_1, \dots, x_n] + hK_1[x_1, \dots, x_n],$$

hence

$$1 \notin gK[x_1, \dots, x_n] + hK[x_1, \dots, x_n],$$

so that the same property will hold for K.

- Let $f(x) = g(x)h(x)$ and $d = \deg f(x)$. According to **Exercise 9.3**, we may suppose that f has the form

$$f(x_1, \dots, x_n) = x_1^d + A_{d-1}(x_2, \dots, x_n)x_1^{d-1} + \cdots + A_0(x_2, \dots, x_n).$$

It follows that the factors g and h of f, considered as polynomials in x_1, with coefficients in $K[x_2, x_3, \dots, x_n]$, may be supposed to be monic. In particular, $\deg_{x_1} f = \deg f^+$.
In the sequel, we denote by W^+ the leading homogeneous component of the arbitrary polynomial W. Since the result asked for is true if the polynomials g and h are not coprime, we may suppose that $\gcd(g, h) = 1$. Suppose by contradiction that there exist $u(x_1, \dots, x_n)$ and $v(x_1, \dots, x_n) \in K[x_1, \dots, x_n]$ such that $ug + vh = 1$. Performing the Euclidean division of v by g in $K[x_2, \dots, x_n][x_1]$, we obtain $v = gq + r$, where $r \neq 0$ and $\deg_{x_1} r < \deg_{x_1} g$. Since $(u+qh)g + rh = 1$, then $g^+ \mid r^+h^+$. Since $\deg_{x_1} r^+ = \deg_{x_1} r < \deg_{x_1} g = \deg_{x_1} g^+$, then some non constant factor of g^+ divides h^+. Therefore f^+ is not square free, contradicting the hypotheses.

Exercise 9.7.
Let $a, b \in \mathbb{Z}$ such that $(a, b) \neq (0, 0)$. Show that the polynomial $f(x, y) = y^3 - x^2y^2 - 3x^2y + x^4y - ax^2 + b$ is irreducible over $\mathbb{Q}$.

Solution 9.7.

Suppose that $f(x,y) = g(x,y)h(x,y)$ is a non trivial factorization of f in $\mathbb{Q}[x,y]$. Since f is monic in y, we must have $\deg_y g \geq 1$ and $\deg_y h \geq 1$. Without loss of generality, we may suppose that $\deg_y g = 1$, $\deg_y h = 2$ and $g(x,y) = cy - P(x)$, where $c \in \mathbb{Q}$ and $P(x) \in \mathbb{Q}[x]$, $\deg P \geq 1$, then we may suppose that $c = 1$. It follows that $f(x, P(x)) = 0$ and therefore $P(x) \mid ax^2 - b$ in $\mathbb{Q}[x]$. Since this condition is impossible if $a = 0$, we suppose that $a \neq 0$. If $\deg P = 1$, set $P(x) = \lambda x + \mu$ with $\lambda \neq 0$. The identity $f(x, P(x)) = 0$, immediately implies $\lambda = 0$, a contradiction. If $\deg P = 2$, then $P(x) = \lambda(ax^2 - b)$ with $\lambda \in \mathbb{Q}^\star$. Using again the same identity, we get the equation $(\lambda a)^3 - (\lambda a)^2 + (\lambda a) = 0$, which has no non zero rational solution λa. In particular, such a factorization $f = gh$ cannot exist, and therefore f must be irreducible.

Exercise 9.8.

Let K be a field, $\alpha_1, \ldots, \alpha_m$ be non zero, algebraic and separable over K of degree $d_1, \ldots, d_m$ respectively and $u_1, \ldots, u_m$ be algebraically independent variables. Set $\overrightarrow{u} = (u_1, \ldots, u_m)$ and $\overrightarrow{\alpha} = (\alpha_1, \ldots, \alpha_m)$.

(1) Show that $\alpha_1, \ldots, \alpha_m$ are algebraic, separable over $K(\overrightarrow{u})$ and $[K(\overrightarrow{u}, \overrightarrow{\alpha}) : K(\overrightarrow{u})] = [K(\overrightarrow{\alpha}) : K]$.

(2) Show that $\gamma = u_1\alpha_1 + \cdots + u_m\alpha_m$ is a primitive element of $K(\overrightarrow{u}, \overrightarrow{\alpha})$ over $K(\overrightarrow{u})$.

(3) Let $P(\overrightarrow{u}, x) \in K(\overrightarrow{u})[x]$ be the minimal polynomial of γ over $K(\overrightarrow{u})$. Show that $P \in K[\overrightarrow{u}, x]$.

(4) Let $d = d_1 \cdots d_m$. We denote the elements $\alpha_1^{e_1} \cdots \alpha_m^{e_m}$ with $0 \leq e_i \leq d_i - 1$ as follows:

$$\theta_1 = 1, \theta_2 = \alpha_1, \ldots, \theta_{d_1} = \alpha_1^{d_1 - 1}, \theta_{d_1+1} = \alpha_2, \ldots,$$
$$\theta_{2d_1} = \alpha_1^{d_1-1}\alpha_2, \ldots, \theta_d = \alpha_1^{d_1-1} \cdots \alpha_m^{d_m - 1}.$$

(a) Show that for any $j \in \{1, \ldots, d\}$, $\gamma\theta_j$ may be written in the form

$$\gamma\theta_j = \sum_{i=1}^{d} l_{ji}(\overrightarrow{u})\theta_i,$$

where $l_{ji} \in K[\overrightarrow{u}]$ and $\deg l_{ji} = 1$.

(b) Show that $P(\overrightarrow{u}, x)$ divides the polynomial

$$D(\overrightarrow{u}, x) = \begin{vmatrix} l_{11} - x & l_{12} & \ldots & l_{1d} \\ l_{21} & l_{22} - x & \ldots & l_{2d} \\ \ldots & \ldots & \ldots & \ldots \\ l_{d1} & l_{d2} & \ldots & l_{dd} - x \end{vmatrix}.$$

(5) Show that for any $i \in \{1, \ldots, m\}$,
$$\alpha_i = -\frac{\frac{\partial P}{\partial u_i}(\overrightarrow{u}, \gamma)}{\frac{\partial P}{\partial x}(\overrightarrow{u}, \gamma)}.$$
(6) Let $\overrightarrow{u^\star} = (u_1^\star, \ldots, u_m^\star) \in K^m$ such that $\frac{\partial P}{\partial x}(\overrightarrow{u^\star}, \gamma^\star) \neq 0$, where $\gamma^\star = u_1^\star \alpha_1 + \cdots + u_m^\star \alpha_m$. Show that $\gamma^\star$ is a primitive element of $K(\overrightarrow{\alpha})$ over K.

(7) Let
$$\mathcal{F} = \left\{\overrightarrow{u^\star} \in K^m, \frac{\partial P}{\partial x}\left(\overrightarrow{u^\star}, \gamma^\star\right) = 0\right\}.$$

(a) Show that
$$\mathcal{F} = \left\{\overrightarrow{u^\star} \in K^m, \frac{\partial P}{\partial x}\left(\overrightarrow{u^\star}, \gamma^\star\right) = 0, \frac{\partial P}{\partial u_i}\left(\overrightarrow{u^\star}, \gamma^\star\right) = 0, i = 1, \ldots, m\right\}.$$

(b) Let $\mathcal{U} = K^m \setminus \mathcal{F}$. Show that for any $\overrightarrow{u^\star} \in \mathcal{U}$, $P(\overrightarrow{u^\star}, x)$ is irreducible over K.

(8) Let $K = \mathbb{Q}$, $\alpha_1 = 2^{1/2}$ and $\alpha_2 = 2^{1/4}$. Compute $D(u_1, u_2, x)$ and deduce $P(u_1, u_2, x)$.

Solution 9.8.

(1) It is trivial that for any $i \in \{1, \ldots, m\}$, α_i is algebraic and separable over $K(\overrightarrow{u})$. Consider the diagram

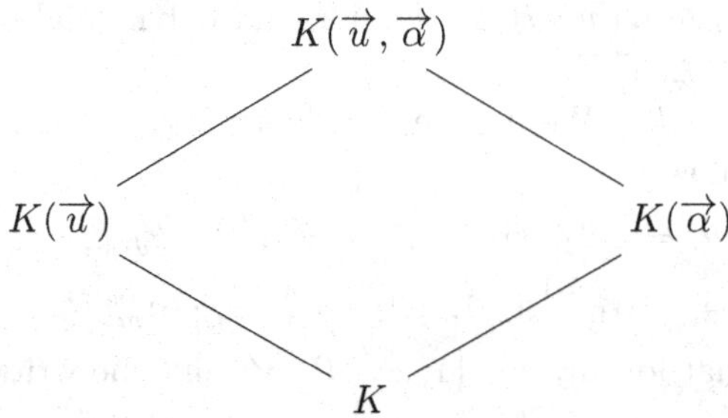

The extensions $K(\overrightarrow{u})$ and $K(\overrightarrow{\alpha})$ are linearly disjoint over K, hence the result on the degrees.

(2) Let Ω be an algebraic closure of $K(\overrightarrow{u})$. Suppose that there exist two $K(\overrightarrow{u})$-isomorphisms $\sigma, \tau : K(\overrightarrow{u}, \overrightarrow{\alpha}) \to \Omega$ such that $\sigma(\gamma) = \tau(\gamma)$, then
$$u_1\sigma(\alpha_1) + \cdots + u_m\sigma(\alpha_m) = u_1\tau(\alpha_1) + \cdots + u_m\tau(\alpha_m),$$
hence $\sigma(\alpha_i) = \tau(\alpha_i)$ for $i = 1, \ldots, m$. Therefore $\sigma = \tau$ and γ is a primitive element of $K(\overrightarrow{u}, \overrightarrow{\alpha})$ over $K(\overrightarrow{u})$.

(3) The polynomial $P(\overrightarrow{u}, x)$ is given by

$$P(\overrightarrow{u}, x) = \prod_{\sigma} \left(x - \sum_{i=1}^{m} u_i \sigma(\alpha_i) \right),$$

where the product runs over the set of $K(\overrightarrow{u})$-embeddings of $K(\overrightarrow{u}, \overrightarrow{\alpha})$ into Ω, hence the result. Notice that any of these embeddings is an extension of one and only one embedding of $K(\overrightarrow{\alpha})$ into Ω fixing the elements of K.

(4) (a) Each θ_j has the form $\theta_j = \alpha_1^{e_1} \cdots \alpha_m^{e_m}$ with $0 \leq e_i \leq d_i - 1$, hence

$$\gamma\theta_j = u_1 \alpha_1^{e_1+1} \cdots \alpha_m^{e_m} + u_2 \alpha_1^{e_1} \alpha_2^{e_2+1} \cdots \alpha_m^{e_m} + u_m \alpha_1^{e_1} \cdots \alpha_m^{e_m+1}.$$

If $e_i + 1 \leq m_i - 1$, then there exists k such that $\alpha_1^{e_1} \cdots \alpha_i^{e_i+1} \cdots \alpha_m^{e_m} = \theta_{i_k}$. If $e_i + 1 = m_i$, then

$$\begin{aligned} \alpha_1^{e_1} \cdots \alpha_i^{e_i+1} \cdots \alpha_m^{e_m} &= \alpha_1^{e_1} \cdots (-a_{d_i-1} \alpha_i^{d_i-1} - \cdots - a_1 \alpha_i - a_0) \cdots \alpha_m^{e_m} \\ &= -a_{d_i-1} \alpha_1^{e_1} \cdots \alpha_i^{d_i-1} \cdots \alpha_m^{e_m} - a_1 \alpha_1^{e_1} \cdots \alpha_i \cdots \alpha_m^{e_m} \\ &\quad - a_0 \alpha_1^{e_1} \cdots \alpha_{i-1}^{e_{i-1}} \alpha_{i+1}^{e_{i+1}} \cdots \alpha_m^{e_m}. \end{aligned}$$

This shows that $\gamma\theta_j$ is a linear combination of the θ_k, $k = 1, \ldots, d$, where the coefficients are linear polynomials in $u_1, \ldots, u_m$ with coefficients in K.

(b) The preceding equations may be written in the form:

$$\begin{aligned} (l_{11} - \gamma)\theta_1 + l_{12}\theta_2 + \cdots + l_{1d}\theta_d &= 0 \\ l_{21}\theta_1 + (l_{22} - \gamma)\theta_2 + \cdots + l_{2d}\theta_d &= 0 \\ \cdots\cdots\cdots\cdots\cdots\cdots\cdots\cdots\cdots\cdots\cdots\cdots \\ l_{d1}\theta_1 + l_{d2}\theta_2 + \cdots + (l_{dd} - \gamma)\theta_d &= 0. \end{aligned}$$

This implies that the homogeneous system

$$\begin{aligned} (l_{11} - \gamma)x_1 + l_{12}x_2 + \cdots + l_{1d}x_d &= 0 \\ l_{21}x_1 + (l_{22} - \gamma)x_2 + \cdots + l_{2d}x_d &= 0 \\ \cdots\cdots\cdots\cdots\cdots\cdots\cdots\cdots\cdots\cdots\cdots\cdots \\ l_{d1}x_1 + l_{d2}x_2 + \cdots + (l_{dd} - \gamma)x_d &= 0 \end{aligned}$$

has a non trivial solution, namely $\theta_1, \ldots, \theta_d$. Therefore the determinant D of this system is zero, i.e. $D(u_1, \ldots, u_m, \gamma) = 0$. Since $P(\overrightarrow{u}, x)$ is the minimal polynomial of γ over $K(\overrightarrow{u})$, then $P(\overrightarrow{u}, x) \mid D(\overrightarrow{u}, x)$. It is possible to compute $D(\overrightarrow{u}, x)$ without using the above determinant. Indeed we have

$$D(\overrightarrow{u}, x) = \prod_{(\sigma_1, \ldots, \sigma_m)} (x - u_1\sigma_1(\alpha_1) - \ldots - u_m\sigma_m(\alpha_m)),$$

where the product runs over all the m-tuples $(\sigma_1, \ldots, \sigma_m)$ and σ_i is a K isomorphism of $K(\alpha_i)$ into Ω.

(5) We fix $j \in \{1, \ldots, m\}$ and we differentiate the identity

$$P\left(\overrightarrow{u}, \sum_{i=1}^{m} u_i \alpha_i\right) = 0$$

relatively to u_j. We obtain

$$\alpha_j \frac{\partial P}{\partial x}(\overrightarrow{u}, \gamma) + \frac{\partial P}{\partial u_j}(\overrightarrow{u}, \gamma) = 0. \qquad \text{(Eq 1)}$$

Since $K(\overrightarrow{u}, \overrightarrow{\alpha})/K(\overrightarrow{u})$ is separable then $\frac{\partial P}{\partial x}(\overrightarrow{u}, \gamma) \neq 0$, hence

$$\alpha_j = -\frac{\frac{\partial P}{\partial u_j}(\overrightarrow{u}, \gamma)}{\frac{\partial P}{\partial x}(\overrightarrow{u}, \gamma)}.$$

(6) Since $\frac{\partial P}{\partial x}(\overrightarrow{u^\star}, \gamma^\star) \neq 0$, then the preceding identity shows that for any $j = 1, \ldots, m$, $\alpha_j \in K(\gamma^\star)$, hence $K(\overrightarrow{\alpha}) = K(\gamma^\star)$ and $\gamma^\star$ is a primitive element of $K(\overrightarrow{\alpha})$.

(7) (a) Easy application of equations **(Eq 1)**.

(b) Since $P(\overrightarrow{u^\star}, \gamma^\star) = 0$, then $\operatorname{Irr}(\gamma^\star, K) \mid P(\overrightarrow{u^\star}, x)$. These two polynomials being monic in x, to get their equality it is sufficient to show that they have the same degree. We have

$$\begin{aligned}
\deg_x P(\overrightarrow{u^\star}, x) &= \deg_x P(\overrightarrow{u}, x) \\
&= [K(\overrightarrow{u}, \gamma) : K(\overrightarrow{u})] \\
&= [K(\overrightarrow{u}, \overrightarrow{\alpha}) : K(\overrightarrow{u})] \\
&= [K(\overrightarrow{\alpha}) : K] \\
&= [K(\gamma^\star) : K],
\end{aligned}$$

hence the result. Here are some explanations of the preceding equalities. The first one is due to the fact that P is monic in x. Since γ is a root of P, we obtain the second one. The third equality comes from the fact that γ is a primitive element of $K(\overrightarrow{u}, \overrightarrow{\alpha})$ over $K(\overrightarrow{u})$. From **(1)**, we deduce the fourth equality. The last equality is due to the fact that $\gamma^\star$ is a primitive element of $K(\overrightarrow{\alpha})$ over K.

(8) We have

$$\begin{aligned}
D(u_1, u_2, x) &= (x + u_1\sqrt{2} + u_2 2^{1/4})(x + u_1\sqrt{2} - u_2 2^{1/4}) \\
&\quad .(x - u_1\sqrt{2} + u_2 i 2^{1/4})(x - u_1\sqrt{2} - u_2 i 2^{1/4}) \\
&\quad .(x - u_1\sqrt{2} - u_2 2^{1/4})(x - u_1\sqrt{2} + u_2 2^{1/4}) \\
&\quad .(x + u_1\sqrt{2} - u_2 i 2^{1/4})(x + u_1\sqrt{2} + u_2 i 2^{1/4}) \\
&= (x^2 + 2u_1^2)^2 - 2(2u_1 x - u_2^2)^2 \\
&\quad .(x^2 + 2u_1^2)^2 - 2(2u_1 x + u_2^2)^2 \\
&= P(u_1, u_2, x) Q(u_1, u_2, x)
\end{aligned}$$

with

$$P(u_1, u_2, x) = (x^2 + 2u_1^2)^2 - 2(2u_1x - u_2^2)^2$$

and

$$Q(u_1, u_2, x) = (x^2 + 2u_1^2)^2 - 2(2u_1x + u_2^2)^2.$$

Question. May it happen that there exists $(u_1^\star, \ldots, u_m^\star, x^\star) \in K^{m+1}$ such that

$$\frac{\partial P}{\partial x}(\overrightarrow{u^\star}, x^\star) = 0, \frac{\partial P}{\partial u_i}(\overrightarrow{u^\star}, x^\star) = 0$$

for $i = 1, \ldots, m$ and such that one of the following conditions holds?

(i) There exists i such that $u_m^\star \notin K$.
(ii) For any i, $u_m^\star \in K$ but $x^\star \neq \sum_{i=1}^m u_i^\star \alpha_i$.

Exercise 9.9.
Let K be a field, $f_i(x_1, \ldots, x_n) \in K[x_1, \ldots, x_n]$ for $i = 1, \ldots, n$ be algebraically independent over K.

(1) Suppose that x_i, for $i = 1, \ldots, n$, is integral over $K[f_1, \ldots, f_n]$, show that

$$K(f_1, \ldots, f_n) \cap K[x_1, \ldots, x_n] = K[f_1, \ldots, f_n].$$

(2) Let $f(x, y) = x^2y^2$ and $g(x, y) = xy(x + y)$ and suppose that the characteristic of the field K is not equal to 2.
   (a) Show that f and g are algebraically independent over K.
   (b) Show that x is algebraic over $K(f, g)$, but not integral over $K[f, g]$.
   (c) Show that $K(f, g) \cap K[x, y] \neq K[f, g]$.

Solution 9.9.

(1) Let $h(x_1, \ldots, x_n)$ be an element of the intersection of the two rings, then h is integral over $K[f_1, \ldots, f_n]$ and belongs to the fraction field of this last ring. This ring is isomorphic to $K[x_1, \ldots, x_n]$, hence factorial. Therefore $K[f_1, \ldots, f_n]$ is integrally closed and then $h \in K[f_1, \ldots, f_n]$. The other inclusion is trivial.
(2) (a) Suppose that there exists $\phi(u, v) \in K[u, v] \backslash K$ such that $\phi(f, g) = 0$. Let $\phi^+(u, v)$ be the the leading form of $\phi(u, v)$. We may write $\phi^+(u, v)$ in the form

$$\phi^+(u, v) = u^e v^k \prod_{i=1}^m (u - \theta_i v),$$

where e, k are non negative integers, the product may be empty and the θ_i are algebraic over K. Since f and g are homogeneous, then $\phi^+(f,g)$ is the leading form of $\phi(f,g)$, hence $\phi^+(f,g) = 0$. We deduce that $f = 0$ or $g = 0$ or $f = \theta_i g$ for some $i \in \{1, \ldots, m\}$ which is impossible.

(b) We have $xy = \epsilon\sqrt{f}$, with $\epsilon = \pm 1$ and $x + y = g/\epsilon\sqrt{f}$. Hence x and y are roots of the polynomial $u^2 - g/(\epsilon\sqrt{f})u + \epsilon\sqrt{f}$. Therefore

$$2x = g/\epsilon\sqrt{f} \pm \sqrt{g^2/f - 4\epsilon\sqrt{f}}.$$

We deduce that

$$\left(2x - g/\epsilon\sqrt{f}\right)^2 = g^2/f - 4\epsilon\sqrt{f},$$

hence $\epsilon\sqrt{f}x^2 = xg - f$ and then

$$fx^4 - g^2x^2 + 2fgx - f^2 = 0.$$

This means that x is algebraic over $K(f,g)$ and is a root of the polynomial

$$\phi(z) = fz^4 - g^2z^2 + 2fgz - f^2.$$

We can use another method to find again this polynomial. Since x is a common root of the polynomials

$$\phi_1(z) = x^2z^2 - f \quad \text{and} \quad \phi_2(z) = xz^2 + x^2z - g,$$

then $\mathrm{Res}_z(\phi_1(z), \phi_2(z)) = 0$. Using standard formulas on the resultant, we obtain

$$\begin{aligned}\mathrm{Res}_z(\phi_1(z), \phi_2(z)) &= x^4\left(x\frac{f}{x^2} + x^2\frac{\epsilon\sqrt{f}}{x} - g\right)\left(x\frac{f}{x^2} - x^2\frac{\epsilon\sqrt{f}}{x} - g\right)\\ &= x^2\left(f^2 + g^2x^2 - 2fgx - fx^4\right).\end{aligned}$$

Since $x \neq 0$ then $\phi(x) = 0$. It remains to show that x is not integral over $K[f,g]$. It is enough to show that $\phi(z)$ is irreducible over $K[f,g]$. This equivalent to prove that the polynomial

$$\psi(u,v,z) = uz^4 - v^2z^2 + 2uvz - u^2$$

is irreducible over K. We consider this polynomial as a polynomial in v of degree 2 and we find that its discriminant is equal to $4uz^6$. Therefore ψ is irreducible over K.

(c) We have

$$g^2/f = (x+y)^2 \in K(f,g) \cap K[x,y].$$

Suppose that $g^2/f \in K[f,g]$, then there exists $\rho(u,v) \in K[u,v]$ such that $g^2 = f\rho(f,g)$. Since f and g are algebraically independent over K, we may compare the degrees in f of the two sides of this identity and get a contradiction. Therefore $K[f,g]$ is strictly contained in $K(f,g) \cap K[x,y]$.

Exercise 9.10.

Let K be field, $d \geq 2$ be an integer and $x_1, \ldots, x_n$ be algebraically independent variables. Set $\vec{x} = (x_1, \ldots, x_n)$. For any $F(\vec{x}) \in K[\vec{x}]$, denote by $S_d(F)$ the univariate polynomial defined by $S_d(F) = F(y, y^d, \ldots, y^{d^{n-1}})$.

(1) Show that the map $S_d : F \to S_d(F)$ is a surjective morphism of K-algebras from $K[\vec{x}]$ into $K[y]$.
(2) Show that the restriction of S_d to the set of polynomials $F(\vec{x})$, such that $\deg_{x_i} < d$ for $i = 1, \ldots, n$, is injective.
(3) Let Ω be an algebraically closed field containing K and $F(\vec{x}) \in K[\vec{x}]$. Suppose that F is reducible over Ω. Show that F is reducible over $\overline{K}$, where $\overline{K}$ is the algebraic closure of K contained in Ω.
(4) Factorize $F(x_1, x_2) = 6x_1^3 + 3x_2x_1^2 + x_2^2 + 2x_2x_1 + 6x_1 + 3x_2$ into irreducible factors in $\mathbb{Q}[x_1, x_2]$.

Solution 9.10.

(1) Clearly S_d is a morphism of algebras. Let $f(y) \in K[y]$, then $f(y) = S_d(f(x_1))$, thus S_d is surjective.
(2) Since the image of a monomial by S_d is a monomial, it is sufficient to prove that any monomial ay^i is the image of one monomial $M = ax_1 \cdots x_n \in K[\vec{x}]$.
Suppose that $S_d(M) = ay^i$, then $ay^i = ay^{i_1 + di_2 + \cdots + d^{n-1}i_n}$, hence $i = i_1 + di_2 + \cdots + d^{n-1}i_n$. This implies that $i_1, \ldots, i_n$ are the digits in base d of i. Since the representation in base d exists and is unique, then the restriction of S_d is injective and also bijective by **(1)**.
(3) Choose a total order (for example the lexicographic order) in the set of monomials $ax_1 \cdots x_n \in K[\vec{x}]$. In order to prove the result, we may suppose that F is monic relatively to this order. Suppose that $F(\vec{x}) = F_1(\vec{x})F_2(\vec{x})$ is a non trivial factorization of $F(\vec{x})$ in $\Omega[\vec{x}]$. We may suppose that F_1 and F_2 are monic. Let d be a positive integer such

that $d > \deg_{x_i} F$ for $i = 1, \ldots, n$, then
$$F\left(y, y^d, \ldots, y^{d^{n-1}}\right) = F_1\left(y, y^d, \ldots, y^{d^{n-1}}\right) F_2\left(y, y^d, \ldots, y^{d^{n-1}}\right).$$
This is a factorization in $\Omega[y]$ of a polynomial in one variable with coefficients in K. Moreover, by (2), it is a non trivial factorization. The coefficients of $F_1(y, y^d, \ldots, y^{d^{n-1}})$ are symmetric functions of a set of roots of $F(y, y^d, \ldots, y^{d^{n-1}})$. Since these roots are elements of $\bar{K}$, the coefficients of $F_1(y, y^d, \ldots, y^{d^{n-1}})$ belong to $\bar{K}$. The same claim is true for $F_2(y, y^d, \ldots, y^{d^{n-1}})$. It follows that $F_1(\vec{x})$ and $F_2(\vec{x}) \in \overline{K}[\vec{x}]$.

(4) Here we let $d = 3$ and we get
$$F(y, y^3) = y^6 + 3y^5 + 2y^4 + 9y^3 + 6y = y(y^5 + 3y^4 + 2y^3 + 9y^2 + 6).$$
We factorize
$$\phi(y) = y^5 + 3y^4 + 2y^3 + 9y^2 + 6$$
over $\mathbb{Q}$. It is easy to verify that ϕ does not vanish for ±1, ±2, ±3, ±6, so that $\phi(y)$ has no factor of degree 1. It follows that if $\phi(y)$ is reducible over $\mathbb{Q}$, then $\phi(y) = g(y)h(y)$, with $g(y)$ and $h(y) \in \mathbb{Z}[y]$, $\deg g = 2$ and $\deg h = 3$. Since $\phi(y) \equiv y^2(y^3 + y^2 + 1) \pmod 2$, then
$$g(y) = y^2 + 2ay + 2b \quad \text{and} \quad h(y) = y^3 + (2c+1)y^2 + 2dy + 2e + 1,$$
where $a, b, c, d, e \in \mathbb{Z}$. Identifying the coefficients of x^i, for $i = 0, \ldots, 4$, in the identity $\phi(y) = g(y)h(y)$, we obtain the following equations
$$\begin{aligned} b(2e+1) &= 3, \\ a(2e+1) + 2bd &= 0, \\ e + 2ad + b(2c+1) &= 4, \\ d + b + a(2c+1) &= 1, \\ a + c &= 1. \end{aligned}$$
The first equation shows that $2e + 1 = \epsilon$, $b = 3\epsilon$ or $2e + 1 = 3\epsilon$, $b = \epsilon$ with $\epsilon = \pm1$.

- If $2e + 1 = \epsilon$ and $b = 3\epsilon$.
  The second equation implies $a = -6d$. The last equation gives $c = 1 + 6d$. Substituting these values of a, b, c, e, in terms of d and e in the fourth equation leads to $72d^2 - 17d + 1 - 3\epsilon = 0$. The discriminant of this quadratic equation is equal to $\Delta = 1 + 2883\epsilon$. If $\epsilon = -1$, then $\Delta < 0$ and the quadratic equation has no integral (nor real) solution. We conclude that $\phi(y)$ is irreducible over $\mathbb{Q}$. If $\epsilon = 1$, then $\Delta = 5 \cdot 173$. It follows that $\Delta \equiv 0 \pmod 5$ but $\Delta \not\equiv 0 \pmod{5^2}$. We are led to the same conclusion as before.

- If $2e+1=3\epsilon$ and $b=\epsilon$.
  By the same method as above, we find that d satisfies the following equation $8d^2-8\epsilon d+27(1-\epsilon)=0$. Its discriminant is equal to 64 if $\epsilon=1$ and -52 if $\epsilon=-1$. If $\epsilon=-1$, then $\phi(y)$ is irreducible over $\mathbb{Q}$. If $\epsilon=1$, then the roots of the quadratic equation are given by $d=0$ or $d=1$. If $d=0$, the second equation implies that $a=0$. The last equation gives $c=1$. The fourth one implies $b=1$. It follows that $g(y)=y^2+2$ and $h(y)=y^3+3y^2+3$. One verifies that these polynomials satisfy the relation $\phi(y)=g(y)h(y)$. Notice that $g(y)$ and $h(y)$ are irreducible over $\mathbb{Q}$, so that we have found the factorization of $\phi(y)$ into irreducible factors over $\mathbb{Q}$. We now have
  $$F(y,y^3)=y(y^2+2)(y^3+3y^2+3).$$
  Suppose that $F(x_1,x_2)=F_1(x_1,x_2)F_2(x_1,x_2)$ is a non trivial factorization of F in $\mathbb{Q}[x_1,x_2]$, then
  $$F(y,y^3)=F_1(y,y^3)F_2(y,y^3)=y(y^2+2)(y^3+3y^2+3).$$
  Without loss of generality, we must consider the following cases
  (i) $F_1(y,y^3)=y$, $F_2(y,y^3)=(y^2+2)(y^3+3y^2+3)$.
  (ii) $F_1(y,y^3)=y^2+2$, $F_2(y,y^3)=y(y^3+3y^2+3)$.
  (iii) $F_1(y,y^3)=y(y^2+2)$, $F_2(y,y^3)=(y^3+3y^2+3)$.

In the first case, $F_1(x_1,x_2)=x_1$. Since $x_1\nmid F(x_1,x_2)$, then we must reject this case. In the second case $F_1(x_1,x_2)=x_1^2+2$. Since $x_1^2+2\nmid F(x_1,x_2)$, this case does not hold. In the third case, we get

$$F_1(x_1,x_2)=x_2+2x_1 \quad\text{and}\quad F_2(x_1,x_2)=x_2+3x_1^2+3.$$

It is now easy to verify that these polynomials satisfy the relation $F(x_1,x_2)=F_1(x_1,x_2)F_2(x_1,x_2)$.

Exercise 9.11.
Let K be a field.

(1) Let $g(x_1,\dots,x_n)\in K[x_1,\dots,x_n]$ and t a new variable. Suppose that $g(tx_1,\dots,tx_n)=t^m u(x_1,\dots,x_n)$, where m is a positive integer and u is a polynomial with coefficients in K. Show that $m=\deg g$, $u=g$ and g is homogeneous.
(2) Let $f(x_1,\dots,x_n)$ an homogeneous polynomial with coefficients in K. Suppose that $f=gh$ is a non trivial factorization of f in $K[x_1,\dots,x_n]$. Show that g and h are homogeneous.

Solution 9.11.

(1) Set

$$g(x_1,\ldots,x_n)=\sum a_{i_1,\ldots,i_n}x_1^{i_1}\cdots x_n^{i_n} \quad\text{and}$$
$$u(x_1,\ldots,x_n)=\sum b_{i_1,\ldots,i_n}x_1^{i_1}\cdots x_n^{i_n}.$$

Then

$$g(tx_1,\ldots,tx_n)=\sum\left(t^{\sum i_j}\right)a_{i_1,\ldots,i_n}x_1^{i_1}\cdots x_n^{i_n}=t^m\sum b_{i_1,\ldots,i_n}x_1^{i_1}\cdots x_n^{i_n}.$$

It follows that for any $(i_1,\ldots,i_n)$, $i_1+\cdots+i_n=m$ and $a_{i_1,\ldots,i_n}=b_{i_1,\ldots,i_n}$, which implies that $\deg g=m$, $u=g$ and g is homogeneous.

(2) • **First proof.** Suppose that g is not homogeneous. Let g^+ and g^- (resp. h^+ and h^-) be the homogeneous parts of g (resp. h) of highest and lowest degree. We have $g^+\neq g^-$ but we may have $h^+=h^-$. We also have $f=g^+h^+$ and $g^-h^-=0$, therefore $h^-=0$. This implies $h=0$ which is a contradiction.

• **Second proof.** Let t a new variable then, from the identity $f=gh$, we deduce that

$$t^d f(x_1,\ldots,x_n)=g(tx_1,\ldots,tx_n)h(tx_1,\ldots,tx_n),$$

where $d=\deg f$. We consider this identity as a factorization of $t^d f$ in $K[x_1,\ldots,x_n][t]$. We deduce that

$$g(tx_1,\ldots,tx_n)=t^k u(x_1,\ldots,x_n) \quad\text{and}$$
$$h(tx_1,\ldots,tx_n)=t^l v(x_1,\ldots,x_n),$$

where $k+l=d$ and u,v are polynomials in $x_1,\ldots,x_n$. We conclude from **(1)** that g and h are homogeneous.

Exercise 9.12.
Let $f(X,Y)\in\mathbb{Q}[X,Y]$ be irreducible and let α be a root of f in an algebraic closure of $\mathbb{Q}(X)$. Let $u(X,Y)=u_n(X)Y^n+\cdots+u_0(X)\in\mathbb{Q}[X,Y]$, $\beta=u(X,\alpha)$ and let $g(X,Y)\in\mathbb{Q}[X,Y]$ be the unique irreducible polynomial(up to a multiplication by a constant) satisfying $g(X,\beta)=0$. Suppose that $n\geq 1$ and that $u_n(X)$ has no rational root. For any $h(X,Y)$ denote by $V(h)$ the set $V(h)=\{(a,b)\in\mathbb{C}^2, h(a,b)=0\}$. If $V(g)\cap\mathbb{Q}^2$, (resp. $V(g)\cap\mathbb{Z}^2$) is finite, show that $V(f)\cap\mathbb{Q}^2$, (resp. $V(f)\cap\mathbb{Z}^2$) is finite.

Solution 9.12.
We have $g(X,u(X,\alpha))=g(X,\beta)=0$, hence

$$g(X,u(X,Y))=f(X,Y)q(X,Y)/D(X),$$

where $D(X) \in \mathbb{Q}[X]$, $q(X,Y) \in \mathbb{Q}[X,Y]$. Moreover, we may suppose that D is monic and $\gcd(D(X), q_0(X), \dots, q_m(X)) = 1$, where $q_0(X), \dots, q_m(X)$ denote the coefficients of q as a polynomial in Y. Let $P(X)$ be an irreducible factor of $D(X)$, then since $g(X, u(X,Y)) \in \mathbb{Q}[X,Y]$, we conclude that $P(X)$ divides the content (in $\mathbb{Q}[X]$) of $f(X,Y)$, hence $D(X) = 1$ and

$$g(X, u(X,Y)) = f(X,Y)q(X,Y).$$

Let $(a,b) \in \mathbb{Q}^2$ (resp. $(a,b) \in \mathbb{Z}^2$) such that $f(a,b) = 0$, then $g(a, u(a,b)) = 0$. Set $c = u(a,b)$, then $(a,c) \in V(g) \cap \mathbb{Q}^2$ (resp. $(a,c) \in V(g) \cap \mathbb{Z}^2$). The hypotheses implies that a and c have a finite number of values. To determine b, we must solve the equation

$$u(a,b) = u_n(a)b^n + \cdots + u_0(a) = c.$$

For a given c and a given a not a root of $u_n(X)$, this equation has a finite number of solutions b in $\mathbb{Q}$ and in $\mathbb{Z}$ and the proof is complete.

Exercise 9.13.
Let $f(x)$ and $g(x) \in \mathbb{Q}[x]$ be non constant polynomials.

(1) Suppose that $g(\mathbb{Q}) \subset f(\mathbb{Q})$. Show that there exists $P(x) \in \mathbb{Q}[x]$ such that $g(x) = f(P(x))$.
(2) If $g(\mathbb{Q}) = f(\mathbb{Q})$, deduce that there exist a and $b \in \mathbb{Q}$, $a \neq 0$ such that $g(x) = f(ax+b)$.
(3) Show that a similar result to **(1)** does not hold over a finite field.

Solution 9.13.

(1) Let $F(X,Y) = f(X) - g(Y)$ and let $F(X,Y) = F_1(X,Y) \cdots F_r(X,Y)$ be its factorization into irreducible factors over $\mathbb{Q}$. Clearly the leading coefficient of $F_i(X,Y)$ for $i = 1, \dots, r$ as a polynomial in X and also as a polynomial in Y is a constant. Let A be the set of elements y of $\mathbb{Q}$ such that for any $i \in \{1, \dots, r\}$, $F_i(X,y)$ is irreducible over $\mathbb{Q}$, then by Hilbert's irreducibility theorem [Schinzel (2000), Chap. 4.4], A is infinite. Let $y \in A$, then $g(y) \in f(\mathbb{Q})$, hence there exists $x \in \mathbb{Q}$ such that $f(x) - g(y) = 0$. Therefore there exists $i_0 \in \{1, \dots, r\}$ such that for infinitely $y \in \mathbb{Q}$, the polynomial $F_{i_0}(X,y)$ is irreducible over $\mathbb{Q}$ and has a root $x \in \mathbb{Q}$. Set $F_{i_0}(X,Y) = \sum_{i=0}^{n} a_i(Y)X^i$, where $n \geq 1$, $a_i(Y) \in \mathbb{Q}[Y]$ for $i = 1, \dots, r$ and $a_n(Y) = c \in \mathbb{Q}$. For $y \in A$, $F_{i_0}(X,y)$ is irreducible and has a rational root, hence $n = 1$ and $F_{i_0}(X,Y) = cX + a_0(Y)$. We now have

$$f(X) - g(Y) = (cX + a_0(Y))G(X,Y),$$

with $G(X,Y) \in \mathbb{Q}[X,Y]$. Substituting for X the value $-a_0(Y)/c$ yields $f(-a_0(Y)/c) = g(Y)$ and the proof is complete.

(2) Easy.

(3) Consider the polynomials $f(X) = X^q$ and $g(X) = X^q - X$ over $\mathbb{F}_q$, then clearly

$$\{0\} = g(\mathbb{F}_q) \subset f(\mathbb{F}_q) = \mathbb{F}_q.$$

If there exists $P(X) \in \mathbb{F}_q[X]$ such that $g(X) = f(P(X))$, then $\deg P = 1$. Set $P(X) = aX + b$, then

$$f(aX+b) = aX^q + b \neq g(X).$$

Exercise 9.14.

Let K be a field, $f(x) \in K[x]$ of degree $d \geq 1$. We say that f is functionally decomposable over K if there exist two non constant polynomials $u(x), v(x) \in K[x]$ such that $f(x) = u(v(x))$. In this case $u(x)$ (resp. $v(x)$) is called the left composition factor (resp. right composition factor) of $f(x)$.

(1) Show that the following propositions are equivalent.

 (i) f is functionally decomposable over K.

 (ii) There exists $h(x) \in K[x]\backslash K$, $\deg h < \deg f$ such that $(h(x)-h(y)) \mid (f(x) - f(y))$ in $K[x,y]$.

(2) Let $h(x) \in K[x]$ be non constant. Suppose that $\deg h < \deg f$. Show that the following propositions are equivalent.

 (i) $h(x)$ is a right composition factor of $f(x)$.

 (ii) There exists $g(x) \in K[x]$ such that $h(x) - t$ divides $f(x) - g(t)$ in $K[x,t]$.

 (iii) For any $a \in K$, there exists $b \in K$ such that $(h(x)-a) \mid (f(x)-b)$.

 (iv) There exist a positive integer $r > \deg f / \deg h$, distinct elements of K, $a_1, \ldots, a_r$ and elements of K, $b_1, \ldots, b_r$ not necessarily distinct such that for any $i \in \{1, \ldots, r\}$, $h(x) - a_i \mid f(x) - b_i$.

 (v) There exist $a \in K$, a positive integer k and a polynomial $g(x) \in K[x]$ such that

$$\mathrm{Res}_x(f(x) - z, h(x) - t) = a(z - g(t))^k.$$

 (vi) There exists a polynomial $g(x) \in K[x]$ such that the remainder of the Euclidean division of $f(x) - z$ by $h(x) - t$ in $K[t,z][x]$ is equal to $g(t) - z$.

Solution 9.14.

(1)
- $(i) \Rightarrow (ii)$. If $f(x) = u(v(x))$, where u and v are non constant, then $v(x) - v(y) \mid f(x) - f(y)$ in $K[x, y]$.
- $(ii) \Rightarrow (i)$. Let $d = \deg f$, $e = \deg h$ and $D = \lfloor d/e \rfloor$. The $h(x)$-adic expansion of $f(x)$ [Lang (1965), Th. 9, Chap. 5.5] has the form:

$$f(x) = \sum_{i=0}^{D} f_i(x)(h(x))^i,$$

where $f_i(x)$ is a polynomial with coefficients in K of degree at most $e-1$ for $i = 0, \ldots, D$. We have

$$\sum_{i=0}^{D} f_i(x)(h(x))^i - \sum_{i=0}^{D} f_i(y)(h(y))^i \equiv 0 \pmod{h(x) - h(y)}.$$

Since

$$h(x) \equiv h(y) \pmod{h(x) - h(y)},$$

then

$$\sum_{i=0}^{D} \big(f_i(x) - f_i(h)\big)(h(y))^i \equiv 0 \pmod{h(x) - h(y)}.$$

Therefore, there exists $A(x, y) \in K[x, y]$ such that

$$\sum_{i=0}^{D} \big(f_i(x) - f_i(h)\big)(h(y))^i = A(x, y)\big(h(x) - h(y)\big).$$

Comparing the degrees of the two sides of this identity leads to $A(x, y) = 0$ and then to

$$\sum_{i=0}^{D} f_i(x)(h(y))^i = \sum_{i=0}^{D} f_i(y)(h(y))^i.$$

Substituting 0 for x in this identity, we obtain two $h(y)$-adic expansions of the same polynomial in $K[y]$. It follows that $f_i(y) \in K$ for $i = 0, \ldots, D$. Set $f_i(y) = b_i$ and $g_i(y) = \sum_{i=0}^{D} b_i x^i$. Then

$$f(y) = \sum_{i=0}^{D} b_i(h(c))^i = g(h(x),$$

which means that $h(x)$ is a right composition factor of $f(x)$.

(2) We prove the following implications.

$$(i) \Rightarrow (ii) \Rightarrow (iii) \Rightarrow (iv) \Rightarrow (i) \Leftrightarrow (v) \text{ and } (ii) \Leftrightarrow (vi).$$

The implications

$$(i) \Rightarrow (ii) \Rightarrow (iii) \Rightarrow (iv)$$

are easy and will be omitted.

- $(iv) \Rightarrow (i)$.
The conditions contained in (iv) express the fact that $f(x)$ is a solution of the following system of congruence equations:

$$f(x) \equiv b_i \pmod{h(x) - a_i}, \quad i = 1, \ldots, r.$$

By the Chinese Remainder Theorem, a solution of this system is given by $f_0(x) = \sum_{i=1}^{r} b_i Y_i(x)$, where $Y_i(x)$ is a polynomial satisfying the conditions:

$$Y_i(x) \equiv 1 \pmod{h(x) - a_i} \quad \text{and} \quad Y_i(x) \equiv 0 \quad \left(\operatorname{mod} \prod_{j \neq i} (h(x) - a_j)\right).$$

We may take

$$Y_i(x) = \prod_{j \neq i} (h(x) - a_j) / \prod_{j \neq i} (a_i - a_j).$$

We thus have

$$f(x) = f_0(x) + u(x) \prod_{i=1}^{r} (h(x) - a_i),$$

where $u(x)$ is a polynomial. We clearly have $\deg f_0 < r \deg h$ and by **(iv)**, $\deg f < r \deg h$, hence

$$f(x) = f_0(x) = \sum_{i=1}^{r} \frac{b_i}{\prod_{j \neq i} (a_i - a_j)} \prod_{j \neq i} (h(x) - a_j).$$

Set

$$c_i = \frac{b_i}{\prod_{j \neq i} (a_i - a_j)}, \quad g_i(x) = c_i \prod_{j \neq i} (x - a_j) \quad \text{and} \quad g(x) = \sum_{i=1}^{r} g_i(x),$$

then

$$f(x) = \sum_{i=1}^{r} g_i(h(x)) = g(h(x)),$$

hence $h(x)$ is a right composition factor of $f(x)$.

- $(i) \Rightarrow (v)$.
  Using **Exercise 8.6(4)**, we obtain
  $$\text{Res}_x(f(x) - z, h(x) - t) = a\big(H(z,t)\big)^k,$$
  where $a \in K$, k is a positive integer and $H(z,t)$ is an irreducible polynomial over K satisfying $H(f(x), h(x)) = 0$. This polynomial H satisfying this property is unique up to multiplication by a constant in K. Since $h(x)$ is a right composition factor of $f(x)$, there exists $g(x) \in K[x]$ such that$f(x) - g(h(x)) = 0$, hence $H(z,t) = z - g(t)$ and the proof is complete.
- $(v) \Rightarrow (i)$.
  Substitute $f(u)$ for z and $h(u)$ for t, where u is a new variable, in the identity
  $$\text{Res}_x(f(x) - z, h(x) - t) = a(z - g(t))^k$$
  and obtain
  $$\text{Res}_x(f(x) - f(u), h(x) - h(u)) = a(f(u) - g(h(u)))^k.$$
  Since the polynomials for which we compute the resultant have the common root $x = u$, then the resultant vanishes, hence $f(u) = g(h(u))$ and $h(x)$ is a right composition factor of $f(x)$.
- $(ii) \Rightarrow (vi)$.
  Since $h(x) - t \mid f(x) - g(t)$, there exists a polynomial $q(x,t)$ such that $f(x) - g(t) = (h(x) - t)q(t,x)$, hence
  $$f(x) - z = (h(x) - t)q(t,x) + g(t) - z.$$
- $(vi) \Rightarrow (ii)$.
  Clear.

Exercise 9.15.
Let $u(x)$, $v(x)$ be two non constant polynomials with coefficients in a factorial ring A, p be a prime of A and K the fraction field of A.

(1) Suppose that $u(x)$ and $v(x)$ satisfy the Eisenstein's irreducibility criteria for the prime p. Show that the same property holds for $u(v(x))$.
(2) Deduce that if the polynomial $f(x)$ with coefficients in A is p-Eisenstein, then for any positive integer k, $f^{[k]}$ is irreducible over K, where $f^{[k]}$ is the k-th iterate (for the composition of functions) of $f(x)$.
(3) Let $u_1, \ldots, u_n, x$ be algebraically independent variables over K. Let $F(x) = x^n + u_1 x^{n-1} + \cdots + u_n \in K(u_1, \ldots, u_n)[x]$. Show that for any positive integer k, $F^{[k]}(x)$ is irreducible over K, where $F^{[k]}$ is the k-th iterate of F.

Solution 9.15.

(1) Set

$$u(x) = \sum_{i=0}^{n} u_i x^i \quad \text{and} \quad v(x) = \sum_{i=0}^{m} v_i x^i,$$

where u_i and $v_i \in A$. Then

$$u(x) \equiv u_n x^n \pmod{p} \quad \text{and} \quad v(x) \equiv v_m x^m \pmod{p}.$$

Therefore $u(v(x)) \equiv u_n v_m^n x^{mn} \pmod{p}$. It follows that all the coefficients of $u(v(x))$ except the leading one are congruent to 0 modulo p. We have

$$u(v(0)) = u_n v_0^n + \cdots + u_1 v_0 + u_0 \equiv u_0 \pmod{p^2},$$

hence the result.

(2) The conclusion for the iterates of a polynomial is trivial.

(3) Consider the unique morphism of rings

$$\Phi : K[u_1, \ldots, u_n, x] \to K[T, x]$$

such that $\Phi(u_i) = T$, for $i = 1, \ldots, n$ and $\Phi(x) = x$, then

$$\Phi(F(x)) = x^n + Tx^{n-1} + \cdots + T,$$

hence $\Phi(F(x))$ is T-Eisenstein. It follows that $\Phi(F)^{[k]}$ is T-Eisenstein, hence irreducible over K. Since $\Phi(F^{[k]}) = \Phi(F)^{[k]}$, then $F(x)^{[k]}$ is irreducible.

Exercise 9.16.

(1) Let K be field, E be an extension of K, $f(x)$, $g(x)$ and $h(x) \in K(x)$. If $K(f, g) = K(h)$, show that $E(f, g) = E(h)$.

(2) Suppose that f nor g is constant. Show that there exists $P(z, w) \in K[z, w]$, irreducible such that $P(f, g) = 0$. Show that $P(z, w)$ is absolutely irreducible.

Solution 9.16.

(1) Since $E(f, g)$ is the smallest field containing E, f, g and since $E(h)$ contains E, f, g, then $E(f, g) \subset E(h)$. On the other hand, we have $h \in K(f, g) \subset E(f, g)$, hence $E(h) \subset E(f, g)$. Therefore $E(f, g) = E(h)$.

(2) We have

$$K \subset K(f) \subset K(f,g) \subset K(x).$$

Clearly the transcendence degree over K of $K(x)$ and of $K(f)$ is equal to 1. Therefore the same property is true for $K(f,g)$. It follows that this last field is an algebraic extension of $K(f)$. Let $Q(f,w) \in K(f)[w]$ be the minimal polynomial of g over $K(f)$. Consider the set of the denominators of all the coefficients of this polynomial and let $d(f)$ be their least common multiple. Then clearly $P(z,w) := d(z)Q(z,w)$ is irreducible in $K[z,w]$ and $P(f,g) = 0$. Let $\bar{K}$ be an algebraic closure of K. By Luroth's Theorem [Schinzel (2000), Th. 2, Chap. 1.1], there exists $h(x)$ in $K[x]$ such that $K(f,g) = K(h)$. By **(1)**, we have $\bar{K}(f,g) = \bar{K}(h)$. Let $\bar{P}(z,w) \in \bar{K}[z,w]$, irreducible such that $\bar{P}(f,g) = 0$. Clearly $\bar{P}(z,w) \mid P(z,w)$ in $\bar{K}[z,w]$.
Consider the following diagram and the similar one when K is replaced by $\bar{K}$.

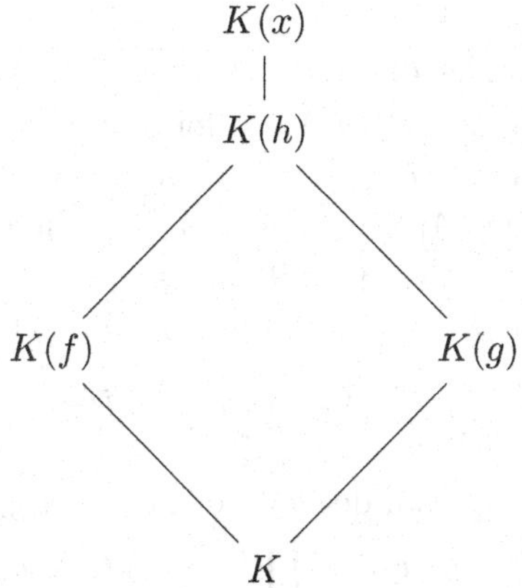

We have

$$\begin{aligned}\deg_z P(z,w) &= [K(f,g) : K(g)] \\ &= \deg h / \deg g \\ &= [\bar{K}(f,g) : \bar{K}(g)] \\ &= \deg_z \bar{P}(z,w).\end{aligned}$$

Therefore $P(z,w)$ and $\bar{P}(z,w)$ are equal up to a multiplication by a constant in $\bar{K}$ and this implies that $P(z,w)$ is absolutely irreducible.

Exercise 9.17.
Let K be a field, Ω be an algebraic closure of K and $F(x_1,\dots,x_n) \in K[x_1,\dots,x_n]$. Let A be a Cartesian subset of Ω^n, i.e. a set of the form $A = A_1 \times A_2 \times \cdots \times A_n$. Suppose that $F(a_1,\dots,a_n) = 0$ for any $(a_1,\dots,a_n) \in A$.

(1) If $|A_i| > \deg_{x_i} F$, show that $F = 0$.
(2) Deduce that there exist $h_1,\dots,h_n \in K[x_1,\dots,x_n]$ such that $F(x_1,\dots,x_n) = \sum_{i=1}^n h_i g_i$, $\deg h_i \leq \deg F - \deg g_i$, where $g_i \in K[x_i]$ and $g_i(x_i) = \prod_{b\in A_i}(x_i - b)$ for $i = 1,\dots,n$.

Solution 9.17.

(1) We proceed by induction on n. If $n = 1$, the statement is trivial. Suppose that the result holds for $n-1$ indeterminates. Write $F(x_1,\dots,x_n)$ in the form
$$F(x_1,\dots,x_n) = F_0 + F_1 x_n + \cdots + F_m x_n^m,$$
where $F_i(x_1,\dots,x_{n-1}) \in K[x_1,\dots,x_{n-1}]$ for $i = 0,\dots,m$. Let $(a_1,\dots,a_{n-1}) \in A_1 \times A_2 \times \cdots \times A_{n-1}$. Consider the univariate polynomial
$$f(x_n) = F_0(a_1,\dots,a_{n-1}) + F_1(a_1,\dots,a_{n-1})x_n + \cdots + F_m(a_1,\dots,a_{n-1})x_n^m.$$
It satisfies the condition $f(a_n) = 0$ for any $a_n \in A_n$, hence $f(x_n) = 0$. It follows that for any $i = 0,\dots,m$, we have $F_i(a_1,\dots,a_{n-1}) = 0$ for any $(a_1,\dots,a_{n-1}) \in A_1 \times A_2 \times \cdots \times A_{n-1}$. Therefore, by the inductive hypothesis, $F_i(x_1,\dots,x_{n-1}) = 0$ for $i = 0,\dots,m$.

(2) We first show that $F(x_1,\dots,x_n)$ may be written in the form
$$F(x_1,\dots,x_n) = \sum_{i=1}^{n} h_i g_i + \bar{F}(x_1,\dots,x_n),$$
where $h_i \in K[x_1,\dots,x_n]$, $\deg h_i \leq \deg F - \deg g_i$,
$$g_i(x_i) = \prod_{b\in A_i}(x_i - b) \in K[x_i]$$
for $i = 1,\dots,n$ and $\deg_{x_i} \bar{F} < \deg g_i$. If $n = 1$, the result follows from the euclidean division of F by g_1. Suppose that $n \geq 2$. Divide F by g_1 in $K[x_2,\dots,x_n][x_1]$. We have $F = g_1 h_1 + \bar{F}_1$ where $\bar{F}_1$ and $h_1 \in K[x_2,\dots,x_n][x_1]$ with $\deg_{x_1} \bar{F}_1 < \deg g_1$, $\deg_{x_i} \bar{F}_1 \leq \deg_{x_i} F$ for $i \neq 1$ and $\deg_{x_i} h_1 \leq \deg_{x_i} F - \deg g_1$. We do the same operation with $\bar{F}_1(x_1,\dots,x_n)$ instead of F and g_2 instead of g_1. We obtain $\bar{F}_1 = g_2 h_2 + \bar{F}_2$ with similar conditions on the degrees of the polynomials. We iterate the process and obtain the announced identity with $\bar{F} = \bar{F}_n$. We apply (1) to $\bar{F}$ and conclude that $\bar{F} = 0$ and $F(x_1,\dots,x_n) = \sum_{i=1}^n h_i g_i$.

Exercise 9.18.
Let K be a field, $f(x,y) \in K[x,y)$ and $f^+(x,y)$ be the leading homogeneous form of f. Suppose that f^+ is square free. We say that two non trivial factorizations

$$f^+(x,y) = u_1(x,y)u_2(x,y) \quad \text{and} \quad f^+(x,y) = v_1(x,y)v_2(x,y)$$

of f in $K[x,y)$ are essentially the same if $v_i = \lambda_i u_i$ for $i = 1,2$ or $v_1 = \lambda_1 u_2$ and $v_2 = \lambda_2 u_1$, where λ_1 and $\lambda_2 \in K^\star$. In the opposite case we say that the factorizations are essentially different. Let d be the number of essentially different factorizations of f^+ in $K[x,y)$

(1) Let

$$V_0(f) = \{\lambda \in K, f(x,y) - \lambda \text{ is reducible over } K\}.$$

Show that $|V_0(f)| \leq d$.

(2) Determine these values of $\lambda \in V_0(f)$ when $K = \mathbb{Q}$ and $f(x,y) = x^4 - y^4 - x^2 - 5y^2$.

(3) Show that the result in **(1)** does not hold if we remove the square free condition on f^+.

(4) Suppose that f^+ has no factor of degree 1 over K. Let

$$V_1(f) = \{(a,b,c) \in K^3, f(x,y) - (ax + by + c) \text{ is reducible over } K\}.$$

Show that $|V_1(f)| \leq d$.

Solution 9.18.

(1) We define a one to one map ϕ from $V_0(f)$ into the set of the distinct factorizations of f^+, which will be denoted by $\mathcal{F}(f^+)$. Let $\lambda \in V_0(f)$ and

$$f(x,y) - \lambda = g_\lambda(x,y)h_\lambda(x,y)$$

a non trivial factorization of $f - \lambda$ over K. We may suppose that $\deg g = k \geq l = \deg h$. Set

$$f(x,y) = \sum_{i=0}^{n} f_i(x,y),$$

$$g_\lambda(x,y) = \sum_{i=0}^{k} g_i(x,y) \quad \text{and}$$

$$h_\lambda(x,y) = \sum_{i=0}^{l} h_i(x,y),$$

where $k + l = n$, $f_i(x, y)$, (resp. $g_j(x, y)$) (resp. $h_t(x, y)$) is 0 or homogeneous of degree i (resp. j), (resp. t). The existence of the factorization is equivalent to the solubility of the following system of equations:

$$
\begin{aligned}
f_n &= g_k h_l \\
f_{n-1} &= g_k h_{l-1} + g_{k-1} h_l \\
&\cdots\cdots\cdots\cdots\cdots\cdots \\
f_k &= g_k h_0 + g_{k-1} h_1 + \cdots + g_{k-l} h_l \\
f_{k-1} &= g_{k-1} h_0 + g_{k-2} h_1 + \cdots + g_{k-l-1} h_l \\
&\cdots\cdots\cdots\cdots\cdots\cdots \\
f_k &= g_k h_0 + g_{k-1} h_1 + \cdots + g_{k-l} h_l \\
f_{k-1} &= g_{k-1} h_0 + g_{k-2} h_1 + \cdots + g_{k-l-1} h_l \\
&\cdots\cdots\cdots\cdots\cdots\cdots \\
f_l &= g_l h_0 + g_{l-1} h_1 + \cdots + g_0 h_l \\
f_t &= g_t h_0 + g_{t-1} h_1 + \cdots + g_0 h_t \quad \text{for} \quad 1 \le t \le l-2 \\
f_0 - \lambda &= g_0 h_0.
\end{aligned}
$$

These equations may be written in the form:

$$f_n = g_k h_l \tag{Eq 0}$$

$$\frac{f_{n-1}}{f_n} = \frac{h_{l-1}}{h_l} + \frac{g_{k-1}}{g_k} \tag{Eq 1}$$

$$\cdots\cdots = \cdots\cdots\cdots\cdots\cdots\cdots$$

$$\frac{f_k}{f_n} = \frac{h_0}{h_l} + \frac{g_{k-1}}{g_k}\frac{h_1}{h_l} + \cdots \frac{g_{k-l}}{g_k} \tag{Eq n-k}$$

$$\frac{f_{k-1}}{f_n} = \frac{g_{k-1}}{g_k}\frac{h_0}{h_l} + \frac{g_{k-2}}{g_k}\frac{h_1}{h_l} + \cdots + \frac{g_{k-l-1}}{g_k} \tag{Eq n-(k-1)}$$

$$\cdots\cdots = \cdots\cdots\cdots\cdots\cdots\cdots$$

$$\frac{f_l}{f_n} = \frac{g_l}{g_k}\frac{h_0}{h_l} + \frac{g_{l-1}}{g_k}\frac{h_1}{h_l} + \cdots + \frac{g_0}{g_k} \tag{Eq n-l}$$

$$\frac{f_t}{f_n} = \frac{g_t}{g_k}\frac{h_0}{h_l} + \frac{g_{t-1}}{g_k}\frac{h_1}{h_l} + \cdots + \frac{g_0}{g_k}\frac{h_t}{h_l} \quad \text{for} \quad 0 < t < l \tag{Eq n-t}$$

$$f_0 - \lambda = g_0 h_0. \tag{Eq n}$$

Now for any $\lambda \in V_0(f)$, let $\phi(\lambda)$ be the factorization **(Eq 0)** of f^+ arising from the factorization of $f - \lambda$. We show that this map is one to one. Given the factorization **(Eq 0)** of f^+, we show that there exists at most one value of λ such that the preceding system has a solution. **(Eq 1)** shows that expanding $\frac{f_{n-1}}{f_n}$ into partial fractions, this determines uniquely g_{k-1} and h_{l-1}. The equations **(Eq 1)**–**(Eq n-k)** determine uniquely $g_{k-1}, \ldots, g_{k-l}$ and $h_{l-1}, \ldots, h_0$. The equations **(Eq n-(k-1))**–**(Eq n-1)** determine uniquely $g_{k-l-1}, \ldots, g_0$. The equations **(Eq n-l-1)**–**(Eq n-1)** may be compatible or not with the preceding. In the positive case the last equation determines uniquely λ. In the negative case there is no value of λ corresponding to the given factorization of f^+.

(2) Here

$$f^+(x,y) = x^4 - y^4 = (x-y)(x+y)(x^2+y^2)$$

and the distinct factorizations of f^+ over $\mathbb{Q}$ are given by:

$$f^+(x,y) = \big((x-y)(x^2+y^2)\big)(x+y) \qquad \text{(Eq a)}$$

$$f^+(x,y) = \big((x+y)(x^2+y^2)\big)(x-y) \qquad \text{(Eq b)}$$

$$f^+(x,y) = (x^2-y^2)(x^2+y^2). \qquad \text{(Eq c)}$$

It follows that $|V_0(f)| \leq 3$. We compute explicitly the values of λ if any.

- Case **(Eq a)**.
  We have

$$f - \lambda = (g_3 + g_2 + g_1 + g_0)(h_1 + h_0),$$
$$g_3 = (x-y)(x^2+y^2),$$
$$h_1(x,y) = x + y$$

  and

$$f_3 = g_3h_0 + g_2h_1 = 0$$
$$f_2 = g_2h_0 + g_1h_1 = -x^2 - 5y^2$$
$$f_1 = g_1h_0 + g_0h_1 = 0$$
$$f_0 - \lambda = g_0h_0 = -\lambda.$$

  The first equation reads $h_0/h_1 + g_2/g_3 = 0$, which implies $g_2 = h_0 = 0$. The second equation gives

$$g_1/g_3 = (-x^2 - 5y^2)/(x^4 - y^4),$$

  which is clearly impossible. Therefore there is no λ corresponding to case (1).
- Case **(Eq b)**.
  Similar computations show that there is no λ corresponding to case **(2)**.

- Case **(Eq c)**.
Here we have,
$$f-\lambda=(g_2+g_1+g_0)(h_2+h_1+h_0),$$
$$g_2=x^2-y^2,$$
$$h_2(x,y)=x^2+y^2$$
and
$$f_3=g_2h_1+g_1h_2=0$$
$$f_2=g_2h_0+g_1h_1+g_0h_2=-x^2-5y^2$$
$$f_1=g_1h_0+g_0h_1=0$$
$$f_0-\lambda=g_0h_0=-\lambda.$$
The first equation reads $h_1/h_2+g_1/g_2=0$, which implies $g_1=h_1=0$. The second gives
$$h_0/h_2+g_0/g_2=(-x^2-5y^2)/(x^4-y^4).$$
We have
$$\begin{aligned}(-x^2-5y^2)/(x^4-y^4)&=\frac{-1}{y^2}\frac{X^2+5}{(X^2-1)(X^2+1)}\\&=\frac{-1}{y^2}\left(\frac{3}{X^2-1}-\frac{2}{X^2+1}\right),\end{aligned}$$
where $X=x/y$. Hence
$$h_0/h_2+g_0/g_2=2/h_2-3/g_2.$$
Therefore $h_0=2$ and $g_0=-3$. The third equation is satisfied and the fourth gives $\lambda=6$. We deduce that $V_0(f)=\{6\}$ and
$$f(x,y)-6=x^4-y^4-x^2-5y^2-6=(x^2-y^2-3)(x^2+y^2+2).$$

(3) Let $f(x,y)=(x+y)^2$, then for any $\lambda\in\mathbb{Q}$ which is a square, $f(x,y)-\lambda$ is reducible over $\mathbb{Q}$ so that $V_0(f)$ is infinite.

(4) Let $(a,b,c)\in K^3$, then $f-(ax+by+c)$ is reducible over K if and only if the equations in (2) except the two last of them which are replaced by
$$f_1-(ax+by)=g_1h_0+g_0h_1$$
$$f_0-c=g_0h_0,$$
are satisfied. The equations **(Eq 1)**–**(Eq n-l)** determine $g_0,\ldots,g_k$ and $h_0,\ldots,h_l$ and since $2\le l\le k$, then the next equations are compatible or not with the preceding. In the affirmative case, equations **(Eq n-1)** and **(Eq n)** determine uniquely (a,b) and c respectively. In the negative case there is no value of λ corresponding to this factorization of f^+.

Exercise 9.19.
Let $\{x_1, \ldots, x_n\}$ be a set of variables. The n-tuple $(x_1, \ldots, x_n)$ will be denoted $\vec{x}$. For $k = 1, \ldots, n$ let $s_k(\vec{x})$ be the elementary symmetric function of degree k in the variables $x_1, \ldots, x_n$. For any $k \geq 1$ let $\sigma_k(\vec{x}) = \sum_{i=1}^n x_i^k$. Let (h, k) and $F_{(h,k)}(\vec{x}) = \sum x_{i_1}^k x_{i_2} \cdots x_{i_h}$, where the indices $i_1, \ldots, i_h$ are distinct. Show that

$$F_{(h,k)}(\vec{x}) = \sigma_{k-1}(\vec{x}) s_h(\vec{x}) - F_{(h+1,k-1)}(\vec{x}).$$

Deduce that

$$F_{(h,k)}(\vec{x}) = \sigma_{k-1}(\vec{x}) s_h(\vec{x}) - \sigma_{k-2}(\vec{x}) s_{h+1}(\vec{x}) + \cdots .$$

Solution 9.19.
We have

$$\begin{aligned}\sigma_{k-1}(\vec{x}) s_h(\vec{x}) &= (x_1^{k-1} + \cdots + x_1^{k-1}) \left(\sum x_{i_1} \cdots x_{i_h}\right) \\ &= F_{(h,k)}(\vec{x}) + F_{(h+1,k-1)}(\vec{x}),\end{aligned}$$

hence

$$F_{(h,k)}(\vec{x}) = \sigma_{k-1}(\vec{x}) s_h(\vec{x}) - F_{(h+1,k-1)}(\vec{x}).$$

We deduce that, for any $1 \leq t \leq \inf(n-h, k-1)$,

$$\begin{aligned}F_{(h,k)}(\vec{x}) &= \sigma_{k-1}(\vec{x}) s_h(\vec{x}) - F_{(h+1,k-1)}(\vec{x}) \\ &= \sigma_{k-1}(\vec{x}) s_h(\vec{x}) - \sigma_{k-2}(\vec{x}) s_{h+1}(\vec{x}) \\ &\quad + \cdots + (-1)^{t-1} \sigma_{k-t}(\vec{x}) s_{h+t-1} + (-1)^t F_{(h+t,k-t)}(\vec{x}).\end{aligned}$$

Let $t = \inf(n-h, k-1)$. If $t = n-h \leq k-1$, then

$$\begin{aligned}F_{(h,k)}(\vec{x}) &= \sigma_{k-1}(\vec{x}) s_h(\vec{x}) - \sigma_{k-2}(\vec{x}) s_{h+1}(\vec{x}) \\ &\quad + \cdots + (-1)^{n-h-1} \sigma_{h+k-n}(\vec{x}) s_{n-1}(\vec{x}) \\ &\quad + (-1)^{n-h} F_{(n,h+k-n)}(\vec{x}) \\ &= \sigma_{k-1}(\vec{x}) s_h(\vec{x}) - \sigma_{k-2}(\vec{x}) s_{h+1}(\vec{x}) \\ &\quad + \cdots + (-1)^{n-h-1} \sigma_{h+k-n}(\vec{x}) s_{n-1}(\vec{x}) \\ &\quad + (-1)^{n-h} s_n(\vec{x}) \sigma_{h+k-1}(\vec{x}).\end{aligned}$$

If $t = k-1 \leq n-h$, then

$$\begin{aligned}F_{(h,k)}(\vec{x}) &= \sigma_{k-1}(\vec{x}) s_h(\vec{x}) - \sigma_{k-2}(\vec{x}) s_{h+1}(\vec{x}) + \cdots + \\ &\quad (-1)^{k-2} \sigma_1(\vec{x}) s_{h+k-2}(\vec{x}) + (-1)^{k-1} F_{(h+k-1,1)}(\vec{x}) \\ &= \sigma_{k-1}(\vec{x}) s_h(\vec{x}) - \sigma_{k-2}(\vec{x}) s_{h+1}(\vec{x}) + \cdots \\ &\quad + (-1)^{k-2} \sigma_1(\vec{x}) s_{h+k-2}(\vec{x}) + (-1)^{k-1} s_{h+k-1}(\vec{x}).\end{aligned}$$

Exercise 9.20.
Let K be a field, $f(t)$, $g(t)$, $A(t)$ and $B(t)$ be non constant polynomials with coefficients in K. If $A(f(x)) - A(g(y))$ divides $B(f(x)) - B(g(y))$, show that $A(x) - A(y)$ divides $B(x) - B(y)$.

Solution 9.20.
Performing the Euclidean division in $K[y][x]$, we obtain

$$B(x) - B(y) = (A(x) - A(y))Q(x, y) + R(x, y),$$

where $Q(x, y)$, $R(x, y) \in K(x, y])$ and $\deg_x R < \deg A$. Substituting $f(x)$ for x and $g(y)$ for y, we get

$$B(f(x)) - B(g(y)) = (A(f(x)) - A(g(y)))Q(f(x), g(y)) + R(f(x), g(y)).$$

Moreover

$$\deg_x R(f(x), g(y)) = \deg_x R(x, y) \deg f < \deg A \deg f = \deg(A(fx)).$$

Hence $R(f(x), g(y)) = 0$. We show that $R(x, y) = 0$. Suppose that this is not true. We may write the identity $R(f(x), g(y)) = 0$ in the form

$$a_n(f(x))g(y)^n + a_{n-1}(f(x))g(y)^{n-1} + \cdots + a_0(f(x)) = 0$$

with $a_n(x) \neq 0$, thus also $a_n(f(x)) \neq 0$. Pick an element x_0, in the algebraic closure of K, say Ω, such that $a_n(f(x_0)) \neq 0$ and substitute it for x in the preceding identity. We obtain:

$$a_n(f(x_0))g(y)^n + a_{n-1}(f(x_0))g(y)^{n-1} + \cdots + a_O(f(x_0)) = 0.$$

This implies $g(y)$ is algebraic over Ω, hence over K, which is a contradiction. We conclude that $R(x, y) = 0$ as desired.

Exercise 9.21.
Let K be a field. The polynomial $g_1(t) \in K[t]$ is said to be a right composition factor of $g(t) \in K[t]$ if there exists $G(t) \in K[t]$ such that $g(t) = G(g_1(t))$. Let $f_1(t)$ and $f(t)$ be non constant polynomials with coefficients in K. Suppose that $f_1(x) - f_1(y)$ divides $f(x) - f(y)$ in $K[x, y]$. The purpose of the exercise is to prove that $f_1(t)$ is a right composition factor of $f(t)$.

(1) If $f_1'(t) \neq 0$, show that $f_1(t)$ is a right composition factor of $f(t)$.
(2) If the characteristic of K is 0, show that $f_1(t)$ is a right composition factor of $f(t)$.
(3) If the characteristic of K is a prime number p, let e and d be non negative integers such that $f_1(x) = f_2(x^{p^e})$ and $f(x) = f_3(x^{p^d})$, where $f_2(x)$ and $f_3(x)$ are polynomials with coefficients in K such that $f_2'(x) \neq 0$ and $f_3'(x) \neq 0$.

(a) Show that $e \le d$ and $f_2(x) - f_2(y) \mid f_3(x^{p^s}) - f_3(y^{p^s})$, where $s = d - e$.

(b) Deduce, from **(1)**, that $f_2(t)$ is a right composition factor of $f_3(t^{p^s})$.

(c) Conclude that $f_1(t)$ is a right composition factor of $f(t)$.

(4) Conversely, if $f_1(t)$ is a right composition factor of $f(t)$, show that $f_1(x) - f_1(y)$ divides $f(x) - f(y)$ in $K[x, y]$.

Solution 9.21.

(1) We prove the result by induction on the degree of $f_1(x)$. If $\deg f_1 = 1$, then $f(t) = F(f_1(t))$, where $F(t) = f(f_1^{-1}(t))$ and f_1^{-1} is the unique polynomial such that $f_1(f_1^{-1}(t)) = t$. Thus the result is obvious in this case. Suppose that $\deg f_1 \ge 2$ and that the result holds for all polynomials $g_1(t)$such that $\deg g_1(t) < \deg f_1(t)$. We distinguish two cases.

- Suppose that

$$f_1(x) - f_1(y) = L_1(x, y) \cdots L_r(x, y),$$

where $L_i(x, y) = a_i x + b_i y + c_i$, a_i, b_i, $c_i \in K$ and $a_i b_i \neq 0$. By **Exercise 4.6**, $K(t)/K(f_1(t))$ is normal. Moreover since $f_1'(t) \neq 0$, then $K(t)/K(f_1(t))$ is separable, thus Galois. Let $q(x, y) \in K[x, y]$ such that

$$f(x) - f(y) = (f_1(x) - f_1(y))q(x, y),$$

then

$$f(x) - f(t) = (f_1(x) - f_1(t))q(x, t).$$

Let $\sigma \in \mathrm{Gal}(K(t), K(f_1(t)))$, then we may extend σ to $K(t, x)$ by setting $\sigma(x) = x$. Applying σ to the preceding identity, we get

$$f(x) - f(\sigma(t)) = (f_1(x) - f_1(t))q(x, \sigma(t)).$$

It follows that $f_1(x) - f_1(t)$ divides both

$$f(x) - f(t) \quad \text{and} \quad f(x) - f(\sigma(t)),$$

thus $f_1(x) - f_1(t)$ divides $f(t) - f(\sigma(t))$. We conclude that $f(t) = f(\sigma(t))$ for any $\sigma \in \mathrm{Gal}(K(t), K(f_1(t)))$ and then $f(t) \in K(f_1(t)))$. Set $f(t) = A(f_1(t))/B(f_1(t))$, where $A(t)$, $B(t) \in K[t]$ and $\gcd(A(t), B(t)) = 1$. By Bezout's Theorem, there exist $a(t)$ and $b(t) \in K[t]$ such that

$$a(t)A(t) + b(t)B(t) = 1.$$

Substituting $f_1(t)$ for t, we obtain

$$a(f_1(t))A(f_1(t)) + b(f_1(t))B(f_1(t)) = 1.$$

Since $B(f_1(t)) \mid A(f_1(t))$, then $B(f_1(t)) \in K$, thus $B(t) \in K$. We conclude that $f_1(t)$ is a right composition factor of $f(t)$.

- Suppose that $f_1(x) - f_1(y)$ has an irreducible factor $\phi(x, y)$, of degree at least 2 in $K[x, y]$ and let u and v be variables such that $\phi(u, v) = 0$. Notice that

$$f_1(u) = f_1(v) \quad \text{and} \quad f(u) = f(v).$$

Consider the following diagram of fields.

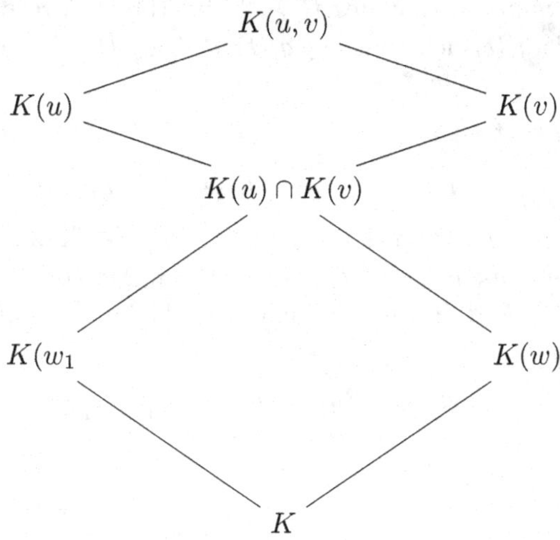

Here

$$w_1 = f_1(u) = f_1(v) \quad \text{and} \quad w = f(u) = f(v).$$

Since $K(u) \cap K(v)$ is an intermediate field between K and $K(u)$, then, by Luroth's Theorem [Schinzel (2000), Th. 2, Chap. 1.1], $K(u) \cap K(v) = K(w_2)$. Moreover since this field contains $w_1 = f_1(u) \in K[u]$, then we may suppose that $w_2 \in K[u]$ (see Exercise 8.10(1)). Set $w_2 = f_2(u)$, where $f_2(t) \in K[t]$. We then have

$$w_1 = g_1(w_2) \quad \text{and} \quad w = g(w_2),$$

where $g(t)$ and $g_1(t)$ are rational functions with coefficients in K. We show that indeed, $g_1(t)$ is a polynomial. A similar proof works for $g(t)$ and will be omitted. Set $w_1 = A(f_2(u))/B(f_2(u))$, where $A(t)$, $B(t) \in K[t]$ and $\gcd(A(t), B(t)) = 1$. By Bezout's Theorem, there exist $a(t)$ and $b(t) \in K[t]$ such that $a(t)A(t) + b(t)B(t) = 1$. Substituting $f_2(t)$ for t, we obtain

$$a(f_2(t))A(f_2(t)) + b(f_2(t))B(f_2(t)) = 1.$$

Since $B(f_2(u)) \mid A(f_2(u))$, then $B(f_2(u)) \in K$, thus $B(u) \in K$. We conclude that $f_1(u) = g_1(f_2(u))$, where $g_1(t)$ is a polynomial with coefficients in K. Similarly, we get $f(u) = g(f_2(u))$, with $g(t) \in K[t]$. Since u is transcendental over K, then we may substitute x or y as well, for u, in these expressions of $f_1(u)$ and $f(u)$. We obtain

$$f_1(x) = g_1(f_2(x)), \quad f(x) = g(f_2(x))$$

and

$$f_1(y) = g_1(f_2(y)), \quad f(y) = g(f_2(y)).$$

Our assumption implies that $g_1(f_2(x)) - g_1(f_2(y))$ divides $g(f_2(x)) - g(f_2(y))$, hence by **Exercise 9.20**,

$$g_1(x) - g_1(y) \quad \text{divides} \quad g(x) - g(y).$$

In order to use the inductive hypothesis, we compare $\deg f_1(u)$ and $\deg g_1(u)$. Since $f_1(u) = g_1(f_2(u))$, then $\deg f_1(u) \geq \deg g_1(u)$. Suppose that here we have equality. Then $\deg f_2(u) = 1$, hence $K(u) = K(u) \cap K(v)$, which implies $K(u) \subset K(v)$. Set $u = C(v)/D(v)$, where $C(t)$ and $D(t)$ are coprime polynomials with coefficients in K. Then $D(v)u - C(v) = 0$. this implies that $\phi(x, y) = D(y)x - C(y)$. But it is easy to see that, being a divisor of $f_1(x) - f_1(y)$, then ordering the terms of $\phi(x, y)$ in the ascending powers of x (resp. y), the leading coefficient is constant in both cases. This implies that $D(t) \in K^\star$, thus $u = h(v)$, where $h(t)$ is a polynomial with coefficients in K. We have $f_1(v) = f_1(u) = f_1(h(v))$. We deduce that $\deg h(t) = 1$ and then $\deg \phi(x, y) = 1$, which contradicts our assumption. We conclude that $\deg f_1(u) > \deg g_1(t)$, so that we may apply the result for $g_1(t)$ and get that $g_1(t)$ is a right composition factor of $g(t)$. Set $g(t) = F(g_1(t))$, where $F(t)$ is a polynomial with coefficients in K, then

$$f(t) = g(f_2(t)) = F(g_1(f_2(t))) = F(f_1(t))$$

and the proof is complete.

(2) Since K is of characteristic 0 and $f_1(t)$ is non constant, then $f_1'(t) \neq 0$, thus **(1)** applies and we conclude that $f_1(t)$ is a right composition factor of $f(t)$.

(3)(a) Suppose that $e > d$ and let $r = e - d$. Then $r \geq 1$ and since

$$f_2(x^{p^e}) - f_2(y^{p^e}) \quad \text{divides} \quad f_3(x^{p^d}) - f_3(y^{p^d}),$$

we get

$$f_2(x^{p^r}) - f_2(y^{p^r}) \quad \text{divides} \quad f_3(x) - f_3(y).$$

Since $r \geq 1$, then $(x - y)^2 \mid f_3(x) - f_3(y)$. We deduce that $f_3'(x) = f_3'(y) = 0$, which is a contradiction. We conclude that $e \leq d$. Since

$$f_2(x^{p^e}) - f_2(y^{p^e}) \quad \text{divides} \quad f_3(x^{p^{s+e}}) - f_3(y^{p^{s+e}}),$$

then

$$f_2(x) - f_2(y) \quad \text{divides} \quad f_3(x^{p^s}) - f_3(y^{p^s}).$$

(b) Since $f_2'(t) \neq 0$, then, by **(1)**, $f_2(t)$ is a right composition factor of $f_3(t^{p^s})$.

(c) According to the result of **(b)**, let $F(t) \in K[t]$ such that $f_3(t^{p^s}) = F(f_2(t))$. Then substituting t^{p^e} for t, we get $f(t) = F(f_1(t))$, that is $f_1(t)$ is a right composition factor of $f(t)$.

(4) Set $f(t) = F(f_1(t))$, then

$$f(x) - f(y) = F(f_1(x)) - F(f_1(y)).$$

This shows that $f_1(x) - f_1(y)$ is a divisor of $f(x) - f(y)$ in $K[x, y]$.

Chapter 10

Integral elements, Algebraic number theory

Exercise 10.1.
Let K be a field, x be an indeterminate and Ω be an algebraic closure of $K(x)$. For any positive integer n let $x^{1/n}$ denotes one fixed root of the polynomial $y^n - x$ in Ω. Denote by $\mathbb{Q}^+$ the set of non negative rational numbers. Let A be the subring of Ω whose elements $f(x)$ have the form $f(x) = \sum_{r\in\mathbb{Q}^+} a_r x^r$, where $a_r \in K$, $a_r = 0$ for all r except finitely many of them.

(1) For any $f(x) \in A$, $f(x) = \sum_{r\in\mathbb{Q}^+} a_r x^r$, let $S(f) = \{r \in \mathbb{Q}^+, a_r \neq 0\}$. Define the degree of $f(x)$ as follows. If $f(x) = 0$ set $\deg f = -\infty$. If $f(x) \neq 0$, set $\deg f = \max_{r\in S(f)} r$. Let $g(x)$ and $h(x)$ be elements of A. Show that $\deg gh = \deg g + \deg h$ and $\deg(g+h) \leq \max(\deg g, \deg h)$.
(2) Show that A is an integral domain and $A^\star = K^\star$.
(3) Show that A is integrally closed.
(4) Show that A is integral over $K[x]$.
(5) Let $f(x) \in K[x]$. If $f(x)$ is irreducible in A, show that $f(x)$ is irreducible in $K[x]$. Show that the converse is false.
(6) Let $f(x) \in A$ such that $\deg f > 0$. If $f(x)$ is irreducible in A, show that it is prime.
(7) Let $a \in K^\star$ and r be a positive rational number. Show that any divisor (in A) of ax^r has the form bx^s, where $b \in K^\star$, s is a non negative rational number and $s \leq r$.
(8) Show that x cannot be written in the form $x = up_1 \cdots p_k$, where $u \in A^\star$ and $p_1, \ldots, p_k$ are irreducible elements of A.
(9) Let B be a Noetherian ring. Show that any $b \in B \setminus B^\star$, $b \neq 0$, may be expressed as a product of irreducible elements of B.
(10) Deduce from **(8)** and **(9)** that A is not Noetherian.

(11) Show that A is not local.
Hint. One may use the following result. A commutative ring B is local if and only if $B \setminus B^\star$ is an ideal.

Solution 10.1.

(1) If $f(x) = 0$ or $g(x) = 0$ then the assertion on the degrees of $f(x)g(x)$ and $f(x) + g(x)$ are satisfied. If $f(x) \neq 0$ and $g(x) \neq 0$, set $f(x) = \sum_{r \leq r_0} a_r x^r$ and $g(x) = \sum_{s \leq s_0} b_s x^s$ with $a_{r_0} \neq 0$ and $b_{s_0} \neq 0$. Then $f(x)g(x) = \sum a_r b_s x^{r+s}$, with $a_{r_0} b_{s_0} \neq 0$ and $r + s < r_0 + s_0$ if $(r, s) \neq (r_0, s_0)$, hence $\deg fg = \deg f + \deg g$. We omit the proof for the degree of $f(x) + g(x)$.

(2) Obvious.

(3) Let $f(x) \in \mathrm{Frac}(A)$. Suppose that $f(x)$ is integral over A, that is, there exist a positive integer n and $a_0(x), \ldots, a_{n-1}(x) \in A$ such that $f(x)^n + a_{n-1}(x)f(x)^{n-1} + \cdots + a_0(x) = 0$. There exists a positive integer N such that $f(x) \in K(x^{1/N})$ and $a_i(x) \in K[x^{1/N}]$. Set $f(x) = g(x^{1/N}) = u(x^{1/N})/v(x^{1/N})$ with $u(y), v(y) \in K[y]$ and $\gcd(u(y), v(y)) = 1$. Substituting x^N for x in the integral equation satisfied by $f(x)$, we obtain: $g(x)^n + a_{n-1}(x^N)g(x)^{n-1} + \cdots + a_0(x^N) = 0$. It follows that $g(x)$ is integral over $K[x]$. Since $g(x) \in K(x)$ and $K[x]$ is integrally closed, then $g(x) \in K[x]$, hence $v(x) \mid u(x)$. We deduce that $v(x) \in K^\star$ and then $f(x) \in A$.

(4) Let $f(x) = \sum_{r \leq r_0} a_r x^r$ be an element of A. To obtain that $f(x)$ is integral over $K[x]$, it is sufficient to prove that x^r is integral over $K[x]$. Set $r = p/q$, where p and q are positive integers, then it is sufficient to prove that $x^{1/q}$ is integral over $K[x]$. This claim is true since $x^{1/q}$ is a root of the monic polynomial $y^q - x$ with coefficients in $K(x)$.

(5) The first part of the statements of **(5)** is obvious and will be omitted. We prove the second part by finding a counterexample. Let $f(x) = x$, then $f(x)$ is irreducible in $K[x]$ but reducible in A since $f(x) = (x^{1/2})^2$.

(6) Suppose that $f(x)$ divides $a(x)b(x)$ in A, then $a(x)b(x) = f(x)g(x)$, where $g(x) \in A$. Let N be a positive integer such that $f(x)$, $g(x)$, $a(x)$, $b(x) \in K[x^{1/N}]$, then $f(x^N)$, $g(x^N)$, $a(x^N)$, $b(x^N) \in K[x]$ and we have $a(x^N)b(x^N) = f(x^N)g(x^N)$. Since $f(x)$ is irreducible in A, then $f(x^N))$ is irreducible in A, hence irreducible in $K[x]$ by **(5)**. The preceding identity implies that $f(x^N)$ divides one, say $a(x^N)$, among the polynomials $a(x^N)$ and $b(x^N)$ in $K[x]$. Let $c(x) \in K[x]$ such that $a(x^N) = c(x)f(x^N)$. Substituting $x^{1/N}$ for x, we get $a(x) =$

$c(x^{1/N})f(x)$, which proves that $f(x)$ divides $a(x)$ in A, establishing that $f(x)$ is prime in A.

(7) Let $f(x) \in A$ be a divisor of ax^r and let $g(x) \in A$ such that $ax^r = f(x)g(x)$. Let N be a positive integer such that ax^r, $f(x)$, $g(x) \in K[x^{1/N}]$. Then we get a factorization of ax^{rN} in $K[x]$: $ax^{rN} = f(x^N)g(x^N)$. It follows that $f(x^N) = bx^t$, where $b \in K^\star$ and t is a positive integer such that $t \leq rN$. Let $s = t/N$, then $f(x) = bx^s$.

(8) Suppose that $x = up_1 \cdots p_k$, where $u \in A^\star$ and $p_1, \ldots, p_k$ are irreducible elements of A. Then by **(7)** $p_i = b_i x^{s_i}$, where $s_i \leq r$. But $b_i x^{s_i} = b_i(x^{s_i/2})^2$ is reducible in A. Thus x cannot be expressed as a product of irreducible elements of A.

(9) Let

$$T = \{t \in B \setminus B^\star,\, t \neq 0,\, t \text{ is not a product of irreducibles of } A\}.$$

Suppose that $T \neq \emptyset$. Consider the set $\mathcal{I}$ of ideals tB for $t \in T$. By the ascending chain condition, this set of ideals contains a maximal element, say t_0B. Since $t_0B \in T$, then t_0 is not irreducible, thus $t_0 = t_1t_2$, where t_1 nor t_2 is a unit and $t_1t_2 \neq 0$. Clearly $t_1 \in T$ or $t_2 \in T$. Without loss of generality, we may suppose that $t_1 \in T$. We now have $t_0B \subsetneq t_1B$, contradicting the maximality of t_0B in $\mathcal{I}$. We conclude that $T = \emptyset$, that is every non zero element of B, not a unit, may be expressed as a product of irreducible elements of B.

(10) By **(9)**, if one finds an element $f(x) \in A \setminus A^\star$, $f(x) \neq 0$, which cannot expressed as a product of irreducibles, then one may conclude that A is not Noetherian. By **(8)**, x is such an element. Thus A is not Noetherian.

Remark. We may prove that A is not Noetherian in the following way. Consider the chain of ideals

$$xA \subset x^{1/2}A \subset \cdots \subset x^{1/2^k}A \subset x^{1/2^{k+1}}A \subset \cdots.$$

It is seen that this chain never terminates. For if $x^{1/2^k}A = x^{1/2^{k+1}}A$, then $x^{1/2^k} \mid x^{1/2^{k+1}}$, which is a contradiction to **(7)**.

(11) We use the hint. Here $A^\star = K^\star$ and then

$$A \setminus A^\star = \{0\} \cup \{f(x) \in A, \deg f > 0\}.$$

Since x, $x+1 \in A \setminus A^\star$ and $x+1-x \notin A \setminus A^\star$, then A is not local.

Exercise 10.2.

Define the polynomials $\binom{x}{n}$ for $n \in \mathbb{N}$ by $\binom{x}{0} = 1$ and $\binom{x}{n} = \frac{x(x-1)\cdots(x-(n-1))}{n!}$ for $n \geq 1$.

(1) Show that these polynomials constitute a basis of $\mathbb{Q}[x]$ over $\mathbb{Q}$.
(2) Let
$$A = \{f(x) \in \mathbb{Q}[x],\ f(\mathbb{Z}) \subset \mathbb{Z}\}.$$
Show that A is a free $\mathbb{Z}$-module and $\mathcal{B} = \{\binom{x}{n},\ n \in \mathbb{N}\}$ is one of its bases.
(3) Show that A is a subring of $\mathbb{Q}[x]$ and $A^\star = \{-1, 1\}$.
(4) Show that $\mathrm{Frac}(A) = \mathbb{Q}(x)$.
(5) Show that A is integrally closed.
(6) Show that A is not Noetherian.
(7) Show that for $n \geq 1$, $\binom{x}{n}$ is irreducible in A. Give an example of an element of A which has at least two distinct factorizations.
(8) For any $f(x) \in A \backslash \{0\}$, denote by $d(f)$, the integer $d(f) = \gcd_{a\in\mathbb{Z}} f(a)$. Let $f(x)$ and $g(x) \in A \setminus \{0\}$. Show that
   (i) $d(f)d(g) \mid d(fg)$.
   (ii) $d(f^m) = d(f)^m$ for any $m \geq 1$.

   Show that we may have $d(f)d(g) \neq d(fg)$.
(9) Let $f(x) = a_0 + a_1 x + \cdots + a_n\binom{x}{n}$ with $a_i \in \mathbb{Z}$ and $a_n \neq 0$. Show that
$$d(f) = \gcd(f(0), f(1), \ldots, f(n)) = \gcd(a_0, a_1, \ldots, a_n).$$

Solution 10.2.

(1) Suppose that $a_0 + a_1 x + \cdots + a_n\binom{x}{n} = 0$, with $a_0, a_1, \ldots, a_n \in \mathbb{Q}$. Set $f(x) = a_0 + a_1 x + \cdots + a_n\binom{x}{n}$. From $f(0) = 0$, we get $a_0 = 0$. Suppose that $a_j = 0$ for any $j \leq k - 1$, then $0 = f(k) = a_k$. To prove that $\mathcal{B}$ generates the vector space, it is sufficient to show that x^j, for $j \geq 0$ is a linear combination of the elements of $\mathcal{B}$ with rational coefficients. Since
$$n!\binom{x}{n} = x(x-1)\cdots(x-(n-1)) = x^n + \text{ lower degree terms},$$
the conclusion follows by induction.
(2) Obviously A is a $\mathbb{Z}$-module. A similar reasoning as in **(1)**, shows that $\mathcal{B}$ is a basis of A over $\mathbb{Z}$.
(3) It is clear that the product of two elements of A is an element of A. Let $f(x) \in A^\star$ and let $g(x) \in A$ such that $f(x)g(x) = 1$, then $\deg f = \deg g = 0$ and then $f(x) = \pm 1$.
(4) From $\mathbb{Z}[x] \subset A \subset \mathbb{Q}[x]$, we conclude that
$$\mathbb{Q}(x) = \mathrm{Frac}(\mathbb{Z}[x]) \subset \mathrm{Frac}(A) \subset \mathrm{Frac}(\mathbb{Q}[x]) = \mathbb{Q}(x),$$
thus $\mathrm{Frac}(A) = \mathbb{Q}(x)$.

(5) Let $f(x) \in \mathbb{Q}(x)$ be integral over A. Write $f(x)$ in the form $f(x) = u(x)/v(x)$ with $u(x)$ and $v(x) \in \mathbb{Z}[x]$ and $\gcd(u(x), v(x)) = 1$. Representing $u(x)$ and $v(x)$ in the form $u(x) = \operatorname{cont}(u)u_1(x)$ and $v(x) = \operatorname{cont}(v)v_1(x)$, where $u_1(x)$ and $v_1(x) \in \mathbb{Z}[x]$, it is seen that $\gcd(\operatorname{cont}(u), \operatorname{cont}(v)) = 1$ and $\gcd(u_1(x), v_1(x)) = 1$. Since $f(x)$ is integral over A, there exist $a_0(x), \ldots, a_{n-1}(x)$ such that

$$f(x)^n + a_{n-1}(x)f(x)^{n-1} + \cdots + a_0(x) = 0.$$

Since $\mathbb{Z}$ is integrally closed, this equation shows that for any $m \in \mathbb{Z}$ such that $v(m) \neq 0$, we have $f(m) \in \mathbb{Z}$. Let C be a positive real number greater than all the absolute values of the roots of $v(x)$. We distinguish two cases.

- $\operatorname{cont}(v) = 1$. Here $\operatorname{cont}(v) \mid \operatorname{cont}(u)$ and $v(m) \mid u(m)$ for any integer $m > C$. **Exercise 1.29** shows that $v(x) \mid u(x)$ in $\mathbb{Z}[x]$, which in turn implies that $f(x) \in \mathbb{Z}[x]$ and then $f(x) \in A$.
- $\operatorname{cont}(v) \neq 1$. We have $\operatorname{cont}(v_1) \mid \operatorname{cont}(u)$ and $v_1(m) \mid u(m)$ for any integer $m \geq C$. **Exercise 1.29** shows that $v_1(x) \mid u(x)$ in $\mathbb{Z}[x]$. Since $\operatorname{cont}(v_1) = 1$ and $\mathbb{Z}[x]$ is a unique factorization domain, then $v_1(x) \mid u_1(x)$ in $\mathbb{Z}[x]$. Our assumptions on u_1 and v_1 implies that $v_1(x) = \pm 1$. We may suppose that $v_1(x) = 1$ and then $f(x) = \frac{\operatorname{cont}(u)u_1(x)}{d}$, where d is a positive integer such that $\gcd(\operatorname{cont}(u), d) = 1$. Since $f(m) \in \mathbb{Z}$ for any $m > C$, then $d \mid u_1(m)$ for any $m > C$. Set $u_1(x) = b_0 + b_1 x + \cdots + b_k \binom{x}{k}$ with $b_j \in \mathbb{Z}$ for $j = 0, \ldots, k$. We claim and we omit the proof that, since $u_1(x) \in \mathbb{Z}[x]$, $b_j \equiv 0 \pmod{j!}$. We may write $u_1(x)$ in the form

  $$u_1(x) = c_0 + c_1 x + \cdots + c_k x(x-1)\cdots(x-(k-1))$$

  with $c_j \in \mathbb{Z}$ for $j = 0, \ldots, k$. Let q be a positive integer such that $dq > C$ and let $j \in \{0, \ldots, k\}$. We have $0 \equiv u_1(dq) \pmod{d} \equiv c_0 \pmod{d}$, that is $c_0 \equiv 0 \pmod{d}$. Suppose by induction that $i!c_i \equiv 0 \pmod{d}$ for any $i = 0, \ldots, j-1$. We have

  $$\begin{aligned} u_1(dq+j) &= c_k(dq+j)(dq+j-1)\cdots(dq+j-(k-1)) \\ &\quad + \cdots + c_j(dq+j)(dq+j-1)\cdots(dq+1) \\ &\quad + \cdots + c_1(dq+j) + c_0 \equiv 0 \pmod{d}, \end{aligned}$$

hence

$$\begin{aligned}u_1(dq+j) &\equiv c_0 + c_1 j + c_2 j(j-1) \\ &\quad + \cdots + c_{j-1} j(j-1)\cdots 2 + c_j j! \pmod d \\ &\equiv c_0 + 1!c_1 \frac{j}{1!} + 2!c_2 \frac{j(j-1)}{2} \\ &\quad + \cdots + (j-1)!c_{j-1} \frac{j(j-1)\cdots 2}{(j-1)!} + c_j j! \pmod d.\end{aligned}$$

The inductive hypothesis implies that $j!c_j \equiv 0 \pmod d$. We now have

$$\begin{aligned}f(x) &= \frac{\mathrm{cont}(u)u_1(x)}{d} \\ &= \mathrm{cont}(u)\frac{c_0 + c_1 x + \cdots + c_k x(x-1)\cdots(x-(k-1))}{d} \\ &= \mathrm{cont}(u)\left(\frac{c_0}{d} + \frac{c_1}{d}x + \frac{2!c_2}{d}\binom{x}{2} + \cdots + \frac{k!c_k}{d}\binom{x}{k}\right).\end{aligned}$$

This shows that $f(x) \in A$.

(6) For any non negative integer j let $P_j(x) = \binom{x}{j}$. Let I_1 be the ideal of A generated by $P_1(x)$. For any prime number p let I_p be the ideal generated by $P_1(x), P_2(x), \ldots, P_{p-1}(x), P_p(x)$, then

$$I_1 \subset I_2 \subset \ldots \subset I_p \subset \ldots.$$

We show that this sequence of ideals is not stationary. Let p_1 and p_2 be two consecutive primes with $p_1 < p_2$. We are going to show that $P_2(x) \notin I_{p_1}$, which will imply that $I_{p_1} \subsetneq I_{p_2}$. Suppose that $P_2(x) \in I_{p_1}$, then $P_{p_2}(x) = a_1(x)P_1(x) + \cdots + a_{p_1}(x)P_{p_1}(x)$ with $a_j(x) \in A$ for $j = 1, \ldots, p_1$. Put $x = p_2$ in this identity. We get

$$1 = a_1(p_2)P_1(p_2) + a_2(p_2)P_2(p_2) + \cdots + a_{p_1}(p_2)P_{p_1}(p_2). \qquad \text{(Eq 1)}$$

For any $j = 1, \ldots, p_1$, we have $P_j(p_2) = \frac{p_2(p_2-1)\cdots(p_2-(j-1))}{j!}$. Since $j \le p_1 < p_2$, then $P_j(p_2) \equiv 0 \pmod{p_2}$. Now **(Eq 1)** reads $1 \equiv 0 \pmod{p_1}$, which is a contradiction.

(7) Let n be a positive integer. Suppose that $\binom{x}{n} = g(x)h(x)$ with $g(x)$ and $h(x) \in A$. Let $k = \deg g$ and $m = \deg h$, then $k + m = n$. We may suppose that $k \le m$. Let b_k (resp. c_m) be the coefficient of $\binom{x}{k}$ (resp. $\binom{x}{m}$) in the representation of $g(x)$ (resp. $h(x)$) in the basis $\mathcal{B}$. Equating the leading coefficients of the two sides of the above identity,

we get $1/n! = (b_k/k!)(c_m/m!)$. If $k = 0$ and $m = n$, then $b_0c_n = 1$, hence $b_0 = \pm 1$, thus $g(x) \in A^\star$. Suppose that $k \neq 0$, then

$$(n-k)!k! = n!b_kc_m = n(n-1)\cdots(n-(k-1))(n-k)!,$$

hence $k! = n(n-1)\cdots(n-(k-1))b_kc_m$. Since $|b_k| \geq 1$, $|c_m| \geq 1$ and $n > k$, we deduce that $|n(n-1)\cdots(n-(k-1))b_kc_m| > k!$, which is a contradiction. Thus $\binom{x}{n}$ is irreducible in A.

For any positive integer n, we have $n\binom{x}{n} = (x-(n-1))\binom{x}{n-1}$. From what was just been proved the element of A in the right side of this identity has exactly 2 irreducible factors, while the element on the left side has $1+r$ irreducible factors, where r is the number of prime numbers distinct or not dividing n, so that in general $1+r > 2$, which means that these factorizations are different.

Even if n is a prime number, that is $r = 1$, these factorizations are different since n is not associate to neither of the factors appearing on the right side of the identity.

(8) (i) For any $a \in \mathbb{Z}$, $d(f) \mid f(a)$ and $d(g) \mid g(a)$, hence $d(f)d(g) \mid f(a)g(a)$, thus $d(f)d(g) \mid d(fg)$.

(ii) By **(i)**, $d(f)^n \mid d(f^n)$. Let p be a prime number such that $p^e \parallel d(f)^m$, then $e = rm$ and $p^r \parallel d(f)$. It follows that for any $a \in \mathbb{Z}$, $p^r \mid f(a)$ and then $prm \mid f(a)^m$. Therefore $pe \mid d(f)^m$.

For the last statement Let $f(x) = x$ and $g(x) = x-1$, then obviously $d(f) = d(g) = 1$ and $d(fg) = 2$, so that $d(f)d(g) \neq d(fg)$ in this case.

(9) We have

$$\begin{aligned} f(0) &= a_0 \\ f(1) &= a_0 + a_1 \\ f(2) &= a_0 + 2a_1 + a_2 \\ \cdots &= \cdots\cdots\cdots\cdots\cdots\cdots\cdots\cdots \\ f(n) &= a_0 + na_1 + n(n-1)/2a_2 + \cdots + \binom{n}{n-1}a_{n-1} + a_n. \end{aligned}$$

These equations show that a given positive d divides $f(0)$, $f(1), \ldots, f(n)$ if and only if d divides $a_0, a_1, \ldots, a_n$, thus $\gcd(f(0), f(1), \ldots, f(n)) = \gcd(a_0, a_1, \ldots, a_n)$.

In **Exercise 1.8**, the following result is proved.

Let K be a field of characteristic 0, $f(X)$ be a polynomial of degree n and $m > n$ be an integer. Then

$$f(m) = \sum_{j=0}^{n}(-1)^{n-j}\binom{m}{j}\binom{m-j-1}{n-j}f(j).$$

This shows that $d(f) = \gcd(f(0), f(1), \ldots, f(n))$.

Remark. In **Exercise 10.1(9)** it is proved that in a Noetherian ring any element, except 0 and a unit, may be expressed as a product of irreducibles of the ring. For the ring subject of the present exercise, we may see that any $f(x) \in A$, $f(x) \notin \{0, -1, 1\}$ is a product of irreducible elements of A in the following way. Let $p_1, \ldots, p_r$ be the list (may be empty) of distinct or not of prime numbers dividing $f(x)$ in A. Let $f_1(x) \in A$ such that $f(x) = p_1 \cdots p_r f_1(x)$. If $f_1(x)$ is irreducible in A, the polynomial $f(x)$ is decomposed into a product of irreducible elements of A. If not, let $f_2(x)$ be an irreducible divisor of $f_1(x)$ in A and let $f_3(x) \in A$ such that $f_1(x) = f_2(x)f_3(x)$. We use the same reasoning for $f_3(x)$ instead of $f_1(x)$. Since $\deg f_3(x) < \deg f_1(x)$, this process must terminates after a finite number of steps. At the end of this process, we get a factorization of $f(x)$ into a product of irreducibles of A. This implies that the converse of the property stated in **Exercise 10.1(9)** does not hold, that is we may have a decomposition into irreducibles even if the ring is not Noetherian.

Exercise 10.3.

(1) Let K be a field. An element $F(x) \in K[[x]]$ is said to be rational if there exist $u(x)$ and $v(x)$ in $K[x]$ such that $v(0) \neq 0$ and $F(x) = u(x)/v(x)$. Let $F(x) = \sum_{n=0}^{\infty} a_n x^n \in K[[x]]$. Show that $F(x)$ is rational if and only if there exist two non negative integers s and n_0 and elements $b_0, \ldots, b_s$ of K such that $s \geq 1$, $b_0 \neq 0$ and $b_0 a_n + b_1 a_{n-1} + \cdots + b_s a_{n-s} = 0$, for any $n \geq n_0$.

(2) Let $g(x)$ and $h(x)$ be coprime polynomials with rational coefficients such that $h(0) = 1$. Suppose that the power series $g(x)/h(x)$ has integral coefficients. Show that $g(x)$ and $h(x) \in \mathbb{Z}[x]$.

(3) Let $g(x)$ and $h(x) \in \mathbb{Z}[x]$ be coprime such that $h(0) \neq 0$. Suppose that $g(x)/h(x) \in \mathbb{Z}[[x]]$. Show that $h(0) = \pm 1$.

(4) Let α be an algebraic number of degree n such that $\mathrm{Tr}_{\mathbb{Q}(\alpha)/\mathbb{Q}}(\alpha^m) \in \mathbb{Z}$ for any positive integer m. Show that α is an algebraic integer.

Solution 10.3.

(1) • **Necessity of the condition.** Let $u(x)$ and $v(x) \in K[x]$, with $v(0) \neq 0$, such that $F(x) = u(x)/v(x)$. Set $v(x) = v_0 + v_1 x + \cdots + v_s x^s$ and let $r = \deg u$. Let $n_0 = Max(s, r+1)$, then for $n \geq n_0$, the coefficient of x^n in the series $v(x)F(x)$, which coincide with $u(x)$,

will be 0. Since this coefficient is equal to $v_0a_n + \cdots + v_sa_{n-s}$, then the result follows by putting $b_i = v_i$.

- **Sufficiency of the condition.** Let $v(x) = b_0+b_1x+\cdots+b_sx^s$. The relations on the a_i and the b_j implies that $v(x)F(x)$ is a polynomial, say $u(x)$ with coefficients in K and of degree no greater than n_0-1. Hence $F(x) = u(x)/v(x)$ is rational.

(2) Clearing denominators and dividing out by the greatest common divisor of the set of coefficients of $g(x)$ and $h(x)$, we may write $F(x)$ in the form $F(x) = g(x)/h(x) = u(x)/v(x)$, where $u(x)$ and $v(x)$ are coprime polynomials with integral coefficients. Further, we may suppose that $v(0) > 0$. We may write a Bezout's Identity relating $u(x)$ and $v(x)$ in the form $a(x)u(x) + b(x)v(x) = c$, where c is a positive integer and $a(x)$ and $b(x)$ are polynomials with integral coefficients. It follows that $a(x)F(x) + b(x) = c/v(x)$. The left side of this identity is a series, say $G(x) = \sum_{n=0}^{\infty} q_nx^n$, with integral coefficients. Thus we get an identity of the form $\sum_{n=0}^{\infty} q_nx^n = c/(v_0 + v_1x + \cdots + v_sx^s)$. We may suppose that in this identity, $\gcd_{n\geq 0}(q_n) = 1$. Otherwise, we get an analogue identity satisfying this condition by dividing c and all the coefficients q_n by this greatest common divisor. We show that $v_0 = 1$. Suppose that the contrary holds and let p be a prime factor of v_0. Since $c = v_0q_0$, then $p \mid c$. Let r be the smallest non negative integer such that $p \nmid q_r$. Write the above identity in the form

$$(v_0 + \cdots + v_sx^s)\sum_{n=0}^{r-1} q_nx^n + (v_0 + \cdots + v_sx^s)\sum_{n=r}^{\infty} q_nx^n = c.$$

Since

$$(v_0 + \cdots + v_sx^s)\sum_{n=0}^{r-1} q_nx^n \equiv 0 \pmod{p},$$

then

$$S := (v_0 + \cdots + v_sx^s)\sum_{n=r}^{\infty} q_nx^n \equiv 0 \pmod{p}.$$

We have

$S = v_0q_rx^r + (v_0q_{r+1} + v_1q_r)x^{r+1} + (v_0q_{r+2} + v_1q_{r+1} + v_2q_r)x^{r+2} + \cdots$, hence $v_0q_{r+1} + v_1q_r \equiv 0 \pmod{p}$, $v_0q_{r+2} + v_1q_{r+1} + v_2q_r \equiv 0 \pmod{p}$, and so on. The first congruence shows that $p \mid v_1$. By induction we get $p \mid v_i$ for $i = 0, \ldots, s$, contradicting the assumptions. We conclude that $v_0 = 1$. We now have $g(x)/h(x) = u(x)/v(x)$, with $\gcd(g(x), h(x)) = \gcd(u(x), v(x)) = 1$ and $h(0) = v(0) = 1$, hence $g(x) = u(x)$ and $h(x) = v(x)$. Thus $g(x)$ and $h(x)$ have integral coefficients.

(3) Since $\frac{g(x)/h(0)}{h(x)/h(0)} \in \mathbb{Z}[[x]]$, then $g(x)/h(0)$ and $h(x)/h(0) \in \mathbb{Z}[x]$, hence $h(0)$ divides $\gcd(g(x), h(x))$. We conclude that $h(0) = \pm 1$.

(4) • **First Proof.** Let $f(x) = \mathrm{Irr}(\alpha, \mathbb{Q}) = \prod_{i=1}^{n}(x - \alpha_i)$ and let

$$g(x) = x^n f(1/x) = \prod_{i=1}^{n}(1 - \alpha_i x),$$

then

$$\begin{aligned} g'(x)/g(x) &= \sum_{i=1}^{n} \frac{-\alpha_i}{1 - \alpha_i x} = \sum_{i=1}^{n} -\alpha_i \sum_{k=0}^{\infty} (\alpha_i x)^k \\ &= -\sum_{i=1}^{n}\sum_{k=0}^{\infty} (\alpha_i)^{k+1} x^k = -\sum_{k=0}^{\infty}\left(\sum_{i=1}^{n} \alpha_i^{k+1}\right) x^k \\ &= -\sum_{k=0}^{\infty} \mathrm{Tr}_{\mathbb{Q}(\alpha)/\mathbb{Q}}(\alpha^{k+1}) x^k. \end{aligned}$$

From the assumptions we deduce that $g'(x)/g(x) \in \mathbb{Z}[[x]]$. Obviously $g(0) = 1$, hence we may apply **(2)** and obtain $g(x) \in \mathbb{Z}[x]$. Since $f(x)$ has the same coefficients as $g(x)$, except that they are written in the inverse order, then $f(x) \in \mathbb{Z}[x]$.

• **Second Proof.** Let $K = \mathbb{Q}(\alpha)$ and consider the lattice

$$M = \mathbb{Z} + \mathbb{Z}\alpha + \cdots \mathbb{Z}\alpha^{n-1}.$$

Let L be its dual lattice, that is

$$L = \{\gamma \in K, \mathrm{Tr}_{K/\mathbb{Q}}(\gamma M) \subset \mathbb{Z}\}.$$

By assumptions, we have $\mathbb{Z}[\alpha] \subset L$. Since L is a free $\mathbb{Z}$-module of rank n, then $\mathbb{Z}[\alpha]$ is a free $\mathbb{Z}$-module. In particular, it is finitely generated over $\mathbb{Z}$. By [Lang (1965), Int. 2, Chap. 9.1], α is integral over $\mathbb{Z}$.

Exercise 10.4.
Let K be a field, $f(t)$, $g(t)$, $f_1(t)$ and $g_1(t)$ be non constant polynomials with coefficients in K. Show that the following propositions are equivalent.

(i) There exists $F(t) \in K[t]$, such that $f(t) = F(f_1(t))$ and $g(t) = F(g_1(t))$.
(ii) The polynomial $f_1(x) - g_1(y)$ divides $f(x) - g(y)$ in $K[x, y]$.

Hint. One may use **Exercise 9.20 and Exercise 9.21**.

Solution 10.4.

- $(i) \Rightarrow (ii)$. Obvious.
- $(ii) \Rightarrow (i)$. Let $\phi(x, y) \in K[x, y]$ be an irreducible factor of $f_1(x) - g_1(y)$ and let u, v be variables such that $\phi(u, v) = 0$. Let $w = f(u)$ and $w_1 = f_1(u)$, then $w = g(v)$ and $w_1 = g_1(v)$. Consider the following diagram of fields.

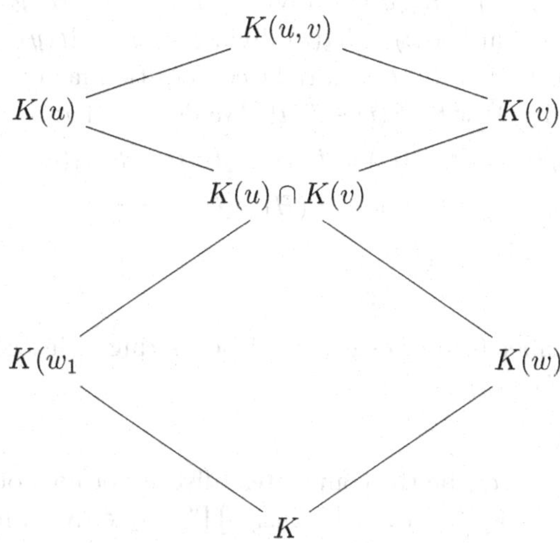

Since $K(u) \cap K(v)$ is an intermediate field between K and $K(u)$ and since this last field is a purely transcendental extension of the first, then, by Luroth's Theorem [Schinzel (2000), Th. 2, Chap. 1.1], $K(u) \cap K(v)$ is purely transcendental over K. Let $w_2 \in K(u) \cap K(v)$ such that $K(u) \cap K(v) = K(w_2)$. Moreover since this intersection contains $w_1 = f_1(u) \in K[u]$, then we may suppose that $w_2 = A(u) \in K[u]$ see **Exercise 8.10(1)**. Using similar argument as above, we may conclude that $K(u) \cap K(v) = K(w_3)$ and $w_3 = B(v) \in K[v]$. We have $w_3 = (aw_2 + b)/(cw_2 + d)$, where a, b, c, $d \in K$ and $ad - bc \neq 0$. It follows that $(aw_2 + b)w_3 - (cw + d) = 0$, hence $(aA(u)) + b)B(v) - (cA(u) + d) = 0$. But it is easy to see that the leading coefficient of $\phi(x, y)$, considered as a polynomial in x or in y, is a constant, hence v is integral over $K[u]$. We deduce, by transitivity, that $B(v)$ is integral over $K[A(u)]$. This implies that, in the preceding equation satisfied by $B(v)$, we have $a = 0$. Thus $B(v) = \lambda A(u) + \mu$ and then $w_3 = \lambda w_2 + \mu$ with $\lambda \in K^\star$ and $\mu \in K$. Therefore we may suppose that $K(u) \cap K(v) = K(w_2)$, where $w_2 = f_2(u) = g_2(v)$ and f_2, g_2 are polynomials with coefficients in K. Set $w_1 = H_1(w_2)$ and $w = H(w_2)$, where H_1 and H are rational functions with coefficients in K. Using the relations $f_1(u) = H_1(f_2(u))$ and $f(u) = H(f_2(u))$, one may prove easily that H_1 and H are polynomials. Since u is transcendental over K, then we have $f_1(x) = H_1(f_2(x))$, $f(x) = H(f_2(x))$, $g_1(y) = H_1(g_2(y))$ and $g(y) = H(g_2(y))$. Our assumptions read as follows $H_1(f_2(x)) - H_1(g_2(y))$ divides $H(f_2(x)) - H(g_2(y))$. **Exercise 9.20** implies that $H_1(x) - H_1(y)$ divides $H(x) - H(y)$. The preceding exercise shows that $H_1(t)$ is a right composition factor of $H(t)$, that is $H(t) = F(H_1(t))$ with $F(t) \in K[t]$. We deduce that

$$f(t) = H(f_2(t)) = F(H_1(f_2(t))) = F(f_1(t)) \quad \text{and}$$
$$g(t) = H(g_2(t)) = F(H_1(g_2(t))) = F(g_1(t))$$

and the proof is complete.

Exercise 10.5.
Let K be a number field of degree n, A be its ring of integers and $\alpha \in A$. Show that $\alpha \mid \mathrm{N}_{K/\mathbb{Q}}(\alpha)$.

Solution 10.5.
Let $\alpha_1 = \alpha, \alpha_2, \ldots, \alpha_n$ be the conjugates (distinct or not) of α over $\mathbb{Q}$. We have $N_{K/\mathbb{Q}}(\alpha) = \alpha \prod_{i=2}^{n} \alpha_i$. Obviously $\prod_{i=2}^{n} \alpha_i \in K$. Since the α_i are algebraic integers, then so is $\prod_{i=2}^{n} \alpha_i$. Therefore $\prod_{i=2}^{n} \alpha_i \in A$. It follows that $\alpha \mid N_{K/\mathbb{Q}}(\alpha)$.

Exercise 10.6.
Let $\theta \in \mathbb{C}$ be an algebraic integer. If θ is a unit, show that $\mathbb{Z}[\theta] = \mathbb{Z}[1/\theta]$.

Solution 10.6.
It is sufficient to prove that $\mathbb{Z}[1/\theta] \subset \mathbb{Z}[\theta]$. Let

$$f(x) = x^n + a_{n-1}x^{n-1} + \cdots a_1 x + \epsilon,$$

where $\epsilon = \pm 1$ and $a_1, \ldots, a_{n-1} \in \mathbb{Z}$, be the minimal polynomial of θ over $\mathbb{Q}$. Then

$$1/\theta = \epsilon^2/\theta = -\epsilon(\theta^{n-1} + a_{n-1}\theta^{n-2} + \cdots a_1) \in \mathbb{Z}[\theta].$$

The stated inclusion, now follows easily.

Exercise 10.7.
Let a and b be non zero, distinct square free rational integers, $K = \mathbb{Q}(\sqrt{a}, \sqrt{b})$ and A be the ring of integers of A. Suppose that $a \equiv b \equiv 1 \pmod 3$. Show that there exists no $\alpha \in A$ such that $A = \mathbb{Z}[\alpha]$.

Solution 10.7.
Suppose that $A = \mathbb{Z}[\alpha]$ for some $\alpha \in A$ and let $f(x)$ be the minimal polynomial of α over $\mathbb{Q}$. Consider the elements of A, $\alpha_1 = (1+\sqrt{a})(1+\sqrt{b})$, $\alpha_2 = (1+\sqrt{a})(1-\sqrt{b})$, $\alpha_3 = (1-\sqrt{a})(1+\sqrt{b})$ and $\alpha_4 = (1-\sqrt{a})(1-\sqrt{b})$. For $i = 1, \dots, 4$, let $f_i(x) \in \mathbb{Z}[x]$ such that $\alpha_i = f_i(\alpha)$. For any $g(x) \in \mathbb{Z}[x]$, denote by $\overline{g(x)}$, the reduced polynomial modulo 3 of $g(x)$. Fix $i \in \{1, \dots, 4\}$. We show that $\overline{f(x)} \mid \overline{f_i(x)f_j(x)}$ for $j \neq i$ and $\overline{f(x)} \nmid \overline{f_i(x)}^n$ for any positive integer n. By **Exercise 2.3**, it is equivalent to prove the following congruences in $\mathbb{Z}[\alpha]$: $f_i(\alpha)f_j(\alpha) \equiv 0 \pmod{p}$ for $i \neq j$ and $f_i(\alpha)^n \not\equiv 0 \pmod{p}$ for $n \geq 1$, that is $\alpha_i\alpha_j \equiv 0 \pmod{p}$ for $i \neq j$ and $\alpha_i^n \not\equiv 0 \pmod{p}$. In any case, for $i \neq j$, $\alpha_i\alpha_j$ may be written in form

$$\alpha_i\alpha_j = (1+\sqrt{a})(1-\sqrt{a})\gamma \quad \text{or}$$
$$\alpha_i\alpha_j = (1+\sqrt{b})(1-\sqrt{b})\gamma,$$

where $\gamma \in A$, thus it is sufficient to prove that $(1+\sqrt{a})(1-\sqrt{a}) \equiv 0 \pmod{p}$ and $(1+\sqrt{b})(1-\sqrt{b}) \equiv 0 \pmod{p}$. This claim is true since by assumptions $a \equiv b \equiv 1 \pmod 3$. To show that $\alpha_i^n \not\equiv 0 \pmod{p}$, it is sufficient to verify that $\mathrm{Tr}_{K/\mathbb{Q}}(\alpha_i^n) \not\equiv 0 \pmod{p}$. Since $\alpha_1, \dots, \alpha_4$ are conjugate over $\mathbb{Q}$, then $\mathrm{Tr}_{K/\mathbb{Q}}(\alpha_i^n) = \mathrm{Tr}_{K/\mathbb{Q}}(\alpha_1^n)$ and we may suppose that $i = 1$. Since $a \equiv b \equiv 1 \pmod 3$, then we have

$$\begin{aligned}\alpha_1^2 - \alpha_1 &= \alpha_1(\alpha_1 - 1)\\ &= (1+\sqrt{a}+\sqrt{b}+\sqrt{ab}(\sqrt{a}+\sqrt{b}+\sqrt{ab}\\ &= a+b+ab+\sqrt{a}(1+2b)+\sqrt{b}(1+2a)+3\sqrt{ab}\\ &\equiv 0 \pmod 3.\end{aligned}$$

We deduce that $\alpha_1^n \equiv \alpha_1 \pmod 3$ and by conjugation, $\alpha_i^n \equiv \alpha_i \pmod 3$. It follows that

$$\mathrm{Tr}_{K/\mathbb{Q}}(\alpha_1^n) = \alpha_1^n + \cdots + \alpha_4^n \equiv \alpha_1 + \cdots + \alpha_4 \equiv 1 \pmod 3.$$

We conclude that $\mathrm{Tr}_{K/\mathbb{Q}}(\alpha_i^n) \not\equiv 0 \pmod 3$.

Fix now $i \in \{1, \dots, 4\}$. Since $\overline{f(x)} \nmid \overline{f_i(x)}^n$ for any positive integer n, then there is an irreducible and monic factor $\overline{g_i(x)}$ of $\overline{f(x)}$ in $\mathbb{F}_3[x]$ such that $\overline{g_i(x)} \nmid \overline{f_i(x)}$. Moreover, for any $j \neq i$, $\overline{g_i(x)} \mid \overline{f_j(x)}$. It follows that $\overline{g_1(x)}, \dots, \overline{g_4(x)}$ are four distinct monic irreducible factors of $\overline{f(x)}$, which is of degree 4. We conclude that $\deg \overline{g_i(x)} = 1$ for $i = 1, \dots, 4$, contradicting the fact that $\mathbb{F}_3[x]$ contains exactly three monic polynomials of degree 1. We deduce that $A \neq \mathbb{Z}[\alpha]$ for any $\alpha \in A$.

Exercise 10.8.
Let $n \geq 3$ be an integer, ξ_n be a primitive n-th root of unity in $\mathbb{C}$ and $\Phi_n(x)$ be the minimal polynomial of ξ_n over $\mathbb{Q}$.

(1) If n is odd show that $\Phi_n(-1) = 1$.
(2) Let $\Psi_n(x)$ be the minimal polynomial of $\xi_n + \xi_n^{-1}$ over $\mathbb{Q}$.
 (a) Show that $\deg \Psi_n = \phi(n)/2$, where ϕ is the Euler's function.
 (b) Show that
$$\Phi_n(x) = x^{\phi(n)/2}\Psi_n(x + x^{-1}). \qquad \text{(Eq 1)}$$
 (c) If $n \not\equiv 0 \pmod 4$, show that $\xi_n + \xi_n^{-1}$ is an algebraic unit.
(3) Let $k \geq 3$ be an integer and $n = 2^k$.
 (a) Let $R(y) = \mathrm{Res}_x(x^2 - yx + 1, x^{2^{k-1}} + 1)$. Show that $R(y)$ is a monic polynomial with integral coefficients and $R(0) = 4$.
 (b) Using the result of **Exercise 3.6**, show that $\Psi_n(0) = \pm 2$. Conclude that $\xi_n + \xi_n^{-1}$ is not an algebraic unit.

Solution 10.8.

(1) We have: $\prod_{\substack{d|n \\ d\neq 1}} \Phi_d(x) = (x^n - 1)/(x - 1)$. Since n is odd, then
$$\prod_{\substack{d|n \\ d\neq 1}} \Phi_d(-1) = ((-1)^n - 1)/(-1 - 1) = 1.$$
Using an inductive reasoning and the fact that $\Phi_3(-1) = 1$, we conclude that $\Phi_n(-1) = 1$.

(2)(a) Let $\theta_n = \xi_n + \xi_n^{-1}$, then the degree of Ψ_n is equal to the number of distinct conjugates of θ_n over $\mathbb{Q}$. Any embedding σ of $\mathbb{Q}(\xi_n)$ into $\mathbb{C}$ is completely determined by the value it takes on ξ_n. Since $\sigma(\xi_n) = \xi_n^s$ with $s \in \{1, \ldots, n-1\}$ and $\gcd(s, n) = 1$, then $\sigma(\theta_n) = \xi_n^s + \xi_n^{-s} = \xi_n^s + \xi_n^{n-s}$. Let σ and τ be two embeddings $\mathbb{Q}(\xi_n)$ into $\mathbb{C}$ and suppose that $\sigma(\theta_n) = \tau(\theta_n)$. Let s and t such that $\sigma(\theta_n) = \xi_n^s + \xi_n^{n-s}$ and $\tau(\theta_n) = \xi_n^t + \xi_n^{n-t}$, then
$$\begin{aligned} \sigma(\theta_n) = \tau(\theta_n) &\Leftrightarrow \xi_n^s + \xi_n^{n-s} = \xi_n^t + \xi_n^{n-t} \\ &\Leftrightarrow (\xi_n^{s-t} - 1)(\xi_n^t - \xi_n^{n-s}) = 0 \\ &\Leftrightarrow \xi_n^{s-t} = 1 \quad \text{or} \quad \xi_n^t = \xi_n^{n-s} \\ &\Leftrightarrow t = s \quad \text{or} \quad t = n - s. \end{aligned}$$
We deduce that the number of distinct conjugates of θ_n over $\mathbb{Q}$ is equal to $\phi(n)/2$ and they are given by $\xi_n^s + \xi_n^{n-s}$ with $\gcd(n, s) = 1$ and $1 \leq s \leq (n-1)/2$. We conclude that $deg\Psi_n = \phi(n)/2$.

(b) To prove the identity, notice that the polynomials on both sides are monic and are of the same degree $\phi(n)$. Since $\Phi(x)$ is the minimal polynomial of ξ_n and also the polynomial on the right hand side vanishes for $x = \xi$, the two polynomials are identical.

(c) Let $m = n$ if n is odd and $m = n/2$ if $n \equiv 2 \pmod 4$, then, in any case, m is odd and $m \geq 3$. This implies that ξ_n^2 is a root of $\Phi_m(x)$. It follows that $\xi_n^2 + 1$ is a root of the polynomial $\Phi_m(x-1)$, which is monic. Moreover, by **(1)**, the constant term of this polynomial is equal to $\Phi_m(-1) = 1$. Hence $\xi_n^2 + 1$ is an algebraic unit. Thus $\xi_n + \xi_n^{-1} = (\xi_n^2 + 1)/\xi_n$ is an algebraic unit.

(3)(a) To simplify the forthcoming computations, we set $\lambda = 2^{k-1}$. Since the roots of $x^2 - yx + 1$ are given by $(y \pm \sqrt{y^2-4})/2$, then

$$\begin{aligned}
R(y) &= \operatorname{Res}_x\left(x^2 - yx + 1, x^{2^{k-1}} + 1\right) \\
&= \operatorname{Res}_x\left(x^2 - yx + 1, x^{\lambda} + 1\right) \\
&= \left(\left(\frac{y+\sqrt{y^2-4}}{2}\right)^{\lambda} + 1\right)\left(\left(\frac{y-\sqrt{y^2-4}}{2}\right)^{\lambda} + 1\right) \\
&= 2 + \left(\frac{y+\sqrt{y^2-4}}{2}\right)^{\lambda} + \left(\frac{y-\sqrt{y^2-4}}{2}\right)^{\lambda} \\
&= 2 + \left(\sum_{j=0}^{\lambda}\binom{\lambda}{j}\sqrt{y^2-4}^{\,j}\,y^{\lambda-j}\right)/2^{\lambda} \\
&\quad + \left(\sum_{j=0}^{\lambda}\binom{\lambda}{j}(-1)^j\sqrt{y^2-4}^{\,j}\,y^{\lambda-j}\right)/2^{\lambda} \\
&= 2 + 2\left(\sum_{\substack{j=0 \\ j \text{ even}}}^{\lambda}\binom{\lambda}{j}\sqrt{y^2-4}^{\,j}\,y^{\lambda-j}\right)/2^{\lambda} \\
&= 2 + 2\left(\sum_{i=0}^{\lambda/2}\binom{\lambda}{2i}(y^2-4)^i y^{\lambda-2i}\right)/2^{\lambda}.
\end{aligned}$$

Obviously, the coefficients of $R(y)$ are rational integers. The leading coefficient of $R(y)$ is equal to $\sum_{i=0}^{\lambda/2}\binom{\lambda}{2i}/2^{\lambda-1}$, hence equal to 1. The constant term of $R(y)$ corresponds to $i = \lambda/2$ and then

$$R(0) = 2 + 2(-4)^{\lambda/2}/2^{\lambda} = 4.$$

(b) We have

$$\Phi_{2^k}(x) = (x^{2^k} - 1)/(x^{2^{k-1}} - 1) = x^{2^{k-1}} + 1,$$

hence $\mathrm{Irr}(\xi_n, \mathbb{Q}, x) = x^{2^{k-1}} + 1$. Let $T(y)$ be the characteristic polynomial of $\xi + \xi^{-1}$, then by **(2)**, **(a)**, we have $T(y) = \Psi_n(y)^2$. Using **Exercise 3.6**, we get

$$R(y) = \mathrm{Res}_x(x^{2^{k-1}} + 1, yx - x^2 - 1) = \mathrm{N}_{\mathbb{Q}(\xi)/\mathbb{Q}}(\xi)T(y) = \Psi_n(y)^2.$$

Using **(3)**, **(a)** and identifying the constant terms in this identity, we obtain $\Psi_n(y) = \pm 2$. We conclude that $\xi_n + \xi_n^{-1}$ is not an algebraic unit.

Exercise 10.9.
Let α be an algebraic integer of degree n and $\alpha_1 = \alpha, \alpha_2, \ldots, \alpha_n$ be its conjugates over $\mathbb{Q}$.

(1) If $|\alpha_i| = 1$ for $i = 1, \ldots, n$, show that α is a root of unity.
(2) If $|\alpha_i| \leq 1$ for $i = 1, \ldots, n$, show that $\alpha = 0$ or α is a root of unity.
(3) Let $A(x)$ and $B(x)$ be relatively prime polynomials in $\mathbb{Z}[x]$ with $B(0) \neq 0$. Suppose that the series $F(x) = A(x)/B(x)$ belongs to $\mathbb{Z}[[x]]$ and that its radius of convergence is equal to 1. Show that the poles of $F(x)$ are roots of unity.
(4) Let $F(x) \in \mathbb{Z}[[x]] \setminus \mathbb{Z}[x]$ and let R be its radius of convergence. Suppose that $F(x)$ is algebraic over $\mathbb{Q}[x]$. Show that $R = 1$ and the poles of F are roots of unity or $R < 1$.

Solution 10.9.

(1) • **First proof.** For any integer $m \geq 1$, let $f_m(x) = \prod_{i=1}^{n}(x - \alpha_i^m)$. Since the α_i^m are conjugate algebraic integers, then $f_m(x) \in \mathbb{Z}[x]$. Let A be the set of these polynomial $f_m(x)$ and B the set whose elements are the roots of any of these polynomials. We show that A is finite, which in turn will imply that B is finite. For any $1 \leq h \leq n$, let $s_h(x_1, \ldots, x_n)$ be the elementary symmetric function of degree h

of $x_1, \ldots, x_n$. Then

$$\begin{aligned}|s_h(\alpha_1^m, \ldots, \alpha_n^m)| &= \left|\sum_{1\le i_1<\cdots<i_h} \alpha_{i_1}^m \cdots \alpha_{i_h}^m\right| \\ &\le \sum_{1\le i_1<\cdots<i_h\le n} |\alpha_{i_1}|\cdots|\alpha_{i_h}| \\ &= \binom{n}{h} \le \binom{n}{\lfloor (n-1)/2\rfloor} \\ &:= C.\end{aligned}$$

It follows that for any $m \ge 1$, the coefficients of $f_m(x)$ belong to the set $\{-C, -C+1, \ldots, 0, 1, \ldots, C\}$. Thus A is finite and so is B. This implies that the map $\phi : \mathbb{N} \setminus \{0\} \to B$, such that $\phi(m) = \alpha^m$ is not injective. We conclude that there exist positive integers $m_1 < m_2$ such that $\alpha^{m_1} = \alpha^{m_2}$, thus $\alpha^{m_1}(\alpha^{m_2-m_1} - 1) = 0$. Since $\alpha \ne 0$, then α is a root of unity.

- **Second proof.** Let $f(x) = x^n - a_1x^{n-1} - \cdots - a_n$ be the minimal polynomial of α over $\mathbb{Q}$. Consider the recurrence sequence (u_m) defined by $u_i = 0$ for $0 \le i \le n-2$, $u_{n-1} = 1$ and

$$u_m = a_1u_{m-1} + \cdots + a_nu_{m-n}$$

for $m \ge n$. According to **Exercise 2.28**, there exist $c_1, \ldots, c_n \in \mathbb{Q}$ such that $u_m = \sum_{i=1}^n c_i\alpha_i^m$ for $m \ge 0$. Our assumptions implies that $|u_m| \le \sum_{i=1}^n |c_i|$. Let $N = \lceil \sum_{i=1}^n |c_i| \rceil$, then $|u_m| \le N$ for any $m \ge 0$. Consider the set A of n-tuples of rational integers $(u_m, u_{m+1}, \ldots, u_{m+n-1})$, where m runs in the set

$$B = \{0, 1, \ldots, (2N+1)^n\}.$$

Since $|A| \le (2N+1)^n$ and $|B| = (2N+1)^n + 1$, then there exist two integers m_1 and m_2 such that $0 \le m_1 < m_2 \le (2N+1)^n$ and

$$(u_{m_1}, u_{m_1+1}, u_{m_1+n-1}) = (u_{m_2}, u_{m_2+1}, u_{m_2+n-1}).$$

Since the sequence (u_n) obeys to a linear homogeneous relation, then $u_{m_1+j} = u_{m_2+j}$ for any $j \ge 0$. Let $k = m_2 - m_1$, then the preceding identity takes the form

$$\sum_{i=1}^n c_i\alpha_i^{m_1+j} = \sum_{i=1}^n c_i\alpha_i^{m_1+j+k}$$

for $j \ge 0$. It follows that $\sum_{i=1}^n c_i\alpha_i^{m_1+j}(\alpha_i - k) = 0$ for $j \ge 0$. This shows that $x_i = c_i(\alpha_i^k - 1)$, $i = 1, \ldots, n$ is a solution of the system of linear equations $\sum_{i=1}^{n} \alpha_i^{m_1+j} x_i = 0$, $j = 0, \ldots, n-1$. The determinant of this system is a Vandermonde determinant and it is non zero. Thus $x_i = 0$ for $i = 1, \ldots, n$. Since $u_{n-1} \neq 0$, then there exists $i_0 \in \{1, \ldots, n\}$ such that $c_{i_0} \neq 0$. We now have $\alpha_{i_0}^k = 1$, thus α is a root of unity.

(2) Suppose that $\alpha \neq 0$. Since $|\alpha_i| \leq 1$ for $i = 1, \ldots, n$, then $|\mathrm{N}_{\mathbb{Q}(\alpha)/\mathbb{Q}}(\alpha)| = |\alpha_1| \cdots |\alpha_n| \leq 1$. Since this norm is a non zero integer, then $|\mathrm{N}_{\mathbb{Q}(\alpha)/\mathbb{Q}}(\alpha)| = 1$. It follows that $|\alpha_i| = 1$ and then **(1)** applies.

(3) Set $B(x) = b_s x^s + \cdots + b_0$. By **Exercise 10.2**, we conclude that $b_0 = \pm 1$. Let $\beta_1, \ldots, \beta_s$ be the roots of $B(x)$ in $\mathbb{C}$, then $|\beta_1 \cdots \beta_s| = 1/|b_s|$. By assumption, the poles of $F(x)$ must lay outside or on the unit circle, thus $\beta_i \geq 1$, for $i = 1, \ldots, s$. It follows that $1/|b_s| \geq 1$. Since b_s is a non zero integer, then $b_s = \pm 1$, hence $|\beta_1 \cdots \beta_s| = 1$. Therefore $|\beta_i| = 1$ for $i = 1, \ldots, s$. 1) shows that β_i is a root of unity for $i = 1, \ldots, s$.

(4) Set $F(x) = \sum_{n=0}^{\infty} a_n x^n$. Since $F(x) \notin \mathbb{Z}[x]$, then for any non negative integer n, there exists an integer $n' > n$ such that $a_{n'} \neq 0$. It follows that $\limsup_{n\to\infty} |a_n|^{1/n} \geq 1$, hence $R = \frac{1}{\limsup_{n\to\infty}} |a_n|^{1/n} \leq 1$. If $R < 1$, the proof is complete. Suppose that $R = 1$; we will show that the poles of F are roots of unity. Since $F(x)$ is algebraic over $\mathbb{Q}(x)$, there exists $\phi(x, y) \in \mathbb{Z}[x, y]$, irreducible such that $\phi(x, F(x)) = 0$. This implies that there exist $B_0(x), \ldots, B_m(x) \in \mathbb{Z}[x]$ such that $B_0(x)B_m(x) \neq 0$ and

$$B_m(x)F(x)^m + \cdots + B_1(x)F(x) + B_0(x) = 0.$$

Let $G(x) = B_m(x)F(x)$. Then $G(x) \in \mathbb{Z}[[x]]$ and it satisfies an equation of the form $G(x)^m + \cdots + C_1(x)G(x) + C_0(x) = 0$, where $C_i(x) \in \mathbb{Z}[x]$ for $i = 1, \ldots, m-1$. Moreover $G(x)$ has the same radius of convergence as $F(x)$, that is 1. From the equation satisfied by $G(x)$, it is seen that $G(x)$ has no pole z with $|z| < \infty$. Set $G(x) = \sum_{n=0}^{\infty} b_n x^n$. Since $\lim_{n\to\infty} |b_n z_0^n| = 0$ for any $|z_0| < 1$, then $\lim_{n\to\infty} \frac{1}{2}|b_n| = 0$, hence $\lim_{n\to\infty} |b_n| = 0$. Since $b_n \in \mathbb{Z}$ for any $n \geq 0$, there exists n_0 such that $b_n = 0$ for $n > n_0$. It follows that $G(x) \in \mathbb{Z}[x]$ and then $F(x) = G(x)/B_m(x) \in \mathbb{Q}(x)$. We may write $F(x)$ in the form $F(x) = A(x)/B(x)$, where the polynomials A and B satisfies the conditions in **(3)**. Therefore the conclusion in **(3)** applies and we conclude that the poles of $F(x)$ are roots of unity.

Exercise 10.10.
Let θ be a root in $\mathbb{C}$ of $g(y) = y^3 + y + 1$, $K = \mathbb{Q}(\theta)$ and $F(x,y) = (x+1)y^3 + x^2y + x^3$. Show that $F(x,y)$ is irreducible over $K(x)$.

Solution 10.10.

- **First proof.** Suppose that the polynomial F is reducible over $K(x)$, then it has a root $f(x)$ in $K(x)$. This root has the form $f(x) = \lambda x^{e_1}$ or $f(x) = \lambda x^{e_2}/(x+1)$, where $\lambda \in K^\star$, $1 \le e_1 \le 3$ and $0 \le e_2 \le 3$. In the first case, we have

$$\lambda^3 x^{3e_1}(x+1) + \lambda x^{e_1+2} + x^3 = 0.$$

It is clear that the leading term on the left side of this identity is equal to $\lambda^3 x^{3e_1+1}$, thus $\lambda = 0$ which is a contradiction. In the second case, we have

$$\lambda^3 x^{3e_2} + \lambda x^{e_2+2}(x+1) + x^3(x+1)^2 = 0.$$

If $e_2 \ge 2$, then the leading term of the left side of the identity is equal to $\lambda^3 x^{3e_2}$, thus $\lambda = 0$ which is a contradiction. If $e_2 \le 1$, then the leading term in the left side of the identity is equal to x^5. This also is a contradiction. We conclude that F is irreducible over $K(x)$.
- **Second proof.** Suppose that $F(x,y)$ is reducible in $K(x)[y]$, then $F(1,y)$ is reducible in $K[y]$, that is the polynomial $h(y) = 2y^3 + y + 1$ is reducible over K. This implies that $h(y)$ has a root, say μ, in K. Since $g(y)$ and $h(y)$ are irreducible over $\mathbb{Q}$, then $K = \mathbb{Q}(\theta) = \mathbb{Q}(\mu) = \mathbb{Q}(2\mu)$. The minimal polynomial of 2μ is given by $h_1(y) = y^3 + 4y + 4$. Easy computations give the following values of the discriminants of $g(x)$ and $h_1(x)$: $\mathrm{Disc}(g(x)) = -31$ and $\mathrm{Disc}(h_1(x)) = -43.4^2$. The first value shows that the rational prime 31 is ramified in K, while the second leads to the opposite conclusion. We conclude that F is irreducible over $K(x)$.

Exercise 10.11.
Let K be a field, $f(x)$ be an element of $K((x))$. Suppose that $f(x)$ is algebraic over $K(x)$.

(1) If $f(x)$ is integral over $K[x]$, show that $f(x) \in K[[x]]$.
(2) Show that the converse of **(1)** is false.

Solution 10.11.

(1) Suppose that $f(x)$ is integral over $K[x]$ and $f(x) \notin K[[x]]$, then

$$f(x) = a_{-\mu}x^{-\mu} + a_{-\mu+1}x^{-\mu+1} + \cdots + a_0 + \cdots ,$$

where μ is a positive integer, the coefficients a_j belong to K and $a_{-\mu} \neq 0$. Write the algebraic equation satisfied by $f(x)$ in the form

$$f(x)^n + b_{n-1}(x)f(x)^{n-1} + \cdots + b_0(x) = 0,$$

where the $b_j(x)$ are polynomials with coefficients in K. The term of the series on the left side of this equation, having the lowest exponent is given by $a^n_{-\mu}x^{-n\mu}$, which implies that $a_{-\mu} = 0$, thus a contradiction.

(2) Let K be any field and $f(x) = 1/(1-x)$, then $f(x)$ is algebraic of degree 1 over $K(x)$, but not integral over $K[x]$. We have

$$f(x) = 1 + x + \cdots + x^k + \cdots ,$$

hence $f(x) \in K[[x]]$.

Exercise 10.12.
Let $p \geq 3$ be a prime number and for $k \in \{1, \ldots, p-1\}$. Let s_k be the elementary symmetric function of degree k of $\{1, 2, ..., p-1\}$.

(1) Let $f(x) = \prod_{i=1}^{p-1}(x-i)$. Show that $f(x) \equiv x^{p-1} - 1 \pmod p$. Deduce that $s_k \equiv 0 \pmod p$ for $k = 1, 2, \ldots, p-2$.
(2) Show that $f(-x) = f(x+p)$. Deduce that if k is odd and $3 \leq k \leq p-2$, then $s_k \equiv 0 \pmod{p^2}$.
(3) Let $k \in \{3, \ldots, p-2\}$ such that k is odd. Show that $s_k \equiv 0 \pmod{p^3}$ if and only if $s_{k-1} \equiv 0 \pmod{p^2}$.
(4) Let $n = (p-1)/2$. Show that $f(x)$ may be written in the form

$$f(x) = a_0 + a_1x(x-p) + a_2x^2(x-p)^2 + \cdots + a_nx^n(x-p)^n,$$

where $a_0, \ldots, a_n \in \mathbb{Z}$, $a_0 = (p-1)!$ and $a_n = 1$. Deduce that $f'(p/2) = 0$.

Solution 10.12.

(1) Since $(\mathbb{Z}/p\mathbb{Z})^*$ is cyclic of order $p-1$, then for any $x \in \mathbb{Z}$, $x \not\equiv 0 \pmod p$, $x^{p-1} \equiv 1 \pmod p$. Therefore any $\bar{x} \in (\mathbb{Z}/p\mathbb{Z})^*$ is a root of $x^{p-1}-1$. It follows that $x^{p-1}-1 \equiv \prod_{i=1}^{p-1}(x-i) \pmod p$. We conclude that $s_k \equiv 0 \pmod p$ for $k = 1, \ldots, p-2$.

(2) We have

$$\begin{aligned} f(x+p) &= \prod_{i=1}^{p-1}(x+p-i) \\ &= (-1)^{p-1}\prod_{i=1}^{p-1}(-x-(p-i)) \\ &= \prod_{j=1}^{p-1}(-x-j) \\ &= f(-x). \end{aligned}$$

Since

$$f(x) = x^{p-1} - s_1x^{p-2} + s_2x^{p-3} + \cdots - s_{p-2}x + (p-1)!,$$

then the above identity takes the form:

$$\begin{aligned} &x^{p-1} + s_1x^{p-2} + s_2x^{p-3} + \cdots + s_{p-2}x \\ &\quad = (x+p)^{p-1} - s_1(x+p)^{p-2} \\ &\qquad + s_2(x+p)^{p-3} - \cdots - s_{p-2}(x+p). \end{aligned} \qquad \text{(Eq 1)}$$

For k odd, $3 \leq k \leq p-2$, we identify the coefficient of x^{p-k-1} is this identity and we obtain:

$$\begin{aligned} s_k = -s_k &+ s_{k-1}\binom{p-k}{1}p \\ &- s_{k-2}\binom{p-k+1}{2}p^2 \\ &+ \cdots - s_1\binom{p-2}{p-1-k}p^{k-1} \\ &+ \binom{p-1}{p-1-k}p^k. \end{aligned} \qquad \text{(Eq 2)}$$

Using **(1)** we conclude that $s_k \equiv -s_k \pmod{p^2}$, thus $s_k \equiv 0 \pmod{p^2}$, as $p \neq 2$.

(3) Let k be an odd integer such that $3 \leq k \leq p-2$. By **(Eq 2)**,

$$\begin{aligned} 2s_k &= s_{k-1}\binom{p-k}{1}p - s_{k-2}\binom{p-k+1}{2}p^2 + \cdots \\ &\equiv s_{k-1}\binom{p-k}{1}p \pmod{p^3}. \end{aligned}$$

It follows that $s_k \equiv 0 \pmod{p^3}$ if and only if $s_{k-1} \equiv 0 \pmod{p^2}$.

(4) Let σ be the unique automorphism of $\mathbb{Q}(x)$ such that $\sigma(x) = -x + p$, then σ is of order 2. Let K be the invariant field of the group G generated by σ, then by Artin's Theorem [Lang (1965), Th. 2, Chap. 8.1], $\mathbb{Q}(x)$ is a Galois extension of K and $\mathrm{Gal}(\mathbb{Q}(x), K) = G$. Let $g(x) = x(x-p)$, then obviously $g(x) \in K$, hence $\mathbb{Q}(g) \subset K \subset \mathbb{Q}(x)$. Since $[\mathbb{Q}(x) : \mathbb{Q}(g)] = 2 = [\mathbb{Q}(x) : K]$, then $K = \mathbb{Q}(g)$. By **(2)**, $f(x) \in K$, hence $f(x) = \frac{u(g(x))}{v(g(x))}$, where $u(x), v(x) \in \mathbb{Z}[x]$ and $\gcd(u(x), v(x)) = 1$. Using a Bezout's identity, $a(x)u(x) + b(x)v(x) = 1$, we conclude that

$$a(g(x))u(g(x)) + b(g(x))v(g(x)) = 1,$$

hence $v(g(x)) \in \mathbb{Q}$ and then v is constant. It follows that $f(x) = u(g(x))$, where $u(x) \in \mathbb{Q}[x]$. Write $u(x)$ in the form $u(x) = u_1(x)/d$ where d is a positive integer relatively prime with $\mathrm{cont}(u_1)$. We have $df(x) = u_1(g(x))$. Since $f(x)$ and $g(x)$ are monic, then the leading coefficient of $u_1(x)$ is equal to d, hence $f(x) = (g(x)-\alpha_1)\cdots(g(x)-\alpha_n)$, where $\alpha_1, ..., \alpha_n$ are the roots (distinct or not) of $u_1(x)$. For any $i \in \{1, ..., n\}$ it is seen that every root of $g(x) - \alpha_i$ is a root of $f(x)$, thus an algebraic integer. It follows that α_i is an algebraic integer for any $i \in \{1, ..., n\}$. Therefore $u_1(x)$ is monic and then $\alpha = 1$. We conclude that

$$f(x) = u_1(g(x)) = a_0 + a_1 g(x) + \cdots + a_n g(x)^n,$$

where $a_0, ..., a_n \in \mathbb{Z}$ and $a_n = 1$. Identifying the constant terms in the two sides of this identity, we get $a_0 = (p-1)!$ We deduce that

$$f'(x) = u_1'(g(x))g'(x) = u_1'(g(x))(2x - p),$$

hence $f'(p/2) = 0$.

Exercise 10.13.
A given monic, irreducible polynomial $h(x) \in \mathbb{Z}[x]$ is said to satisfy the property **(P)**, if for any prime number p, the reduction modulo p of $h(x)$ is reducible over $\mathbb{F}_p$.

(1) Let $h_1(x), h_2(x) \in \mathbb{Z}[x]$, irreducible such that $\mathbb{Q}(\alpha) = \mathbb{Q}(\beta)$ for some roots α, β of $h_1(x)$ and $h_2(x)$ respectively. Show that $h_1(x)$ satisfies **(P)** if and only if $h_2(x)$ does.
Hint. One may use the result that if in a given number field some non ramified prime p has a prescribed splitting type, then there exist infinitely many prime numbers having the same splitting type.

(2) Let $f(x)$ and $g(x)$ be two monic polynomials with integral coefficients. Let (θ, m) (resp. (ϕ, n)) be the couple formed by a root and the degree of $f(x)$ (resp. $g(x)$). Suppose that $f(x)$ and $g(x)$ are irreducible over $\mathbb{Q}$ and that the fields $\mathbb{Q}(\theta)$ and $\mathbb{Q}(\phi)$ are linearly disjoint over $\mathbb{Q}$. Suppose furthermore that one of the following conditions holds:

(i) $\gcd(m, n) > 1$.

(ii) $\gcd(m, n) = 1$ and the splitting field of $g(x)$ is not linearly disjoint from $\mathbb{Q}(\theta)$ over $\mathbb{Q}$.

Let γ be an algebraic integer which is a primitive element of $\mathbb{Q}(\theta, \phi)$ and let $h(x)$ be its minimal polynomial over $\mathbb{Q}$. Show that $h(x)$ satisfies the property **(P)**.

(3) Show that **(2)**, **(i)** applies for $f(x) = x^2 - d_1$ and $g(x) = x^2 - d_2$, where d_1 and d_2 are distinct square free integers. Deduce a polynomial $h(x)$ of degree 4 satisfying **(P)**.

(4) Show that (2), (ii) applies for $f(x) = x^2 + x + 1$ and $g(x) = x^3 - 2$.

(5) On studying the example $f(x) = x^2 - 2$, $g(x) = x^3 - x + 1$ show that the condition on linear disjointness stated in **(ii)** is necessary.

Solution 10.13.

(1) It is sufficient to prove that $h_1(x)$ satisfies **(P)** implies $h_2(x)$ satisfies **(P)**. By contradiction suppose that $h_1(x)$ satisfies **(P)** while $h_2(x)$ does not. Then, there exists a prime p such that the reduction of $h_2(x)$ modulo p is irreducible over $\mathbb{F}_p$. This implies that p is inert in $\mathbb{Q}(\beta)$. By a result recalled above, there exists an infinite set T of prime numbers which are inert in $\mathbb{Q}(\beta)$. Since $h_1(x)$ satisfies **(P)**, then for any $l \in T$, the reduction of $h_1(x)$ is reducible over $\mathbb{F}_l$. Therefore all these primes l divide the index of α, which is impossible since this index is finite.

(2) Write γ in the form

$$\gamma = \gamma'/d = u(\theta, \phi)/d,$$

where γ' is an algebraic integer, d is a rational integer and $u(x, y)$ is a polynomial with integral coefficients. By (1), we may replace γ by γ' and thus suppose $d = 1$. Fix a prime p and denote as usual by $\overline{u}(x)$ the reduced polynomial of $u(x)$ modulo p. Let $\Theta_1, \ldots, \Theta_m$ (resp. $\Phi_1, \ldots, \Phi_n$) be the roots of $\overline{f}(x)$ (resp. $\overline{g}(x)$ in an algebraic closure of $\mathbb{F}_p$ and let $d = \gcd(m, n)$. Suppose first that we are in the case (i). Set $\gamma_1 = \overline{u}(\Theta_1, \Phi_1)$. Since γ_1 is a root of $\overline{h}(x)$, the minimal polynomial of γ_1 over $\mathbb{F}_p$ divides $\overline{h}(x)$. Since $\Theta_1 \in \mathbb{F}_{p^{m_1}}$ and $\Phi_1 \in \mathbb{F}_{p^{n_1}}$, where

$1 \leq m_1 \leq m$ and $1 \leq n_1 \leq n$, we deduce that $\gamma_1 \in \mathbb{F}_{p^{m_1}} \cdot \mathbb{F}_{p^{n_1}} = \mathbb{F}_{p^k}$ where $k = \text{lcm}(m_1, n_1)$. If $m_1 < m$ or $n_1 < n$, then $k \leq m_1 n_1 < mn$. If $m_1 = m$ and $n_1 = n$ then $k = \text{lcm}(m, n) = mn/d < mn$. In all cases we have $k < mn$, hence the minimal polynomial of γ_1 over $\mathbb{F}_p$ is of degree smaller than mn. This implies that $\overline{h(x)}$ is reducible over $\mathbb{F}_p$. Suppose now that we are in the case **(ii)**. Since $[\mathbb{F}_p(\Theta_1, \Phi_1) : \mathbb{F}_p] < mn$ in the case where $\overline{g(x)}$ is reducible over $\mathbb{F}_p$, we may suppose that $\overline{g(x)}$ is irreducible over $\mathbb{F}_p$. We have $[\mathbb{F}_p(\Phi_1) : \mathbb{F}_p] \leq n$. Let E be the splitting field of $g(x)$ over $\mathbb{Q}$ and let $f_1(x)$ be the minimal polynomial of θ over E. Then $\deg f_1 < m$. Reducing modulo p, and writing each Φ_i as a polynomial in Φ_1 with coefficients in $\mathbb{F}_p$ shows that $[\mathbb{F}_p(\Phi_1, \Theta_1) : \mathbb{F}_p(\Phi_1)] < m$, hence $\overline{h(x)}$ is reducible over $\mathbb{F}_p$.

(3) Here clearly the fields $\mathbb{Q}(\sqrt{d_1})$ and $\mathbb{Q}(\sqrt{d_2})$ are linearly disjoint over $\mathbb{Q}$ and $\gcd(m, n) = 2$. Therefore the conditions of (2), (i) are satisfied. The minimal polynomial $h(x)$ of $\gamma = \sqrt{d_1} + \sqrt{d_2}$ satisfies **(P)**. Straightforward computations show that $h(x) = x^4 - 2(d_1 + d_2)x^2 + (d_1 - d_2)^2$.

(4) Here also the fields $\mathbb{Q}(\theta)$ and $\mathbb{Q}(\phi)$ are linearly disjoint over $\mathbb{Q}$. We have $\gcd(m, n) = 1$ and the fields $\mathbb{Q}(j)$, $\mathbb{Q}(2^{1/3}, j)$ are not linearly disjoint over $\mathbb{Q}$, so that the conditions stated in **(2)**, **(ii)** are satisfied.

(5) As in **(4)** the fields $\mathbb{Q}(\theta)$ and $\mathbb{Q}(\phi)$ are linearly disjoint over $\mathbb{Q}$ and $\gcd(m, n) = 1$. The splitting field L of $g(x)$ is $\mathbb{Q}(\beta, \sqrt{-23})$. It is linearly disjoint from $\mathbb{Q}(\theta)$ since $f(x)$ has no root in L. Therefore one of the conditions in (2), (ii) is not fulfilled. That this condition cannot be omitted may be seen through the minimal polynomial $h(x)$ of $\alpha\beta$. Easy computations shows that $h(x) = x^6 - 4x^4 + 4x^2 - 8$ and $h(x)$ reduced modulo 3 is irreducible. Therefore $h(x)$ does not satisfy **(P)**.

Exercise 10.14.

Let d be a square free positive integer such that $d \equiv 1 \pmod 4$, $K = \mathbb{Q}(\sqrt{d})$, A be its ring of integers, $\omega = (1 + \sqrt{d})/2$ and $\epsilon = a + b\omega$ be a fundamental unit of K. Show that $(\mathbb{Z}[\sqrt{d}])^\star = A^\star$ in the following cases

(i) $d \equiv 1 \pmod 8$.

(ii) $d \equiv 5 \pmod 8$ and $b \equiv 0 \pmod 2$.

Show that $(\mathbb{Z}[\sqrt{d}])^\star = \{\pm(\epsilon^3)^n, n \in \mathbb{Z}\}$ if $d \equiv 5 \pmod 8$ and $b \equiv 1 \pmod 2$.

Solution 10.14.

It is sufficient to prove that $\epsilon \in \mathbb{Z}[\sqrt{d}]$ in the cases **(i)**, **(ii)** and $\epsilon \notin \mathbb{Z}[\sqrt{d}]$, $\epsilon^2 \notin \mathbb{Z}[\sqrt{d}]$ and $\epsilon^3 \in \mathbb{Z}[\sqrt{d}]$ in the last case.

- **case** $d \equiv 1 \pmod 8$. We have $\epsilon = a + b\omega = a + b/2 + \sqrt{d}b/2$. Since $\mathrm{N}_{K/\mathbb{Q}}(\epsilon) = \pm 1$, then

$$(a + b/2)^2 - db^2/4 = a^2 + ab + b^2(1 - d)/4 = \pm 1.$$

Since $(1 - d)/4 \equiv 0 \pmod 2$, then $a^2 + ab \equiv 1 \pmod 2$. It follows that $a \equiv 1 \pmod 2$ and $b \equiv 0 \pmod 2$, thus $\epsilon \in \mathbb{Z}[\sqrt{d}]$.

- **case** $d \equiv 5 \pmod 8$. As above, we have

$$a^2 + ab + b^2(1 - d)/4 = \pm 1,$$

hence $a \equiv 1 \pmod 2$, and $b \equiv 0 \pmod 2$ or $a \equiv 0 \pmod 2$, and $b \equiv 1 \pmod 2$ or $a \equiv 1 \pmod 2$, and $b \equiv 1 \pmod 2$. In the first case $\epsilon = a + b\omega = a + b/2 + \sqrt{d}b/2 \in \mathbb{Z}[\sqrt{d}]$. In the last cases, it is clear that $\epsilon \notin \mathbb{Z}[\sqrt{d}]$. We show that $\epsilon^3 \in \mathbb{Z}[\sqrt{d}]$. Since ϵ^3 is a unit of A, then

$$\epsilon^3 = x + y\omega = x + y(1 + \sqrt{d})/2 = x + y/2 + \sqrt{d}y/2 = (a + b\omega)^3.$$

Thus to conclude that $\epsilon^3 \in \mathbb{Z}[\sqrt{d}]$, it is sufficient to show that $y \equiv 0 \pmod 2$. Using the relations

$$\omega^2 = (d - 1)/4 + \omega \quad \text{and}$$
$$\omega^3 = (d - 1)/4 + \omega(d + 3)/4,$$

we get $y = b^3(d + 3)/4 + 3ab(a + b)$, hence $y \equiv 0 \pmod 2$ for the last cases. Now since $\epsilon^3 \in \mathbb{Z}[\sqrt{d}]$ and $\epsilon \notin \mathbb{Z}[\sqrt{d}]$, then $\epsilon^2 \notin \mathbb{Z}[\sqrt{d}]$.

Exercise 10.15.
Let A be a Dedekind ring, K be its fraction field, $\alpha \in K^\star$ and $\mathcal{P}$ be a prime ideal of A. Show that there exist a and $b \in A$ such that $\alpha = a/b$ and $(a, \mathcal{P}) = 1$ or $(b, \mathcal{P}) = 1$.

Solution 10.15.
We have $(\alpha) = IJ^{-1}$, where I and J are ideals of A, which we may suppose to be coprime. Let $c \in I$ such that $((c)I^{-1}, \mathcal{P}) = 1$ and let $Q = (c)I^{-1}$, then $(\alpha) = (c)(b)^{-1}$, where $(b) = JQ$. Hence $\alpha = \epsilon c/b$ where ϵ is a unit in A. Set $a = \epsilon c$. If $\mathcal{P} \mid \mathcal{I}$, then $\mathcal{P} \nmid J$ and $\mathcal{P} \nmid Q$, hence $\mathcal{P} \nmid b$. If $\mathcal{P} \nmid I$, then $\mathcal{P} \nmid a$, hence the result.

Exercise 10.16.
Let K be a number field, A be its ring of integers and I be a non zero ideal of A. Show that $(A : I) = (I^{-1} : A)$.

Solution 10.16.
Let $\mathcal{A}$, $\mathcal{B}$ and $\mathcal{C}$ be fractional ideals of K such that $\mathcal{B} \subset \mathcal{A}$, then the A-modules $\mathcal{AC}/\mathcal{BC}$ and $\mathcal{A}/\mathcal{B}$ are isomorphic. Apply this result with $\mathcal{A} = I^{-1}$, $\mathcal{B} = A$ and $\mathcal{C} = I$ and obtain $I^{-1}I/AI \simeq I^{-1}/A$, thus $A/I \simeq I^{-1}/A$. The equality of indices now follows.

Exercise 10.17.
Let K and E be number fields such that $K \subset E$, A and B be their respective rings of integers.

(1) Let I and J be ideals of A. Suppose that $IB \mid JB$. Show that $I \mid J$.
(2) Let I be an ideal of A. Show that $IB \cap A = I$.
(3) Find an example of an ideal J of B for which $(J \cap A)B \neq J$.

Solution 10.17.

(1) Let $\mathcal{P}_1, \ldots, \mathcal{P}_r$ be the list of the prime ideals of A dividing I or J (or both). Set $I = \mathcal{P}_1^{e_1} \cdots \mathcal{P}_r^{e_r}$ and $J = \mathcal{P}_1^{h_1} \cdots \mathcal{P}_r^{h_r}$, where the exponents e_i, h_i are non negative integers. We have $\mathcal{P}_i B = \prod_{j=1}^{t_i} \mathcal{Q}_{ij}^{k_{ij}}$, where the $\mathcal{Q}_{ij}$ are prime ideals of B, hence
$$IB = \prod_{j=1}^{t_1} \mathcal{Q}_{1j}^{e_1 k_{1j}} \cdots \prod_{j=1}^{t_r} \mathcal{Q}_{rj}^{e_r k_{rj}} \quad \text{and}$$
$$JB = \prod_{j=1}^{t_1} \mathcal{Q}_{1j}^{h_1 k_{1j}} \cdots \prod_{j=1}^{t_r} \mathcal{Q}_{rj}^{h_r k_{rj}}.$$
Since $IB \mid JB$, then $e_i k_{ij} \leq h_i k_{ij}$, hence $e_i \leq h_i$ for any i, which implies that $I \mid J$.
(2) Apply **(1)** with $J = IB \cap A$. Then
$$JB = IB \cap AB = IB \cap B = IB,$$
thus $IB \mid JB$ and $JB \mid IB$. It follows that $J = I$, that is $IB \cap A = I$.
(3) Let $K = \mathbb{Q}$, E be a number field in which there is a prime ideal $\mathcal{Q}$ lying over $p\mathbb{Z}$, for some prime number p. Suppose that the splitting of p in E is given by $pB = \mathcal{Q}^e \prod \mathcal{Q}_i^{e_i}$ with $e \geq 2$ or $\prod$ is non empty. Then
$$(\mathcal{Q} \cap A)B = pB = \mathcal{Q}^e \prod \mathcal{Q}_i^{e_i} \neq \mathcal{Q}.$$

Exercise 10.18.
Let K be a number field of degree n and A be its ring of integers. Let p be a prime number, $\mathcal{P}$ be a prime ideal of A lying over $p\mathbb{Z}$ and let a be a rational integer. Suppose that $\mathcal{P} \mid a$. Show that $p \mid a$.

Solution 10.18.

Let f be the residual degree of $\mathcal{P}$ over $p\mathbb{Z}$. Since $aA = \mathcal{P}I$, where I is an ideal of A, then

$$|a^n| = |\mathrm{N}_{K/\mathbb{Q}}(a)| = \mathrm{N}_{K/\mathbb{Q}}(\mathcal{P})\,\mathrm{N}_{K/\mathbb{Q}}(I) = p^f\,\mathrm{N}_{K/\mathbb{Q}}(I).$$

This shows that $p \mid a$.

Exercise 10.19.

Let K be a number field, A be its ring of integers. Let I and J be ideals of A satisfying the following condition

(C) For any prime number p, $I + pA = \mathcal{P}^e$ and $J + pA = \mathcal{P}^h$, where $\mathcal{P}$ is a prime ideals of A and e, h are non negative integers.

(1) Show that $\mathrm{N}_{K/\mathbb{Q}}(\gcd(I,J)) = \gcd(\mathrm{N}_{K/\mathbb{Q}}(I), \mathrm{N}_{K/\mathbb{Q}}(J))$.
(2) Show that the preceding identity no longer holds if we omit the condition **(C)**.

Solution 10.19.

(1) Write I and J in the form $I = \mathcal{P}_1^{e_1}\cdots\mathcal{P}_r^{e_r}$, $J = \mathcal{P}_1^{h_1}\cdots\mathcal{P}_r^{h_r}$, where $\mathcal{P}_1, \dots \mathcal{P}_r$ are prime ideals of A and the exponents e_i, h_i are non negative integers such that $e_i + h_i \geq 1$. The condition **(C)** implies that $\mathcal{P}_i \cap \mathbb{Z} \neq \mathcal{P}_j \cap \mathbb{Z}$ if $i \neq j$. For $i = 1, \dots, r$, let p_i be the prime number lying under $\mathcal{P}_i$ and let f_i such that $\mathrm{N}_{K/\mathbb{Q}}(\mathcal{P}_i) = p_i^{f_i}$, then the p_i are distinct and we have $\gcd(I,J) = \mathcal{P}_1^{inf(e_1,h_1)}\cdots\mathcal{P}_r^{inf(e_r,h_r)}$, hence

$$\begin{aligned}\mathrm{N}_{K/\mathbb{Q}}(\gcd(I,J)) &= \mathrm{N}_{K/\mathbb{Q}}\left(\mathcal{P}_1^{\inf(e_1,h_1)}\right)\cdots\mathrm{N}_{K/\mathbb{Q}}\left(\mathcal{P}_r^{\inf(e_r,h_r)}\right)\\ &= p_1^{f_1\inf(e_1,h_1)}\cdots p_r^{f_r\inf(e_r,h_r)}.\end{aligned}$$

On the other hand, we have

$$\begin{aligned}\mathrm{N}_{K/\mathbb{Q}}(I) &= \mathrm{N}_{K/\mathbb{Q}}\left(\mathcal{P}_i^{e_1}\right)\cdots\mathrm{N}_{K/\mathbb{Q}}\left(\mathcal{P}_r^{e_r}\right)\\ &= p_1^{f_1e_1}\cdots p_r^{f_re_r} \quad \text{and}\\ \mathrm{N}_{K/\mathbb{Q}}(J) &= \mathrm{N}_{K/\mathbb{Q}}\left(\mathcal{P}_1^{h_1}\right)\cdots\mathrm{N}_{K/\mathbb{Q}}\left(\mathcal{P}_r^{h_r}\right)\\ &= p_1^{f_1h_1}\cdots p_r^{f_rh_r}.\end{aligned}$$

It follows that

$$\gcd(\mathrm{N}_{K/\mathbb{Q}}(I), \mathrm{N}_{K/\mathbb{Q}}(J)) = p_1^{inf(f_1e_1,f_1h_1)}\cdots p_r^{\inf(f_re_r,f_rh_r)}.$$

Since $\inf(ab,ac) = a\inf(b,c)$ for any positive integers a, b, c, then $\mathrm{N}_{K/\mathbb{Q}}(\gcd(I,J)) = \gcd(\mathrm{N}_{K/\mathbb{Q}}(I), \mathrm{N}_{K/\mathbb{Q}}(J))$.

(2) Let $\mathcal{P}$ and $\mathcal{Q}$ be distinct prime ideals of A such that $\mathcal{P}\cap\mathbb{Z} = \mathcal{Q}\cap\mathbb{Z} = p\mathbb{Z}$ and $\mathrm{N}_{K/\mathbb{Q}}(\mathcal{P}) = \mathrm{N}_{K/\mathbb{Q}}(\mathcal{Q}) = p^f$, with $f \geq 1$. Let $I = \mathcal{P}^{e_1}\mathcal{Q}^{e_2}$ and $J = \mathcal{P}^{h_1}\mathcal{Q}^{h_2}$ with $1 \leq e_1 < h_1$ and $1 \leq h_2 < e_2$, then $\gcd(I,J) = \mathcal{P}^{e_1}\mathcal{Q}^{h_2}$, hence $\mathrm{N}_{K/\mathbb{Q}}(\gcd(I,J)) = p^{f(e_1+h_2)}$. On the other hand, we have

$$\mathrm{N}_{K/\mathbb{Q}}(I) = p^{f(e_1+e_2)} \quad \text{and}$$
$$\mathrm{N}_{K/\mathbb{Q}}(J) = p^{f(h_1+h_2)}.$$

The assumptions on e_1, e_2, h_1 and h_2 show that

$$f(e_1+h_2) \neq f(e_1+e_2) \quad \text{and}$$
$$f(e_1+h_2) \neq f(h_1+h_2),$$

hence

$$\mathrm{N}_{K/\mathbb{Q}}(\gcd(I,J)) \neq \gcd(\mathrm{N}_{K/\mathbb{Q}}(I), \mathrm{N}_{K/\mathbb{Q}}(J)).$$

Here is an explicit example satisfying the preceding conditions:

$$K = \mathbb{Q}(i), \quad A = \mathbb{Z}[i],$$
$$\mathcal{P} = (1+i)A, \quad \mathcal{Q} = (1-i)A,$$
$$I = (1-i)(1+i)^2A = 2(1+i)A \quad \text{and}$$
$$J = (1-i)^2(1+i)A = 2(1-i)A.$$

Exercise 10.20.
Let K be a number field of degree n, A be its ring of integers. Let k be a positive integer and let I be the ideal of A generated by $\alpha_1,\dots,\alpha_k$. Show that $\mathrm{N}_{K/\mathbb{Q}}(I)$ divides $\gcd_{i=1}^{k}\big(\mathrm{N}_{K/\mathbb{Q}}(\alpha_i)\big)$. Show that the converse does not hold.

Solution 10.20.
Let p be a prime factor of $\mathrm{N}_{K/\mathbb{Q}}(I)$ and let $e = \nu_p\big(\mathrm{N}_{K/\mathbb{Q}}(I)\big)$. Let $\mathcal{P}_1,\dots,\mathcal{P}_r$ be the prime ideals of A lying over $p\mathbb{Z}$, then $I = {\mathcal{P}_1}^{h_1}\cdots{\mathcal{P}_r}^{h_r}$, with $pA + J = A$ and $\sum_{i=1}^{r} h_i f_i = e$. Here f_i is the residual degree of $\mathcal{P}_i$. Since $I = \alpha_1 A + \dots + \alpha_k A$, then for any $i \in \{1,\dots,r\}$, ${\mathcal{P}_i}^{h_i}$ divides $\alpha_j A$ for $j = 1,\dots,k$. We deduce that for $j = 1,\dots,k$, $\prod {\mathcal{P}_i}^{h_i}$ divides $\alpha_i A$ for any $i \in \{1,\dots,r\}$. Thus using the norm, we conclude that $p^{\sum h_i f_i}$ divides $\mathrm{N}_{K/\mathbb{Q}}(\alpha_i)$, for any $i \in \{1,\dots,r\}$ that is p^e divides $\gcd\big(\mathrm{N}_{K/\mathbb{Q}}(\alpha_i)\big)$.

For the converse, let $f(x) = x^3 - 9x - 6$, $\theta \in \mathbb{C}$ be a root of $f(x)$ and $K = \mathbb{Q}(\theta)$. Let $I = (\theta, \theta-1) = A$, then $\mathrm{N}_{K/\mathbb{Q}}(I) = 1$. It is easy to see that the minimal polynomial of $\theta - 1$ is given by $g(x) = x^3 + 3x^2 - 6x - 14$, thus $\mathrm{N}_{K/\mathbb{Q}}(\theta-1) = 14$. Since $\mathrm{N}_{K/\mathbb{Q}}(\theta) = 6$, then $\gcd\big(\mathrm{N}_{K/\mathbb{Q}}(\theta), \mathrm{N}_{K/\mathbb{Q}}(\theta-1)\big) = 2$.

Exercise 10.21.
Let K be a number field of degree n, A be its ring of integers and $\alpha_1, \ldots, \alpha_k$ be non zero elements of A. Let I be the ideal of A generated by these elements. Show that $I^{-1} = \sum_{i=1}^{k} \alpha_i^{-1} A$ if and only if α_i is associate to α_1 for $i = 1, \ldots, k$.

Solution 10.21.

- **Necessity of the condition.** Since α_1^{-1} and $\alpha_i^{-1} \in I^{-1}$, then $\alpha_i^{-1}\alpha_1 \in A$ and $\alpha_1^{-1}\alpha_i \in A$. Set $\alpha_i^{-1}\alpha_1 = a$ and $\alpha_1^{-1}\alpha_i = b$, where a and $b \in A$, then $\alpha_1 = a\alpha_i$ and $\alpha_i = b\alpha_i$, which shows that α_1 and α_i are associate.
- **Sufficiency of the condition.** Set $\alpha_i = \epsilon_i \alpha_1$, where $\epsilon_i \in A^{\star}$ for $i = 1, \ldots, k$. Then

$$I = (\epsilon_1 \alpha_1, \ldots, \epsilon_k \alpha_1) = \alpha_1 A(\epsilon_1, \ldots, \epsilon_k) = \alpha_1 A.$$

Let $J = \alpha_1^{-1} A$, then obviously, $I^{-1} = J$. Since $\alpha_i^{-1} A = \alpha_1^{-1} A$, then we now have $\sum_{i=1}^{k} \alpha_i^{-1} A = \alpha_1^{-1} A = I^{-1}$.

Exercise 10.22.
Give an example of two irreducibles, non associate α and β in the ring of integers A of some number field such that $\gcd(\alpha A, \beta A) \neq A$.

Solution 10.22.
Let $K = \mathbb{Q}(\sqrt{-14})$ and $A = \mathbb{Z}[\sqrt{-14}]$. Let $\alpha = 2 + \sqrt{-14}$ and $\beta = \alpha' = 2 - \sqrt{-14}$, then $N_{K/\mathbb{Q}}(\alpha) = N_{K/\mathbb{Q}}(\beta) = 18$. In A, 2 is ramified and 3 splits, say $2A = \mathcal{P}^2$ and $3A = \mathcal{Q}\mathcal{Q}'$, where $\mathcal{P}$, $\mathcal{Q}$ and $\mathcal{Q}'$ are prime ideals of A of the first degree. It is easy to see that none of these prime ideals is principal. Since $3 \nmid \alpha$, then $\alpha A = \mathcal{P}\mathcal{Q}^2$ and $\beta A = \mathcal{P}(\mathcal{Q}')^2$. It follows that $\gcd(\alpha A, \beta A) = \mathcal{P} \neq A$. Suppose that α is reducible in A, say $\alpha = \alpha_1\alpha_2$, where α_1, nor α_2 is a unit of A, then $\alpha_1 A \alpha_2 A = \mathcal{P}\mathcal{Q}^2$. Hence, we may suppose that, $\alpha_1 A = \mathcal{P}$ and $\alpha_2 A = \mathcal{Q}^2$ or $\alpha_1 A = \mathcal{Q}$ and $\alpha_2 A = \mathcal{P}\mathcal{Q}$. In any case $\mathcal{P}$ or $\mathcal{Q}$ would be principal, which is a contradiction. Thus α is irreducible, which in turn implies, by conjugation, the same property for β.

Exercise 10.23.
Let K be a field, and A be its ring of integers. Let α and π be elements of A such that π is prime and $\pi \mid \alpha$. Let $\alpha = \epsilon \alpha_1^{e_1} \cdots \alpha_s^{e_s}$ be any factorization of α into powers of irreducibles of A, with $\epsilon \in A^{\star}$. Show that there exists a unique $i \in \{1, \ldots, s\}$ such that π is an associate to α_i.

Solution 10.23.
Since π is prime and $\pi \mid \epsilon\alpha_1^{e_1}\cdots\alpha_s^{e_s}$, then there exists $i \in \{1,\ldots,s\}$ such that $\pi \mid \alpha_i$. Since α_i is irreducible, then π is an associate to α_i. Since α_i and α_j are not associate for $i \neq j$, then the index i is unique.

Exercise 10.24.
Let K be a number field and A be its ring of integers.

(1) Let $\mathcal{Q}_1,\ldots,\mathcal{Q}_k$ be (distinct or not) prime ideals of A with $k \geq 2$. Suppose that for any $1 \leq l \leq \lfloor k/2 \rfloor$, $\mathcal{Q}_{j_1}\cdots\mathcal{Q}_{j_l}$ is not principal but $\mathcal{Q}_1\cdots\mathcal{Q}_k$ is principal, say $\mathcal{Q}_1\cdots\mathcal{Q}_k = \alpha A$. Show that α is irreducible in A.

(2) Let $\alpha \in A$ and let $\alpha A = \mathcal{P}_1\cdots\mathcal{P}_k$ be the factorization of αA into a product of (distinct or not) prime ideals of A. Let $\mathrm{P} = \{\mathcal{P}_1,\ldots,\mathcal{P}_k\}$. We consider all the partitions $\mathrm{P} = \dot{\cup}_{i=1}^t C_i$ of P which satisfy the following conditions.
(C) For any $i \in \{1,\ldots,t\}$, $\prod_{\mathcal{P}\in C_i}\mathcal{P}$ is principal but $\prod_{\mathcal{P}\in J}\mathcal{P}$ is not for any $\emptyset \subsetneq J \subsetneq C_i$.
We include here the possibility $t = 1$, that is $\prod_{\mathcal{P}\in \mathrm{P}}\mathcal{P}$ is principal but $\prod_{\mathcal{P}\in J}\mathcal{P}$ is not for any $\emptyset \subsetneq J \subsetneq \mathrm{P}$.
Show that we may find all the partitions of P satisfying **(C)** in the following way. Start with any $\mathcal{P}_{i_1} \in \mathrm{P}$. If it is principal then set $C_1 = \{\mathcal{P}_{i_1}\}$. If not adjoin to it $\mathcal{P}_{i_2}, \mathcal{P}_{i_3},\ldots,\mathcal{P}_{i_l}$ until the product $\mathcal{P}_{i_1}\cdots\mathcal{P}_{i_l}$ is principal but any product of h ideals among them, with $h < l$, is never principal. Set $C_1 = \{\mathcal{P}_{i_1},\ldots,\mathcal{P}_{i_l}\}$. If $C_1 = \mathrm{P}$, then the process is finished. If not start with any $\mathcal{P}_{j_1} \in \mathrm{P}\setminus C_1$ and construct C_2 in the same way as it was done for C_1. We then define $C_3, C_4, \ldots$.

(3) Let $\mathrm{P} = \dot{\cup}_{i=1}^t C_i$ be a partition of P satisfying the conditions **(C)** and let, for any $i \in \{1,\ldots,t\}$, $\alpha_i \in A$ such that $\prod_{\mathcal{P}\in C_i}\mathcal{P} = \alpha_i A$. Show that α_i is irreducible.

(4) Let $\mathrm{P} = \dot{\cup}_{i=1}^t C_i$ be a partition of P satisfying the conditions **(C)** and let, for any $i \in \{1,\ldots,t\}$, $\alpha_i \in A$ such that $\prod_{\mathcal{P}\in C_i}\mathcal{P} = \alpha_i A$. Show that $\alpha = \epsilon\alpha_1\cdots\alpha_t$, where $\epsilon \in A^\star$, is a factorization of α into a product of irreducibles of A.

(5) Show that the map ϕ which maps a given partition $\mathrm{P} = \dot{\cup}_{i=1}^t C_i$ of P satisfying the conditions **(C)** onto the factorization into irreducibles $\alpha = \epsilon\alpha_1\cdots\alpha_t$, obtained in **(4)**, is one to one and onto.

(6) Let $\theta \in A$, $\theta = \theta_1\cdots\theta_s$ be one of its factorizations into irreducibles of A and let $\theta A = \mathcal{P}_1\cdots\mathcal{P}_r$ be the factorization of θA into a product of prime ideals of A. Show that $r \geq s$ and that $r = s$ if and only if $\theta_1,\ldots,\theta_s$ are prime.

(7) Let $K = \mathbb{Q}(\sqrt{-14})$ and $A = \mathbb{Z}[\sqrt{-14}]$. Determine all the distinct factorizations of 90 into a product of irreducibles of A.

Solution 10.24.

(1) Suppose that α is reducible in A, say $\alpha = \beta\gamma$, then $\beta A \gamma A = \mathcal{Q}_1 \cdots \mathcal{Q}_k$, hence

$$\beta A = \mathcal{Q}_{j_1} \cdots \mathcal{Q}_{j_l} \quad \text{and} \quad \gamma A = \mathcal{Q}_{j_l+1} \cdots \mathcal{Q}_{j_k},$$

with $1 \leq l < k$. Permuting β and γ, if necessary, we may suppose that $l \leq \lfloor k/2 \rfloor$. $\mathcal{Q}_{j_1} \cdots \mathcal{Q}_{j_l}$ is then principal with $l \leq \lfloor k/2 \rfloor$ contradicting the assumptions. Thus α is irreducible.

(2) The integer l exists since $\prod_{\mathcal{P} \in \mathrm{P}} \mathcal{P}$ is principal. Since

$$\prod_{\mathcal{P} \in \mathrm{P}} \mathcal{P} = \prod_{\mathcal{P} \in C_1} \mathcal{P} \prod_{\mathcal{P} \in \mathrm{P} \setminus C1} \mathcal{P},$$

then $\prod_{\mathcal{P} \in \mathrm{P} \setminus C1} \mathcal{P}$ is principal so that C_2 is defined. We define similarly $C_3, C_4 \ldots$. Since P is finite the process must stop and we obtain a partition of P satisfying **(C)**. It is also clear that, starting with a given partition satisfying **(C)**, we may recover it by using the described process.

(3) This follows directly from **(1)**.

(4) We have

$$\begin{aligned} \alpha A &= \left(\prod_{\mathcal{P} \in C_1} \mathcal{P} \right) \cdots \left(\prod_{\mathcal{P} \in C_t} \mathcal{P} \right) \\ &= (\alpha_1 A) \cdots (\alpha_t A) \\ &= (\alpha_1 \cdots \alpha_t) A, \end{aligned}$$

hence $\alpha = \epsilon \alpha_1 \cdots \alpha_t$, where ϵ is a unit of A. We thus get a factorization of α into a product of irreducibles of A.

(5) Let $\mathrm{P} = \cup_{i=1}^{s} C_i$ and $\mathrm{P} = \cup_{j=1}^{t} D_j$ be two partitions of P satisfying **(C)**. For any $(i,j) \in \{1, \ldots, s\} \times \{1, \ldots, t\}$, let α_i and $\beta_j \in A$ such that

$$\prod_{\mathcal{P} \in C_i} \mathcal{P} = \alpha_i A \quad \text{and} \quad \prod_{\mathcal{P} \in D_j} \mathcal{P} = \beta_j A.$$

Suppose that these partitions have the same image under ϕ, then $s = t$ and reordering if necessary the β_i, we have α_i is associate to β_i. Let $\epsilon_i \in A^\star$ such that $\beta_i = \epsilon_i \alpha_i$. Then

$$\prod_{\mathcal{P} \in D_i} \mathcal{P} = \beta_i A = \epsilon_i \alpha_i A = \prod_{\mathcal{P} \in C_i} \mathcal{P},$$

hence $D_i = C_i$. It follows that ϕ is one to one. Let $\alpha = \alpha_1 \cdots \alpha_h$ be a factorization of α into a product of irreducibles of A. Then $\alpha A = (\alpha_1 A) \cdots (\alpha_h A)$. This implies that there exists a partition $\mathrm{P} = \cup_{i=1}^{k} C_i$ of P such that $\prod_{\mathcal{P} \in C_i} \mathcal{P} = \alpha_i A$ for $i = 1, \ldots, h$. This shows that the given factorization of α is the image of a partition under ϕ, that is ϕ is onto.

(6) Since $\mathrm{P} = \cup_{i=1}^{k} C_i$, then $r = \sum_{i=1}^{s} |C_i|$, hence $r \geq s$. We have: $r = s$ if and only if $|C_i| = 1$ for any i, that is if and only if each $\mathcal{P}_i$ is principal, thus if and only if α_i is prime.

(7) The splitting of the rational primes in A is given by the following rules. 2 and 7 are ramified. In particular, $2A = \mathcal{P}^2$. Are inert in A the odd primes p such that $(\frac{-14}{p}) = -1$. The odd primes such that $(\frac{-14}{p}) = 1$ splits as products of two prime ideals of A of residual degree equal to 1. For instance, 3 and 5 are such primes. Set $3A = \mathcal{P}_3\mathcal{P}_3^{'}$ and $5A = \mathcal{P}_5\mathcal{P}_5^{'}$. We have

$$90 = 2.3^2.5 = \mathcal{P}^2\mathcal{P}_3^2(\mathcal{P}_3^{'})^2\mathcal{P}_5\mathcal{P}_5^{'}.$$

We consider this as a product of 8 prime ideals of A. We form all the partitions of the set of these ideals satisfying **(C)**. The following table of the small values of the norm map shows that 2, 3 and 5 are not norms of elements of A.

Table 1

a	1	2	3	0	1	4	2	3	4
b	0	0	0	1	1	0	1	1	1
$a^2 + 14b^2$	1	4	9	14	15	16	18	23	30

It follows that none of the eight prime ideals of A appearing in the factorization of $90A$ is principal. This means that there is no class of cardinality 1 in any partition. We look at the classes C_i (if any) of cardinality 2. We have

$$2A = \mathcal{P}^2, \quad 3A = \mathcal{P}_3\mathcal{P}_3^{'} \quad \text{and} \quad 5A = \mathcal{P}_5\mathcal{P}_5^{'},$$

which gives 3 classes. We have

$$\begin{aligned} \mathrm{N}_{K/\mathbb{Q}}(\mathcal{P}_3\mathcal{P}_5) &= N_{K/\mathbb{Q}}\left(\mathcal{P}_3^{'}\mathcal{P}_5^{'}\right) \\ &= \mathrm{N}_{K/\mathbb{Q}}\left(\mathcal{P}_3\mathcal{P}_5^{'}\right) \\ &= \mathrm{N}_{K/\mathbb{Q}}\left(\mathcal{P}_3^{'}\mathcal{P}_5\right) \\ &= 15 = \mathrm{N}_{K/\mathbb{Q}}\left(1 + \sqrt{-14}\right). \end{aligned}$$

We may suppose that the ideals lying over $3\mathbb{Z}$ or $5\mathbb{Z}$ are labeled such that $\mathcal{P}_3\mathcal{P}_5 = (1+\sqrt{-14})A$ and then by conjugation, $\mathcal{P}_3'\mathcal{P}_5' = (1-\sqrt{-14})A$. The products $\mathcal{P}_3\mathcal{P}_5'$ and $\mathcal{P}_3'\mathcal{P}_5$ are not principal because the factorization of $(1+\sqrt{-14})A$, (resp. $(1-\sqrt{-14})A$) into a product of prime ideals is unique. We next consider

$$\mathcal{P}\mathcal{P}_3, \quad \mathcal{P}\mathcal{P}_3', \quad \mathcal{P}\mathcal{P}_5, \quad \mathcal{P}\mathcal{P}_5', \quad \mathcal{P}_3^2 \quad \text{and} \quad (\mathcal{P}_3')^2.$$

Let I be any of these ideals. If it was principal, say $I = (a+b\sqrt{-14})$, then $\mathrm{N}_{K/\mathbb{Q}}(I) = a^2+14b^2$. But 6, 9, 10, and 15 are not norms of elements of A (see the table), thus none of these products of 2 ideals is principal. For the classes of cardinality 3, we must look at the following products of ideals, obtained by adjoining one ideal to the preceding products of 2 ideals which where not principal:

$$\mathcal{P}\mathcal{P}_3\mathcal{P}_5', \quad \mathcal{P}\mathcal{P}_3'\mathcal{P}_5, \quad \mathcal{P}\mathcal{P}_3^2, \quad \mathcal{P}(\mathcal{P}_3')^2 \quad \text{and}$$
$$\mathcal{P}_3^2\mathcal{P}_5, \quad \mathcal{P}_3^2\mathcal{P}_5', \quad (\mathcal{P}_3')^2\mathcal{P}_5, \quad (\mathcal{P}_3')^2\mathcal{P}_5'.$$

We have

$$\mathrm{N}_{K/\mathbb{Q}}(\mathcal{P}\mathcal{P}_3\mathcal{P}_5') = \mathrm{N}_{K/\mathbb{Q}}(\mathcal{P}\mathcal{P}_3'\mathcal{P}_5) = 30 = \mathrm{N}_{K/\mathbb{Q}}(4 \pm \sqrt{-14}),$$

hence

$$\mathcal{P}\mathcal{P}_3\mathcal{P}_5' = (4+\epsilon_1\sqrt{-14})A \quad \text{and}$$
$$\mathcal{P}\mathcal{P}_3'\mathcal{P}_5 = (4-\epsilon_1\sqrt{-14})A,$$

where $\epsilon = \pm 1$. We have

$$\mathrm{N}_{K/\mathbb{Q}}(\mathcal{P}\mathcal{P}_3^2) = \mathrm{N}_{K/\mathbb{Q}}(\mathcal{P}(\mathcal{P}_3')^2) = 18 = \mathrm{N}_{K/\mathbb{Q}}(2 \pm \sqrt{-14}),$$

hence

$$\mathcal{P}\mathcal{P}_3^2 = (2+\epsilon_2\sqrt{-14})A \quad \text{and} \quad \mathcal{P}(\mathcal{P}_3')^2 = (2-\epsilon_2\sqrt{-14})A,$$

with $\epsilon_2 = \pm 1$. We determine the values of ϵ_1 and ϵ_2 as follows. We have

$$\mathcal{P}(\mathcal{P}_3')^2\mathcal{P}_3\mathcal{P}_5 = (\mathcal{P}\mathcal{P}_3'\mathcal{P}_5)(\mathcal{P}_3\mathcal{P}_3'),$$

hence

$$(2-\epsilon_2\sqrt{-14})(1+\sqrt{-14}) = (4-\epsilon_1\sqrt{-14})(3)A.$$

It follows that

$$(2+14\epsilon_2+\sqrt{-14}(2-\epsilon_2))A = (12-3\epsilon_1\sqrt{-14})A,$$

thus

$$(2+14\epsilon_2+\sqrt{-14}(2-\epsilon_2)) = \pm(12-3\epsilon_1\sqrt{-14}).$$

Clearly the sign $-$ must be rejected and then we conclude that $\epsilon_2=-1$ and $\epsilon_1=1$. We have proved that the first four products of prime ideals are principal. The norm of the remaining products is equal to 45 and since this integer is not the norm of an integer of A, then these products are not principal. To obtain a class C_i of cardinality 4, we must start with one of the preceding products of three ideals which was not principal and adjoin to it one ideal so that the resulting product is principal but the products of 2 or 3 of them is not principal. It appears that no class of cardinality 4 exists.
We reproduce here the list of all the classes which were found:

$$\begin{aligned}
&\mathcal{P}^2 = 2A, \quad \mathcal{P}_3\mathcal{P}_3' = 3A, \quad \mathcal{P}_5\mathcal{P}_5' = 5A,\\
&\mathcal{P}_3\mathcal{P}_5 = (1+\sqrt{-14})A, \quad \mathcal{P}_3'\mathcal{P}_5' = (1-\sqrt{-14})A,\\
&\mathcal{P}\mathcal{P}_3\mathcal{P}_5' = (4+\sqrt{-14})A, \quad \mathcal{P}\mathcal{P}_3'\mathcal{P}_5 = (4-\sqrt{-14})A,\\
&\mathcal{P}\mathcal{P}_3^2 = (2-\sqrt{-14})A, \quad \mathcal{P}(\mathcal{P}_3')^2 = (2+\sqrt{-14})A.
\end{aligned}$$

They are all of cardinality 2 or 3. The only ways of expressing the integer 8 as a sum of two's and three's is $8=3+3+2$ or $2+2+2+2$. Thus to get a factorization into irreducibles (a partition), we must select four classes of cardinality 2 or two classes of cardinality 3 and one of cardinality 2. There result the following factorizations, the first two correspond to the first selection of classes.

$$\begin{aligned}
(\mathcal{P}^2)(\mathcal{P}_3\mathcal{P}_3')^2(\mathcal{P}_5\mathcal{P}_5') &= 2.3^2.5A\\
(\mathcal{P}^2)(\mathcal{P}_3\mathcal{P}_3')(\mathcal{P}_3\mathcal{P}_5)(\mathcal{P}_3'\mathcal{P}_5') &= 2.3(1+\sqrt{-14})(1-\sqrt{-14})A\\
(\mathcal{P}\mathcal{P}_3^2)(\mathcal{P}(\mathcal{P}_3')^2)(\mathcal{P}_5\mathcal{P}_5') &= 5(2-\sqrt{-14})(2+\sqrt{-14})A\\
(\mathcal{P}\mathcal{P}_3\mathcal{P}_5')(\mathcal{P}\mathcal{P}_3'\mathcal{P}_5)(\mathcal{P}_3\mathcal{P}_3') &= 3(4+\sqrt{-14})(4-\sqrt{-14})A\\
(\mathcal{P}\mathcal{P}_3\mathcal{P}_5')(\mathcal{P}(\mathcal{P}_3')^2)(\mathcal{P}_3\mathcal{P}_5) &= (4+\sqrt{-14})(2+\sqrt{-14})(1+\sqrt{-14})A\\
(\mathcal{P}\mathcal{P}_3'\mathcal{P}_5)(\mathcal{P}\mathcal{P}_3^2)(\mathcal{P}_3'\mathcal{P}_5') &= (4-\sqrt{-14})(2-\sqrt{-14})(1-\sqrt{-14})A.
\end{aligned}$$

The left side of each of these equalities is equal to the ideal $90A$. The right side gives a generator, say γ, of this ideal, hence $90=\pm\gamma$ (since the units in A are 1 and -1). We deduce that

$$\begin{aligned}
90 &= 2.3^2.5 \\
&= 2.3(1+\sqrt{-14})(1-\sqrt{-14}) \\
&= 5(2-\sqrt{-14})(2+\sqrt{-14}) \\
&= 3(4+\sqrt{-14})(4-\sqrt{-14}) \\
&= -(4+\sqrt{-14})(2+\sqrt{-14})(1+\sqrt{-14}) \\
&= -(4-\sqrt{-14})(2-\sqrt{-14})(1-\sqrt{-14}).
\end{aligned}$$

We have seen that the product of the ideals in a given class is a principal ideal generated by an irreducible, thus in these factorizations, the factors are irreducible in A.

Exercise 10.25.
Let α and β be algebraic integers over $\mathbb{Q}$, $f(x)$ and $g(x)$ be their respective minimal polynomials. Show that $f(\beta)$ is an algebraic unit if and only if $g(\alpha)$ is an algebraic unit.

Solution 10.25.
By symmetry, it is sufficient to show that if $f(\beta)$ is a unit, then so is $g(\alpha)$. Set $\gamma = f(\beta)$, then γ is a root of an equation of the form $\gamma^n + \cdots + a_1\gamma + \epsilon$, where $\epsilon = \pm 1$, and $a_1, \ldots, a_{n-1} \in \mathbb{Z}$. It follows that

$$1/\gamma = -\epsilon(\gamma^{n-1} + \cdots + a_1),$$

hence $1/f(\beta) \in \mathbb{Z}[f(\beta)] \subset \mathbb{Z}[\beta]$. Therefore there exists $h(x) \in \mathbb{Z}[x]$ such that $f(\beta)h(\beta) = 1$. We deduce that there exists $u(x) \in \mathbb{Z}[x]$ such that $f(x)h(x) - 1 = g(x)u(x)$. Indeed $u(x)$ is the quotient in the Euclidean division of $f(x)h(x) - 1$ by $g(x)$. Since these polynomials have integral coefficients and $g(x)$ is monic, then the quotient has integral coefficients and then $u(x) \in \mathbb{Z}[x]$. Substituting α for x, in the preceding identity, yields $u(\alpha)g(\alpha) = -1$, hence $g(\alpha)$ is a unit.

Exercise 10.26.

(1) Let $r \geq 2$ be an integer and $c_{11}, \ldots, c_{1r} \in \mathbb{Z}$ such that $\gcd(c_{11}, \ldots, c_{1r}) = d$. Show that there exists an $r \times r$ matrix $A = (a_{ij})$ with integral coefficients such that $a_{1j} = c_{1j}$ for $j = 1, \ldots, r$ and $\operatorname{Det} A = d$.
(2) Let K be a number field of degree n. Deduce from **(1)** that K has an integral basis of the form $\{1, \omega_1, \ldots, \omega_{n-1}\}$.

Solution 10.26.

(1) • **First proof.** The proof is by induction on r. Suppose that $r = 2$. By Bezout's identity, there exist $a_{21}, a_{22} \in \mathbb{Z}$ such that $a_{11}a_{22} - a_{21}a_{12} = d$ and the claim is obvious in this case. Suppose that $r \geq 3$, the result is true for $r-1$ and let $d' = \gcd(c_{11}, \dots, c_{1r-1})$, then $d = \gcd(d', c_{1r})$. By the inductive hypothesis, we may find elements $a_{ij} \in \mathbb{Z}$, $i, j = 1, \dots r-1$ such that $a_{1j} = c_{1j}$ and $\mathrm{Det}(a_{ij}) = d'$. If x and y are two integers to be determined later, we have

$$\begin{vmatrix} a_{11} & a_{12} & \cdots & a_{1(r-1)} & a_{1r} \\ a_{21} & a_{22} & \cdots & a_{2(r-1)} & 0 \\ \cdots & \cdots & \cdots & \cdots & \cdots \\ a_{(r-1)1} & a_{(r-1)2} & \cdots & a_{(r-1)(r-1)} & 0 \\ xa_{11}/d' & xa_{12}/d' & \dots & xa_{1(r-1)}/d' & y \end{vmatrix} = yd' - a_{1r}(x/d')d' = yd' - xa_{1r}.$$

Now it is possible to determine x and $y \in \mathbb{Z}$ such that $yd' - xa_{1r} = d$ and the proof is complete.

• **Second proof.** We first prove the result when $d = 1$. Let M be a free $\mathbb{Z}$-module of rank r. Let $\{e_1, \dots, e_r\}$ be a basis of M. Let $\alpha \in M$ such that $\alpha = \sum_{j=1}^{r} c_{1j}e_j$. To get the result it is sufficient to prove that it is possible to complete α for obtaining a basis of M. We prove the following steps.

- ⋆ $M/\mathbb{Z}\alpha$ is torsion free.
- ⋆ Any finitely generated, torsion free, $\mathbb{Z}$-module is free.
- ⋆ The exact sequence $0 \to \mathbb{Z}\alpha \xrightarrow{i} M \xrightarrow{s} M/\mathbb{Z}\alpha \to 0$ splits. Here i and s are the canonical injection and surjection respectively and the other maps are the trivial ones.

We begin with the first point. Let $\lambda \in \mathbb{Z} \setminus \{0\}$ and $\bar{m} \in M/\mathbb{Z}\alpha$. Suppose that $\lambda\bar{m} = \bar{0}$. Then $\lambda \sum_j m_j e_j = k \sum_j c_{1j}e_j$. It follows that $\lambda \mid k$, say $k = \lambda q$ and then $m = \sum_j m_j e_j = q\alpha$ which means $\bar{m} = \bar{0}$. For the second point, let N be a finitely generated, torsion free, $\mathbb{Z}$-module generated by n elements, say $v_1, \dots, v_n$, and not by fewer elements. We prove that there is no non trivial relation $\sum_{i=1}^{n} a_i v_i$ with $a_i \in \mathbb{Z}$, so that $v_1, \dots, v_n$ are linearly independent over $\mathbb{Z}$. Assume that the contrary holds and among all sets of n generators and all such relations, choose one for which $\sum |a_i|$ is minimal.

Case one and only one of the a_i is non zero.

We may Suppose that $a_1 \neq 0$. Since $a_1 v_1 = 0$ and N is torsion free, then we have a contradiction.

Case at least two of the a_i are non zero.
We may suppose that $a_1 \neq 0$ and $a_2 \neq 0$. If a_1 and a_2 have opposite signs, we may replace a_1 by $-a_1$ and v_1 by $-v_1$ and get a new system of generators with a similar relation as the preceding one and obeying to the minimality condition. Therefore, multiplying the relation by -1 if necessary, we may suppose that $a_1 \geq a_2 > 0$. Clearly $v_1, v_2 + v_1, v_3, \dots, v_n$ still generate N. Here we have

$$(a_1 - a_2)v_1 + a_2(v_2 + v_1) + a_3v_3 \cdots + a_nv_n = 0.$$

The sum of the absolute values of the coefficients is given by

$$|a_1 - a_2| + |a_2| + \cdots + |a_n| = |a_1| + |a_3| + \cdots + |a_n|,$$

so it is smaller than the preceding one which is minimal. In this case also we have reached a contradiction. Thus $v_1, \dots, v_n$ are linearly independent over $\mathbb{Z}$.
We prove the last point. $M/\mathbb{Z}\alpha$ is torsion free by the first point. Since it is finitely generated, then it is free over $\mathbb{Z}$. Let $s \leq r$ be its rank. We may suppose that $\{\overline{e_1}, \dots, \overline{e_s}\}$ is a basis. To show that the short exact sequence

$$0 \to \mathbb{Z}\alpha \xrightarrow{i} M \xrightarrow{s} M/\mathbb{Z}\alpha \to 0$$

splits, we prove that there exists a morphism of $\mathbb{Z}$-modules $f : M/\mathbb{Z}\alpha \to M$ such that $sof = Id_{M/\mathbb{Z}\alpha}$. The unique morphism of modules f such that $f(\overline{e_i}) = e_i$ for $i = 1, \dots, s$ satisfies the conditions. Therefore the short exact sequence splits and $M \simeq \mathbb{Z}\alpha \bigoplus M/\mathbb{Z}\alpha$. It follows that α may be completed for obtaining a basis of M and the case $d = 1$ is finished.
Suppose now that $d \geq 2$. Let $b_{1j} = c_{1j}/d$. Then by what have been proved it is possible to find a matrix $A = (a_{ij})$ such that $\operatorname{Det} A = 1$ and $a_{1j} = b_{1j}$. Multiply the first row of A by d and obtain a matrix B with $\operatorname{Det} B = d \operatorname{Det} A = d$ and the matrix B has the required properties.
Remark. Indeed the result proved in the second point is valid if we replace $\mathbb{Z}$ by any principal domain, [Ribenboim (2001), Th. 1, Chap. 6.2]. The third point may be proved by using a general result: If $0 \to E \to F \to G \to 0$ is a short exact sequence of modules and if G is free, then the sequence splits, [Ribenboim (2001), Lemma 1, Chap. 6.2].

(2) Let $\{\gamma_1, \dots, \gamma_n\}$ be an integral basis of K and let $c_{11}, \dots, c_{1n} \in \mathbb{Z}$ such that $1 = c_{11}\gamma_1 + \cdots + c_{1n}\gamma_n$. Since $\gcd(c_{11}, \dots, c_{1n}) = 1$, then by (1), there exists a $n \times n$ matrix $B = (b_{ij})$ such that $b_{1j} = c_{1j}$ for $j = 1, \dots, n$ and $\operatorname{Det} B = 1$. For $i = 1, \dots, n$, let $\omega_i = \sum_{j=1}^{n} b_{ij}\gamma_j$, then $\begin{pmatrix} \omega_1 \\ \omega_2 \\ \cdot \\ \cdot \\ \cdot \\ \omega_n \end{pmatrix} = B \begin{pmatrix} \gamma_1 \\ \gamma_2 \\ \cdot \\ \cdot \\ \cdot \\ \gamma_n \end{pmatrix}$. Since B in invertible and B^{-1} has integral coefficients, then $\omega_1, \dots, \omega_n$ is an integral basis. Since $\omega_1 = 1$, the proof is complete.

Exercise 10.27.
Let $(\alpha_1, \dots, \alpha_m) \in \mathbb{C}^m$ such that for any $(i, j) \in \{1, \dots, m\}^2$, we have $\alpha_i\alpha_j = \sum_{k=1}^{m} a_{ijk}\alpha_k$, where $a_{ijk} \in \mathbb{Z}$ for all ijk. Show that $\alpha_1, \dots, \alpha_m$ are algebraic integers.

Solution 10.27.
We have

$$\begin{aligned} \alpha_1^2 &= a_{111}\alpha_1 + \cdots + a_{11m}\alpha_m \\ \alpha_1\alpha_2 &= a_{121}\alpha_1 + \cdots + a_{12m}\alpha_m \\ &\cdots\cdots\cdots\cdots\cdots\cdots\cdots\cdots \\ \alpha_1\alpha_m &= a_{1m1}\alpha_1 + \cdots + a_{1mm}\alpha_m, \end{aligned}$$

where the coefficients a_{ijk} are integers. We may express these equations in the form:

$$\alpha_1 \begin{pmatrix} \alpha_1 \\ \alpha_2 \\ \cdots \\ \alpha_m \end{pmatrix} = M \begin{pmatrix} \alpha_1 \\ \alpha_2 \\ \cdots \\ \alpha_m \end{pmatrix},$$

where M is the matrix with integral coefficients given by

$$M = \begin{pmatrix} a_{111} & a_{112} & \cdots & a_{11m} \\ a_{121} & a_{122} & \cdots & a_{12m} \\ \cdots & \cdots & \cdots & \cdots \\ a_{1m1} & a_{1m2} & \cdots & a_{1mm} \end{pmatrix}.$$

This implies that α_1 is an eigenvalue of the matrix M. It follows that α_1 is a root of the polynomial $f(x) = \operatorname{Det}(M - xI)$. Since this polynomial has integral coefficients and since the leading coefficient is equal to $(-1)^m$, then α_1 is an algebraic integer. The same proof is valid for $\alpha_2, \dots, \alpha_n$.

Exercise 10.28.
Let K be a number field, A be its ring of integers and $\theta \in A$ such that $A \neq \mathbb{Z}[\theta]$. Let p be a prime number and $S = \{\alpha + pA, \alpha \in \mathbb{Z}[\theta]\}$.

(1) Show that S is a subring of A/pA.
(2) Consider the following identities

(i) $S = \mathbb{Z}[\theta]/pA$.
(ii) $S = \mathbb{Z}[\theta]/p\mathbb{Z}[\theta]$.

Show that **(i)** has no meaning and **(ii)** is false. Show that S is a quotient of some ring by an ideal.

Solution 10.28.

(1) Obvious.
(2) Since pA is not contained in $\mathbb{Z}[\theta]$, then $\mathbb{Z}[\theta]/pA$ has no meaning. The zero elements of S and of $\mathbb{Z}[\theta]/p\mathbb{Z}[\theta]$ are equal to pA and $p\mathbb{Z}[\theta]$ respectively. Since $\mathbb{Z}[\theta] \subsetneq A$, then $p\mathbb{Z}[\theta]p \subsetneq A$, thus **(ii)** is false. We claim and we omit the proof that $S = (\mathbb{Z}[\theta] + pA)/pA$.

Exercise 10.29.
Let p be a prime number, K be a number field of degree n over $\mathbb{Q}$ and A be its ring of integers. Let θ be an element of A. It is known that $\mathrm{Disc}(\theta)$ may be written in the form $\mathrm{Disc}(\theta) = I(\theta)^2 \,\mathrm{Disc}(K)$, where $I(\theta)$ is a non negative integer called the index of θ. Moreover $I(\theta) \neq 0$ if and only if θ is primitive over $\mathbb{Q}$. In this case $I(\theta) = (A : \mathbb{Z}[\theta])$, see [Marcus (1977), Exercise 27, Chap. 1]. Show that the following assertions are equivalent.

(i) $p \mid I(\theta)$.
(ii) There exist an integer m, $1 \leq m \leq n-1$ and $a_0, a_1, \ldots, a_{m-1} \in \mathbb{Z}$ such that $\theta^m + a_{m-1}\theta^{m-1} + \cdots + a_0 \equiv 0 \pmod{pA}$.
(iii) $\mathbb{F}_p[\theta] \subsetneq A/pA$.

Solution 10.29.

$\star$ *(i)* $\Leftrightarrow$ *(ii)*.

- **First proof.** Let $\{\omega_1, \ldots, \omega_n\}$ be an integral basis. For $j = 0, \ldots, n-1$, write θ^j in the form $\theta^j = \sum_{i=1}^{n} c_i^j \omega_j$. These equations may be formulated as follows

$$\begin{pmatrix} 1 \\ \theta \\ \cdot \\ \cdot \\ \cdot \\ \theta^{n-1} \end{pmatrix} = M \begin{pmatrix} \omega_1 \\ \omega_2 \\ \cdot \\ \cdot \\ \cdot \\ \omega_n \end{pmatrix}, \text{ where } M \text{ is the matrix } M = (c_i^j)_{0\le j\le n-1}^{1\le i\le n}.$$

Let σ_j, for $j = 1, \dots, n$, be the distinct embeddings of K into $\mathbb{C}$. Then from the above identity of matrices, we get

$$\begin{pmatrix} 1 \\ \sigma_j(\theta) \\ \cdot \\ \cdot \\ \cdot \\ \sigma_j(\theta)^{n-1} \end{pmatrix} = M \begin{pmatrix} \sigma_j(\omega_1) \\ \sigma_j(\omega_2) \\ \cdot \\ \cdot \\ \cdot \\ \sigma_j(\omega_n) \end{pmatrix}.$$

All these n equations may be summarized in the following single one $\left(\sigma_j(\theta^i)\right)_{0\le i\le n-1}^{1\le j\le n} = M\left(\sigma_j(\omega_i)\right)_{1\le i\le n}^{1\le j\le n}$. Taking determinants of both sides and squaring, we obtain

$$\operatorname{Disc}(1, \theta, \dots, \theta^{n-1}) = \operatorname{Det}(M)^2 \operatorname{Disc} K,$$

hence $I(\theta) = |\operatorname{Det}(M)|$. It follows that $p \mid I(\theta)$ if and only if $\operatorname{Det}(M) \equiv 0 \pmod p$, if and only if the columns of M are linearly dependent over $\mathbb{F}_p$. This is equivalent to the existence of $a_0, \dots, a_{n-1} \in \mathbb{Z}$, not all 0 such that

$$a_{n-1}\theta^{n-1} + a_1\theta + \cdots + a_0 \equiv 0 \pmod{pA},$$

that is if and only if there exist an integer m, $1 \le m \le n-1$ and $a_0, a_1, \dots, a_{m-1}$ such that $\theta^m + a_{m-1}\theta^{m-1} + \cdots + a_0 \equiv 0 \pmod{pA}$.

- **Second proof.** We have

$p \mid I(\theta) \Leftrightarrow p \mid (A : \mathbb{Z}[\theta])$

$\Leftrightarrow$ There exists $\gamma \in A$ whose order in $A/\mathbb{Z}[\theta]$ is equal to p

$\Leftrightarrow$ There exists $\gamma \in A \setminus \mathbb{Z}[\theta]$, $\quad p\gamma \in \mathbb{Z}[\theta]$

$\Leftrightarrow$ There exists $\gamma \in A \setminus \mathbb{Z}[\theta]$, $\quad \gamma = \left(\sum_{i=0}^{n-1} a_i\theta^i\right)/p$ with $a_i \in \mathbb{Z}$

$\Leftrightarrow$ There exist integers $m, a_0, a_1, \dots, a_{m-1}$, with $1 \le m \le n-1$ such that $\theta^m + a_{m-1}\theta^{m-1} + \cdots + a_0 \equiv 0 \pmod{pA}$.

⋆ (*ii*) ⇔ (*iii*). (*ii*) means that $1, \theta, \dots, \theta^{n-1}$ are linearly dependent over $\mathbb{F}_p$. Since the dimension of A/pA over $\mathbb{F}_p$ is equal to n, then (*iii*) has the same meaning. We thus get the equivalence of (*ii*) and (*iii*).

Exercise 10.30.

Let K be a number field of degree n, A be its ring of integers, $\theta \in A$ and $f(x)$ be its characteristic polynomial over $\mathbb{Q}$.

(1) Let p be a prime number. For any $u(x) \in \mathbb{Z}[x]$, denote by $\overline{u(x)}$ the polynomial with coefficients in $\mathbb{F}_p$ obtained from $u(x)$ by reducing modulo p its coefficients. Let $\mathcal{I} = \{u(x)) \in \mathbb{F}_p[x], u(\theta) \equiv 0 \pmod{p}\}$. Show that $\mathcal{I}$ is a non zero principal ideal of $\mathbb{F}_p[x]$. Let $u_0(x)$ be a monic generator of $\mathcal{I}$ and let $M_\theta(x)$ be any monic polynomial with integral coefficients such that $\overline{M_\theta(x)} = u_0(x)$. Call this polynomial the minimal polynomial modulo p of θ. Show that $\overline{M_\theta(x)} \mid \overline{f(x)}$ in $\mathbb{F}_p[x]$. Recall from **Exercise 10.29** that $p \mid I(\theta)$ if and only if $\deg M_\theta(x) < n$. This equivalence may be used to answer the questions hereafter.

(2) Let $pA = \mathcal{P}_1^{e_1} \cdots \mathcal{P}_r^{e_r}$ be the splitting of p in A. Write $f(x)$ in the form

$$f(x) = \prod_{j=1}^{s} f_j(x)^{h_j} + pg(x), \qquad \text{(Eq 1)}$$

where $g(x)$, $f_1(x), \dots, f_s(x)$ are polynomials with integral coefficients, $\deg g < n$, $f_1(x), \dots, f_s(x)$ are monic, distinct and irreducible over $\mathbb{F}_p$.

(a) For any $i \in \{1, \dots, r\}$, show that there exists a unique $j = j(i) \in \{1, \dots, s\}$ such that $f_j(\theta) \equiv 0 \pmod{\mathcal{P}_i}$.

(b) Show that $M_\theta(x) \equiv \prod_{j=1}^{s} f_j(x)^{h'_j}(x) \pmod{p}$, where $1 \le h'_j \le h_j$ for $j = 1, \dots, s$.

(c) Deduce that the map $\phi : \{1, \dots, r\} \to \{1, \dots, s\}$ defined by $\phi(i) = j(i)$, where $j(i)$ is the unique integer in $\{1, \dots, s\}$ such that $f_{j(i)}(\theta) \equiv 0 \pmod{\mathcal{P}_i}$, is surjective.

(d) Using **Exercise 10.29**, show that $p \mid I(\theta)$ if and only if there exists $j \in \{1, \dots, s\}$ such that $h'_j < h_j$.

(3) Let $j \in \{1, \dots, s\}$ and let $J = \{i \in \{1, \dots, r\}$ such that $f_j(\theta) \equiv 0 \pmod{\mathcal{P}_i}\}$. Show that $h'_j = max_{i \in J} \lceil e_i / \nu_{\mathcal{P}_i}(f_j(\theta)) \rceil$.

(4) Let $j \in \{1, \dots, s\}$ such that $h_j \ge 2$. Show that $h'_j < h_j$ if and only if $\overline{f_j(x)} \mid \overline{g(x)}$ in $\mathbb{F}_p[x]$.

(5) Show that the following conditions are equivalent.

(i) $p \mid I(\theta)$.

(ii) There exists $j \in \{1, \ldots, s\}$ such that $h_j \geq 2$ and $\overline{f_j(x)} \mid \overline{g(x)}$ in $\mathbb{F}_p[x]$.

(6) In each of the following cases, determine $I(\theta)$, $\mathrm{Disc}(K)$, an integral basis and for any prime p dividing $\mathrm{Disc}(\theta)$, its splitting in K.

(i) $f(x) = x^3 - 4$.
(ii) $f(x) = x^3 + 4x + 8$.

Solution 10.30.

(1) Obviously $\mathcal{I}$ is an ideal of $\mathbb{F}_p[x]$, hence principal generated by some polynomial $u_0(x)$. Since $\overline{f(x)} \in \mathcal{I}$, then $\mathcal{I} \neq (0)$. Therefore $u_0(x) \neq 0$ and we may suppose that $u_0(x)$ is monic. Since $\overline{M_\theta(x)} = u_0(x)$ and $\overline{f(x)} \in \mathcal{I}$, then $\overline{M_\theta(x)} \mid \overline{f(x)}$ in $\mathbb{F}_p[x]$.

(2)(a) Since $\overline{M_\theta(x)} \mid \overline{f(x)}$ in $\mathbb{F}_p[x]$, then $\overline{M_\theta(x)} \mid \prod_{j=1}^{s} \overline{f_j(x)}^{h_j}$. Let $i \in \{1, \ldots, r\}$ and let $\psi(x) \in \mathbb{F}_p[x]$ be the minimal polynomial of $\theta + \mathcal{P}_i$ over $\mathbb{F}_p$. Since $\psi(x) \mid \overline{M_\theta(x)}$, then $\psi(x)$ divides $\prod_{j=1}^{s} \overline{f_j(x)}^{h_j}$, hence there exists $j = j(i) \in \{1, \ldots, s\}$ such that $\psi(x) = \overline{f_j(x)}$. From $\psi(\theta + \mathcal{P}_i) = 0$, we conclude that $f_j(\theta) \equiv 0 \pmod{\mathcal{P}_i}$. We show that this index $j = j(i)$ is unique. Suppose that $f_j(\theta) \equiv f_{j'}(\theta) \equiv 0 \pmod{\mathcal{P}_i}$ for $j \neq j'$. Since $\overline{f_j(x)}$ and $\overline{f_{j'}}(x)$ are coprime, then there exist $u(x)$ and $v(x) \in \mathbb{F}_p[x]$ such that $u(x)\overline{f_j(x)} + v(x)\overline{f_{j'}}(x) = 1$. Substituting θ for x, we get $0 \equiv 1 \pmod{p}$, which is a contradiction. Thus the index $j(i)$ is unique.

(b) Since $\overline{M_\theta(x)} \mid \prod_{j=1}^{s} \overline{f_j(x)}^{h_j}(x)$, then $M_\theta(x) \equiv \prod_{j=1}^{s} f_j(x)^{h'_j}(x) \pmod{p}$, where $0 \leq h'_j \leq h_j$ for $j = 1, \ldots, s$. We show that, in fact, $h'_j \geq 1$ for $j = 1, \ldots, s$. Let $\gamma = M_\theta(\theta)$. Then $\gamma \in A$ and $\gamma \equiv 0 \pmod{p}$, hence γ satisfies an equation of the form:

$$\gamma^n - pa_{n-1}\gamma^{n-1} + p^2 a_{n-2}\gamma^{n-2} + \cdots + (-1)^n p^n a_0 = 0,$$

where $a_i \in \mathbb{Z}$ for $i = 0, \ldots, n-1$. It follows that

$$M_\theta(\theta)^n - pa_{n-1}M_\theta(\theta)^{n-1} + p^2 a_{n-2}M_\theta(\theta)^{n-2} + \cdots + (-1)^n p^n a_0 = 0.$$

We conclude that there exists $q(x) \in \mathbb{Z}[x]$ such that

$$M_\theta(x)^n - pa_{n-1}M_\theta(x)^{n-1} + \cdots + (-1)^n p^n a_0 = fx)q(x).$$

We deduce that $M_\theta(x)^n \equiv fx)q(x) \pmod{p}$. From this, it is seen that any prime factor of $\overline{f(x)}$ in $\mathbb{F}_p[x]$ is a divisor of $\overline{M_\theta(x)}$. Thus $h'_j \geq 1$ for $j = 1, \ldots, s$.

(c) Suppose that there exists $j_0 \in \{1,\ldots,s\}$ such that for all $i \in \{1,\ldots,r\}$, $f_{j_0}(\theta) \not\equiv 0 \pmod{\mathcal{P}_i}$. Then $\prod_{\substack{j=1,\ldots,s\\ j\neq j_0}} f_j^{h'_j}(\theta) \equiv 0 \pmod{p}$, contradicting the property established in (b), that $h'_{j_0} \geq 1$. Thus the map ϕ is surjective.

(d) In **Exercise 10.29**, it is proved that $p \mid I(\theta)$ if and only if $\deg M_\theta(x) < n$, hence by **(b)** if and only if there exists $j \in \{1,\ldots,s\}$ such that $h'_j < h_j$.

(3) Let $h''_j = max_{i\in J}\lceil e_i/\nu_{\mathcal{P}_i}(f_j(\theta))\rceil$. Let $i \in J$ then

$$\nu_{\mathcal{P}_i}(f_j(\theta))^{h'_j} = h'_j\nu_{\mathcal{P}_i}(f_j(\theta) \geq e_i,$$

hence $h'_j \geq \frac{e_i}{\nu_{\mathcal{P}_i}(f_j(\theta))}$. It follows that $h'_j \geq \lceil e_i/\nu_{\mathcal{P}_i}(f_j(\theta))\rceil$ and then $h'_j \geq h''_j$. We prove the reverse inequality. For any $i \in J$, we have $\frac{e_i}{\nu_{\mathcal{P}_i}(f_j(\theta))} \leq \lceil e_i/\nu_{\mathcal{P}_i}(f_j(\theta))\rceil$, hence $\frac{e_i}{\nu_{\mathcal{P}_i}(f_j(\theta))} \leq h''_j$. It follows that for any $i \in J$, $e_i \leq h''_j\nu_{\mathcal{P}_i}(f_j(\theta))$. We conclude that $f_j(\theta)^{h''_j} \equiv 0 \pmod{\mathcal{P}_i^{e_i}}$. Therefore $h_j' \leq h_j''$.

(4) Substituting θ for x in **(Eq 1)**, we obtain the following identity.

$$\frac{\prod_{j=1}^s f_j(\theta)^{h'_j}}{p}\prod_{j=1}^s f_j(\theta)^{h_j-h'_j} = -g(\theta).$$

If $h_k > h'_k$ for some $k \in \{1,\ldots,s\}$, then $h_k \geq 2$ and the left side of this identity is an integer, hence $f_k(\theta) \mid g(\theta)$ in A. By **(2)(a)**, there exists $i \in \{1,\ldots,r\}$ such that $f_k(\theta) \equiv 0 \pmod{\mathcal{P}_i}$. It follows that $g(\theta) \equiv 0 \pmod{\mathcal{P}_i}$. Since $\overline{f_k(x)}$ is the minimal polynomial of $\theta + \mathcal{P}_i$ over $\mathbb{F}_p$, then $\overline{f_k(x)} \mid \overline{g(x)}$ in $\mathbb{F}_p[x]$. Conversely suppose that there exists $k \in \{1,\ldots,s\}$ such that $h_k \geq 2$ and $\overline{f_k(x)} \mid \overline{g(x)}$ in $\mathbb{F}_p[x]$. Let $J = \{i \in \{1,\ldots,r\}$ such that $f_k(\theta) \equiv 0 \pmod{\mathcal{P}_i}\}$. By **(3)**, we have $h'_k = max_{i\in J}\lceil e_i/\nu_{\mathcal{P}_i}(f_k(\theta))\rceil$. Let $t \in J$ such that this maximum is reached, that is $h'_k = \lceil e_t/\nu_{\mathcal{P}_t}(f_k(\theta))\rceil$.

- If $\nu_{\mathcal{P}_t}(f_k(\theta)) \geq e_t$, then $h'_k = 1$. Since $h_k \geq 2$, then, by **(2)(d)**, the result is proved in this case.
- If $\nu_{\mathcal{P}_t}(f_k(\theta)) < e_t$, let $q(x)$ and $r(x) \in \mathbb{Z}[x]$ such that $g(x) = f_k(x)q(x) + pr(x)$. Using **(Eq 1)**, we obtain

$$\prod_{j=1}^s f_j(\theta)^{h_j} = -pf_k(\theta)q(\theta) - p^2r(\theta).$$

Let A and B be the left and right side of this identity respectively. We have

$$h'_k = \lceil e_t/\nu_{\mathcal{P}_t}(f_k(\theta))\rceil < 1 + e_t/\nu_{\mathcal{P}_t}(f_k(\theta)),$$

hence $h_k'\nu_{\mathcal{P}_t}(f_k(\theta)) < e_t + \nu_{\mathcal{P}_t}(f_k(\theta))$. Thus

$$\begin{aligned}\nu_{\mathcal{P}_t}(A) &= h_k'\nu_{\mathcal{P}_t}(f_k(\theta)) + (h_k - h_k')\nu_{\mathcal{P}_t}(f_k(\theta)) \\ &< e_t + \nu_{\mathcal{P}_t}(f_k(\theta)) + (h_k - h_k')\nu_{\mathcal{P}_t}(f_k(\theta)).\end{aligned}$$

On the other hand, we have $\nu_{\mathcal{P}_t}(B) \geq e_t + \nu_{\mathcal{P}_t}(f_k(\theta))$. It follows that

$$e_t + \nu_{\mathcal{P}_t}(f_k(\theta)) < e_t + \nu_{\mathcal{P}_t}(f_k(\theta)) + (h_k - h_k')\nu_{\mathcal{P}_t}(f_k(\theta)),$$

which implies that $h_k' < h_k$.

(5) ⋆ *(i)* ⇒ *(ii)*. By **(2), (d)**, there exists $j \in \{1, \ldots, s\}$ such that $h_j' < h_j$. It follows that $h_j \geq 2$ and then by **(4)**, $\overline{f_j(x)} \mid \overline{g(x)}$ in $\mathbb{F}_p[x]$.

⋆ *(ii)* ⇒ *(i)*. By **(4)**, $h_j' < h_j$, hence by **(2), (d)** $p \mid I(\theta)$.

(6) (i) Here $f(x) = x^3 - 4$, hence $\mathrm{Disc}(f) = \mathrm{Disc}(\theta) = -3^3 2^4$. This shows that 3 is ramified in A. In this case, **(Eq 1)** reads:

$$f(x) = (x-1)^3 + 3(x^2 - x - 1)$$

and its is seen that $\overline{x-1} \nmid x^2 - x - 1$ in $\mathbb{F}_3[x]$, hence $3 \nmid I(\theta)$. Therefore, for the prime 3, $M_\theta(x) = f(x)$. Dedekind's Theorem (see [Marcus (1977), Th. 27, Chap. 3]) shows that 3 is totally ramified. For the prime 2, we have $f(x) = x^3 - 2.2$ and $\overline{x} \mid \overline{2}$ in $\mathbb{F}_2[x]$, hence $2 \mid I(\theta)$. Obviously $\theta/2$ is not integral since its minimal polynomial is given by $f_1(x) = x^3 - 1/2$. Let $\mu = \theta^2$, then the minimal polynomial of μ is given by $f_2(x) = x^3 - 4^2$. From this it is seen that $\theta^2/2$ is integral, (thus $M_\theta(x) = x^2$ for the prime 2) but $\theta^2/4$ is not. Recall that for a number field E and for any algebraic integer $\gamma \in E$ which is primitive over $\mathbb{Q}$, there exists an integral basis of the form $\{1, f_1(\gamma)/d_1, \ldots, f_{n-1}(\gamma)/d_{n-1}\}$, where $f_i(x)$ is a monic polynomial of degree i with integral coefficients and the d_i are positive integers satisfying the conditions: For any integer $d_i' > d_i$ and any monic polynomial $g_i(x)$ of degree i with integral coefficients, $g_i(\theta)/d_i'$ is not integral. Moreover $I(\gamma) = d_1 \cdots d_{n-1}$ [Marcus (1977), Th. 13, Exrcises 38–40, Chap. 2]. Applying this theorem for our case, we conclude that $\{1, \theta, \theta^2/2\}$ is an integral basis, $I(\theta) = 2$ and $\mathrm{Disc}(K) = 2^2.3^3$. To see how the prime 2 splits in A, we compute the minimal polynomial of $\theta^2/2$. It is given by $f_3(x) = x^3 - 2$, hence by **(5)**, $2 \nmid I(\theta^2/2)$ and then 2 is totally ramified in A.

(ii) Here $f(x) = x^3 + 4x + 8$. We have $\mathrm{Disc}(f) = -4.4^3 - 27.8^2 = -2^6.31$. This shows immediately that $31 \nmid I(\theta)$ and then 31 is ramified in A. Since $f(x) \equiv (x+3)^2(x-6) \pmod{31}$, then Dedekind's Theorem

[Marcus (1977), Th. 27, Chap. 3] implies that the splitting of 31 in A is given by $31 = \mathcal{P}_1^2\mathcal{P}_2$, where $\mathcal{P}_1$ and $\mathcal{P}_2$ are prime ideals of A both of residual degree equal to 1. We now look at the prime 2. We have $f(x) = x^3 + 2(2x+4)$. Since $\overline{x} \mid \overline{2x+4}$ in $\mathbb{F}_2[x]$, then $2 \mid I(\theta)$. The element $\theta/2$ is a root of $f_1(x) = x^3 + x + 1$, hence it is an algebraic integer. We conclude that for the prime 2, $M_\theta(x) = x$. It is seen that $\theta/4$ is not integral. Obviously $\theta^2/4$ is integral. From the value of the discriminant of θ and the theorem cited above, we conclude that $\{1, \theta/2, \theta^2/4\}$ is an integral basis, $I(\theta) = 2^3$ and $\mathrm{Disc}(K) = -31$. It is easy to see that the minimal polynomial of $\theta^2/4$ over $\mathbb{Q}$ is given by $f_2(x) = x^3 + 2x^2 + x - 1$. Since $\overline{f_2(x)}$ is irreducible over $\mathbb{F}_2$, then the prime 2 is inert in A.

Exercise 10.31.
Let K be a number field of degree n, A be its ring of integers and p be a prime number. Let $\theta \in A$ be a primitive element, $I(\theta)$ be its index. Show that the following assertions are equivalent

(i) $p \nmid I(\theta)$.
(ii) $A = \mathbb{Z}[\theta] + pA$.
(iii) $A \subset \mathbb{Z}_{(p)}[\theta]$, where $\mathbb{Z}_{(p)}$ is the localized ring of $\mathbb{Z}$ in the prime ideal $p\mathbb{Z}$.

Solution 10.31.
We show the implications $(i) \Rightarrow (ii) \Rightarrow (iii) \Rightarrow (i)$.

- $(i) \Rightarrow (ii)$. Consider the following diagram.

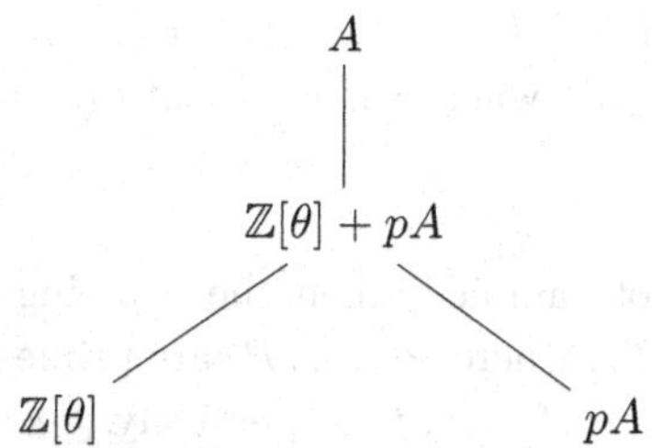

We have $I(\theta) = (A : \mathbb{Z}[\theta])$ and $(A : pA) = p^n$, hence $(A : \mathbb{Z}[\theta] + pA)$ divides both $I(\theta)$ and $(A : pA)$. It follows that $(A : \mathbb{Z}[\theta] + pA) = 1$ and then $\mathbb{Z}[\theta] + pA = A$.
- $(ii) \Rightarrow (iii)$. Obviously $\mathbb{Z}[\theta]$ and pA are contained in $\mathbb{Z}_{(p)}[\theta]$, hence $A \subset \mathbb{Z}_{(p)}[\theta]$.

- $(iii) \Rightarrow (i)$. It is known that A has a basis over $\mathbb{Z}$ of the form
$$\{1, f_1(\theta)/d_1, \ldots, f_{n-1}(\theta)/d_{n-1}\},$$
where $f_i(x)$ is a monic polynomial of degree i, with integral coefficients and $d_1, \ldots, d_{n-1}$ are positive integers such that $d_i \mid d_{i+1}$, for $i = 1, \ldots, n-2$. Moreover $I(\theta) = d_1 \cdots d_{n-1}$ [Marcus (1977), Th. 13 and Exercise 40 Chap. 1]. Since $f_i(\theta)/d_i \in \mathbb{Z}_{(p)}[\theta]$, then $p \nmid d_i$, hence $p \nmid I(\theta)$.

Exercise 10.32.
Let K be a number field, A be its ring of integers, $\theta \in A$ and $g(x) = \mathrm{Irr}(\theta, \mathbb{Q}, x)$. Let p be a prime number such that p is not ramified in K and $p \nmid I(\theta)$. Let $g_1(x), \ldots, g_r(x) \in \mathbb{Z}[x]$ be distinct monic irreducible polynomials such that $g(x) \equiv g_1(x) \cdots g_r(x) \pmod p$.

(1) Suppose that $g_i(x) \not\equiv x \pmod p$ for any $i \in \{1, \ldots, r\}$ and let d_i be the order of $g_i(x)$ over $\mathbb{F}_p$.
   (a) Show that the sequence $(\theta^n)_{n \geq 0}$, considered modulo p, is periodic and its period $T_p(\theta)$ is equal to $\mathrm{lcm}_{i=1}^r(d_i)$.
   (b) Show that $T_p(\theta)$ divides $\mathrm{lcm}_{i=1}^r(p^{f_i} - 1)$.
   (c) Show that there exists a primitive element $\gamma \in A$ such that $T_p(\gamma) = \mathrm{lcm}_{i=1}^r(p^{f_i} - 1)$.
   (d) Compute $T_3(\theta)$ and $T_7(\theta)$ when θ is a root of $g(x) = x^3 + 4x + 1$.

(2) Suppose that $g_1(x) \equiv x \pmod p$ and let d_i be the order of $g_i(x)$ for $i = 2, \ldots, r$.
   (a) Show that the sequence $(\theta^n)_{n \geq 1}$, considered modulo p, is periodic and its period $T_p(\theta)$ is equal to $\mathrm{lcm}_{i=2}^r(d_i)$.
   (b) Compute $T_2(\alpha)$ when α is a root of $h(x) = x^3 - 5x^2 + 5x - 2$.

Solution 10.32.

(1)(a) Since p is not ramified, then the splitting of p in A has the form $pA = \mathcal{P}_1 \cdots \mathcal{P}_r$, where $\mathcal{P}_1, \ldots, \mathcal{P}_r$ are prime ideals of A with residual degrees equal to $f_1, \ldots, f_r$ respectively. Set $T = \mathrm{lcm}_{i=1}^r(d_i)$. For any $i = 1, \ldots, r$, we have $\theta^{d_i} \equiv 1 \pmod{\mathcal{P}_i}$, hence $\theta^T \equiv 1 \pmod{\mathcal{P}_i}$. It follows that $\theta^T \equiv 1 \pmod{\prod_{i=1}^r \mathcal{P}_i}$, that is $\theta^T \equiv 1 \pmod{pA}$. We deduce that for any non negative integer n, $\theta^{n+T} \equiv \theta^n \pmod{pA}$. Let $S \geq 1$ be an integer such that $\theta^S \equiv 1 \pmod{pA}$, then $\theta^S \equiv 1 \pmod{\mathcal{P}_i}$ for any $i \in \{1, \ldots, r\}$. It follows that the order of $\theta + \mathcal{P}_i$ in $A/\mathcal{P}_i$ divides S, that is $d_i \mid S$ for any $i \in \{1, \ldots, r\}$. Hence $T =$

$\text{lcm}_{i=1}^r(d_i)$ divides S. This implies that the period of the sequence $(\theta^n)_{n\geq 0}$, modulo p, is equal to $T = T_p(\theta)$.

(b) We have $f_i = [A/\mathcal{P}_i : \mathbb{F}_p]$, hence the order of $\theta + \mathcal{P}_i$ in $A/\mathcal{P}_i$ divides $|A/\mathcal{P}_i^\star| = p^{f_i} - 1$. We deduce that $T = \text{lcm}_{i=1}^r(d_i)$ divides $\text{lcm}_{i=1}^r(p^{f_i} - 1)$.

(c) For any $i \in \{1, \dots, r\}$ let $h_i(x) \in \mathbb{Z}[x]$ be monic of degree f_i such that $\overline{h_i(x)}$ has its order over $\mathbb{F}_p$ equal to $p^{f_i} - 1$. Suppose that the $h_i(x)$ are chosen so that if $f_i = f_j$ and $i \neq j$, then $\overline{h_i(x)} \neq \overline{h_j(x)}$. For any $i \in \{1, \dots, r\}$ let $\gamma_i \in A$ such that $\gamma_i + \mathcal{P}_i$ is a root of $\overline{h_i(x)}$. Let $\gamma \in A$ such that $\gamma \equiv \gamma_i \pmod{\mathcal{P}_i}$. Then by **Exercise 10.30**, γ is a primitive element of K and $p \nmid \gamma$. By **(a)**, $T_p(\gamma) = \text{lcm}_{i=1}^r(p^{f_i} - 1)$.

(d) The factorization of $g(x)$ modulo 3 is given by

$$g(x) \equiv (x-1)(x^2 + x - 1),$$

hence $3A = \mathcal{P}_1\mathcal{P}_2$, where $\mathcal{P}_1$ and $\mathcal{P}_2$ are prime ideals of A with residual degrees equal to 1 and 2 respectively. It follows that $d_1 = 1$ and $d_2 \mid 3^2 - 1$. Obviously $d_2 \neq 1$. If $d_2 = 2$, then $\theta^2 \equiv 1 \pmod{3A}$, hence $\theta^3 \equiv \theta \pmod{3A}$. But $\theta^3 = -4\theta - 1 \not\equiv \theta \pmod{3A}$, thus $d_2 \neq 2$. The same reasoning shows that $d_2 \neq 4$. It follows that $T_3(\theta) = \text{lcm}(d_1, d_2) = 8$.

In $\mathbb{F}_7[x]$, $\overline{g(x)}$ is irreducible, hence $7A$ is prime. It follows that $T_7(\theta) \mid 7^3 - 1$. Since $7^3 - 1 = 2.3^2.19$, then the possible values of $T_7(\theta)$ are the followings: $1, 2, 3, 6, 9, 18, 19, 2.19, 3.19, 6.19, 3^2.19, 7^3 - 1$. Obviously we may exclude the values $1, 2$. We have

$$\begin{aligned}
&\theta^3 \equiv 3\theta - 1 \pmod{7A}, \theta^4 \equiv 3\theta^2 - \theta, \theta^5 \equiv -\theta^2 + 2\theta - 3 \pmod{7A},\\
&\theta^6 \equiv 2\theta^2 + \theta + 1 \pmod{7A}, \theta^9 \equiv \theta^2 - \theta \pmod{7A},\\
&\theta^{18} \equiv 3\theta^2 + 2 \pmod{7A}, \theta^{19} \equiv -3\theta - 3 \pmod{7A},\\
&\theta^{2.19} \equiv 2\theta^2 - 3\theta + 2 \pmod{7A}, \theta^{3.19} \equiv 3\theta^2 - \theta \pmod{7A},\\
&\theta^{6.19} \equiv \theta - 1 \pmod{7A}, \theta_{3^2.19}(\theta) \equiv -3\theta - 3 \pmod{7A},
\end{aligned}$$

hence $T_7(\theta) = 7^3 - 1$.

(2)(a) Since p is not ramified, then the splitting of p in A has the form $pA = \mathcal{P}_1 \cdots \mathcal{P}_r$, where $\mathcal{P}_1, \dots, \mathcal{P}_r$ are prime ideals of A with residual degrees equal to $f_1, \dots, f_r$ respectively, where $f_1 = 1$. Set $T = \text{lcm}_{i=2}^r(d_i)$. For any $i = 2, \dots, r$, we have $\theta^{d_i} \equiv 1 \pmod{\mathcal{P}_i}$, hence $\theta^T \equiv 1 \pmod{\mathcal{P}_i}$. It follows that $\theta^T \equiv 1 \pmod{\prod_{i=2}^r \mathcal{P}_i}$, that is $\theta^T - 1 \equiv 0 \pmod{\prod_{i=2}^r \mathcal{P}_i}$ and then $\theta^{T+1} \equiv \theta \pmod{p}$. The proof may be completed by using similar arguments as in **(1)(a)**.

(b) We have $h(x) \equiv x(x^2+x+1) \pmod 2$ and the polynomial x^2+x+1 is irreducible over $\mathbb{F}_2$. The order of this polynomial is equal 3, hence $T_2(\alpha) = 3$. This means that $\theta^{n+3} \equiv \theta^n \pmod{2A}$ for $n \geq 1$.

Exercise 10.33.
Let $f(x) = x^3 - 2x^2 - 1$, θ be a root of $f(x)$ in $\mathbb{C}$, $K = \mathbb{Q}(\theta)$ and A be the ring of integers of K.

(1) Show that $f(x)$ is irreducible over $\mathbb{Q}$ and that $\{1, \theta, \theta^2\}$ is a basis of A over $\mathbb{Z}$.
(2) Show that $2A = \mathcal{P}_1\mathcal{P}_2$, where $\mathcal{P}_1$ and $\mathcal{P}_2$ are prime ideals of A having their residual degrees equals to 1 and 2 respectively. Show that $\mathcal{P}_1 = (\theta - 1)A$ and $\mathcal{P}_2 = (\theta^2 + \theta + 1)A$.
(3) Show that $3A$ is prime in A.
(4) Let $I = 3(\theta - 1)^2 A + 4(\theta^2 + \theta + 1)A$. Determine a basis of I^{-1} over $\mathbb{Z}$.
(5) Show that $36 \in I$ and find $\alpha \in A$ such that $I = 36A + \alpha A$.

Solution 10.33.

(1) Since the polynomial $f(x)$ has no root in $\mathbb{Z}$, then it is irreducible over $\mathbb{Q}$. According to **Exercise 10.5**, we have $\mathbb{Z}[1/\theta] = \mathbb{Z}[\theta]$, thus it is equivalent to show that $\{1, 1/\theta, 1/\theta^2\}$ is a basis of A over $\mathbb{Z}$. This claim is true since the minimal polynomial of $1/\theta$ is given by $F(x) = x^3+2x-1$ and the discriminant of this polynomial, being equal to -59, is square free.
(2) Since $f(x) = (x-1)(x^2+x+1) - 2x^2$, then, by **Exercise 10.30**, $2 \nmid I(\theta)$ and $2A = \mathcal{P}_1\mathcal{P}_2$, where $\mathcal{P}_1$ and $\mathcal{P}_2$ are prime ideals of A having their residual degrees equal to 1 and 2 respectively. Since $\theta - 1$ is a root of the polynomial $f(x+1)$ and since the constant term of this polynomial is equal to $f(1)$, that is to -2, then $\mathrm{N}_{K/\mathbb{Q}} = 2$, then $\mathcal{P}_1 = (\theta - 1)A$. Let $\gamma = \theta^2 + \theta + 1$. By the method used in **Exercise 5.30(3)(b)**, we find that the characteristic polynomial (here equal to the minimal polynomial) $g(x)$ of γ is given by $g(x) = x^3 + 3x^2 - 4$. Moreover, since $g(x) = x^2(x+1) + 2(x^2 - 2)$, then by **Exercise 10.30**, $2 \mid I(\gamma)$ and $\mathcal{P}_2 = \gamma A = (\theta^2 + \theta + 1)A$.
(3) The reduced polynomial of $f(x)$ modulo 3 has no root in $\mathbb{F}_3$, hence it is irreducible over this field and then $3A$ is a prime ideal of A.
(4) Let $x \in K$, then

$$\begin{aligned} x \in I^{-1} &\Leftrightarrow xI \subset A \\ &\Leftrightarrow x\big(3(\theta-1)^2\big) \in A \quad \text{and} \quad x\big(4(\theta^2+\theta+1)\big) \in A. \end{aligned}$$

Set $x = (x_0 + x_1\theta + x_2\theta^2)/d$ with $x_0, x_1, x_2 \in \mathbb{Z}$ and d is a positive integer. Using the relations $\theta^3 = 2\theta^2 + 1$, $\theta^4 = 4\theta^2 + \theta + 2$, we find the following identities

$$3x(\theta - 1)^2 = \frac{\theta^2(x_0 + x_2) + \theta(-2x_0 + x_1 + x_2) + (x_0 + x_1)}{d} \quad \text{and}$$

$$4x(\theta^2 + \theta + 1) = \frac{\theta^2(x_0 + 3x_1 + 7x_2) + \theta(x_0 + x_1 + x_2) + (x_0 + x_1 + 3x_2)}{d}.$$

We deduce that $x \in I^{-1}$ if and only if d divides the following integers

$$3(x_0 + x_2), \quad 3(-2x_0 + x_1 + x_2), \quad 3(x_0 + x_1),$$
$$4(x_0 + 3x_1 + 7x_2), \quad 4(x_0 + x_1 + x_2), \quad 4(x_0 + x_1 + 3x_2).$$

(a) Suppose that $x \in I^{-1}$ and d is odd, then since d divides the three last integers in the above list, we conclude that d divides each of the integers $x_0+3x_1+7x_2$, $x_0+x_1+x_2$, $x_0+x_1+3x_2$. This implies that d divides x_1, x_2, x_3 which means that $x \in A$.

(b) Suppose that $x \in I^{-1}$ and $d \equiv 2 \pmod 4$. Let $\delta = \gcd(d, 12)$ and set $d = \delta d'$, where d' is odd and $\delta = 2$, or 6. From the conditions on d described above we deduce in particular that d divides each of the following integers $x_0 + 3x_1 + 7x_2$, $x_0 + x_1 + x_2$, $x_0 + x_1 + 3x_2$. It is easy to show successively that $d' \mid x_2$, $d' \mid x_1$ and $d' \mid x_0$. Let $y_0 = x_0/d'$, $y_1 = x_1/d'$ and $y_2 = x_2/d'$, then δ divides each of the followings integers

$$3(y_0 + y_2), \quad 3(-2y_0 + y_1 + y_2), \quad 3(y_0 + y_1),$$
$$4(y_0 + 3y_1 + 7y_2), \quad 4(y_0 + y_1 + y_2), \quad 4(y_0 + y_1 + 3y_2).$$

- If $\delta = 2$, we get $y_0 \equiv y_1 \equiv y_2 \pmod 2$. Set $y_1 = y_0 + 2z_1$ and $y_2 = y_0 + 2z_2$, then we conclude that

$$\begin{aligned} x &= (x_0 + x_1\theta + x_2\theta^2)/(\delta d') \\ &= \left(y_0 + (y_0 + 2z_1)\theta + (y_0 + 2z_2)\theta^2\right)/2 \\ &= y_0(1 + \theta + \theta^2)/2 + z_1\theta + z_2\theta^2. \end{aligned}$$

- If $\delta = 6$, we get $2 \mid y_0 + y_2, y_1 + y_2, y_0 + y_1$, that is $y_2 \equiv y_1 \equiv y_0 \pmod 2$. We also obtain that $3 \mid y_0 + y_2, y_0 + y_1 + y_2, y_0 + y_1$, which implies $y_2 \equiv y_1 \equiv y_0 \equiv 0 \pmod 3$. From these conditions modulo 2 and modulo 3, we may set $y_0 = 3z_0$, $y_1 = 3z_0+6z_1$, $y_2 = 3z_0 + 6z_2$. As above, we obtain $x = z_0(1+\theta+\theta^2)/2 + z_1\theta + z_2\theta^2$.

(c) Suppose that $x \in I^{-1}$ and $d \equiv 0 \pmod 4$. As above, let $\delta = \gcd(d, 12)$ and set $d = \delta d'$, where d' is a positive integer and $\delta = 4$, or 12. We know that $\delta d'$ divides each of the followings integers

$$3(x_0 + x_2), \quad 3(-2x_0 + x_1 + x_2), \quad 3(x_0 + x_1),$$
$$4(x_0 + 3x_1 + 7x_2), \quad 4(x_0 + x_1 + x_2), \quad 4(x_0 + x_1 + 3x_2).$$

- If $\delta = 4$, then $3 \nmid d'$ and conclude that d' divides each of the integers

$$(x_0 + x_2), \quad (-2x_0 + x_1 + x_2), \quad (x_0 + x_1),$$
$$(x_0 + 3x_1 + 7x_2), \quad (x_0 + x_1 + x_2), \quad (x_0 + x_1 + 3x_2).$$

It follows that d' divides each of x_0, x_1, x_2. Let $y_0 = x_0/d'$, $y_1 = x_1/d'$ and $y_2 = x_2/d'$, then 4 divides each of the followings integers

$$3(y_0 + y_2), \quad 3(-2y_0 + y_1 + y_2), \quad 3(y_0 + y_1),$$
$$4(y_0 + 3y_1 + 7y_2), \quad 4(y_0 + y_1 + y_2), \quad 4(y_0 + y_1 + 3y_2),$$

thus 4 divides each of the integers

$$(y_0 + y_2), \quad (-2y_0 + y_1 + y_2), \quad (y_0 + y_1).$$

It follows that $y_1 \equiv -y_0 \pmod 4$ and $y_2 \equiv -y_0 \pmod 4$. Set $y_1 = -y_0 + 4z_1$ and $y_2 = -y_0 + 4z_2$, then $x = y_0(1 - \theta - \theta^2)/4 + z_1\theta + z_2\theta^2$.

- If $\delta = 12$, similar computations as above show that d' divides x_0, x_1, x_2. Setting $y_0 = x_0/d'$, $y_1 = x_1/d'$ and $y_2 = x_2/d'$, we get $y_2 \equiv y_1 \equiv y_0 \equiv 0 \pmod 3$ and $y_1 \equiv -y_0 \pmod 4$, $y_2 \equiv -y_0 \pmod 4$. Set $y_0 = 3z_0$, $y_1 = -3z_0 + 12z_1$ and $y_2 = -3z_0 + 12z_2$, then $x = z_0(1 - \theta - \theta^2)/4 + z_1\theta + z_2\theta^2$.

We have proved that x is a linear combination of

$$\{1, \theta, \theta^2\} \quad \text{or} \quad \{(1 + \theta + \theta^2)/2, \theta, \theta^2\} \quad \text{or} \quad \{(1 - \theta - \theta^2)/4, \theta, \theta^2\}.$$

Since $(1 + \theta + \theta^2)/2 = 2(1 - \theta - \theta^2)/4 + \theta + \theta^2$, then in any case x is a linear combination of the last set of generators. On the other hand since $A \subset I^{-1}$, then θ and $\theta^2 \in I^{-1}$. To obtain the same conclusion for $(1 - \theta - \theta^2)/4$, it is sufficient to verify that

$$(1 - \theta - \theta^2)/4(3\theta^2 - 6\theta + 3) \quad \text{and}$$
$$(1 - \theta - \theta^2)/4(4\theta^2 + 4\theta + 4) \in A.$$

We have

$$(1-\theta-\theta^2)/4(3\theta^2-6\theta+3) = -3(\theta^2+\theta) \in A \quad \text{and}$$
$$(1-\theta-\theta^2)/4(4\theta^2+4\theta+4) = (1-\theta-\theta^2)(\theta^2+\theta+1) \in A,$$

hence the result. We conclude that $\{(1-\theta-\theta^2)/4, \theta, \theta^2\}$ is a basis of I^{-1} over $\mathbb{Z}$.

(5) We have $36\theta = 3(4\theta^2+4\theta+4) - 4(3\theta^2-6\theta+3)$. Since θ is a unit, then $36 \in A$. Set $\mathcal{B} = 36I^{-1}$, then we have

$$\begin{aligned} \alpha A + 36A = I &\Leftrightarrow \alpha II^{-1} + 36II^{-1} = II^{-1} \\ &\Leftrightarrow \alpha I^{-1} + \mathcal{B} = I \\ &\Leftrightarrow \alpha I^{-1} + \mathcal{P} = I, \quad \text{for any} \quad \mathcal{P} \mid \mathcal{B} \\ &\Leftrightarrow \alpha I^{-1} \not\subset \mathcal{P}, \quad \text{for any} \quad \mathcal{P} \mid \mathcal{B} \\ &\Leftrightarrow \alpha \notin I\mathcal{P}, \quad \text{for any} \quad \mathcal{P} \mid \mathcal{B}. \end{aligned}$$

The factorizations of $3(\theta-1)^2$ and $4(\theta^2+\theta+1)$ into products of prime ideals of A are given by $3(\theta-1)^2 = 3A\mathcal{P}_1^2$ and $4(\theta^2+\theta+1) = \mathcal{P}_1^2\mathcal{P}_2^3$, hence $I = \mathcal{P}_1^2$ and then $\mathcal{B} = (3A)^2\mathcal{P}_2^2$. The conditions above on α read $\alpha \notin I\mathcal{P}_2$ and $\alpha \notin 3AI$. Clearly $\alpha = (\theta-1)^2$ works.

Exercise 10.34.

Let K be a field of characteristic 0, Ω be an algebraic closure of K and let $\mathcal{U}$ be the set of elements $\beta \in \Omega$ for which there exists $\alpha \in \Omega^*$ such that $\beta = \alpha/\alpha'$, where α' is a conjugate of α over K. Call the elements of $\mathcal{U}$, $\mathcal{U}$-numbers.

(1) Let F be a finite extension of $\mathbb{Q}$ and k be a positive integer. Show that for any prime number p, except finitely of them, $x^k - p$ is irreducible over F.

(2) Let $\beta \in \Omega$. Suppose that $\beta^k \in \mathcal{U}$ for some positive integer k. Show that $\beta \in \mathcal{U}$.

(3) Let $\beta \in \Omega$, N_β be the normal closure of $K(\beta)$ over K, $G_\beta = \mathrm{Gal}(N_\beta, K)$. For any $\sigma \in G_\beta$, let e_σ be the smallest positive integer such that $\sigma^{e_\sigma}(\beta) = \beta$ and let $P(\sigma, \beta) = \prod_{i=0}^{e_\sigma - 1} \sigma^i(\beta)$.

(a) Show that the following conditions are equivalent.

(i) $\beta \in \mathcal{U}$.

(ii) There exists $\sigma \in G_\beta$ such that $P(\sigma, \beta)$ is a root of unity.

(b) If the equivalent conditions (i) and (ii) hold and $P(\sigma, \beta)$ is a k-th root of unity, show that β may be written in the form $\beta = \alpha/\alpha'$, where α' is a conjugate of α over K and $\alpha^k \in N_\beta$.

(c) Show that (ii) is equivalent to the following assertion.

(iii) There exists $\sigma \in G_\beta$ such that $\prod_{i=0}^{n-1} \sigma^i(\beta)$ is a root of unity where n is the order of σ.

(4) Let $E = K(\beta)$ and suppose that E/K is cyclic. Show that $\beta \in \mathcal{U}$ if and only if $\mathrm{N}_{E/K}(\beta) = 1$.

Solution 10.34.

(1) Let p be a prime number and $f(x) = x^k - p \in F[x]$. Then, by [Lang (1965), Th. 16, Chap. 8.9], $f(x)$ is reducible over F if and only if there exists a prime $l \mid k$ such that $p \in F^l$ or $4|k$ and $p \in -4F^4$. For any prime number l such that $l \mid k$, let $\mathcal{P}_l = \{p \quad \text{prime}, \quad p \in F^l\}$. If $4 \mid k$, let $\mathcal{P}_4 = \{p \quad \text{prime}, \quad p \in -4F^4\}$. To get that the set of primes p, for which $x^k - p$ is reducible over F, it is sufficient to prove that $\mathcal{P}_l$ is finite for any prime factor l of k and that $\mathcal{P}_4$ is finite. Suppose that $\mathcal{P}_l$ is infinite for some prime factor l of k, then as it is said above, for any $p \in \mathcal{P}_l$, there exists $\alpha \in F$ such that $p = \alpha^l$, that is F contains the field $\mathbb{Q}(p^{1/l})$. Since the number of subfields of F is finite, there exist two distinct prime numbers p_1 and p_2 belonging to $\mathcal{P}_l$, such that $p_1 \neq l$, $p_2 \neq l$ and $\mathbb{Q}(p_1^{1/l}) = \mathbb{Q}(p_2^{1/l})$.
Claim. Let p and l be distinct prime numbers and $K = \mathbb{Q}(p^{1/l})$. Then except possibly for l, the only prime number which is ramified in K is p.
Proof. Let $\alpha = p^{1/l}$ and $g(x) = \mathrm{Irr}(\alpha, \mathbb{Q}) = x^l - p$. We have

$$\mathrm{Disc}(g) = (-1)^{l(l-1)/2} N_{K/\mathbb{Q}}(g^{'}(\alpha)) = \pm(1) l^l p^{l-1}.$$

This shows that the only prime, excluding l, which can be ramified in K is p. We show that effectively p is ramified in K. Using the result of **Exercise 10.30**, it is seen that $p \nmid I(\alpha)$ and then p is ramified in K. We deduce from this claim that p_i is ramified in $\mathbb{Q}(p_i^{1/l})$ but not in $\mathbb{Q}(p_j^{1/l})$ for $i, j \in \{1, 2\}$, $i \neq j$, contradicting the equality of the fields. We have proved that $\mathcal{P}_l$ is finite. The same proof, mutadis mutandis, works for $\mathcal{P}_4$ and will be omitted.

(2) Let $\gamma \in \Omega$ such that $\beta^k = \frac{\gamma}{\gamma^{'}}$, where $\gamma^{'}$ is a conjugate of γ over K. Let $\rho \in \Omega$ such that $\rho^k = \gamma$. Let $\sigma : K(\gamma) \to \Omega$ be the unique K-embedding such that $\sigma(\gamma) = \gamma^{'}$ and let $\hat{\sigma} : K(\rho) \to \Omega$ be its extension to $K(\rho)$. Since $\rho^k = \gamma$, then $\hat{\sigma}(p)^k = \hat{\sigma}(\gamma) = \sigma(\gamma) = \gamma^{'}$. Therefore $\beta^k = (\frac{\rho}{\hat{\sigma}(\rho)^k})^k = \frac{\rho^k}{\hat{\sigma}(\rho)^k}$. We deduce that $\beta = \epsilon_k \rho/\hat{\sigma}(\rho)$, where ϵ_k is a k-th root of unity in Ω. Let $F = K(\rho, \epsilon_k)$ and let p be a prime number

such that $x^k - p$ is irreducible over F. Let $p^{1/k}$ be a root of $x^k - p$ in Ω, then we may write β in the form $\beta = \frac{\rho p^{1/k}}{\hat{\sigma}(\rho)\epsilon_k^{-1}p^{1/k}}$. Consider the following diagram of fields.

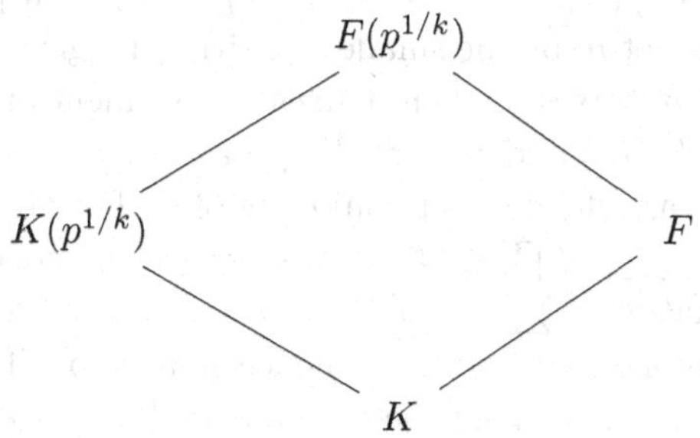

Since F and $K(p^{1/k})$ are linearly disjoint over K, there exists a K-embedding $\tau : F(p^{1/k}) \to \Omega$ such that

$$\tau(p^{1/k}) = \epsilon_k^{-1}p^{1/k} \quad \text{and}$$
$$\tau(\rho) = \hat{\sigma}(\rho).$$

We deduce that $\tau(\rho p^{1/k}) = \hat{\sigma}(\rho)\epsilon_k^{-1}p^{1/k}$, hence $\beta = \rho p^{1/k}/\tau(\rho(p^{1/k})$, thus $\beta = \alpha/\alpha'$ where $\alpha = \rho p^{1/k}$ and $\alpha' = \tau(\rho p^{1/k})$ is a conjugate of α.

(3) • $(i) \Rightarrow (ii)$. Let $\alpha \in \Omega$ such that $\beta = \frac{\alpha}{\tau(\alpha)}$ with $\tau \in \mathrm{Gal}(N_\alpha, K)$, then $\beta \in N_\alpha$, hence $N_\beta \subset N_\alpha$. Let σ be the restriction of τ to N_β, then $\sigma \in G_\beta$. We consider $P(\sigma,\beta) = \prod_{i=0}^{e_\sigma - 1}\sigma^i(\beta)$. Let n be the smallest positive integer such that $\tau^n(\alpha) = \alpha$. Since

$$\sigma^n(\beta) = \tau^n(\beta) = \tau^n\left(\frac{\alpha}{\tau(\alpha)}\right) = \frac{\tau^n(\alpha)}{\tau^{n+1}(\alpha)} = \frac{\alpha}{\tau(\alpha)} = \beta,$$

then $m \mid n$. Set $n = mk$. On the one hand we have

$$\prod_{i=1}^{n}\sigma^{i-1}\left(\frac{\alpha}{\tau(\alpha)}\right) = \prod_{i=1}^{n}\sigma^{i-1}(\beta) = (\beta_1, \cdots, \beta_m)^k = P(\sigma,\beta)^k.$$

On the other hand we have

$$\prod_{i=1}^{n}\sigma^{i-1}\left(\frac{\alpha}{\tau(\alpha)}\right) = \prod_{i=1}^{n}\tau^{i-1}(\alpha)/\prod_{i=1}^{n}\tau^{i}(\alpha) = \tau^0(\alpha)/\tau^n(\alpha) = 1,$$

hence $P(\alpha,\beta)^k = 1$. It follows that $P(\sigma,\beta)$ is a root of unity.

- *(ii)* $\Rightarrow$ *(i)*. Let $\sigma \in G_\beta$ such that $P(\sigma, \beta)$ is a k-th root of unity for some positive integer k and let e be the smallest positive integer such $\sigma^e(\beta) = \beta$. Let N be the Galois closure of $K(\beta)$ over K, $F = \mathrm{Inv}(\sigma)$ and $n = [N : F]$, then, $| < \sigma > | = [N : F] = n$. Therefore $e \mid n$. We deduce that $\prod_{i=0}^{n-1} \sigma_i(\beta)^k = 1$. Let $\gamma \in N_\beta$ be a primitive element over K and let m be the smallest positive integer such that $\sigma^m(\gamma) = \gamma$, then as σ^m fixes γ, then it fixes any element of N_β, thus $m = n$. Let $\beta_i = \sigma^{i-1}(\beta)$ and $\gamma_i = \sigma^{i-1}(\gamma)$ for $i = 1, \ldots, n$ then $\prod_{i=1}^n \beta^i = 1$ and the γ_i are the distinct conjugate of γ. For any $j \in \{1, 2, \ldots, s\}$, let $\mu_j = \sum_{i=1}^n \gamma_i^j \prod_{l=1}^i \beta_l^k$. Consider the homogeneous system of linear equations: $\sum_{i=1}^n \gamma_i^j x_i = 0, j = 1, \ldots, n$. Its determinant is a Vandermonde determinant, hence non zero. Therefore it has a unique solution, namely the trivial one. It follows that there exists $j \in \{1, \ldots, n\}$ such that $\mu_j \neq 0$. Otherwise $x_1 = \cdots = x_n = 0$, contradicting the fact that $\prod_{l=1}^n \beta_l^k \neq 0$. For this index j, we have
$$\begin{aligned}\beta^k\sigma(\mu_j) &= \beta^k\sigma(\gamma_1^j\beta_1^k + \gamma_2^j\beta_1^k\beta_2^k + \cdots + \gamma_n^j\beta_1^k\cdots\beta_n^k) \\ &= \beta^k(\gamma_2^j\beta_2^k + \gamma_3^j\beta_2^k\beta_3^k + \cdots + \gamma_n^j\beta_2^j\cdots\beta_n^k + \gamma_1^j\beta_2^k\cdots\beta_n^k\beta_1^k).\end{aligned}$$
Since $\beta_1^k \cdots \beta_n^k = 1$, then $\beta^k\sigma(\mu_j) = \mu_j$. Therefore $\beta_1^k = \mu_j/\sigma(\mu_j) \in \mathcal{U}$.
Using **(2)**, we conclude that $\beta \in \mathcal{U}$.

(c) *(ii)* $\Leftrightarrow$ *(iii)*. Since $\sigma^n = \mathrm{Id}_{N_\beta}$, then $\sigma^n(\beta) = \beta$, hence $e_\sigma \mid n$. Set $n = e_\sigma \cdot q$ where q is a positive integer, then $\prod_{i=0}^{n-1} \sigma^i(\beta) = P(\sigma, \beta)$, hence $\prod_{i=0}^{n-1} \sigma^i(\beta)$ is a root of unity if and only if $P(\sigma, \beta)$ is.

(4) • **Sufficiency of the condition.**
Let σ be a generator of $\mathrm{Gal}(E, K)$, $n = [E : K]$, then
$$1 = \mathrm{N}_{E/K}(\beta) = \prod_{i=0}^{n-1} \sigma^i(\beta),$$
hence by **(3) (iii)**, $\beta \in \mathcal{U}$.

- **Necessity of the condition.**
Let $\alpha \in E$ such that $\beta = \frac{\alpha}{\tau(\alpha)}$, where $\tau \in \mathrm{Gal}(E, K)$, then
$$\mathrm{N}_{E/K}(\beta) = \mathrm{N}_{E/K}(\alpha)/\mathrm{N}_{E/K}(\tau(\alpha)) = 1.$$

Exercise 10.35.
Let K be a number field, A be its ring of integers and p be a prime number. Let $\mathcal{P}$ be a prime ideal of A such that $\mathrm{N}_{K/\mathbb{Q}}(\mathcal{P}) = p^f$, where f is a positive integer. Let $\phi(x) \in \mathbb{Z}[x]$ be monic of degree f such that its reduction modulo p is irreducible.

(1) Show that there exists $\theta_1 \in A$ such that $\nu_{\mathcal{P}}(\phi(\theta_1)) = 1$.
(2) Let $k \geq 1$ be an integer. Show that $A/\mathcal{P}^k \simeq \mathbb{F}_p[x]/\phi^k(x)\mathbb{F}_p[x]$. Deduce that $|(A/\mathcal{P}^k)^*| = p^{f(k-1)}(p^f - 1)$.
(3) Show that there exists $\theta \in A$ such that $\nu_{\mathcal{P}}(\phi(\theta)) = 1$ and $\nu_{\mathcal{P}'}(\phi(\theta)) = 0$ for any prime ideal $\mathcal{P}'|p$ and $\mathcal{P}' \neq \mathcal{P}$. Deduce that

$$\mathcal{P} = pA + \phi(\theta)A = (p, \phi(\theta)).$$

(4) Let $\{\omega_1, \ldots, \omega_n\}$ be a basis of A over $\mathbb{Z}$ and let $\vec{u} = (u_1, \ldots, u_n)$ be an n-tuple of independent variables over $\mathbb{Q}$. Let $\eta = \sum_{i=1}^{n} u_i\omega_i$.
   (a) Show that $\eta \equiv \sum_{j=0}^{f-1} L_j(\vec{u})\theta^j \pmod{\mathcal{P}}$, where $L_j(\vec{u})$ is a linear form with coefficients in $\mathbb{F}_p$ for $j = 0, 1, \ldots, f-1$.
   (b) Let $H(\vec{u}, x) = \prod_{k=0}^{f-1}(x - \sum_{j=0}^{f-1} L_j(\vec{u})\theta^{jp^k})$. Show that $H(\vec{u}, x) \in \mathbb{F}_p[\vec{u}, x]$ and $H(\vec{u}, x)$ is irreducible over $\mathbb{F}_p$.
   (c) Let $\hat{H}(\vec{u}, x) \in \mathbb{Z}[\vec{u}, x]$ be a monic lift (in x) of $H(u, x)$ and let $\alpha_1, \ldots, \alpha_d \in A$ be all the non zero coefficients of the monomials $u_1^{i_1} \cdots u_n^{i_n}$ in $H(\vec{u}, \eta)$. Show that $\mathcal{P} = (p, \alpha_1, \ldots, \alpha_d)$.

Solution 10.35.

(1) Let $\theta_1 \in A$ such that $\theta_1 + \mathcal{P}$ is a primitive element of $A/\mathcal{P}$ over $\mathbb{F}_{\mathcal{P}}$ and $\phi(x)$ is its minimal polynomial over $\mathbb{F}_p$, then $\phi(\theta_1) \equiv 0 \pmod{\mathcal{P}}$. If $\phi(\theta_1) \not\equiv 0 \pmod{\mathcal{P}^2}$, the proof is complete. Otherwise, let $\tau \in \mathcal{P} \setminus \mathcal{P}^2$ and $\mu = \theta_1 + \tau$, then $\phi(\mu) \equiv 0 \pmod{\mathcal{P}}$ and $\phi(\mu) \not\equiv 0 \pmod{\mathcal{P}^2}$. Therefore replacing θ_1 by μ if necessary, we may suppose that $\nu_{\mathcal{P}}(\phi(\theta_1)) = 1$.
(2) Consider the map $\Psi : \mathbb{F}_p[x] \to A/\mathcal{P}^k$ such that $\Psi(g(x)) = g(\theta_1) + \mathcal{P}^k$. Obviously Ψ is a morphism of rings. Let $g(x) \in \mathbb{F}_p[x]$. If $\phi^k(x) \mid g(x)$, then obviously $g(\theta_1) \equiv 0 \pmod{\mathcal{P}^k}$. We prove the converse by induction on k. The claim is true for $k = 1$. Since $\phi(x)$ is the minimal polynomial of $\theta_1 + \mathcal{P}$ in $A/\mathcal{P}$. Suppose the claim is true for $k-1$ and suppose that $g(\theta_1) \equiv 0 \pmod{\mathcal{P}^k}$. Let $g(x) = \phi(x)q(x) + r(x)$, where $q(x), r(x) \in \mathbb{F}_p[x]$ and $\deg r < \deg \phi$. Since $r(\theta_1) \equiv 0 \pmod{\mathcal{P}}$, then $r(x) = 0$, that is $g(x) = \phi(x)q(x)$. Since $\nu_{\mathcal{P}}(\phi(\theta_1)) = 1$, then $q(\theta_1) \equiv 0 \pmod{\mathcal{P}^{k-1}}$. Therefore $\phi^{k-1}(x)|q(x)$ by the inductive hypothesis. It follows that $\phi^k(x) \mid g(x)$. We deduce that $\operatorname{Ker}\Psi = \phi^k(x)\mathbb{F}_p[x]$ and then $\mathbb{F}_p[x]/\phi^k(x)\mathbb{F}_p[x] \simeq \operatorname{Im}\Psi$. Let $f = \deg \phi = [A/\mathcal{P} : \mathbb{F}_p]$, then

$$|\mathbb{F}_p[x]/\phi^f(x)\mathbb{F}_p[x]| = p^{kf} = |A/\mathcal{P}^k|,$$

hence $\operatorname{Im}\Psi = A/\mathcal{P}^k$ and the isomorphism is established. The elements of the form

$$a_0(x) + a_1(x)\phi(x) + \cdots + a_{k-1}(x)\phi^{k-1}(x),$$

where $a_i(x) \in \mathbb{F}_p[x]$, $\deg a_i < \deg \phi = f$, for $i = 0, \ldots, k-1$, constitute a complete set of representatives of the cosets of the elements of $\mathbb{F}_p[x]$ modulo $\phi^k(x)\mathbb{F}_p[x]$. Any element of this form is a unit in $\mathbb{F}_p[x]/\phi^k(x)\mathbb{F}_q$ if and only if $a_0(x) \neq 0$. Therefore the number of units is equal to $(p^f - 1)(p^f)^{k-1} = (p^f - 1)p^{f(k-1)}$.

(3) Let I be the ideal of A such that $pA = \mathcal{P}^e I$ and $\mathcal{P} \nmid I$ and let $\gamma \in I$ then $\gamma \notin \mathcal{P}$, and γ is a unit in $A/\mathcal{P}^2$, hence $\gamma^{(p^f-1)p^f} \equiv 1 \pmod{\mathcal{P}^2}$. Let $\theta = \theta_1\gamma^{(p^f-1)p^f}$, then $\theta \equiv \theta_1 \pmod{\mathcal{P}^2}$, hence $\nu_{\mathcal{P}}(\phi(\theta)) = 1$.
On the other hand let $\mathcal{P}'$ be a prime ideal of A such that $\mathcal{P}' \mid I$. We have $\phi(\theta) = \theta^f + a_{f-1}\theta^{f-1} + \cdots + a_0$, where $a_0, \ldots, a_{f-1} \in \mathbb{F}_p$ and $a_0 \neq 0$ hence $\phi(\theta) \equiv a_0 \pmod{\mathcal{P}'} \not\equiv 0 \pmod{\mathcal{P}'}$. It is clear that the only ideal of A which divides both pA and $\phi(\theta)A$ is equal to $\mathcal{P}$, hence $\mathcal{P} = pA + \phi(\theta)A$.

(4)(a) For $i = 1, ..., n$, we have $\omega_i \equiv \sum_{j=0}^{f-1} a_i^j\theta^j \pmod{\mathcal{P}}$, where $a_i^j \in \mathbb{Z}$, hence $\eta \equiv \sum_{i=1}^n u_i \sum_{j=0}^{f-1} a_i^j\theta^j \pmod{\mathcal{P}}$, thus

$$\eta \equiv \sum_{j=0}^{f-1}\sum_{i=1}^{n}(u_i a_i^j)\theta^j \pmod{\mathcal{P}}.$$

Set $L_j(\vec{u}) = \sum_{i=1}^n u_i a_i^j$, then $\eta \equiv \sum_{j=0}^{f-1}(\vec{u})\theta^j \pmod{\mathcal{P}}$.

(b) Consider the following diagram of fields:

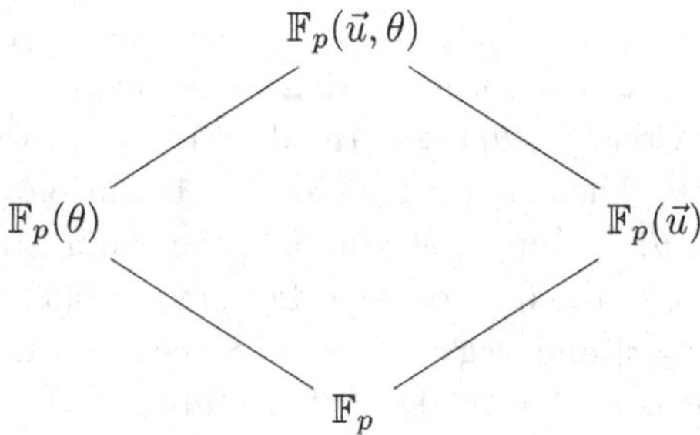

Since $\mathbb{F}_p(\theta)$ is algebraic over $\mathbb{F}_p$ and $\mathbb{F}_p(\vec{u})$ is purely transcendental over $\mathbb{F}_p$, then

$$[\mathbb{F}_p(\vec{u}, \theta) : \mathbb{F}_p(\vec{u})] = [\mathbb{F}_p(\theta) : \mathbb{F}_p] = f.$$

Clearly the characteristic polynomial over $\mathbb{F}_p(\vec{u})$ of η is given by

$$H(\vec{u},x) = \prod_{k=0}^{f-1}\left(x - \sum_{j=0}^{f-1} L_j(\vec{u})\theta^{jp^k}\right).$$

Let $\vec{u}^\star = (u_1^*, ..., u_n^*) \in \mathbb{Z}^n$ such that $\theta = \sum_{i=1}^n u_i^* w_i$, that is $L_1(\vec{u}^*) = 1$ and $L_j(\vec{u}^*) = 0$ for $j \neq 1$, then $H(u^*, x) = \phi(x)$. Since this last polynomial is irreducible over $\mathbb{F}_p$, then so is $H(\vec{u}, x)$.

(c) Let $\vec{u}^* = (u_1^*, ..., u_n^*)$ such that $\theta = \sum_{i=1}^n u_i^* w_i$, then $\hat{H}(u^*, x) \equiv \phi(x) \pmod p$, hence

$$\mathcal{P} = (p, \phi(\theta)) = (p, \hat{H}(u^*, \theta)).$$

It follows that $\mathcal{P} = (p, \alpha_1, \ldots, \alpha_d)$ because if a prime ideal $\mathcal{Q}$ lying over $p\mathbb{Z}$, divides $\alpha_1, \ldots, \alpha_d$ then it divides $\hat{H}(u^*, \theta)$ which is a contradiction.

Exercise 10.36.

Let $f(x) \in \mathbb{Z}[x]$ be monic irreducible of degree n, θ be a root of f, $K = \mathbb{Q}(\theta)$ and A be the ring of integers of K. Let $\mathfrak{P}$ be the set of the prime numbers p such that $p \mid I(\theta)$.

(1) Let $\omega_0, \omega_1, \ldots, \omega_{n-1}$ be an integral basis of the form $\omega_i = f_i(\theta)/d_i$, where $f_i(x)$ is a monic polynomial with integral coefficients of degree i and $d_0, \ldots, d_{n-1}$ are positive integers such that $1 = d_0 \mid d_1 \mid \ldots \mid d_{n-1}$ and $I(\theta) = \prod_{i=0}^{n-1} d_i$. This basis exists by [Marcus (1977), Th. 13 and Exercise 40, Chap. 1]. Let $B = \mathbb{Z}[\theta]$ and for any $p \in \mathfrak{P}$ let

$$B_p = \{\gamma \in A, \text{ there exists}\, h \in \mathbb{N}, p^h\gamma \in B\}.$$

Show that $\{f_i(\theta)/p^{\nu_p(d_i)}, i = 0, \ldots, n-1\}$ is a $\mathbb{Z}$-basis of B_p.

(2) For any $p \in \mathfrak{P}$, let $\{1, f_1^{(p)}/p^{h_1(p)}, \ldots, f_{n-1}^{(p)}/p^{h_{n-1}(p)}\}$ be a $\mathbb{Z}$-basis of B_p, where $f_i^{(p)}(x)$ is a monic polynomial of degree i with integral coefficients and $h_1(p), \ldots, h_{n-1}(p)$ are non negative integers such that $h_1(p) \le \cdots \le h_{n-1}(p)$. Show that for any $i \in \{1, \ldots, n-1\}$, there exists a monic polynomial $f_i(x) \in \mathbb{Z}[x]$, of degree i such that for any $p \in \mathfrak{P}$, $f_i(x) \equiv f_i^{(p)}(x) \pmod{p^{h_i(p)}}$. Show that $\{1, \frac{f_1(\theta)}{\prod_{p\in\mathfrak{P}} p^{h_1(p)}}, \ldots, \frac{f_{n-1}(\theta)}{\prod_{p\in\mathfrak{P}} p^{h_{n-1}(p)}}\}$ is an integral basis of K.

(3) Let μ be a root of $g(x) = x^3 - 5x^2 - 12x - 64$, $E = \mathbb{Q}(\mu)$, A be the ring of integers of E and $B = \mathbb{Z}[\mu]$.

(a) Show that $g(x)$ is irreducible over $\mathbb{Q}$ and $\mathrm{Disc}(g) = -4^2 \cdot 5^2 \cdot 503$.

(b) Show that $I(\mu) = 20$.

(c) Show that $\{1, \mu, (\mu^2 - \mu)/4\}$ (resp. $\{1, \mu, (\mu^2 + 2\mu + 2)/5\}$) is a $\mathbb{Z}$-basis of B_2 (resp. B_5) and deduce an integral basis of E.

Solution 10.36.

(1) Let $\gamma \in B_p$, $\gamma = \sum_{i=0}^{k} \lambda_i f_i(\theta)/d_i$, where $1 \leq k \leq n-1$, $\lambda_i \in \mathbb{Z}$ and $\lambda_k \neq 0$. We prove by induction on k that λ is a linear combination of $1, f_1(\theta)/p^{\nu_p(d_1)}, \ldots, f_{n-1}(\theta)/p^{\nu_p(d_{n-1})}$ with integral coefficients. The result is trivial for $k = 0$. Suppose that the result is true for any $l < k$. Let h be a non negative integer such that $p^h\gamma \in B$, then there exists $b_0, \ldots, b_k \in \mathbb{Z}$ such that

$$\sum_{i=0}^{k} \lambda_i p^h f_i(\theta)/d_i = \sum_{i=0}^{k} b_i \theta^i.$$

Setting $d_k = p^{\nu_p(d_k)} d'_k$, we deduce in particular that

$$\lambda_k p^h = b_k p^{\nu_p(d_k)} d'_k.$$

This shows that $d'_k \mid \lambda_k$. We have

$$\lambda_k f_k(\theta)/d_k = (\lambda_k/d'_k) f_k(\theta)/p^{\nu_p(d_k)} \in B_p,$$

hence $\sum_{i=0}^{k-1} \lambda_i f_i(\theta)/d_i \in B$. From our assumptions, it follows that

$$\sum_{i=0}^{k-1} \lambda_i f_i(\theta)/d_i = \sum_{i=0}^{k-1} \mu_i f_i(\theta)/p^{\nu_p(d_i)},$$

where $\mu_i \in \mathbb{Z}$ for $i = 1, \ldots, k-1$. We conclude that

$$\gamma = \sum_{i=0}^{k} \mu_i f_i(\theta)/p^{\nu_p(d_i)}$$

with $\mu_k = \lambda_k/d'_k$. Therefore

$$\left\{1, f_1(\theta)/p^{\nu_p(d_1)}, \ldots, f_{n-1}(\theta)/p^{\nu_p(d_{n-1})}\right\}$$

is a generating set of B_p. Obviously, this set is free over $\mathbb{Z}$. Therefore it is a $\mathbb{Z}$-basis of B_p.

(2) The existence of the polynomials $f_i(x)$ is established in **Exercise 1.30.** We know that A has a $\mathbb{Z}$-basis of the form 1, $g_1(\theta)/d_1, \ldots, g_{n-1}(\theta)/d_{n-1}$, where $g_i(x)$ is a monic polynomial of degree i with integral coefficients and $d_1, \ldots, d_{n-1}$ are positive integers such that $d_1 \mid d_2 \mid d_{n-1}$ and $I(\theta) = d_1 d_2 \cdots d_{n-1}$ [Marcus (1977), Th. 13 and Exercise 40 Chap. 2]. Moreover any of the polynomials $g_i(x)$ may be replaced by any monic polynomial $h_i(x)$ of degree i with

integral coefficients satisfying $h_i(\theta) \equiv 0 \pmod{d_i}$ [Marcus (1977), Exercise 39, Chap. 2]. For our purpose we take $h_i(x) = f_i(x)$. The proof will be complete if it is shown that $d_i = \prod_{p\in\mathfrak{P}} p^{h_i(p)}$ and $f_i(\theta) \equiv 0$ $(\text{mod } \prod_{p\in\mathfrak{P}} p^{h_i(p)})$ for $i = 1, \ldots, n-1$. These congruences are obvious and we omit their proof. Let $p \in \mathfrak{P}$. Suppose that $p^{l_i(p)} \mid\mid d_i$, then

$$(d_i/p^{l_i(p)})g_i(\theta) = g_i(\theta)/p^{l_i(p)} \in B_p,$$

hence $l_i(p) \leq h_i(p)$. On the other hand, $f_i(\theta)/p^{h_i(p)} \in A$, hence $h_i(p) \leq l_i(p)$. It follows that $d_i = \prod_{p\in\mathfrak{P}} p^{h_i(p)}$ for $i = 1, \ldots, n-1$.

(3)(a) The reduction modulo 3 of $g(x)$ is irreducible over $\mathbb{F}_3$, hence $g(x)$ is irreducible over $\mathbb{Q}$. We omit the details of the computations for the discriminant of $g(x)$. One may apply the general formula for the discriminant of a polynomial of degree 3. Alternatively one may proceed as follows. We have $\text{Disc}(g) = -\text{N}_{E/\mathbb{Q}}(g'(\mu)$. We compute the characteristic polynomial, say $u(x)$, of $g'(\mu)$ by the method used in **Exercise 2.7(4)**, then $\text{Disc}(g) = b_0$, where b_0 is the constant coefficient of $u(x)$.

(b) Since $\text{Disc}(g) = -4^2 \cdot 5^2 \cdot 503$, then $503 \nmid I(\mu)$, $0 \leq \nu_2(I(\mu)) \leq 2$ and $0 \leq \nu_5(I(\mu)) \leq 1$. We have $g(x) = x^2(x-1) - 4(x^2 - 3x - 16)$ and $x \mid 2(x^2 - 3x - 16)$ in $\mathbb{F}_2[x]$, hence, by **Exercise 10.30**, $2 \mid I(\mu)$. We compute the minimal polynomial $h(x)$ of $\alpha = (\mu^2 - \mu)/4$ by using the method of **Exercise 2.7(4)**, we find $h(x) = x^3 - 5x^2 - 34x - 72$. Therefore $\alpha \in A$ and $\nu_2(I(\mu)) = 2$.
We also have $g(x) = (x-1)(x-2)^2 - 5(x^2 + 2x + 12)$ and $x - 2 \mid (x^2 + 2x + 12)$ in $\mathbb{F}_5[x]$, hence $5 \mid I(\mu)$. Therefore $\nu_5(I(\mu)) = 1$. We deduce that $I(\mu) = 20$.

(c) It is clear that $B_2 = \mathbb{Z} + \mathbb{Z}\mu + \mathbb{Z}\mu(\mu-1)/2$, hence $\{1, \mu, \mu(\mu-1)/4\}$ is a $\mathbb{Z}$-basis of B_2. Similarly $B_5 = \mathbb{Z} + \mathbb{Z}\mu + \mathbb{Z}(\mu^2 + 2\mu + 2)/5$, hence $\{1, \mu, (\mu^2 + 2\mu + 2)/5\}$ is a $\mathbb{Z}$-basis of B_5. To compute an integral basis, we use the method of **(2)**. We compute an integral basis of the form $\{1, \mu, g_2(\mu)/20\}$, where $g_2(x)$ is a monic polynomial of degree 2, with integral coefficients and satisfying the following conditions: $g_2(x) \equiv x^2 - x \pmod{4\mathbb{Z}[x]}$ and $g_2(x) \equiv x^2 + 2x + 2 \pmod{5\mathbb{Z}[x]}$. We easily find a solution, namely $g_2(x) = x^2 + 7x - 8$. Therefore $\{1, \mu, (\mu^2 + 7\mu - 8)/20\}$ is an integral basis.

Exercise 10.37.

Let A be an integral domain, K be its fraction field and $f(x) = a_0x^n + \cdots + a_n \in A[x] \setminus A$. Let θ be a root of f in an algebraic closure of K. For $i = 1, \ldots, n-1$, let $u_i(\theta) = a_0\theta^i + a_1\theta^{i-1} + \cdots + a_i$.

(1) Show that $\gamma_k := u_k(\theta)$ is integral over A.
(2) Show that $\theta\gamma_k$ is integral over A.
(3) Let $B = A + A\gamma_1 + \cdots + A\gamma_{n-1}$. Show that B is a ring.

Solution 10.37.

(1) • **First proof.**
Consider the polynomial $R_k(y) = \operatorname{Res}_x\left(y - u_k(x), \frac{1}{a_0} f(x)\right)$. Then

$$R_k(y) = \begin{vmatrix} -a_0 & -a_1 & \dots & y - a_k & 0 & \dots & 0 \\ 0 & -a_0 & \dots & -a_{k-1} & y - a_k & \dots & 0 \\ \dots & \dots & \dots & \dots & \dots & \dots & \dots \\ 0 & 0 & \dots & 0 & -a_0 & \dots & y - a_k \\ 1 & a_1/a_0 & \dots & a_n/a_0 & 0 & \dots & 0 \\ 0 & 1 & \dots & a_{n-1}/a_0 & a_n/a_0 & \dots & 0 \\ \dots & \dots & \dots & \dots & \dots & \dots & \dots \\ 0 & 0 & \dots & 0 & 1 & \dots & a_n/a_0 \end{vmatrix}.$$

Denote by b_i^j the coefficient appearing in the above determinant at the intersection of the i-th row with the j-th column. $R_k(y)$ is a sum of $n + k$ terms. One of them is

$$\pm b_1^{k+1} b_2^{k+2} \cdots b_n^{k+n} b_{n+1}^1 b_{n+2}^2 \cdots b_{n+k}^k = \pm(y - a_k)^n (1)^k.$$

From this it appears that $R_k(y)$ is a polynomial in y of degree n and its leading coefficient is equal to ± 1. Moreover $u_k(\theta)$ is a root of this polynomial. Hence to get the result it is sufficient to prove that $R_k(y) \in A[y]$. We may write $f(x)$ in the form

$$f(x) = x^{n-k} u_k(x) + \hat{f}(x),$$

where $\hat{f}(x)$ is a polynomial of degree at most $n - (k+1)$ with coefficients in A. We have

$$\begin{aligned} R_k(y) &= \operatorname{Res}_x\left(y - u_k(x), \frac{1}{a_0} f(x)\right) \\ &= \operatorname{Res}_x\left(y - u_k(x), \frac{1}{a_0}(x^{n-k} u_k(x) + \hat{f}(x))\right) \\ &= (-a_0)^n \prod_\xi \frac{1}{a_0}(\xi^{n-k} u_k(\xi) + \hat{f}(\xi)) \\ &= (-1)^n a_0^{n-k} \prod_\xi (\xi^{n-k} u_k(\xi) + \hat{f}(\xi)) \\ &= (-1)^i (-a_0)^{n-k} \prod_\xi (\xi^{n-k} y + \hat{f}(\xi)) \\ &= (-1)^i \operatorname{Res}_x\left(y - u_k(x), y x^{n-k} + \hat{f}(x)\right)). \end{aligned}$$

Here ξ denotes any root of $u_k(x) - y$ in an algebraic closure of $K(y)$. The polynomials in x, for which we compute the resultant, in the last equality above, have their coefficients in $A[y]$, hence $R_k(y) \in A[y]$.

- **Second proof.**

  We have

$$\begin{aligned}
\gamma_k &= a_0\theta^k + a_1\theta^{k-1} + \cdots + a_{k-1}\theta + a_k \\
\theta\gamma_k &= a_0\theta^{k+1} + a_1\theta^k + \cdots + a_{k-1}\theta^2 + a_k\theta \\
&\cdots\cdots\cdots\cdots\cdots\cdots \\
\theta^{n-k-1}\gamma_k &= a_0\theta^{n-1} + a_1\theta^{n-2} + \cdots + a_{k-1}\theta^{n-k} + a_k\theta^{n-k-1} \\
\theta^{n-k}\gamma_k &= -a_{k+1}\theta^{n-k-1} - a_{k+2}\theta^{n-k-2} - \cdots + a_{n-1}\theta - a_n \\
&\cdots\cdots\cdots\cdots\cdots\cdots \\
\theta^{n-1}\gamma_k &= -a_{k+1}\theta^{n-2} - a_{k+2}\theta^{n-3} - \cdots - a_{n-1}\theta^k - a_n\theta^{k-1}.
\end{aligned}$$

This shows that $1, \theta, \ldots, \theta^{n-1}$ is a non zero solution of the following homogeneous system of equations:

$$\begin{aligned}
(a_k - \gamma_k)x_0 + a_{k-1}x_1 + \cdots + a_0x_k &= 0 \\
(a_k - \gamma_k)x_1 + a_{k-1}x_2 + \cdots + a_0x_{k+1} &= 0 \\
\cdots\cdots\cdots\cdots\cdots\cdots& \\
(a_k - \gamma_k)x_{n-k-1} + a_{k-1}x_{n-k} + \cdots + a_0x_{n-1} &= 0 \\
\gamma_kx_{n-k} + a_{k+1}x_{n-k-1} + \cdots + a_nx_0 &= 0 \\
\cdots\cdots\cdots\cdots\cdots\cdots& \\
\gamma_kx_{n-1} + a_{k+1}x_{n-2} + \cdots + a_nx_{k-1} &= 0.
\end{aligned}$$

The determinant of this system must be zero. Therefore γ_k is a root of a polynomial with coefficients in A whose leading coefficient is $(-1)^k$, hence γ_k is integral over A.

(2) It is easy to verify that $\gamma_{k+1} = \theta\gamma_k + a_{k+1}$, hence $\theta\gamma_k = \gamma_{k+1} - a_{k+1}$. It follows, by **(1)**, that $\theta\gamma_k$ is integral over A.

(3) Clearly B is an additive group. For completing the proof it is sufficient to show that $\gamma_i\gamma_j \in B$ for all $i, j \in \{1, \ldots, n-1\}$. We use a formula relating γ_i with γ_{i-1} proved in **(2)**. We obtain

$$\gamma_i\gamma_j = (\theta\gamma_{i-1} + a_i)\gamma_j = \gamma_{i-1}(\gamma_{j+1} - a_{j+1}) + a_i\gamma_j$$

and the conclusion follows by induction on i.

Exercise 10.38.
Let A be an integral domain, K be its fraction field and E be an extension of K. Let $f(x) \in K[x]$ be monic and non constant such that if the characteristic p is positive, then $p \nmid \deg f$. Suppose that $f(x) = g(h(x))$, where $g(x)$ and $h(x)$ are monic non constant polynomials with coefficients in E such that $h(0) = 0$.

(1) Show that $g(x)$ and $h(x) \in K[x]$.
(2) If $f(x) \in A[x]$, show that the coefficients of $g(x)$ and $h(x)$ belong to the integral closure of A in K.
(3) Let $f_1(x) \in K[x]$. Suppose that $f_1(x) = g_1(h_1(x))$, where $g_1(x)$ and $h_1(x)$ are non constant polynomials with coefficients in K. Show that there exist $g(x), h(x) \in K[x]$, non constant such that, h is monic, $h(0) = 0$, $\deg g_1 = \deg g$, $\deg h_1 = \deg h$ and $f_1(x) = g(h(x))$.
(4) Show that the conclusion in **(1)** does not (necessarily) hold if we omit the condition on the characteristic of K.

Solution 10.38.

(1) Let $m = \deg g$, $n = \deg h$,

$$g(x) = x^m + b_1x^{m-1} + \cdots + b_m,$$
$$h(x) = x^n + c_1x^{n-1} + \cdots + c_{n-1}x \quad \text{and}$$
$$f(x) = x^{mn} + a_1x^{mn-1} + \cdots + a_{mn},$$

where $a_1, \ldots, a_{mn} \in K$ and $b_1, \ldots, b_m, c_1, \ldots, c_{n-1} \in E$. Identifying the coefficients of $x^{mn-1}, \ldots, x^{mn-(n-1)}$ in the identity $f(x) = g(h(x))$, we obtain the following equations:

$$a_1 = mc_1 \quad \text{and}$$
$$a_i = mc_i + F_i(c_1, \ldots, c_{i-1}) \quad \text{for} \quad i = 2, \ldots, n-1,$$

where $F_i \in \mathbb{Z}[x_1, ..., x_{i-1}]$ is homogeneous of degree i. Since $p \nmid degf$, then $p \nmid m$ and $c_1 = a_1/m \in K$.
By induction, we conclude that $c_1, c_2, \ldots, c_{n-1} \in K$. We next identify in the same identity the coefficients of $x^{mn-n}, \ldots, x^{mn-nm}$ and we obtain the following equations:

$$b_1 + G_1(c_1, \ldots, c_{n-1}) = a_n$$
$$b_i + G_i(b_1, \ldots, b_{i-1}, c_1, \ldots, c_{n-1}) = a_{ni} \quad \text{for} \quad i = 2, \ldots, m,$$

where $G_i \in \mathbb{Z}[y_1, \ldots, y_{i-1}, x_1, \ldots, x_{n-1}]$ is homogeneous of degree $m-i$. By induction we see that $b_i \in K$ for $i = 1, \ldots, m$. Therefore $g(x)$ and $h(x) \in K[x]$.

(2) Let $\alpha_1, \ldots, \alpha_m$ be the roots of $g(x)$ is an algebraic closure of K, then $f(x) = (h(x) - \alpha_1) \cdots (h(x) - \alpha_m)$. Let $i \in \{1, \ldots, m\}$ and β be any root of $h_i(x) = h(x) - \alpha_i$, then β is a root of $f(x)$. Since $f(x) \in A[x]$, we conclude that β is integral over A. Therefore the coefficients of $h_i(x)$ are integral over A. By **(1)**, $h(x) \in K[x]$, hence the coefficients of $h(x)$ belong to the integral closure of A in K. Now $\alpha_1, \ldots, \alpha_m$ are integral over A, hence all the coefficients of $g(x)$ are integral over A. By **(1)** these coefficients are in K, hence belong to the integral closure of A in K.

(3) Let $a, b \in E$ such that the polynomial $h(x) = ah(x) + b$ is monic and has a zero constant coefficient. Let $g(x) = g_1(\frac{x-b}{a})$, then these polynomials g and h satisfy the required conditions.

(4) Let $A = K = \mathbb{F}_2$ and let α be a root of $u(x) = x^2 + x + 1$ in an algebraic closure of $\mathbb{F}_2$. Let $f(x) = x^4 + x \in \mathbb{F}_2[x]$, then $f(x) = g(h(x))$, with

$$g(x) = x^2 + (\alpha + 1)x \quad \text{and} \quad h(x) = x^2 + \alpha x.$$

These polynomials have their coefficients in $\mathbb{F}_2(\alpha)$ but not in $\mathbb{F}_2$. Notice that $f(x) = g_1(h_1(x))$, where $g_1(x) = h_1(x) = x^2 + x$.

Exercise 10.39.

(1) Let L be a field, Ω be an algebraic closure of L, $(a_{ij})_{(i,j)}, (b_{ij})_{(i,j)}$ be two sets of indeterminates with $0 \le i < r, 0 \le j < r$ and $i + j > r$. Let $A = L[(a_{ij}), (b_{ij})]$. Set $f((a_{ij}), x, y) = \sum_{(i,j)} a_{ij} x^i y^j$ and $g((b_{ij}), x, y) = \sum b_{ij} x^i y^j$. Let $\phi(x, y) \in L[x, y]$ such that $\deg \phi = r$. Let I be the ideal of A generated by the finite set B formed by the coefficients of the polynomial $f(x, y)g(x, y) - \phi(x, y) \in A[x, y]$. Show that the following conditions are equivalent.

   (i) $\phi(x, y)$ is irreducible over Ω.
   (ii) I has no zero in Ω.
   (iii) $I = A$.

(2) Let L be a number field, $\mathcal{O}$ be its ring of integers, $\phi(x, y) \in L[x, y]$ absolutely irreducible. Show that for almost all prime ideals $\mathcal{P}$ of $\mathcal{O}$, $\phi(x, y) \pmod{\mathcal{P}}$ is irreducible over an algebraic closure of $\mathcal{O}/\mathcal{P}$.

Solution 10.39.

(1) $\phi(x, y)$ is irreducible over Ω if and only if for any families $(a^*_{ij})_{(i,j)}$ and $(b^*_{ij})_{(i,j)}$ of elements of Ω, $\phi(x, y) - f((a^*_{ij}), x, y)g((a^*_{ij}), x, y) \neq 0$ if and only if (ii).

(*ii*) and (*iii*) are equivalent by Hilbert's Nullstellensatz [Lang (1965), Chap. 10.2].

(2) Let $r = \deg \phi(x, y)$ and let A, B, I be the sets defined in **(1)**. Since ϕ is absolutely irreducible then $I = A$ (by **iii**), hence there exist a positive integer k, $\lambda_1, \dots, \lambda_k \in A$ and $\mu_1, \dots, \mu_k \in B$ such that $1 = \sum_{i=1}^{k} \lambda_i \mu_i$. Any of these λ_i and any of these μ_i has the form $\frac{u((a_{ij}),(b_{ij}))}{d}$, where d is a positive integer and $u((a_{ij}), (b_{ij}))$ is a polynomial in the variables a_{ij}, b_{ij} with coefficients in $\mathcal{O}$. We first exclude the prime ideals of A which divides some denominator of a coefficient of ϕ. Exclude also the primes $\mathcal{P}$ for which $\deg \bar{\phi} < r$, where $\bar{\phi}$ is the reduced polynomial of ϕ modulo $\mathcal{P}$. Finally exclude the primes $\mathcal{P}$ which divide any denominator d of λ_i or μ_i in the preceding representation. Clearly the excluded primes are finite in number. For any other prime $\mathcal{P}$, we define $\bar{A}, \bar{B}, \bar{I}$ similarly as A, B, I but now for the field $\mathcal{O}/\mathcal{P}$ and for the polynomial $\bar{\phi}(x, y)$ of degree r. If we reduce modulo $\mathcal{P}$ the above Bezout's identity, we obtain $\bar{1} = \sum_{i=1}^{k} \bar{\lambda}_i \bar{\mu}_i$, hence $\bar{I} = \bar{A}$. Therefore $\bar{\phi}$ is absolutely irreducible accordingly to **(1)**.

Exercise 10.40.

Let $f(x) = x^n + \cdots + a_1 x + a_0 \in \mathbb{Z}[x]$ be irreducible, θ be a root of f and $K = \mathbb{Q}(\theta)$. Let k be a positive integer with $k < n$ and p be a prime number such that $p^k \parallel a_0$ and $p^{k+1-i} \mid a_i$ for $i = 1, \dots, k$. Show that p ramifies in K.

Solution 10.40.

Suppose that p does not ramify in K, then $pA = \mathcal{P}_1 \cdots \mathcal{P}_r$, where A is the ring of integers of K and $\mathcal{P}_1, \dots, \mathcal{P}_r$ are distinct prime ideals of A. We have $a_0 A = \mathcal{P}_1^k \cdots \mathcal{P}_r^k I$, where I is an ideal of A whose norm is coprime with p. Since $N_{K/\mathbb{Q}}(\theta) = \pm a_0$, then the ideal θA is divisible by at least one $\mathcal{P}_i$ say $\mathcal{P}$. As $\mathcal{P}^{k+1-i} \mid a_i$ for $i = 1, \dots, k$, we have

$$\begin{aligned} a_0 &= a_0 - f(\theta) \\ &= (-a_1\theta - \dots - a_k\theta^k) - (a_{k+1}\theta^{k+1} + \cdots + a_n\theta^n) \\ &\equiv 0 \pmod{\mathcal{P}^{k+1}}, \end{aligned}$$

contradicting $p^k \parallel a_0$. Thus p ramifies in K.

Exercise 10.41.

If A and B are sets we write $A \underset{\sim}{\subset} B$ (resp. $A \underset{\sim}{=} B$) to mean that all the elements of A, but finitely of them, are elements of B (resp. A and B have the same elements except a finite number of elements). For any non constant $F(x) \in \mathbb{Z}[x]$ of degree m define $D(F)$ and $D_i(F)$ for $i = 1, \ldots, n$ by

$$D(F) = \{p \text{ prime, there exists } x \in \mathbb{Z}, \quad F(x) \equiv 0 \pmod p\}$$
$$D_i(F) = \{p \text{ prime, the equation } F(x) \equiv 0 \pmod p \text{ has exactly } i \text{ solutions}\}.$$

(1) Show that $D(F)$ is infinite.

(2) Recall the following result from [Marcus (1977), Th. 13, Chap. 1 and Exercise 40, Chap. 1]. Let K be a number field of degree m, A be its ring of integers and $\gamma \in A$ be a primitive element over $\mathbb{Q}$. Then A has a basis over $\mathbb{Z}$ of the form

$$\{\omega_0 = 1, \omega_1 = u_1(\gamma)/d_1, \ldots, \omega_{m-1} = u_{m-1}(\gamma)/d_{m-1}\},$$

where $u_i(x) \in \mathbb{Z}[x]$ and d_i is a positive integer for $i = 1, \ldots, m-1$. Moreover $d_1 \mid d_2 \mid \cdots \mid d_{m-1}$ and $\mathrm{I}(\gamma) = d_1 d_2 \cdots d_{m-1}$. This shows that any element $\phi \in A$ may be expressed in the form $\phi = A(\gamma)/d$, with $A[x] \in \mathbb{Z}[x]$ and d is a positive divisor of $\mathrm{I}(\gamma)$.
Let $f(x)$ and $g(x)$ be non constant polynomials with integral coefficients. Suppose that f and g are monic and irreducible over $\mathbb{Q}$. Let α and β be roots of $f(x)$ and $g(x)$ respectively in $\mathbb{C}$. If $\mathbb{Q}(\alpha) \subset \mathbb{Q}(\beta)$, show that if $p \in D(g)$ and $p \nmid I(\beta)$, then $p \in D(f)$. Deduce that $D(g) \overset{\subset}{\sim} D(f)$. Show that the inclusion may be strict.

(3) Let $f(x)$ and $g(x)$ be non constant polynomials with integral coefficients. Suppose that f and g are monic and irreducible over $\mathbb{Q}$. Let α and β be roots of $f(x)$ and $g(x)$ respectively in $\mathbb{C}$. If $\mathbb{Q}(\alpha) = \mathbb{Q}(\beta)$, show that if $p \nmid \mathrm{I}(\alpha)\,\mathrm{I}(\beta)$, then

$$p \in D(g) \Leftrightarrow p \in D(f) \quad \text{and}$$
$$p \in D_i(g) \Leftrightarrow p \in D_i(f),$$

for $i = 1, \ldots, n$. Deduce that $D(f) \overset{=}{\sim} D(g)$ and $D_i(f) \overset{=}{\sim} D_i(g)$ for $i = 1, \ldots, n$.

(4) Suppose that f is monic irreducible of degree n and let α be one of its roots. Suppose that $\mathbb{Q}(\alpha)/\mathbb{Q}$ is normal. If p is a prime number such that $p \nmid \mathrm{Disc}(f)$, show that $p \in D(f) \Leftrightarrow p \in D_n(f)$. Deduce that $D(f) \overset{=}{\sim} D_n(f)$.
Show that if γ is a root of $F(x) = x^4 + x^2 + x - 2$, then $\mathbb{Q}(\gamma)/\mathbb{Q}$ is not normal.

(5) Let $f(x)$ and $g(x)$ be non constant polynomials with integral coefficients. Suppose that f and g are monic and irreducible over $\mathbb{Q}$. Let α and β be roots of $f(x)$ and $g(x)$ respectively in $\mathbb{C}$. Suppose that $\mathbb{Q}(\beta)/\mathbb{Q}$

is normal and that $\mathbb{Q}(\alpha) \subset \mathbb{Q}(\beta)$. Let $n = \deg f$. Let p be a prime number such that $p \nmid \mathrm{I}(\beta)\,\mathrm{Disc}(f)$. Show that $p \in D(g) \Rightarrow p \in D_n(f)$ and that $D(g) \underset{\sim}{\subset} D_n(f)$.

Deduce that if $h(x)$ is a monic irreducible polynomial over $\mathbb{Q}$ of degree n, then $D_n(h)$ is infinite.

(6) Let $K/\mathbb{Q}$ be a number field of degree n, L be a normal closure of K. Let α and β be primitive elements of K and L respectively which are supposed to be algebraic integers. Let $f(x) = \mathrm{Irr}(\alpha, \mathbb{Q})$ and $g(x) = \mathrm{Irr}(\beta, \mathbb{Q})$. Let p be a prime number such that $p \nmid I(\beta)Disc(f)$. Show $p \in D(g) \Leftrightarrow p \in D_n(f)$ and that $D(g) \underset{\sim}{=} D_n(f)$.

(7) Let $f_1(x) = x^3 - 18x - 6$, $f_2(x) = x^3 - 36x - 78$, $f_3(x) = x^3 - 54x - 150$ and let α_1, α_2, α_3 be a root of f_1, f_2, f_3 respectively. Show that these polynomials are irreducible over $\mathbb{Q}$ and $\mathrm{Disc}(f_1) = \mathrm{Disc}(f_2) = \mathrm{Disc}(f_3) = 2^2.3^5.23$. By looking at the primes 5 and 11, show that the fields $\mathbb{Q}(\alpha_1)$, $\mathbb{Q}(\alpha_2)$ and $\mathbb{Q}(\alpha_3)$ are distinct.

Solution 10.41.

(1) Let $m = \deg F$ and set $F(x) = b_m x^m + b_{m-1}x^{m-1} + \cdots + b_0$, where $b_i \in \mathbb{Z}$ for any $i \in \{0, \ldots, m\}$. If $b_0 = 0$, then the result is clear. Suppose that $b_0 \neq 0$. There exists $x_0 \in \mathbb{Z}$ such that $f(x_0) \neq 1$ and $f(x_0) \neq -1$, hence $D(F) \neq \emptyset$ since it contains a prime divisor of $F(x_0)$. Suppose that this set is finite and let $p_1, \ldots, p_r$ be its elements and set $a = p_1 \cdots p_r$. We have

$$\begin{aligned} F(acx) &= b_m(ab_0x)^m + b_{m-1}(ab_0x)^{m-1} + \cdots + b_1(ab_0x) + b_0 \\ &= b_0\big(a^m b_0^{m-1} b_m x^m + \cdots + b_1 ax + 1\big) \\ &:= b_0 G(x), \end{aligned}$$

hence $D(G) \subset D(F)$. For the same reason as for F, we have $D(G) \neq \emptyset$. Let $p \in D(G)$, then $p \mid a$ and there $x \in \mathbb{Z}$ such that

$$p \mid a^m b_0^{m-1} b_m x^m + \cdots + b_1 ax + 1,$$

hence $p \mid 1$ which is a contradiction. Therefore $D(F)$ is infinite.

(2) Let $n = \deg f$. The element α may be written in the form $\alpha = u(\beta)/d$, where $u(x) \in \mathbb{Z}[x]$ and d is a positive integer dividing $I(\beta)$ and such that d and the coefficients of $u(x)$ are coprime. Since $g(x) \mid f(u(x)/d)$ in $\mathbb{Q}[x]$, there exists $q(x) \in \mathbb{Z}[x]$ such that

$$f(u(x)/d) = g(x)q(x)/d^n. \qquad \text{(Eq 1)}$$

From this identity it is seen that for any prime $p \nmid d$, we have $p \in D(g) \Rightarrow p \in D(f)$, hence $D(g) \lesssim D(f)$.
Suppose that $f(x) = x^2 + 9$ and $g(x) = x^2 + 1$. Let $\alpha = 3i$ and $\beta = i$, then obviously $\mathbb{Q}(\alpha) \subset \mathbb{Q}(\alpha)$. We have $f(0) \equiv 0 \pmod 3$, so that $3 \in D(f)$. On the other hand, we have $g(0) \equiv 1 \pmod 3$ and $g(\pm 1) \equiv -1 \pmod 3$, hence $3 \notin D(g)$. this proves that the inclusion proved above may be strict.

(3) Our assumptions allow us to adjoin to the preceding equation the following one

$$g\left(\frac{v(x)}{d'}\right) = \frac{f(x)h(x)}{(d')^n}, \qquad \text{(Eq 2)}$$

where $v(x)$ and $h(x) \in \mathbb{Z}[x]$. We have already proved that if $p \nmid \mathrm{I}(\beta)$ then $p \in D(g) \Rightarrow p \in D(f)$, thus $D(g) \lesssim D(f)$. Using **(2)**, we obtain similar conclusions when permuting α and β, hence $D(g) \eqsim D(f)$.
Let $p \in D_i(g)$ such that $p \nmid d$ and let $x_1, \ldots, x_i$ be the distinct zeros of $g(x)$ modulo p. By **(Eq 1)**, it is seen that $u(x_i)/d$ is a root of $f(x)$ modulo p. We show that if moreover $p \nmid d'$, these roots are distinct modulo p. We have $\frac{1}{d'}v(u(\beta)/d) = \beta$, hence

$$\frac{1}{d'}v(u(x)/d) = x + g(x)a(x)/(d'd^e), \qquad \text{(Eq 3)}$$

where $a(x) \in \mathbb{Z}[x]$ and e is a non negative integer. Suppose that $u(x_i)/d \equiv u(x_j)/d \pmod p$, then **(Eq 3)** shows that $x_i \equiv x_j \pmod p$. Therefore the number of distinct roots of $f(x)$ modulo p is at least equal to i. Inverting the roles of $f(x)$ and $g(x)$, we obtain $D_i(f) \eqsim D_i(g)$ for $i = 1, \ldots, n$.

(4) The inclusion $D_n(f) \lesssim D(f)$ is trivial. We show the reverse inclusion. Set $f(x) = x^n + a_{n-1}x^{n-1} + \cdots + a_0$, with $a_0, \ldots, a_{n-1} \in \mathbb{Z}$ and let $\alpha_1 = \alpha, \alpha_2 = u_2(\alpha)/d, \ldots, \alpha_n = u_n(\alpha)/d$, where d is a positive integer dividing $\mathrm{I}(\alpha)$ and $u_2(x), \ldots, u_n(x) \in \mathbb{Z}[x]$. Let

$$\begin{aligned} F(x,y) &= (y-x)\left(y - \frac{u_2(x)}{d}\right)\cdots\left(y - \frac{u_{n-1}(x)}{d}\right) \\ &= y^n + \frac{A_{n-1}(x)}{d}y^{n-1} + \cdots + \frac{A_0(x)}{d^{n-1}}, \qquad \text{(Eq 4)} \end{aligned}$$

then $A_i(x) \in \mathbb{Z}[x]$ for $i = 0, \ldots, n-1$ and

$$F(\alpha, y) = f(y) = y^n + \frac{A_{n-1}(\alpha)}{d}y^{n-1} + \cdots + \frac{A_0(\alpha)}{d^{n-1}},$$

hence $A_j(\alpha)/d^{n-j} = a_j$ for $j = 0, \ldots, n-1$. Therefore $f(x)$ divides $A_j(x) - d^{n-j}a_j$ in $\mathbb{Q}[x]$ for $j = 0, \ldots, n-1$. Let $q_j(x) \in \mathbb{Q}[x]$ such that

$A_j(x) - d^{n-j}a_j = f(x)q_j(x)$. Since $f(x)$ is monic then $q_j(x) \in \mathbb{Z}[x]$ for $j = 0, \ldots, n-1$. We have

$$\begin{aligned} F(x,y) &= y^n + \left(a_{n-1} + f(x)\frac{q_{n-1}(x)}{d}\right)y^{n-1} + \cdots + \left(a_0 + f(x)\frac{q_0(x)}{d^{n-1}}\right) \\ &= f(y) + f(x)\frac{G(x,y)}{d^{n-1}}, \qquad \text{(Eq 5)} \end{aligned}$$

with $G(x,y) \in \mathbb{Z}[x,y]$. Let $p \in D(f)$, then there exists $a \in \mathbb{Z}$ such that $f(a) \equiv 0 \pmod p$. If $p \nmid d$, then by By **(Eq 4)** and **(Eq 5)**, we have

$$F(a,y) = (y-a)(y-u_2(a)/d)\cdots(y-u_{n-1}(a))/d \equiv f(y) \pmod p.$$

Since $p \nmid \operatorname{Disc}(f)$, then $a, u_2(a)/d, \ldots, u_{n-1}(a)/d$ are distinct modulo p. Therefore $p \in D_n(f)$.
We apply the result for $F(x) = x^4 + x^2 + x - 2$. We omit the proof that $F(x)$ is irreducible over $\mathbb{Q}$. We have $F(x) \equiv x(x^3+x+1) \pmod 2$ and the factor $x^3 + x + 1$ is irreducible modulo 2. This shows that $F(x)$ is separable modulo 2, thus $2 \nmid \operatorname{Disc}(F)$. Obviously $2 \in D(F)$ but $2 \notin D_4(F)$. We deduce that $\mathbb{Q}(\gamma)/\mathbb{Q}$ is not normal.

(5) Let $p \in D(g)$ such that $p \nmid \operatorname{Disc}(f)\,\mathrm{I}(\beta)$. Set

$$f(x) = x^n + a_{n-1}x^{n-1} + \cdots + a_0.$$

Let $\alpha_1, \ldots \alpha_n$ be the roots of f which are in $\mathbb{Q}(\beta)$. Let $u_i(x) \in \mathbb{Z}[x]$ for $i = 1, \ldots, n$ such that $\alpha_i = u_i(\beta)/d$ where d is a divisor of $I(\beta)$. Let

$$F(x,y) = \prod_{i=1}^{n}(y - u_i(x)/d) = y^n + (B_{n-1}(x)/d)y^{n-1} + \cdots + B_0(x)/d^n.$$

Then

$$F(\beta,y) = \prod_{i=1}^{n}(y-\alpha_i) = f(y) = y^n + (B_{n-1}(\beta)/d)y^{n-1} + \cdots + B_0(\beta)/d^n.$$

It follows that $B_j(\beta)/d^{n-j} - a_j = 0$ for $j = 0, \ldots, n-1$. Therefore $g(x) \mid B_j(x) - a_j d^{n-j}$ in $\mathbb{Z}[x]$. Set $B_j(x) - a_j d^{n-j} = g(x)q_j(x)$ with $q_j(x) \in \mathbb{Z}[x]$. Then

$$\prod_{i=1}^{n}(y - u_i(x)/d) = f(y) + g(x)H(x,y)/d^n,$$

where $H(x,y) \in \mathbb{Z}[x,y]$. Let $x_0 \in \mathbb{Z}$ such that $g(x_0) \equiv 0 \pmod p$. Since $p \nmid d$ and $p \nmid \operatorname{Disc}(f)$ then $f(y)$ factorizes into distinct linear factors modulo p, which implies that $p \in D_n(f)$. We deduce that $D(g) \subsetneqq D_n(f)$.

Let α be a root of h, L be a Galois closure of $\mathbb{Q}(\alpha)$ over $\mathbb{Q}$, β be a primitive element of L and $g(x)$ be the minimal polynomial of β. By the first part of **(5)**, we have $D(g) \lesssim D_n(h)$. Since $D(g)$ is infinite by **(1)**, then so is $D_n(h)$.

(6) By **(5)**, we have $D(g) \lesssim D_n(f)$. We prove the reverse inclusion. Let $p \in D_n(f)$ and let $\alpha_1, \dots \alpha_n$ be the roots of f, then

$$f(x) = \prod_{i=1}^{n}(x - \alpha_i) \equiv \prod_{i=1}^{n}(x - a_i) \pmod p,$$

where the a_i are distinct integers. We deduce that $\sigma_j(\alpha_1, \dots, \alpha_n) \equiv \sigma_j(a_1, \dots, a_n) \pmod p$, where σ_j is the elementary symmetric polynomial of degree j. The primitive element theorem asserts that there exists a primitive element of $L/\mathbb{Q}$ of the form $\sum_{i=1}^{n} c_i\alpha_i$, where $c_i \in \mathbb{Z}$ for $i = 1, \dots, n$. By **(3)**, we may suppose that $\beta = \sum_{i=1}^{n} c_i\alpha_i$. Consider the polynomials

$$\phi(x) = \prod\left(x - \sum_{i=1}^{n} c_i\alpha_{\tau(i)}\right) = \sum d_k x^k \quad \text{and}$$

$$\psi(x) = \prod\left(x - \sum_{i=1}^{n} c_i a_{\tau(i)}\right) = \sum \delta_k x^k,$$

where the products run over the permutations τ of the set $\{1, \dots, n\}$. Since d_k is a symmetric polynomial of $\alpha_1, \dots, \alpha_n$ and δ_k is a symmetric polynomial of $a_1, \dots, a_n$, then $\delta_k \equiv d_k \pmod p$. Therefore $\phi(x) \equiv \psi(x) \pmod p$. Since $\phi(\beta) = 0$, then $g(x) \mid \phi(x)$ in $\mathbb{Z}[x]$, hence $\overline{g(x)} \mid \overline{\phi(x)}$, where $\overline{g(x)}$ (resp. $\overline{\phi(x)}$) denotes the reduced polynomial modulo p of g (resp. ϕ). Therefore $\overline{g(x)} \mid \overline{\psi(x)}$. Since $\overline{\psi(x)}$ is a product of linear factors (distinct or not), so is $\overline{g(x)}$, which implies $p \in D(g)$.

(7) Eisenstein's irreducibility criterion for the prime $p = 2$ applies for the three polynomials, hence they are irreducible. It is known that the discriminant of a polynomial $P(x) = x^3 + ax + b$ is given by $\mathrm{Disc}(P) = -4a^3 - 27b^2$. Using this formula, we obtain

$$\mathrm{Disc}(f_1) = \mathrm{Disc}(f_2) = \mathrm{Disc}(f_3) = 2^2.3^5.23.$$

We have $f_1(x) \equiv (x+1)(x+2)(x-3) \pmod{11}$ and both f_2, f_3 are irreducible modulo 11, hence $11 \in D(f_1)$, $11 \notin D(f_2)$, $11 \notin D(f_3)$. Therefore $D(f_1) \not\subset D(f_2)$ and $D(f_1) \not\subset D(f_3)$. On the other hand, we have $f_3(x) \equiv x(x-2)(x+2) \pmod 5$ and $f_2(x)$ is irreducible modulo 5, hence $5 \in D(f_3)$ and $5 \notin D(f_2)$. Therefore $D(f_3) \not\subset D(f_2)$ and the proof is complete by **(3)**.

Exercise 10.42.
Let $\theta_1, \theta_2, \theta_3$ be the roots in $\mathbb{R}$ of $f(x) = x^3 - 3x + 1$ such that $\theta_1 < \theta_2 < \theta_3$.

(1) Show that $\theta_1 = -1.879...$, $\theta_2 = 0.347...$ and $\theta_3 = 1.529....$
(2) Show that $\theta_2 = \theta_1^2 - 2$ and $\theta_3 = -\theta_1^2 - \theta_1 + 2$.
(3) Let $\alpha_i = \theta_i^2$ for $i = 1, 2, 3$, $a = \theta_1\theta_2\alpha_3 + \theta_2\theta_3\alpha_1 + \theta_3\theta_1\alpha_2$ and $b = \theta_1\theta_2\alpha_3 + \theta_2\theta_3\alpha_2 + \theta_3\theta_1\alpha_1$. Show that among a and b one and only one of them is a rational integer. Compute this integer.

Solution 10.42.

(1) Using a simple calculator, we find

$$f(-1.879) = 0.002925561 > 0, \quad f(-1.88 = -0.004672) < 0,$$
$$f(0.347) = 0.000781923 > 0, \quad f(0.348) = -0.001855808 < 0,$$
$$f(1.529) = -0.012441111 < 0 \quad \text{and} \quad f(1.53) = 0.08423 > 0,$$

hence $\theta_1 = -1.879...$, $\theta_2 = 0.347...$ and $\theta_3 = 1.529....$

(2) Since $f(x)$ is irreducible over $\mathbb{F}_2$, then it is irreducible over $\mathbb{Q}$. Set $\theta_1 = \theta$. Using the identities

$$\theta^3 = 3\theta - 1, \quad \theta^4 = 3\theta^2 - \theta,$$
$$\theta^5 = -\theta^2 + 9\theta - 3 \quad \text{and} \quad \theta^6 = 9\theta^2 - 6\theta + 1,$$

one verifies easily that $f(\theta^2 - 2) = f(-\theta^2 - \theta + 2) = 0$. Obviously θ_1, θ_2 and θ_3 are distinct. Therefore they constitute the complete set of roots of $f(x)$. An alternative way to prove the same assertions is to show that $f(x) \mid f(x^2 - 2)$. In that case it is obvious that $\theta_2 - 2$ is a root of $f(x)$. Moreover the divisibility property of polynomials shows also that $f(\theta_2^2 - 2) = 0$. Straightforward computations shows that $\theta_2^2 - 2 = -\theta^2 - \theta + 2$.

(3) • **First proof.** We have

$$\theta_1^2 = \theta^2$$
$$\theta_2^2 = (\theta^2 - 2)^2 = -\theta^2 - \theta + 4$$
$$\theta_3^2 = (-\theta^2 - \theta + 2)^2 = \theta + 2.$$

Using these relations and also the expressions of θ^3, θ^4, θ^5 and θ^6 as polynomials in θ of degree at most 2, we find $a = 0$ and $b = 3(\theta^2 - 4)$. Thus a is a rational integer and b is not.

• **Second proof.** Using the approximate values of θ_1, θ_2 and θ_3 respectively, we get $a = 0.0029...$ and $b = -11.6039....$ Therefore we may conclude that b is not an integer. The approximate value of

a does not allow us to conclude that a is an integer. From **(2)**, we see that $\mathbb{Q}(\theta)$ is a cyclic extension of $\mathbb{Q}$ of degree 3, whose Galois group is generated by the automorphism σ such that $\sigma(\theta) = \theta_2 = \theta^2 - 2$. It is clear that a is an algebraic integer. Since

$$\sigma(a) = \sigma(\theta_1)\sigma(\theta_2)\sigma(\alpha_3) + \sigma(\theta_2)\sigma(\theta_3)\sigma(\alpha_1) + \sigma(\theta_3)\sigma(\theta_1)\sigma(\alpha_2) = a,$$

then $a \in \mathbb{Q}$. It follows that a is a rational integer and its approximate value shows that $a = 0$.

Exercise 10.43.
Let $f(x) = x^3 - 3x + 1$, θ be a root of $f(x)$ in $\mathbb{C}$, $K = \mathbb{Q}(\theta)$ and A be the ring of integers of K.

(1) Show that $f(x)$ is irreducible over $\mathbb{Q}$. Show that the roots of $f(x)$ are given by $\theta_1 = \theta$, $\theta_2 = \theta^2 - 2$ and $\theta_3 = -\theta^2 - \theta + 2$.
(2) Show that $\mathrm{Disc}(f) = 3^4$ and that $3 \nmid I(\theta)$. Deduce that $A = \mathbb{Z}[\theta]$ and $\mathrm{Disc}(K) = 3^4$.
(3) Show that $2A$ is prime, $3A = \mathcal{P}^3$, where $\mathcal{P}$ is a prime ideal of A with residual degree equal to 1. Show that $\mathcal{P} = (\theta + 1)A$. Show that $\mathcal{D}_K = \mathcal{P}^4 = (3\theta + 3)A$.
(4) Show that $17A = \mathcal{P}_1\mathcal{P}_2\mathcal{P}_3$, where $\mathcal{P}_1$, $\mathcal{P}_2$ and $\mathcal{P}_3$ are prime ideals of A of residual degree equal to 1. Show that $\mathcal{P}_1 = (\theta+3)A$, $\mathcal{P}_2 = (\theta^2+1)A$ and $\mathcal{P}_3 = (-\theta^2 - \theta + 5)A$.
(5) Compute a basis over $\mathbb{Z}$ of $\mathcal{D}_K$.

Solution 10.43.

(1) Since $f(x)$ is irreducible over $\mathbb{F}_2$, then it is irreducible over $\mathbb{Q}$. Using the identities

$$\theta^3 = 3\theta - 1,\ \theta^4 = 3\theta^2 - \theta,\ \theta^5 = -\theta^2 + 9\theta - 3 \text{ and } \theta^6 = 9\theta^2 - 6\theta + 1,$$

one verifies easily that $f(\theta^2 - 2) = f(-\theta^2 - \theta + 2) = 0$. Obviously θ_1, θ_2 and θ_3 are distinct. Therefore they constitute the complete set of roots of $f(x)$. An alternative way to prove the same assertions is to show that $f(x) \mid f(x^2 - 2)$. In that case it is obvious that $\theta_2 - 2$ is a root of $f(x)$. Moreover the divisibility property of polynomials shows also that $f(\theta_2^2 - 2) = 0$. Straightforward computations shows that $\theta_2^2 - 2 = -\theta^2 - \theta + 2$.
(2) We have $\mathrm{Disc}(f) = -4(-3)^3 - 27 = 3^4$. We find that $f(x) \equiv (x+1)^3 \pmod 3$ and $f(x) = (x+1)^3 - 3x(x+2)$. Since $x + 1 \nmid x(x+2)$ in $\mathbb{F}_3[x]$, then, by **Exercise 10.30** $3 \nmid I(\theta)$. It follows that $A = \mathbb{Z}[\theta]$ and $\mathrm{Disc}(K) = 3^4$.

(3) Since $f(x) \equiv x^3 + x + 1 \pmod 2$ and since this last polynomial is irreducible over $\mathbb{F}_2$, then by Dedekind's Theorem [Marcus (1977), Th. 27, Chap. 3], $2A$ is prime, that is the prime 2 is inert in A.
Here also, for the prime 3, Dedekind's Theorem applies since $3 \nmid I(\theta)$ and we conclude that $3A = \mathcal{P}^3$, where $\mathcal{P}$ is a prime ideal of A with residual degree equal to 1. Moreover $\mathcal{P} = (3, \theta + 1)$. We compute the norm of $\theta + 1$. Let $\beta = \theta + 1$, then β is a root of

$$g(x) = (x-1)^3 - 3(x-1) + 1 = x^3 - 3x^2 + 3.$$

We conclude that $\mathrm{N}_{K/\mathbb{Q}}(\theta + 1) = -3$. It follows that $\mathcal{P} = (\theta + 1)A$. Since $A = \mathbb{Z}[\theta]$ and since $\theta - 1$ is a unit, then

$$\mathcal{D}_K = f'(\theta)A = (3\theta^2 - 3)A = 3(\theta-1)(\theta+1)A = 3(\theta+1)A = \mathcal{P}^4.$$

(4) We have $f(x) \equiv (x+3)(x+4)(x-7) \pmod{17}$, hence $17A = \mathcal{P}_1\mathcal{P}_2\mathcal{P}_3$, where $\mathcal{P}_1$, $\mathcal{P}_2$ and $\mathcal{P}_3$ are prime ideals of A of residual degree equal to 1. Moreover

$$\begin{aligned} \mathcal{P}_1 &= (17, \theta + 3), \\ \mathcal{P}_2 &= (17, \theta + 4) \quad \text{and} \\ \mathcal{P}_3 &= (17, \theta - 7). \end{aligned}$$

By the method used in **(3)**, we compute the norm of $\theta + 3$ and we find $\mathrm{N}_{K/\mathbb{Q}}(\theta+3) = 17$. Therefore $\mathcal{P}_1 = (\theta+3)A$. Since $\mathrm{Disc}(K)$ is a square, then K is a Galois extension of $\mathbb{Q}$. It follows that the prime ideals of A lying over $17\mathbb{Z}$ are conjugate. Thus $\mathcal{P}_2 = (\theta_2 + 3)A = (\theta^2 + 1)A$ and $\mathcal{P}_3 = (\theta_3 + 3)A = (-\theta^2 - \theta + 5)A$.

(5) Let

$$\mathbb{Z}[\alpha]^v = \{x \in K, \mathrm{Tr}_{K/\mathbb{Q}}(x\mathbb{Z}[\alpha]) \subset \mathbb{Z}\},$$

then $\mathcal{D} = (\mathbb{Z}[\alpha]^v)^{-1}$. We begin with the determination of a $\mathbb{Z}$-basis of $\mathbb{Z}[\alpha]^v$. Since

$$f(x) = x^3 - 3x + 1 = (x - \theta)(x^2 + \theta x + \theta^2 - 3),$$

then according to [Lang (1965), Prop. 1, Chap. 8.6], $\mathbb{Z}[\alpha]^v$ is generated over $\mathbb{Z}$ by $1/f'(\theta)$, $\theta/f'(\theta)$ and $(\theta^2-3)/f'(\theta)$. We have $1/f'(\theta) = 1/(3\theta^2 - 3) = (2\theta^2 + \theta - 4)/9$. It follows that

$$\begin{aligned} \mathbb{Z}[\alpha]^v &= \mathbb{Z}\frac{(2\theta^2+\theta-4)}{9} + \mathbb{Z}\frac{\theta(2\theta^2+\theta-4)}{9} + \mathbb{Z}\frac{(\theta^2-3)(2\theta^2+\theta-4)}{9} \\ &= \mathbb{Z}(2\theta^2+\theta-4)/9 + \mathbb{Z}(\theta^2+2\theta-2)/9 + \mathbb{Z}(-4\theta^2-2\theta+11)/9. \end{aligned}$$

Let $x = a + b\theta + c\theta^2 \in \mathbb{Z}[\theta]$, then $x \in \mathcal{D}$ if and only if

$$x\frac{(2\theta^2+\theta-4)}{9} \in \mathbb{Z}[\theta],$$
$$x\frac{(\theta^2+2\theta-2)}{9} \in \mathbb{Z}[\theta] \quad \text{and}$$
$$x\frac{(-4\theta^2-2\theta+11)}{9} \in \mathbb{Z}[\theta].$$

Easy calculations show that

$$x\frac{(2\theta^2+\theta-4)}{9} = \frac{u(\theta)}{9},$$
$$x\frac{(\theta^2+2\theta-2)}{9} = \frac{v(\theta)}{9} \quad \text{and}$$
$$x\frac{(-4\theta^2-2\theta+11)}{9} = \frac{w(\theta)}{9}$$

with

$$u(\theta) = \frac{(2a+b+2c)\theta^2+(a+2b+c)\theta-4a-2b-c}{9},$$
$$v(\theta) = \frac{(a+2b+c)\theta^2+(2a+b+5c)\theta-2a-b-2c}{9} \quad \text{and}$$
$$w(\theta) = \frac{(-4a-2b-11c)\theta^2+(-2a-11b-2c)\theta+a+4b+2c}{9}.$$

Therefore, the conditions for x to belong to $\mathcal{D}$ are given by

$$\begin{aligned}
2a+b+2c &\equiv 0 \pmod 9,\\
a+2b+c &\equiv 0 \pmod 9,\\
-4a-2b-c &\equiv 0 \pmod 9,\\
a+2b+c &\equiv 0 \pmod 9,\\
2a+b+5c &\equiv 0 \pmod 9,\\
-2a-b-2c &\equiv 0 \pmod 9,\\
-4a-2b-11c &\equiv 0 \pmod 9,\\
-2a-11b-2c &\equiv 0 \pmod 9 \quad \text{and}\\
a+4b+2c &\equiv 0 \pmod 9.
\end{aligned}$$

The second and the third equations imply that $3a \equiv 0 \pmod 9$, that is $a \equiv 0 \pmod 3$. Set $a = 3a_1$. The first two equations then become $b + 2c \equiv 3a_1 \pmod 9$ and $2b + c \equiv -3a_1 \pmod 9$. We deduce that

$c \equiv 0 \pmod 3$. Set $c = 3c_1$. Then $b \equiv 3(a_1 + c_1) \pmod 9$. Set $b = 3(a_1 + c_1) + 9b_1$, then

$$x = 3a_1 + (3(a_1+c_1)+9b_1)\theta + 3c_1\theta^2 = a_1(3+3\theta) + b_1(9\theta) + c_1(\theta + 3\theta^2).$$

This shows that $\mathcal{D}$ is generated over $\mathbb{Z}$ by $\{3 + 3\theta, 9\theta, \theta + 3\theta^2\}$. This set is obviously free over $\mathbb{Z}$, so it is a basis over $\mathbb{Z}$.

Exercise 10.44.
Let $p \equiv 1 \pmod 8$ be a prime number, $f(x) = x^4 - p$. Let $\theta \in \mathbb{C}$ be a root of $f(x)$, $K = \mathbb{Q}(\theta)$ and A be the ring of integers of K.

(1) Show that $\alpha = (\theta^2 + 1)/2$ and $\beta = (\theta^3 + \theta^2 + \theta + 1)/4$ are algebraic integers.
(2) Show that $\mathrm{Disc}(\theta) = -2^8p^3$.
(3) Show that $p \nmid I(\theta)$.
(4) Show that $2^3 \mid I(\theta)$.
(5) Deduce that $I(\theta) = 2^3$, $\mathrm{Disc}(K) = -2^2p^3$ and $\{1, \theta, \alpha, \beta\}$ is an integral basis of K.
(6) If $p \equiv 1 \pmod{16}$, show that the splitting of the prime 2 in A is given by $(2) = \mathcal{P}_1^2\mathcal{P}_1'\mathcal{P}_1''$, where the prime ideals appearing there have their residual degree equal to 1.

Solution 10.44.

(1) Since $\alpha = (1 \pm \sqrt{p})/2$, then α is an integer of the field $\mathbb{Q}(\sqrt{p})$. The element β is a root of the polynomial

$$g(x) = x^4 - x^3 - \frac{3}{8}(p-1)x^2 - \frac{4}{64}(p-1)^2x - \frac{4}{64}(p-1)^3,$$

hence it is an algebraic integer.
(2) We have

$$\mathrm{Disc}(\theta) = \mathrm{N}_{K/\mathbb{Q}}(f'(\theta)) = \mathrm{N}_{K/\mathbb{Q}}(4\theta^3) = 4^4\big(\mathrm{N}_{K/\mathbb{Q}}(\theta)\big)^3 = -2^8p^3.$$

(3) We have $f(x) \equiv x^4 \pmod p$, $f(x) = x^4 - p(1)$ and $p \nmid 1$, hence $p \nmid I(\theta)$.
(4) We know that the ring of integers of K has a basis of the form

$$\{1, f_1(\theta)/d_1, f_2(\theta)/d_2, f_3(\theta)/d_3\},$$

where for every i, $f_i(x)$ is a monic polynomial of degree i, with integral coefficients and d_i is a positive integer. Moreover $I(\theta) = d_1d_2d_3$ and the coefficients of $f_i(x)$ are determined modulo d_i [Marcus (1977), Th. 13, Exercises 39 and 40, Chap. 2]. Since $f(x) \equiv (x-1)^4 \pmod 2$, then for any prime ideal $\mathcal{P}$ lying over $2\mathbb{Z}$, we have $\theta \equiv 1 \pmod{\mathcal{P}}$. We deduce that if $2 \mid d_1$, then $f_1(\theta) \equiv 0 \pmod 2$, hence $f_1(\theta) \equiv 0 \pmod{\mathcal{P}}$, thus $f_1(x) = x + 1$. Let $\mu = \theta + 1$, then $\theta = \mu - 1$ and $(\mu - 1)^4 - p = 0$. It follows that

$$\mu^4 - 4\mu^3 + 6\mu^2 - 4\mu + 1 - p = 0.$$

From this equation it is seen that $\mu/2$ is not integral. We conclude that $2 \nmid d_1$. We know that $\alpha = (\theta^2 + 1)/2$ is integral, so that $2 \mid d_2$. Does $2^2 \mid d_2$? We must look at

$$f_2(x) = x^2 + 1, x^2 - 1, x^2 + 2x + 1, x^2 + 2x - 1.$$

The minimal polynomial of $\alpha = (\theta^2 + 1)/2$ is given by

$$g(x) = x^2 - x + (1 - p)/4,$$

thus it is seen that $\alpha/2 = (\theta^2 + 1)/4$ is not integral.
Let

$$\gamma = (\theta^2 - 1)/2 = \alpha - 1,$$

then

$$\gamma^2 + \gamma + (1 - p)/4 = 0.$$

This implies that $(\theta^2 - 1)/4$ is not integral.
Let $\rho = (\theta^2 + 2\theta + 1)/2 = \mu^2/2$. We find that ρ satisfies the following equation:

$$\rho^4 - 2\rho^3 + \frac{(35 - p)}{2}\rho^2 + \frac{7 - 3p}{2}\rho + \frac{(1 - p)^2}{16} = 0.$$

We deduce that the characteristic polynomial of $\rho/2$ is given by

$$h(x) = x^4 - x^3 + \frac{(35 - p)}{8}x^2 + \frac{7 - 3p}{16}x + \frac{(1 - p)^2}{16^2}.$$

Since $p \equiv 1 \pmod 8$, then $p \not\equiv 35 \pmod 8$, so that $\rho/2$ is not integral. Let $\nu = (\theta^2 + 2\theta - 1)/2$, then $\nu = \rho - 1$. Using the equation of ρ, we find that of ν:

$$\nu^4 + 2\nu^3 + \frac{(35 - p)}{2}\nu^2 + (40 - 4p)\nu + 6 - 3p + (1 - p)^2/16 + (35 - p)/2 = 0.$$

It follows that $\nu/2$ is a root of the polynomial

$$u(x) = x^4 + x^3 + \frac{(35 - p)}{8}x^2 + \frac{(10 - p)}{2}x + c,$$

where c is an integer. Since $p \equiv 1 \pmod 8$, then $p \equiv 1 \pmod 2$, hence $p \not\equiv 10 \pmod 2$, thus $\nu/2$ is not integral.

We conclude that $2 \mid d_2$, but $2^2 \nmid d_2$. We now discuss the 2-adic valuation of d_3. Since β is integral, then $2^2 \mid d_3$. Does $2^3 \mid d_3$? Since the coefficients of $f_3(x)$ are determined modulo 8, we must consider the following possibilities for $f_3(x)$:

$$f_{31}(x) = x^3 + x^2 + x + 1, \quad f_{32} = x^3 + x^2 + x + 5,$$
$$f_{33}(x) = x^3 + 5x^2 + x + 1, \quad f_{34}(x) = x^3 + 5x^2 + x + 5,$$
$$f_{35}(x) = x^3 + x^2 + 5x + 1, \quad f_{36}(x) = x^3 + x^2 + 5x + 5,$$
$$f_{37}(x) = x^3 + 5x^2 + 5x + 1, \quad f_{38}(x) = x^3 + 5x^2 + 5x + 5.$$

The following table gives the characteristic polynomial of $f_{3i}(\theta)$.

Table 1

$f_{3i}(\theta)$	$\mathrm{Char}(f_{3i}(\theta), \mathbb{Q})$
$f_{31}(\theta)$	$x^4 - 4x^3 + (6 - 6p)x^2 + (-4 + 8p - 4p^2)x + 1 - 3p + 3p^2$ $-p^3$
$f_{32}(\theta)$	$x^4 - 20x^3 + (150 - 6p)x^2 + (-500 + 56p - 4p^2)x + 625$ $-131p + 19p^2 - p^3$
$f_{33}(\theta)$	$x^4 - 4x^3 + (6 - 54p)x^2 + (-4 + 88p - 20p^2)x + 1 - 35p$ $+547p^2 - p^3$
$f_{34}(\theta)$	$x^4 - 20x^3 + (150 - 54p)x^2 + (-500 + 520p - 20p^2)x + 625$ $-1251p + 627p^2 - p^3$
$f_{35}(\theta)$	$x^4 - 4x^3 + (6 - 22p)x^2 + (-4 - 56p - 4p^2)x + 1 - 547p$ $+35p^2 - p^3$
$f_{36}(\theta)$	$x^4 - 20x^3 + (150 - 22p)x^2 + (-500 + 120p - 4p^2)x + 625$ $-675p + 51p^2 - p^3$
$f_{37}(\theta)$	$x^4 - 4x^3 + (6 - 70p)x^2 + (-46360p - 20p^2)x + 1 - 195p$ $+195p^2 - p^3$
$f_{38}(\theta)$	$x^4 - 20x^3 + (150 - 70p)x^2 + (-500 + 200p - 20p^2)x + 625$ $+125p + 275p^2 - p^3$

This table shows that the characteristic polynomial of $f_{3i}(\theta)$ is given by

$$F(x) = x^4 - 4x^3 + \text{ lower degree terms} \quad \text{or}$$
$$F(x) = x^4 - 20x^3 + \text{ lower degree terms}.$$

In the first case the characteristic polynomial of $f_{3i}(\theta)/8$ is given by $G(x) = x^4 - (1/2)x^3 +$ lower degree terms, hence $f_{3i}(\theta)/8$ is not an algebraic integer, thus $2^3 \nmid d_3$ in this case. Similarly we get the same conclusion in the second case. It follows that $I(\theta) = d_1 d_2 d_3 = 2^3$.

Remark. The polynomials appearing in this table suggest that we may avoid computing them explicitly. In fact, finding $\mathrm{Tr}_{\mathbb{Q}(\theta)/\mathbb{Q}}(f_{3i}(\theta))$ is sufficient for our needs.

(5) From **(2)** and **(3)**, we conclude that $p^3 \parallel \mathrm{Disc}(K)$. From **(2)** and **(4)**, we get $2^2 \parallel \mathrm{Disc}(K)$. It follows from **(2)** that $I(\theta) = 2^3$ and $\mathrm{Disc}(K) = -2^2p^3$. Consider the inclusions $\mathbb{Z}[\theta] \subset \mathbb{Z}[\alpha, \beta] \subset A$. It is easy to compute the matrix which expresses the coordinates of θ^i, $i = 0, \dots, 3$, in the basis $\{1, \theta, \alpha, \beta\}$ of K over $\mathbb{Q}$. Its determinant is equal to 8. Therefore $(\mathbb{Z}[\alpha, \beta] : \mathbb{Z}[\theta]) = 8 = I(\theta)$. We deduce that $A = \mathbb{Z}[\alpha, \beta]$ and then $\{1, \theta, \alpha, \beta\}$ is a basis of A over $\mathbb{Z}$.

(6) Obviously, K contains the quadratic field $E = \mathbb{Q}(\sqrt{p}) = \mathbb{Q}(\theta^2)$. Let B be the ring of integers of E. Since $p \equiv 1 \pmod 8$, then $2B = \mathcal{Q}_1\mathcal{Q}_2$, where $\mathcal{Q}_1$ and $\mathcal{Q}_2$ are prime ideals of B having their residual degree equal to 1. From **(5)**, we know that 2 is ramified in A. It follows that the splitting of 2 in A takes one of the following forms.

(i) $2A = \mathcal{P}_1^2\mathcal{P}_1'^2$.
(ii) $2A = \mathcal{P}_1^2\mathcal{P}_2$.
(iii) $2A = \mathcal{P}_1^2\mathcal{P}_1'\mathcal{P}_1''$.

Here all the ideals appearing in the splitting of 2, except $\mathcal{P}_2$, have their residual degree equal to 1. The residual degree of $\mathcal{P}_2$ is equal to 2.

In order to decide which of the splittings (i), (ii) and (iii) is the right one, we show that for any $\gamma \in A$, we have

$$\gamma^2 \equiv \gamma \pmod 2 \quad \text{or} \quad \gamma^2 \equiv \gamma + 1 - \theta \pmod 2.$$

It is easy to prove the following formulas:

$$\theta^2 = 2\alpha - 1 \quad \text{and} \quad \theta^3 = 4\beta - 2\alpha - \theta.$$

Using these relations, we get

$$\beta^2 = \left(\frac{\theta^3 + \theta^2 + \theta + 1}{4}\right)^2 = \beta + \frac{p-1}{8}\alpha + \frac{p-1}{8}\theta + \frac{p-1}{8} \quad \text{and}$$

$$\alpha^2 = \left(\frac{\theta^2 + 1}{2}\right)^2 = \alpha + \frac{p-1}{4}.$$

Since $p \equiv 1 \pmod{16}$, using the preceding identities, we get $\theta^2 \equiv 1 \pmod 2$, $\alpha^2 \equiv \alpha \pmod 2$ and $\beta^2 \equiv \beta \pmod 2$. We deduce that for any $\gamma = a + b\theta + c\alpha + d\beta$, we have
$$\gamma^2 \equiv a + b\theta^2 + c\alpha^2 + d\beta^2 \equiv a + b + c\alpha + d\beta \equiv \gamma + b(1-\theta) \pmod 2.$$
Hence $\gamma^2 \equiv \gamma \pmod 2$ if $b \equiv 0 \pmod 2$ and $\gamma^2 \equiv \gamma + 1 - \theta \pmod 2$ if $b \equiv 1 \pmod 2$.
The element $\delta = \theta - 1$ will be of some help in the proof. Its minimal polynomial is given by $u(x) = x^4 + 4x^3 + 6x^2 + 4x + 1 - p$. From this, it is seen that $\mathrm{N}_{K/\mathbb{Q}}(\delta) = 1 - p \equiv 0 \pmod{16}$ and $(\delta - 1)/2$ is not an algebraic integer.

- Suppose that the splitting (*ii*) holds, that is $2A = \mathcal{P}_1^2\mathcal{P}_2$. From $u(x)$, we see that $\theta \equiv 1 \pmod{\mathcal{P}_1}$ and $\theta \equiv 1 \pmod{\mathcal{P}_2}$. Since $\theta \not\equiv 1 \pmod 2$, then ${\mathcal{P}_1}^2 \nmid \theta - 1$. Since
$$\mathrm{N}_{K/\mathbb{Q}}(\theta - 1) = \mathrm{N}_{K/\mathbb{Q}}(\delta) \equiv 0 \pmod{16},$$
then ${\mathcal{P}_2}^2 \mid \theta - 1$. Choose elements γ_1 and γ_2 of A such that $\gamma_1 \in {\mathcal{P}_1}^2$ and $\gamma_2 + \mathcal{P}_2$ is a primitive element over $\mathbb{F}_2$ of the field $A/\mathcal{P}_2$, which is isomorphic to $\mathbb{F}_4$. By the Chinese remainder theorem, there exists $\gamma \in A$ such that $\gamma \equiv \gamma_1 \pmod{\mathcal{P}_1^2}$ and $\gamma \equiv \gamma_2 \pmod{\mathcal{P}_2}$. Hence $\gamma \equiv 0 \pmod{\mathcal{P}_1^2}$ and $\gamma^2 + \gamma + 1 \equiv 0 \pmod{\mathcal{P}_2}$. Therefore $\gamma^3 + \gamma^2 + \gamma \equiv 0 \pmod 2$. We have seen that $\gamma^2 \equiv \gamma \pmod 2$ or $\gamma^2 \equiv \gamma + 1 - \theta \pmod 2$. If $\gamma^2 \equiv \gamma \pmod 2$, then $\gamma^3 \equiv 0 \pmod 2$, hence $\gamma^3 \equiv 0 \pmod{\mathcal{P}_2}$, which is a contradiction. If $\gamma^2 \equiv \gamma + 1 - \theta \pmod 2$, then $\gamma^3 + 1 - \theta \equiv 0 \pmod 2$. Since $\mathcal{P}_2 \mid 1 - \theta$, then $\mathcal{P}_2 \mid \gamma^3$. Therefore $\mathcal{P}_2 \mid \gamma$, which again is a contradiction. We have shown that the splitting (*ii*) does not hold.
- Suppose that the splitting (*i*) holds, that is $2A = \mathcal{P}_1^2\mathcal{P}_1^{'2}$. We have $\theta \equiv 1 \pmod{\mathcal{P}_1}$ and $\theta \equiv 1 \pmod{\mathcal{P}_1^{'}}$. Since $\theta \not\equiv 1 \pmod 2$, then $\mathcal{P}_1^2 \nmid \theta - 1$ or $\mathcal{P}_1^{'2} \nmid \theta - 1$. Without loss of generality, we may suppose that $\mathcal{P}_1^{'2} \nmid \theta - 1$. Since $\mathrm{N}_{K/\mathbb{Q}}(\theta - 1) \equiv 0 \pmod{16}$, then $\mathcal{P}_1^3 \mid \theta - 1$. Choose $\gamma \in A$ such that $\mathcal{P}_1 \parallel \gamma - 1$. Recall that we proved that any element of A satisfies a congruence equation modulo 2 of two possible forms. Since $\gamma^2 - \gamma = \gamma(\gamma - 1) \not\equiv 0 \pmod 2$, then the first form of these equations does not hold. Suppose that the second form is satisfied, that is $\gamma^2 - \gamma \equiv 1 - \theta \pmod 2$, then $\gamma^2 - \gamma \equiv 1 - \theta \pmod{\mathcal{P}_1^2}$. The $\mathcal{P}_1$-adic valuation of the left side of this congruence equation is equal to 1, while the valuation of the right side is at least equal to 3, hence a contradiction.

We conclude that $2A = \mathcal{P}_1^2\mathcal{P}_1^{'}\mathcal{P}_1^{''}$.

Exercise 10.45.
Let K be a number field with class number $h = 2$, A be its ring of integers.

(1) Let $\rho \in A$ be an irreducible element. Show that ρ is prime or $\rho A = \mathcal{P}_1\mathcal{P}_2$, where $\mathcal{P}_1$ and $\mathcal{P}_2$ are prime ideals of A, distinct or not.
(2) Let $\alpha \in A^\star$. Let $\alpha = \pi_1 \cdots \pi_r$ and $\alpha = \chi_1 \cdots \chi_s$ be two factorizations of α into products of (distinct or not) irreducible elements of A. Let r_1 (resp. s_1) be the number of (distinct or not) prime elements appearing in the first (resp. second) factorization and suppose that these primes are $\pi_1, \ldots, \pi_{r_1}$ and $\chi_1, \ldots, \chi_{s_1}$ respectively. Show that $r_1 = s_1$ and reordering if necessary $\chi_1, \ldots, \chi_{r_1}$, we have χ_i and π_i are associate for $i = 1, \ldots, r_1$. Deduce that $r = s$.

Solution 10.45.

(1) Suppose that ρ is not prime and let $\rho A = \mathcal{P}_1 \cdots \mathcal{P}_r$ be the factorization of ρA as a product of (distinct or not) prime ideals of A, then $r \geq 2$. Suppose that r is odd, say $r = 2k+1$. Since $h = 2$, then $\mathcal{P}_i\mathcal{P}_{i+1} = \alpha_i A$, where $\alpha_i \in A \setminus A^\star$ for $i = 1, 3 \ldots, 2k-1$. It follows that $\mathcal{P}_{2k+1}$ is principal say $\mathcal{P}_{2k+1} = \alpha_{2k+1}A$, thus $\rho = \epsilon\alpha_1\alpha_3 \cdots \alpha_{2k+1}$ with $\epsilon \in A^\star$, contradicting the assumption that ρ is irreducible. We conclude that r is even, say $r = 2k$. As above, mutadis mutandis, we obtain $\rho = \epsilon\alpha_1\alpha_3 \cdots \alpha_{2k-1}$, where $\epsilon \in A^\star$. This implies that $k = 1$; otherwise ρ would be reducible. We conclude that $r = 2$.
(2) We may assume that $r_1 \leq s_1$. Since χ_1 is prime and divides $\pi_1 \cdots \pi_r$, then it divides some π_i, thus it is an associate of this π_i. Since π_j is not prime if $j > r_1$, then $i \leq r_1$. Therefore we may assume that $i = 1$, that is χ_1 and π_1 are associated. Set $\pi_1 = \epsilon_1\pi_1$, then $\pi_2 \cdots \pi_r = \epsilon_1\chi_2 \cdots \chi_s$. Iterating the same reasoning as above, we obtain that χ_j is associate to π_j, for $j = 1, \ldots, r_1$. Set $\pi_j = \epsilon_j\chi_j$ for $j = 1, \ldots, r_1$, then $\pi_{r_1+1} \cdots \pi_r = \epsilon_1 \cdots \epsilon_{r_1}\chi_{r_1+1} \cdots \chi_s$. If $r_1 < s_1$, then the prime element χ_{r_1+1} will divide some irreducible (but not prime) element appearing on the left side of this identity, which is a contradiction. We conclude that $r_1 = s_1$ and χ_j is associate to π_j for $j = 1, \ldots, r_1$. We also have got the following relation $\pi_{r_1+1} \cdots \pi_r = \epsilon_1 \cdots \epsilon_{r_1}\chi_{r_1+1} \cdots \chi_s$, where $\epsilon_1, \ldots, \epsilon_{r_1}$ are units of A and all the factors π_i, χ_j are irreducible elements but not prime. It follows that $\pi_{r_1+1}A \cdots \pi_r A = \chi_{r_1+1}A \cdots \chi_s A$. According to **(1)**, any ideal appearing in this identity is a product of exactly to prime ideals in A. Thanks to the uniqueness of the factorization into prime ideals in A, we get $2(r - r_1) = 2(s - r_1)$, hence $r = s$.

Exercise 10.46.
Let $K = \mathbb{Q}(\sqrt{p})$ where $p = 2$ or p is a prime number congruent to 1 modulo 4.

(1) Let I be an ideal of A such that $I = I'$, where I' is the ideal of A obtained from I by applying the automorphism σ of K such that $\sigma(\sqrt{p}) = -\sqrt{p}$. Show that there exists $c \in \mathbb{Z}$ such that $I = cA$ or $I = c\sqrt{p}A$.
(2) Let ϵ be a fundamental unit of K. Show that $\mathrm{N}_{K/\mathbb{Q}}(\epsilon) = -1$.
(3) Show that the class number h of K is odd.

Solution 10.46.

(1) If a prime ideal $\mathcal{P}$ of A divides I and $e = \nu_{\mathcal{P}}(I)$, then $e = \nu_{\mathcal{P}}(I^{'})$. Notice that in A, there exists one and only one prime ideal ramified in A, namely $\sqrt{p}A$. Write the factorization of I into a product of prime ideals in the form $I = \Pi_1\Pi_2(\sqrt{p}A)^k$, where

$$\Pi_1 = \prod_{\deg \mathcal{P}=2} \mathcal{P}^e \quad \text{and} \quad \Pi_2 = \prod_{\deg \mathcal{P}=1} \mathcal{P}^h \mathcal{P}'^h,$$

then clearly $\Pi_1 = aA$ and $\Pi_2 = bA$, where a and b are positive integers. If k is even, say $k = 2l$, then $(\sqrt{p}A)^k = p^l A$, hence $I = abp^l A$, thus $I = cA$ with $c \in \mathbb{Z}$.
If k is odd, say $k = 2l+1$, then $\sqrt{p}^k A = p^l\sqrt{p}A$, hence $I = abp^l\sqrt{p}A$, thus $I = c\sqrt{p}A$ with $c \in \mathbb{Z}$.

(2) Suppose that $\mathrm{N}_{K/\mathbb{Q}}(\epsilon) = 1$, then by Hilbert's Theorem 90, [Lang (1965), Chap. 8.6], there exists $\gamma \in A$ such that $\epsilon = \gamma/\gamma^{'}$, where $\gamma^{'}$ is the conjugate of γ distinct from γ. We deduce that $\gamma A = \gamma^{'} A$. According to **(1)**, $\gamma = \eta c$ or $\gamma = \eta c\sqrt{p}$ with $c \in \mathbb{Z}$ and η is a unit of A. In the first case, we have $\epsilon = \gamma/\gamma^{'} = \eta/\eta^{'} = \pm\eta^2$. In the second case, we get $\epsilon = \gamma/\gamma^{'} = -\eta/\eta^{'} = \pm\eta^2$. In any case we reached a contradiction since ϵ is a fundamental unit.

(3) Suppose that h is even, then there exists a class $[\mathcal{A}]$ of order 2 in the group of ideal classes, that is $[\mathcal{A}] \neq [A]$ and $[\mathcal{A}]^2 = [A]$. This implies that $[\mathcal{A}]^{-1} = [\mathcal{A}]$. Since $[\mathcal{A}][\mathcal{A}^{'}] = [\mathcal{A}\mathcal{A}^{'}] = [aA]$, with $a \in \mathbb{Z}$, then $[\mathcal{A}^{'}] = [\mathcal{A}^{-1}] = [\mathcal{A}]$. It follows that there exist α and $\beta \in A$ such that $\alpha\mathcal{A} = \beta\mathcal{A}^{'}$. We deduce that $\mathrm{N}_{K/\mathbb{Q}}(\alpha) = \pm\,\mathrm{N}_{K/\mathbb{Q}}(\beta)$.
If the sign in this identity is $+$, then $\mathrm{N}_{K/\mathbb{Q}}(\alpha/\beta) = 1$, which implies, by Hilbert's Theorem 90 [Lang (1965), Chap. 8.6], $\alpha/\beta = \gamma/\gamma^{'}$, with $\gamma \in A$. We deduce that $\gamma^{'}\alpha\mathcal{A} = \gamma\beta\mathcal{A} = \gamma^{'}\beta\mathcal{A}^{'}$, thus $\gamma\mathcal{A} = \gamma^{'}\mathcal{A}^{'} = (\gamma\mathcal{A})^{'}$. We have found in this case an ideal I equivalent to $\mathcal{A}$ such that $I = I^{'}$.

If the sign in this identity is $-$, then $\mathrm{N}_{K/\mathbb{Q}}(\alpha/\epsilon\beta) = 1$, hence $\alpha/\epsilon\beta = \gamma/\gamma^{'}$, with $\gamma \in A$. We deduce that $\gamma^{'}\alpha\mathcal{A} = \gamma\epsilon\beta\mathcal{A} = \gamma^{'}\beta\mathcal{A}^{'}$, thus $\gamma\mathcal{A} = \gamma\epsilon\mathcal{A} = \gamma^{'}\mathcal{A}^{'} = (\gamma\mathcal{A})^{'}$. In this case also we have found an ideal I equivalent to $\mathcal{A}$ such that $I = I^{'}$. By **(1)**, we conclude that I is principal ant then $[\mathcal{A}]$ is of order 1, which is a contradiction.

Exercise 10.47.
Let $f(x) = x^3 - 7x + 10$.

(1) Show that $f(x)$ is irreducible over $\mathbb{Q}$. Show that this polynomial has exactly one real root. Denote it by θ.
(2) Let $K = \mathbb{Q}(\theta)$ and A be the ring of integers of K. Show that $\{1, \theta, \omega\}$ is a basis of A over $\mathbb{Z}$ where $\omega = \frac{\theta^2-\theta+2}{4}$. Deduce that any $\alpha \in A$ may be written in the form $\alpha = \frac{a+b\theta+c\theta^2}{4}$, where $a, b, c \in \mathbb{Z}$, $a \equiv 0 \pmod 2$, $a + 2b \equiv a - 2c \equiv b + c \equiv 0 \pmod 4$ and conversely any element α of this form with these conditions on a, b, c, belongs to A.
(3) Let $\epsilon = \frac{2+\theta-\theta^2}{4} = 1 - \omega$. For any $\alpha \in K$ denote by $\alpha^{'}$ and $\alpha^{''}$ be the images of α by the two non real embeddings of K into $\mathbb{C}$.
Show that $|\theta| < 3.2$, $|\theta^{'}| < 1.8$, $|\theta^{''} - \theta^{'}| < 1.6$, $|\theta^{'} - \theta| < 4.9$, $|\epsilon| < 2.9$ and $|\epsilon^{'}| < 0.6$. Deduce that ϵ is a fundamental unit of A.
(4) Show that $h(K) = 1$.
(5) Show that $A = \mathbb{Z}[\omega]$.

Solution 10.47.

(1) Since $f(x) = x^3 - x + 1 \pmod 3$ and since $x^3 - x + 1$ is irreducible over $\mathbb{F}_3$, then $f(x)$ is irreducible over $\mathbb{Q}$. We omit the proof that $f(x)$ has a unique real root.
(2) It is easy to show that the characteristic polynomial of ω is given by $g(x) = x^3 - 5x^2 + 5x - 2$, hence ω is an algebraic integer. The index of $\mathbb{Z}[\theta]$ in $\mathbb{Z} + \mathbb{Z}\theta + \mathbb{Z}\omega$ is equal to the determinant
$$\Delta = \begin{vmatrix} 1 & 0 & 2 \\ 0 & 1 & -1 \\ 0 & 0 & 4 \end{vmatrix}, \text{ hence } \mathrm{Disc}(1, \theta, \theta^2) = 4^2 \,\mathrm{Disc}(1, \theta, \omega).$$
Since $\mathrm{Disc}(1, \theta, \theta^2) = \mathrm{Disc}(f) = -4^2 \cdot 83$, then $\mathrm{Disc}(1, \theta, \omega) = -83$. Therefore $\mathrm{Disc}(K) = -83$ and $\{1, \theta, \omega\}$ is an integral basis.

Let $\alpha \in A$, then

$$\begin{aligned}\alpha &= x + y\theta + z\omega \\ &= x + y\theta + z\left(\frac{2-\theta+\theta^2}{4}\right) \\ &= \frac{4x+2z+(4y-z)\theta+z\theta^2}{4},\end{aligned}$$

hence α has the form $\alpha = \frac{a+b\theta+c\theta^2}{4}$, where $a = 4x+2z, b = 4y-z, c = z$. Moreover these identities show that $a \equiv 0 \pmod 2$, $a+2b \equiv 0 \pmod 4$ and $b+c \equiv 0 \pmod 4$. Conversely, given $a, b, c \in \mathbb{Z}$ satisfying these congruences, we may compute $x, y, z \in \mathbb{Z}$.

(3) Let $z = x + iy$ with $x, y \in \mathbb{R}$ be a root of $f(x)$, then $f(z) = 0$ is equivalent to the following system of equations

$$\begin{aligned} x^3 - 3xy^2 - 7x + 10 &= 0 \\ y(3x^2 - y^2 - 7) &= 0. \end{aligned}$$

If $y = 0$, then $z = x = \theta$. If $y \neq 0$, then

$$\begin{aligned} 3x^2 - y^2 - 7 = 0 \quad &\text{and} \\ x^3 - 3xy^2 - 7x + 10 = 0&, \end{aligned}$$

hence $y^2 = 3x^2 - 7$ and x is the unique real root of the equation $4x^3 - 7x - 5 = 0$. It is easy to show that the characteristic polynomial of ϵ is equal to $h(x) = x^3+2x^2-2x+1$, hence ϵ is a unit in A. If ϵ is not a fundamental unit then there exist an integer $n \geq 2$ and integers a, b, c such that $\pm\epsilon = (\frac{a+b\theta+c\theta^2}{4})^n$ with $a \equiv 0 \pmod 2$, $a + 2b \equiv a - 2c \equiv b + c \equiv 0 \pmod 4$. Moreover, we may suppose that $c \leq 0$. We deduce that $a + b\theta + c\theta^2 = 4(\pm\epsilon)^{1/n}$ and conjugating this equation, we have

$$\begin{aligned} a + b\theta^{'} + c\theta^{'2} &= 4(\pm\epsilon)^{'1/n} \\ a + b\theta^{''} + c\theta^{''2} &= 4(\pm\epsilon'')^{1/n}. \end{aligned}$$

The determinant of this system of linear equations in a, b, c is given by $\delta = \begin{pmatrix} 1 & \theta & \theta^2 \\ 1 & \theta' & \theta'^2 \\ 1 & \theta'' & \theta''^2 \end{pmatrix} = \sqrt{\mathrm{Disc}(\theta)} = \pm 4i\sqrt{83}$. Solving these equations, we obtain:

$$\begin{aligned} b &= \frac{4}{\delta}(\pm\epsilon)^{1/n}(\theta'^2 - \theta''^2) + (\pm\epsilon')^{1/n}(\theta^{''2} - \theta^2) + (\pm\epsilon^{''1/n})(\theta^2 - \theta^{'2})) \\ c &= \frac{4}{\delta}(\pm\epsilon)^{1/n}(\theta^{''} - \theta') + (\pm\epsilon)^{'1/n}(\theta - \theta'') + (\pm\epsilon)^{''1/n}(\theta^{'} - \theta)). \end{aligned}$$

Since $\epsilon^{''} = \overline{\epsilon^{'}}$, $\theta^{''} = \overline{\theta^{'}}$, $\theta - \theta^{''} = \overline{\theta - \theta^{'}}$, then $|\epsilon^{''}| = |\epsilon^{'}|$, $|\theta^{''}| = |\theta^{'}|$ and $|\theta - \theta^{''}| = |\overline{\theta - \theta^{'}}|$. We deduce that

$$\begin{aligned}|b| &\leq 1/\sqrt{83}(|\epsilon^{1/n}||\theta||\theta^{'} - \theta^{''}| + 2|\epsilon^{'}|^{1/n}|\theta^{'}||\theta^{'} - \theta|) \\ &\leq 1/\sqrt{83}(|\theta||\epsilon|^{1/2}|\theta^{'} - \theta^{''}| + 2|\theta^{'}||\theta^{'} - \theta|) \\ &\leq 1/\sqrt{83}\Big((3.2)\cdot\sqrt{2.9}\cdot(1.6) + 2\cdot(1.8)\cdot(4.9)\Big) < \frac{26.4}{9.11} < 2.89\end{aligned}$$

and

$$\begin{aligned}|c| &\leq 1/\sqrt{83}(|\epsilon|^{1/2}|\theta^{'} - \theta^{''}| + 2|\theta^{'} - \theta|) \\ &< 1/\sqrt{83}\left(\sqrt{2.9}\cdot(1.6) + 2\cdot(4.9)\right) \\ &< \frac{12.536}{9.11} < 1.4.\end{aligned}$$

We conclude that $b = 0$ or $b = \pm 1$ or $b = \pm 2$ and since $c < 0$, then $c = 0$ or $c = -1$. The condition $b + c \equiv 0 \pmod 4$ implies that $(b, c) = (0, 0)$ or $(b, c) = (1, -1)$, since $\epsilon \notin \mathbb{Q}$, we immediately reject the case $(b, c) = (0, 0)$. In the second case, we have $4^n(\pm\epsilon) = (a + \theta - \theta^2)^n$, hence $\mathrm{N}_{K/\mathbb{Q}}(a + \theta - \theta^2) = \pm 64$. But we have $a + \theta - \theta^2 = a - 2 + 4\epsilon$. Let $\mu = a + \theta - \theta^2$, then $\epsilon = \frac{\mu - (a-2)}{4}$. Using the characteristic polynomial of ϵ, we obtain

$$\left(\frac{\mu - (a-2)}{4}\right)^3 + 2\left(\frac{\mu - (a-2)}{4}\right)^2 - 2\left(\frac{\mu - (a-2)}{4}\right) + 1 = 0,$$

hence

$$(\mu - (a-2))^3 + 2\cdot 4\,(\mu - (a-2))^2 - 2\cdot 4^2\,(\mu - (a-2)) + 4^3 = 0.$$

Therefore

$$\mathrm{N}_{K/\mathbb{Q}}(a + \theta - \theta^2) = \mathrm{N}_{K/\mathbb{Q}}(\mu) = (a-2)^3 - 8(a-2)^2 - 32(a-2) - 64.$$

Since $\mathrm{N}_{K/\mathbb{Q}}(\mu) = \pm 64$, then $2|a$ but $4 \nmid a$. Let $d = a - 2$, then d is an integral root of one of the polynomials: $x^3 - 8x^2 - 32x$, $x^3 - 8x^2 - 128$. It is easy to verify that the only possibility is $d = 0$ that is $a = 2$ and then $\pm 4\epsilon = (2 + \theta - \theta^2)^n = (4\epsilon)^n$ which implies $n = 1$, contradicting our assumption $n \geq 2$.

(4) We use the result [Marcus (1977), Cor. 2, Chap. 5] that any class of ideals of A contains an integral ideal I, whose norm satisfies the condition $\mathrm{N}(I) \leq \frac{4}{\pi}\frac{3!}{3^3}\sqrt{|\operatorname{Disc}(K)|} < 2.58$, hence we have to look at the ideals of norm 2. We have $g(x) \equiv x(x^2 + x + 1) \pmod 2$ and $x^2 + x + 1$ is irreducible over $\mathbb{F}_2$, hence $2A = \mathcal{P}_1\mathcal{P}_2$ where $\mathcal{P}_1$ and $\mathcal{P}_2$ are prime

ideals of A of residual degrees 1 and 2 respectively. Therefore there is one and only one ideal of A having its norm equal to 2, namely $\mathcal{P}_1$. Moreover since $\mathrm{N}_{K(\mathbb{Q})}(\omega) = 2$ then $\mathcal{P}_1 = \omega A$ and $\mathcal{P}_1$ is principal. Therefore $h(K) = 1$.

(5) We have $\theta^2 = -2 + \theta + 4\omega$, hence $\omega^2 = (3\theta^2 - 7\theta + 6)/4 = -\theta + 3\omega$, thus $\theta = 3\omega - \omega^2$. This shows that $A = \mathbb{Z}[\theta, \omega] = \mathbb{Z}[\omega]$.
Remark. It is seen that the absolute value of the determinant of the transition matrix from the basis of $\mathbb{K}/\mathbb{Q}$, $\{1, \theta, \omega\}$ to the other basis $\{1, \omega, \omega^2\}$ is equal to 1, hence the second basis is a basis of A.

Exercise 10.48.
Let p be a prime number, d be a square free integer. Suppose that $d \equiv 2 \pmod 4$, $p \equiv 1 \pmod 4$, $\gcd(p, d) = 1$ and $pd < 0$. Let $\Delta = \frac{1+\sqrt{p}}{2}$, $K = \mathbb{Q}(\sqrt{pd})$ and $E = K(\sqrt{p})$. Let R and S be the rings of integers of K and E respectively. The prime number p being ramified in R, let $\mathcal{P}$ be the unique prime ideal of R lying over $p\mathbb{Z}$.

(1) Show that $\mathcal{P}$ is not principal.
(2)(a) If $\{A, B\}$ is a basis of S over R, show that $\{1, \Delta\}$ is also a basis of S over R.
(b) Deduce that S has no basis $\{A, B\}$ over R.

Solution 10.48.

(1) Since $pd \equiv 2 \pmod 4$, then $\{1, \sqrt{pd}\}$ is a basis of R over $\mathbb{Z}$. Suppose that $\mathcal{P}$ is principal generated by $a + b\sqrt{pd}$ with a and $b \in \mathbb{Z}$. Set $\gamma = \sqrt{pd}$. From the relations $\mathcal{P}^2 = pR$ and $(\sqrt{pd})^2 = pd$, we deduce that $p = (a+b\gamma)(u+v\gamma)$ and $\gamma = (a+b\gamma)(x+y\gamma)$ with $(u, v, x, y) \in \mathbb{Z}^4$. Hence

$$au + bvpd = p, \quad av + bu = 0, \quad ax + bypd = 0, \quad ay + bx = 1.$$

Eliminating u between the first two equations, we obtain the following equation $-v(a^2 - dpb^2) = pb$. This equation shows that if $v = 0$, then $b = 0$. If $b = 0$ then $\mathcal{P} = aR$, hence $N_{K/\mathbb{Q}}(\mathcal{P}) = a^2$ contradicting $N_{K/\mathbb{Q}}(\mathcal{P}) = p$. Therefore we may suppose that $v \neq 0$ and $b \neq 0$. Since $dp < 0$, then the absolute value the left side of the above equation satisfies the condition

$$|v||a^2 - dpb^2| = |v|(a^2 - dpb^2) < |pb|,$$

contradicting the same equation. Therefore $\mathcal{P}$ is not principal.

(2)(a) For any $\delta \in E$, denote by $\bar{\delta}$ the image of δ by the unique K-automorphism of E such that $\sqrt{p} \to -\sqrt{p}$. Since 1 and Δ are integral over R, then $1 = \alpha_1 A + \beta_1 B$ and $\Delta = \alpha_2 A + \beta_2 B$ with $(\alpha_1, \beta_1, \alpha_2, \beta_2) \in R^4$. We deduce that $\begin{pmatrix} 1 & 1 \\ \Delta & \bar{\Delta} \end{pmatrix} = \begin{pmatrix} \alpha_1 & \beta_1 \\ \alpha_2 & \beta_2 \end{pmatrix} \begin{pmatrix} A & \overline{A} \\ B & \overline{B} \end{pmatrix}$. Equating the determinants of both sides of this equality and squaring, we obtain:
$p = (\overline{\Delta} - \Delta)^2 = \epsilon^2 \delta$, where $\epsilon = \begin{vmatrix} \alpha_1 & \beta_1 \\ \alpha_2 & \beta_2 \end{vmatrix}$ and $\delta = \begin{vmatrix} A & \overline{A} \\ B & \overline{B} \end{vmatrix}^2$.
It follows that $\mathcal{P}^2 = (\epsilon R)^2 \delta R$. Since $\mathcal{P}$ is not principal and since the factorization of the ideals of R into a product of prime ideals is unique, we conclude that $\epsilon R = R$ and $\mathcal{P}^2 = \delta R$. Therefore ϵ is a unit of R and then $\{1, \Delta\}$ is a basis of S over R.

(b) Suppose that S has a basis $\{A, B\}$ over R, then $\{1, \Delta\}$ is also a basis. Since $\sqrt{d} \in S$, there exist $\alpha, \beta \in R$ such that

$$\sqrt{d} = \sqrt{pd}/\sqrt{p} = \alpha + \beta\Delta.$$

By conjugating over K, we obtain $-\sqrt{pd}/\sqrt{p} = \alpha + \beta\overline{\Delta}$, hence

$$2\sqrt{pd}/\sqrt{p} = \beta(\Delta - \overline{\Delta}) = \beta\sqrt{p}.$$

It follows that $2\sqrt{pd} = \beta p$ and then $2R\sqrt{pd}R = (\beta R) \cdot \mathcal{P}^2$. The right side of this equality is divisible by $\mathcal{P}^2$, while the first is divisible by $\mathcal{P}$ but not $\mathcal{P}^2$ since $p \equiv 1 \pmod 4$ thus $p \neq 2$. We have then reached a contradiction.

Exercise 10.49.
Let K be a cubic extension of $\mathbb{Q}$ such that $\mathrm{Disc}(K) < 0$. Suppose that $K \subset \mathbb{R}$ and let A be the ring of integers of K and ϵ be the fundamental unit of A such that $\epsilon > 1$.

(1) Show that $|\mathrm{Disc}(K)|/4 < \epsilon^3 + 7$.
(2) Deduce that the real root θ of $f(x) = x^3 - 2x^2 - 1$ is a fundamental unit of the ring of integers of the field $\mathbb{Q}(\theta)$.

Solution 10.49.

(1) Let $\sigma : K \to \mathbb{C}$ be one of the complex embeddings, then

$$\mathrm{N}_{K/\mathbb{Q}}(\epsilon) = \epsilon\sigma(\epsilon)\overline{\sigma(\epsilon)} = \epsilon|\sigma(\epsilon)|^2,$$

hence $\mathrm{N}_{K/\mathbb{Q}}(\epsilon) = 1$. Let $\rho = |\sigma(\epsilon)| = \frac{1}{\sqrt{\epsilon}}$. Set $\sigma(\epsilon) = \rho e^{i\theta}, \overline{\sigma(\epsilon)} = \rho e^{-i\theta}$ and $\epsilon = \frac{1}{\rho^2}$. We have $\mathrm{Disc}(1, \epsilon, \epsilon^2) = I(\epsilon)^2\,\mathrm{Disc}(K)$, hence $|\mathrm{Disc}(K)| \leq$

$|\operatorname{Disc}(1,\epsilon,\epsilon^2)|$. We have

$$\begin{aligned}
\operatorname{Disc}(1,\epsilon,\epsilon^2) &= \begin{vmatrix} 1 & \epsilon & \epsilon^2 \\ 1 & \sigma(\epsilon) & \sigma(\epsilon)^2 \\ 1 & \overline{\sigma(\epsilon)} & \overline{\sigma(\epsilon)^2} \end{vmatrix} \\
&= [(\sigma(\epsilon)-\overline{\sigma(\epsilon)})(\epsilon-\sigma(\epsilon))(\epsilon-\overline{\sigma(\epsilon)})]^2 \\
&= 2i\rho(\sin\theta)\left(\frac{1}{\rho^2}-\rho(\cos\theta+i\sin\theta)\left(\frac{1}{\rho^2}-\rho(\cos\theta-i\sin\theta)\right)^2\right. \\
&=-4\rho^2(\sin^2\theta)\left(\left(\frac{1}{\rho^2}-\rho(\cos\theta)\right)^2+\rho^2(\sin^2\theta)\right)\Bigg)^2 \\
&= -4\rho^2(\sin^2\theta)\rho^2\left(\frac{1}{\rho^4}-\frac{2\cos\theta}{\rho}+\rho^2\right)^2 \\
&= -4(\sin^2\theta)\left(\frac{1}{\rho^3}+\rho^3-2(\cos\theta)\right)^2.
\end{aligned}$$

It follows that

$$|\operatorname{Disc}(K)|/4 \le |\operatorname{Disc}(1,\epsilon,\epsilon^2)/4 = (\sin^2\theta)\left(\frac{1}{\rho^3}+\rho^3-2(\cos\theta)\right)^2.$$

Let $c=\cos\theta$ and $z=\rho^3+\frac{1}{\rho^3}$, then $z-2=\frac{\rho^6-2\rho+1}{\rho^3}=\frac{(\rho^3-1)^2}{\rho^3}>0$, hence $z>2$. We deduce that

$$\begin{aligned}
|\operatorname{Disc}(K)|/4 &\le (1-c^2)(z-2c)^2 \\
&= (z^2-4cz+4c^2)(1-c^2) \\
&= z^2-c^2z^2-4cz(1-c^2)+4c^2(1-c^2) \\
&= z^2-(cz+2(1-c^2))^2+4(1-c^2)^2+4c^2(1-c^2) \\
&= z^2+4-\left(cz+2(1-c^2\right)^2-4c^2 < z^2+4=\rho^6+6+\frac{1}{\rho^6} \\
&< \epsilon^3+7.
\end{aligned}$$

(2) We have

$$\begin{aligned}
\operatorname{Disc}(f) &= (-1)^3 N_{K/\mathbb{Q}} f'(\theta) \\
&= -\operatorname{N}_{K/\mathbb{Q}}(3\theta^2-4\theta) \\
&= -\operatorname{N}_{K/\mathbb{Q}}(\theta) N_{K/\mathbb{Q}}(3\theta-4) \\
&= -\operatorname{N}_{K/\mathbb{Q}}(3\theta-4).
\end{aligned}$$

Let $\beta=3\theta-4$, thus $\theta=\frac{\beta+4}{3}$, then $(\frac{\beta+4}{3})^3-2(\frac{\beta+4}{3})^2-1=0$, hence $(\beta+4)^3-6(\beta+4)^2-27=0$. Therefore $\beta^3+6\beta^2-59=0$. We conclude that $\text{Disc}(f) = -59$. Since 59 is a prime number, then $\text{Disc}(K) = -59$ and $\{1, \theta, \theta^2\}$ is a basis of the ring of integer of $K = \mathbb{Q}(\theta)$ over $\mathbb{Z}$. Let ϵ be the fundamental unit of K such that $\epsilon > 1$. By **(1)**, $\epsilon^3 + 7 > |\text{Disc}(K)|/4 > 14.75$, hence $\epsilon > 1.95$ and then $\epsilon^2 > 3.8$. We have $f(2) = -1 < 0$, $f(3) = 8 > 0$, hence $\theta \in]2, 3[$. Since $g(3.8) = 3.8^2 \cdot (1.8) - 1 > \theta > 0$, then $\theta < 3.8 < \epsilon^2$. Since θ is a unit greater than 1, then it is the unique fundamental unit of K which is greater than 1.

Exercise 10.50.
Let $f(x) = x^3 + 4x + 1$.

(1) Show that $f(x)$ is irreducible over $\mathbb{Q}$.
(2) Show that $f(x)$ has exactly one real root. Let θ be this root and let $K = \mathbb{Q}(\theta)$ and A be the ring of integers of K. Show that $\text{Disc}(K) = -283$ and $\{1, \theta, \theta^2\}$ is a basis of A over $\mathbb{Z}$.
(3) Show that $283A = \mathcal{R}^2\mathcal{R}'$, $\mathcal{D}_K = \mathcal{R}$, where $\mathcal{R}$ and $\mathcal{R}'$ are prime ideals of A with residual degree equal to 1. Show that $\mathcal{D}_K$ is principal and determine for it a generator.
(4) Show that $\frac{-1}{\theta}$ is the fundamental unit of A, greater than 1.
Hint. One may use the preceding exercise.
(5) Show that $h(K) = 2$.

Solution 10.50.

(1) Obvious.
(2) We have $f'(x) = 3x^2 + 4 > 0$, hence $f(x)$ is strictly increasing from $-\infty$ to $+\infty$. Therefore the equation $f(x) = 0$ has one and only one real solution. We have

$$\text{Disc}(1, \theta, \theta^2) = \text{Disc}(f) = -4(4^3) - 27 = -283.$$

Since 283 is a prime number then $\text{Disc}(K) = -283$ and $\{1, \theta, \theta^2\}$ is a basis of A over $\mathbb{Z}$.
(3) Since 283 is ramified in K, then $283A = \mathcal{R}^2\mathcal{R}'$ or $283A = \mathcal{R}^3$. In the second case $\mathcal{R}^3 \mid \mathcal{D}_K$ and then $283^2 \mid \text{Disc}\, K$, which is a contradiction. We conclude that $283A = \mathcal{R}^2\mathcal{R}'$. We deduce that $\mathcal{D}_K = \mathcal{R}$. Since $A = \mathbb{Z}[\theta]$, then $\mathcal{D}_K$ is principal and $\mathcal{D}_K = f'(\theta)A = (3\theta^2 + 4)A$.
(4) It is easy to show that the minimal polynomial of $-1/\theta$ is given by $g(x) = x^3 - 4x^2 - 1$, hence $-1/\theta$ is a unit of A. Moreover, we have $g'(x) = 3x^2 - 8x$ and we give the table of variations of the function g.

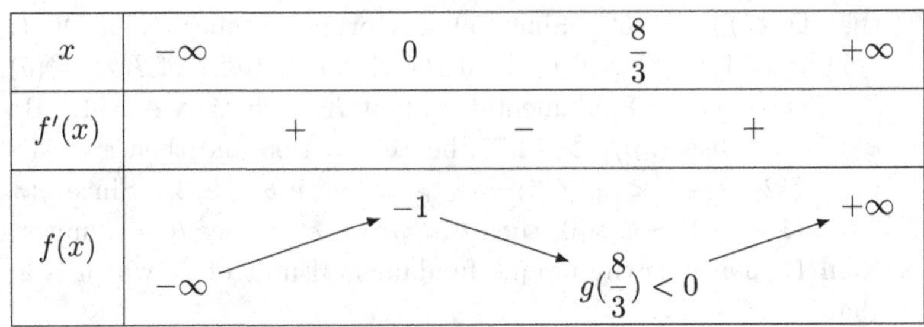

x	$-\infty$		0		$\frac{8}{3}$		$+\infty$
$f'(x)$		$+$		$-$		$+$	
$f(x)$	$-\infty$	$\nearrow$	-1	$\searrow$	$g(\frac{8}{3}) < 0$	$\nearrow$	$+\infty$

We have $g(4) = -1 < 0$ and $g(5) = 24 > 0$, hence $-1/\theta \in]4, 5[$. Let ϵ be the fundamental unit of A such that $\epsilon > 1$ then by **Exercise 10.49**,

$$\epsilon^3 + 7 > |\operatorname{Disc}(K)|/4 = \frac{283}{4} = 70.75,$$

hence $\epsilon^3 > 63.75$. We deduce that $\epsilon^2 > 15.21 > -1/\theta$. Since $-1/\theta$ is a unit > 1 then $-1/\theta = \epsilon^n$ for some positive integer n. Suppose that $n \geq 2$, then $-1/\theta = \epsilon^n \geq \epsilon^2$ contradicting the inequality already proved $-1/\theta < \epsilon^2$. It follows that $n = 1$ and then $-1/\theta$ is a fundamental unit of A.

(5) In order to compute $h(K)$, we use the result [Marcus (1977), Cor. 2, Chap. 5] stating that every class of ideals contains an integral ideal I such that

$$\mathrm{N}_{K/\mathbb{Q}}(I) < \frac{4}{\pi} \frac{3!}{3^3} \sqrt{|\operatorname{Disc} K|} < 5,$$

hence we must look at the ideals of norm 2, 3 and 4 respectively. We have

$$f(x) \equiv x^3 + 1 \equiv (x+1)(x^2 + x + 1) \pmod 2,$$

hence $2A = \mathcal{P}_1\mathcal{P}_2$, where $\mathcal{P}_1$ and $\mathcal{P}_2$ are prime ideals of A of residual degree equal to 1 and 2 respectively. Moreover $\mathcal{P}_1 = (2, \theta + 1)$ and $\mathcal{P}_2 = (2, \theta^2 + \theta + 1)$. We write $I \sim J$ when I and J have the same class. The splitting of 2 in A shows that $\mathcal{P}_1 \sim \mathcal{P}_2$. We have

$$f(x) \equiv x^3 + x + 1 \equiv (x-1)(x^2 + x - 1) \pmod 3,$$

hence $3A = \mathcal{Q}_1\mathcal{Q}_2$ and $\mathcal{Q}_1 \sim \mathcal{Q}_2$, where

$$\mathcal{Q}_1 = (3, \theta - 1) \quad \text{and} \quad \mathcal{Q}_2 = (3, \theta^2 + \theta - 1).$$

It is easy to show that the characteristic polynomial of $\theta^2 + \theta + 1$ is given by $x^3 + 5x + 10x - 12$, hence $(\theta^2 + \theta + 1)A = \mathcal{P}_2\mathcal{Q}_1$. It follows that $\mathcal{Q}_2 \sim \mathcal{Q}_1 \sim \mathcal{P}_2 \sim \mathcal{P}_1$. Therefore $h(K) = 1$ or $h(K) = 2$ according to

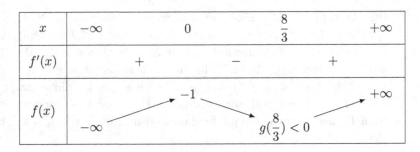

$\mathcal{P}_1$ is principal or not. It is easy to show that the minimal polynomial (and characteristic polynomial) of $\theta+1$ is given by x^3-3x^2+7x-4, hence $\mathrm{N}_{K/\mathbb{Q}}(\theta+1)=4$. It follows that $(\theta+1)A=\mathcal{P}_1^2$ because $\theta+1\equiv 0 \pmod{\mathcal{P}_1}$ but $\theta+1\not\equiv 0 \pmod{\mathcal{P}_2}$. Suppose by contradiction that $\mathcal{P}_1$ is principal and let $\gamma\in A$ such that $\mathcal{P}_1=\gamma A$, then $(\theta+1)A=(\gamma^2A)$, hence there exists an integer n such that $\theta+1=\pm\gamma^2.\epsilon^n$, where ϵ is the fundamental unit of A computed above, that is $\epsilon=-1/\theta$. If n is even, say $n=2k$, then

$$\theta+1=\pm\gamma^2\epsilon^{2k}=\pm(\gamma\epsilon^k)^2:=\pm(a+b\theta+c\theta^2)^2,$$

where $a,b,c\in\mathbb{Z}$.
If n is odd, say $n=2k+1$, then $\theta+1=\pm\gamma^2\epsilon^{2k+1}$, hence

$$\frac{\theta+1}{\epsilon}=\pm(\gamma\epsilon^k)^2=\pm(a+b\theta+c\theta^2)^2.$$

Straightforward computations show that

$$(a+b\theta+c\theta^2)^2=a^2-2bc+(-c^2+2ab-8bc)\theta+(b^2-4c^2+2ac)\theta^2.$$

- case $\theta+1=(a+b\theta+c\theta^2)^2$.
  In this case, we obtain the following equations
  $$a^2-2bc=1,\quad -c^2+2ab-8bc=1,\quad b^2-4c^2+2ac=0.$$
  The third equation shows that $2\mid b$, the same equation then implies $2\mid ac$. The first equation shows that $2\nmid a$, hence $2\mid c$. The second equation then implies $c^2\equiv -1 \pmod 4$ which is a contradiction.
- case $\theta+1=-(a+b\theta+c\theta^2)^2$.
  Here, we obtain the following equations
  $$a^2-2bc=-1,\quad -c^2+2ab-8bc=-1,\quad b^2-4c^2+2ac=0.$$
  The last equation shows that $2\mid b$. The first equation then implies $a^2\equiv -1 \pmod 4$ which is a contradiction.
- case $\frac{\theta+1}{\epsilon}=-\theta^2-\theta=(a+b\theta+c\theta^2)^2$.
  We get the following equations
  $$a^2-2bc=0,\quad -c^2+2ab-8bc=-1,\quad b^2-4c^2+2ac=-1.$$
  The first equation shows that $2\mid a$. The last equation then implies $b^2\equiv -1 \pmod 4$ which is a contradiction.
- case $\frac{\theta+1}{\epsilon}=-\theta^2-\theta=-(a+b\theta+c\theta^2)^2$.
  Here we get the following equations
  $$a^2-2bc=0,\quad -c^2+2ab-8bc=1,\quad b^2-4c^2+2ac=1.$$
  The first equation shows that $2\mid a$ and then $2\mid bc$. The second equation implies $\gcd(b,c)=\gcd(a,c)=1$. We deduce that $2\nmid c$, $2\mid a$ and $2\mid b$. These conditions contradict the third equation.

We conclude that $\mathcal{P}_1$ is not principal and $h(K)=2$.

Exercise 10.51.
Let K be a number field of degree $n \geq 2$ and A be its ring of integers. Suppose that the following property is satisfied.

(P) There exist a prime number p and $\theta \in A$ such that $p \mid \theta^2, p^2 \nmid \theta^2$ and $p^2 \mid \theta^3$.

(1) Show that p is ramified in K.
(2) Show that $n \neq 2$.
(3) Show that **(P)** holds for the prime 2 in the field $\mathbb{Q}(\theta)$, where θ is a root of $f(x) = x^3 - 12x + 12$. Deduce that 2 is totally ramified in $\mathbb{Q}(\theta)$.

Solution 10.51.

(1) Suppose that p is not ramified in K and let $\mathcal{P}$ be a prime ideal of A lying of over $p\mathbb{Z}$. Since $\mathcal{P} \mid \theta^2$, then $\mathcal{P} \mid \theta$, hence $p \mid \theta$, which contradicts the hypothesis $p^2 \nmid \theta^2$. We conclude that p is ramified in K.
(2) Suppose that $n = 2$ and **(P)** holds in K for some prime p, then p is ramified, say $pA = \mathcal{P}^2$. Since $\mathcal{P}^2|\theta^2$, then $\mathcal{P} \mid \theta$. Let $k = \nu_{\mathcal{P}}(\theta)$. Since $p^2 \nmid \theta^2$, then $\mathcal{P}^4 \nmid \theta^2$, hence $k = 1$. We deduce that $\nu_{\mathcal{P}}(\theta^3) = 3$, contradicting the condition $p^2 \mid \theta^3$.
(3) Let $\mu = \theta^2$, then it is easy to show that the minimal polynomial of μ over $\mathbb{Q}$ is given by

$$g(x) = x^3 - 8 \cdot 3x^2 + 4^2 \cdot 3^2 x - 4^2 \cdot 3^2.$$

Therefore it is seen that $\mu/2$ is an algebraic integer but not $\mu/4$. Since $\theta^3 = 4(3\theta - 3)$, then obviously $4 \mid \theta^3$. Since the property **(P)** holds for the prime 2 in K, then 2 is ramified in K.
Suppose that $2A = \mathcal{P}_1^2\mathcal{P}_1^{'}$, where $\mathcal{P}_1$ and $\mathcal{P}_1^{'}$ are prime ideals of A of residual degree equal to 1. Since $\mathcal{P}_1^2\mathcal{P}_1^{'} \mid \theta^2$, then $\mathcal{P}_1 \mid \theta$ and $\mathcal{P}_1^{'} \mid \theta$. Since $\mathcal{P}_1^4\mathcal{P}_1^{'2} \nmid \theta^2$, then $\mathcal{P}_1^4 \nmid \theta^2$. Therefore $\nu_{\mathcal{P}_1}(\theta) = 1$. It follows that $\nu_{\mathcal{P}_1}(\theta^3) = 3$, contradicting the fact that $\mathcal{P}_1^4\mathcal{P}_1^{'2} \mid \theta^3$. We conclude that 2 is totally ramified in A.

Exercise 10.52.
Let $f(x) = x^5 - 5x - 5$, θ be a root of $f(x)$ in $\mathbb{C}, K = \mathbb{Q}(\theta)$ and A be its ring of integers.

(1) Show that $f(x)$ is irreducible over $\mathbb{Q}$. Show that $\text{Disc}(f) = 5^5 \cdot 3^2 \cdot 41$, $\text{Disc}(K) = 5^5 \cdot 41$ and $A = \mathbb{Z} + \mathbb{Z}\theta + \mathbb{Z}\theta^2 + \mathbb{Z}\theta^3 + \mathbb{Z}\omega$, where $\omega = (\theta^4 + \theta^3 + \theta^2 + \theta - 1)/3$.

(2) Show that the characteristic polynomial over $\mathbb{Q}$ of ω is given by $h(x) = x^5 - 5x^4 - 10x^2 - 10x - 3$. Using the factorization of $f(x)$ in $\mathbb{F}_3[x]$ and looking at $h(x)$, show that the splitting of the prime 3 in A has the form $3A = \mathcal{P}_1\mathcal{P}_1^{'}\mathcal{P}_3$, where $\mathcal{P}_1$, $\mathcal{P}_1^{'}$ are prime ideals of A of residual degree equal to 1 and $\mathcal{P}_3$ is a prime ideal of A of residual degree equal to 3. Show that

$$\begin{aligned}\mathcal{P}_1 &= (3,\omega),\ \mathcal{P}_1^{'} = (3,\omega+1),\\ \mathcal{P}_3 &= (3,\omega-1) = (3,\theta^3-\theta^2+1) \quad \text{and}\\ \mathcal{P}_1\mathcal{P}_1^{'} &= (3,\theta-1).\end{aligned}$$

(3) Show that $\mathcal{P}_1$, $\mathcal{P}_1^{'}$ and $\mathcal{P}_3$ are principal and find for each of these ideals a generator.

(4) Show $5A = \mathcal{P}^5$ where $\mathcal{P}$ is a prime ideal of A of residual degree equal to 1. Show that $\mathcal{P} = \theta A$.

(5) Show that $2A = \mathcal{Q}_2\mathcal{Q}_3$, where $\mathcal{Q}_2$, $\mathcal{Q}_3$ are prime ideals of A having their residual degrees equal to 2 and 3 respectively. Show that

$$\mathcal{Q}_2 = (2,\theta^2+\theta+1) \quad \text{and} \quad \mathcal{Q}_3 = (2,\theta^3+\theta^2+1).$$

(6) Show that the ideals $\mathcal{Q}_2$ and $\mathcal{Q}_3$ are principal and determine for each of them a generator.

(7) Determine bases over $\mathbb{Z}$ of $\mathcal{Q}_2$ and $\mathcal{Q}_3$.

Solution 10.52.

(1) Eisenstein's irreducibility theorem shows that $f(x)$ is irreducible over $\mathbb{Q}$. We have

$$\mathrm{Disc}(\theta) = \mathrm{N}_{K/\mathbb{Q}}(f'(\theta)) = \mathrm{N}_{K/\mathbb{Q}}(5(\theta^4-1)) = 5^5\,\mathrm{N}_{K/\mathbb{Q}}(\theta^2-1)\,\mathrm{N}_{K/\mathbb{Q}}(\theta^2+1).$$

It is easy to see that the characteristic polynomial of θ^2 over $\mathbb{Q}$ is given by $g(x) = x^5 - 10x^3 + 25x - 25$. Therefore θ^2+1 (resp. θ^2-1) is a root of $g(x-1)$ (resp. $g(x+1)$). We deduce that $\mathrm{N}_{K/\mathbb{Q}}(\theta^2+1) = 41$ and $\mathrm{N}_{K/\mathbb{Q}}(\theta^2-1) = 3^2$. Therefore $\mathrm{Disc}(\theta) = 5^5\cdot 3^2\cdot 41$. We have $f(x) = x^5 - 5(x+1)$ and since $x \nmid x+1$ in $\mathbb{F}_5[x]$, then by **Exercise 10.30**, $5 \nmid I(\theta)$, thus $\nu_5(\mathrm{Disc}(K)) = 5$. We have

$$f(x) = (x-1)^2(x^3-x^2+1) + 3(x^4-x^3-x-2)$$

and since $x-1 | x^4-x^3-x-2$ in $\mathbb{F}_3[x]$, then, by **Exercise 10.30**, $3 \mid I(\theta)$. From the identity $\mathrm{Disc}(\theta) = I(\theta)^2\,\mathrm{Disc}(K)$, we conclude that $3 \nmid \mathrm{Disc}(K)$. From the same identity we conclude that $41 \mid \mathrm{Disc}(K)$. It follows that $\mathrm{Disc}(K) = 5^5\cdot 41$. Since

$$f(x) \equiv (x-1)^2(x^3-x^2+1) \pmod 3,$$

and since $x^3 - x^2 + 1$ is irreducible over $\mathbb{F}_3$ and $3 \mid I(\theta)$, then $(\theta - 1)(\theta^3 - \theta^2 + 1) \equiv 0 \pmod 3$, that is $\frac{(\theta-1)(\theta^3-\theta^2+1)}{3} \in A$. We have

$$\omega = \frac{\theta^4 + \theta^3 + \theta^2 + \theta - 1}{3} = \frac{(\theta - 1)(\theta^3 - \theta^2 + 1)}{3} + \theta^3,$$

hence $\omega \in A$. Consider the matrix M expressing the coordinates of $1, \theta, \theta^2, \theta^3, \theta^4$ in the basis $\{1, \theta, \theta^2, \theta^3, \omega\}$. It is given by:

$$M = \begin{pmatrix} 1 & 0 & 0 & 0 & 1 \\ 0 & 1 & 0 & 0 & -1 \\ 0 & 0 & 1 & 0 & -1 \\ 0 & 0 & 0 & 1 & -1 \\ 0 & 0 & 0 & 0 & 3 \end{pmatrix}.$$

Its determinant is equal to 3 and we have

$$\mathrm{Disc}(1, \theta, \theta^2, \theta^3, \theta^4) = \mathrm{Det}\, M^2\, \mathrm{Disc}(1, \theta, \theta^2, \theta^3, \omega),$$

hence

$$\mathrm{Disc}(1, \theta, \theta^2, \theta^3, \omega) = 5^5 \cdot 41 = \mathrm{Disc}(K).$$

It follows that $\{1, \theta, \theta^2, \theta^3, \omega\}$ is an integral basis.

(2) Let $\gamma = 3\omega = \theta^4 + \theta^3 + \theta^2 + \theta - 1$. Using the relation $\theta^5 = 5\theta + 5$, we obtain

$$\begin{aligned} \theta\gamma &= \theta^4 + \theta^3 + \theta^2 + 4\theta + 5 \\ \theta^2\gamma &= \theta^4 + \theta^3 + 4\theta^2 + 10\theta + 5 \\ \theta^3\gamma &= \theta^4 + 4\theta^3 + 10\theta^2 + 10\theta + 5 \\ \theta^4\gamma &= 4\theta^4 + 10\theta^3 + 10\theta^2 + 10\theta + 5. \end{aligned}$$

This shows that $1, \theta, \theta^2, \theta^3, \theta^4$ is a non trivial solution of the following homogeneous system of linear equations:

$$\begin{aligned} (-1 - \gamma)x_0 + x_1 + x_2 + x_3 + x_4 &= 0 \\ 5x_0 + (4 - \gamma)x_1 + x_2 + x_3 + x_4 &= 0 \\ 5x_0 + 10x_1 + (4 - \gamma)x_2 + x_3 + x_4 &= 0 \\ 5x_0 + 10x_1 + 10x_2 + (4 - \gamma)x_3 + x_4 &= 0 \\ 5x_0 + 10x_1 + 10x_2 + 10x_3 + (4 - \gamma)x_4 &= 0. \end{aligned}$$

This implies that

$$\begin{vmatrix} -1 - \gamma & 1 & 1 & 1 & 1 \\ 5 & 4 - \gamma & 1 & 1 & 1 \\ 5 & 10 & 4 - \gamma & 1 & 1 \\ 5 & 10 & 10 & 4 - \gamma & 1 \\ 5 & 10 & 10 & 10 & 4 - \gamma \end{vmatrix} = 0.$$

Standard computations on determinants show that the characteristic polynomial of γ is given by

$$h_1(x) = x^5 - 5\cdot 3x^4 - 10\cdot 3^3x^2 - 10\cdot 3^4x - 3^6.$$

It follows that the characteristic polynomial of ω is equal to

$$h(x) = \frac{h_1(3x)}{3^5} = x^5 - 5x^4 - 10x^2 - 10x - 3.$$

We have seen in **(1)** that $3 \mid I(\theta)$ and

$$f(x) \equiv (x-1)^2(x^3 - x^2 + 1) \pmod 3.$$

This shows that there exists one prime ideal of A lying over $3\mathbb{Z}$, having its residual degree equal to 3. This also shows that there exist two prime ideals lying over $3\mathbb{Z}$ of residual degree equal to 1 or one prime ideal of residual degree equal to 2. But looking at $h(x)$, we conclude that $\mathrm{N}_{K/\mathbb{Q}}(\omega) = 3$. Therefore ωA is a prime ideal lying over $3\mathbb{Z}$ with residual degree equal to 1. We conclude that there exist two prime ideals lying over $3\mathbb{Z}$ having their residual degree equal to 1. Therefore $3A = \mathcal{P}_1\mathcal{P}_1'\mathcal{P}_3$. The congruence modulo 3, satisfied by $f(x)$ shows that

$$\mathcal{P}_1\mathcal{P}_1' = (3, \theta - 1) \quad \text{and} \quad \mathcal{P}_3 = (3, \theta^3 - \theta^2 + 1).$$

Since $h(x) \equiv x(x+1)(x-1)^3 \pmod 3$, we conclude that

$$\mathcal{P}_3 = (3, \omega - 1),\ \mathcal{P}_1 = (3, \omega) \quad \text{and} \quad \mathcal{P}_1' = (3, \omega + 1).$$

(3) We have seen above that $\mathrm{N}_{K/\mathbb{Q}}(\omega) = 3$, hence $\mathcal{P}_1 = \omega A$, thus $\mathcal{P}_1$ is principal. Since the characteristic polynomial of $\omega - 1$ over $\mathbb{Q}$ is equal to $h(x+1)$, then $\mathrm{N}_{K/\mathbb{Q}}(\omega - 1) = 3^3$.
Therefore $\mathcal{P}_3 = (\omega - 1)A$, and then $\mathcal{P}_3$ is principal. Since $3A = \mathcal{P}_1\mathcal{P}_3\mathcal{P}_1'$, then $\mathcal{P}_1'$ is also principal. Let μ be a generator of $\mathcal{P}_1'$, then $3A = \omega(\omega-1)\mu A$. It follows that $3 = \omega(\omega - 1)\mu\epsilon$, where ϵ is a unit of A, hence $\mu = \frac{3}{\omega(\omega-1)\epsilon}$. Since μ and $\epsilon\mu$ generates the same ideal, we may suppose that $\epsilon = 1$ and then $\mu = \frac{3}{\omega(\omega-1)}$. Using the relation $\theta^5 = 5\theta + 5$, we find

$$\omega^2 = \frac{2\theta^4 + 5\theta^3 + 8\theta^2 + 11\theta + 7}{3},$$

hence

$$\omega^2 - \omega = \frac{\theta^4 + 4\theta^3 + 7\theta^2 + 10\theta + 8}{3}.$$

We deduce that

$$\frac{3}{\omega^2 - \omega} = \frac{9}{\theta^4 + 4\theta^3 + 7\theta^2 + 10\theta + 8}.$$

By performing three Euclidean divisions, we may compute the *gcd* of the polynomials $x^5 - 5x - 5$ and $x^4 + 4x^3 + 7x^2 + 10x + 8$ and then obtain a Bezout's identity relating these polynomials. Substituting in this identity θ for x, we obtain

$$\frac{27}{\theta^4 + 4\theta^3 + 7\theta^2 + 10\theta + 8} = \theta^4 - 2\theta^3 + 4\theta^2 - 5\theta - 1 = 3\omega - 3\theta^3 + 3\theta^2 - 6\theta.$$

It follows that

$$\mu = \frac{9}{\theta^4 + 4\theta^3 + 7\theta^2 + 10\theta + 8} = \omega - \theta^3 + \theta^2 - 2\theta,$$

thus $\mathcal{P}_1' = (\omega - \theta^3 + \theta^2 - 2\theta)A$.

(4) Since $f(x) = x^5 - 5(x+1)$, then $5 \nmid I(\theta)$ and 5 is totally ramified in A, that is $5A = \mathcal{P}^5$, where $\mathcal{P}$ is a prime ideal of A of residual degree equal to 1. Moreover since $\mathrm{N}_{K/\mathbb{Q}}(\theta) = 5$, then $\mathcal{P} = \theta A$, thus $\mathcal{P}$ is principal.

(5) The factorization of $f(x)$ over $\mathbb{F}_2$ into a product of irreducible factors is given by

$$f(x) \equiv (x^2 + x + 1)(x^3 + x^2 + 1) \pmod 2,$$

hence $2A = \mathcal{Q}_2\mathcal{Q}_3$, where the residual degrees of $\mathcal{Q}_2$ and $\mathcal{Q}_3$ are equal to 2 and 3 respectively. Moreover, we have

$$\mathcal{Q}_2 = (2, \theta^2 + \theta + 1) \quad \text{and} \quad \mathcal{Q}_3 = (2, \theta^3 + \theta^2 + 1).$$

(6) We compute $\mathrm{N}_{K/\mathbb{Q}}(\theta^2 + \theta + 1)$. Set $\beta = \theta^2 + \theta + 1$. Then

$$\begin{aligned}
\theta\beta &= \theta^3 + \theta^2 + \theta \\
\theta^2\beta &= \theta^4 + \theta^3 + \theta^2 \\
\theta^3\beta &= \theta^4 + \theta^3 + 5\theta + 5 \\
\theta^4\beta &= \theta^4 + 5\theta^2 + 10\theta + 5,
\end{aligned}$$

hence $1, \theta, \theta^2, \theta^3, \theta^4$ is a non trivial solution of the following homogeneous system of linear equations

$$\begin{aligned}
(1-\beta)x_0 + x_1 + x_2 &= 0 \\
(1-\beta)x_1 + x_2 + x_3 &= 0 \\
(1-\beta)x_2 + x_3 + x_4 &= 0 \\
5x_0 + 5x_1 + (1-\beta)x_3 + x_4 &= 0 \\
5x_0 + 10x_1 + 5x_2 + (1-\beta)x_4 &= 0.
\end{aligned}$$

It follows that

$$\begin{vmatrix} 1-\beta & 1 & 1 & 0 & 0 \\ 0 & 1-\beta & 1 & 1 & 0 \\ 0 & 0 & 1-\beta & 1 & 1 \\ 5 & 5 & 0 & 1-\beta & 1 \\ 5 & 10 & 5 & 0 & 1-\beta \end{vmatrix} = 0.$$

Straightforward computations show that the characteristic polynomial of β over $\mathbb{Q}$ is given by

$$q(x) = x^5 - 3x^4 + 24x^3 - 11x^2 - 36x + 36,$$

hence $\mathrm{N}_{K/\mathbb{Q}}(\beta) = -3^2 \cdot 2^2$. It follows that

$$\begin{aligned} (\theta^2 + \theta + 1)A &= \mathcal{Q}_2\mathcal{P}_1^2 \quad \text{or} \\ (\theta^2 + \theta + 1)A &= \mathcal{Q}_2\mathcal{P}_1^{'2} \quad \text{or} \\ (\theta^2 + \theta + 1)A &= \mathcal{Q}_2\mathcal{P}_1\mathcal{P}_1^{'}. \end{aligned}$$

Since by **(3)**, $\mathcal{P}_1$ and $\mathcal{P}_1^{'}$ are principal then so are $\mathcal{P}_1^2$, $\mathcal{P}_1^{'2}$ and $\mathcal{P}_1\mathcal{P}_1^{'}$. It follows that $\mathcal{Q}_2$ is principal and then $\mathcal{Q}_3$ is also principal. Moreover $\mathcal{Q}_2$ is generated by one and only one of the following elements:

$$\begin{aligned} &(\theta^2 + \theta + 1)/\omega^2, \\ &(\theta^2 + \theta + 1)/(\omega - \theta^3 + \theta^2 - 2\theta)^2 \quad \text{and} \\ &(\theta^2 + \theta + 1)/\omega(\omega - \theta^3 + \theta^2 - 2\theta), \end{aligned}$$

according to the above possible factorizations of $(\theta^2 + \theta + 1)A$. Indeed, we must find among these three elements the only one which is an algebraic integer. We have seen that $\mathcal{P}_1\mathcal{P}_1^{'} = (3, \theta - 1)$, hence $\mathcal{P}_1\mathcal{P}_1^{'} \mid \theta - 1$. It follows that $\mathcal{P}_1^2\mathcal{P}_1^{'2} \mid (\theta - 1)^2$. From the splitting of the prime 3 it is seen that $\nu_{\mathcal{P}_1}(3\theta) = \nu_{\mathcal{P}_1^{'}}(3\theta) = 1$. Now the identity

$$\theta^2 + \theta + 1 = (\theta - 1)^2 + 3\theta$$

shows that $\mathcal{P}_1^2 \nmid \theta^2 + \theta + 1$ and $\mathcal{P}_1^{'2} \nmid \theta^2 + \theta + 1$. It follows that we have the third possibility for the generator, namely

$$\mathcal{Q}_2 = \frac{(\theta^2 + \theta + 1)}{\omega(\omega - \theta^3 + \theta^2 - 2\theta)} A = (1 + \theta + \omega)A.$$

Since $2A = \mathcal{Q}_2\mathcal{Q}_3$, then $2/(1 + \theta + \omega)$ is a generator of $\mathcal{Q}_3$, thus

$$\mathcal{Q}_3 = (-3 - \theta^2 - \theta^3 + 2\omega)A.$$

(7) We know that $A/\mathcal{Q}_2$ is isomorphic to $\mathbb{F}_{2^2}$, so there exists a surjective morphism of rings $\phi : A \to \mathbb{F}_{2^2}$ such that $\operatorname{Ker}\phi = \mathcal{Q}_2$. To define ϕ it is sufficient to define ϕ on $\mathbb{Z}$ and also the value of $\phi(\theta)$. Obviously for any $a \in \mathbb{Z}$, $\phi(a) = \bar{a}$, where $\bar{a}$ denotes the class of a modulo 2. Since $\theta^2 + \theta + 1 = 0$ in $A/\mathcal{Q}_2$, then $\phi(\theta^2 + \theta + 1) = 0$ in $\mathbb{F}_{2^2}$, hence $\phi(\theta)^2 + \phi(\theta) + 1 = 0$ in $\mathbb{F}_{2^2}$, i.e. $\phi(\theta)$ is a root $x^2 + x + 1$. Set $\rho = \phi(\theta)$. Let $\alpha = a_0 + a_1\theta + a_2\theta^2 + a_3\theta^3 + a_4\omega$ be an element of A, then

$$\begin{aligned}
\alpha \in \mathcal{Q}_2 &\Leftrightarrow \alpha \in \operatorname{Ker}(\phi)\\
&\Leftrightarrow \overline{a_0} + \overline{a_1}\rho + \overline{a_2}\rho^2 + \overline{a_3}\rho^3 + \overline{a_4}(\rho^4 + \rho^3 + \rho^2 + \rho + 1) = 0\\
&\Leftrightarrow \overline{a_0} + \overline{a_2} + \overline{a_3} + \overline{a_4} + (\overline{a_1} + \overline{a_2} + \overline{a_4})\rho = 0\\
&\Leftrightarrow \begin{cases} \overline{a_0} + \overline{a_2} + \overline{a_3} + \overline{a_4} = 0\\ \overline{a_1} + \overline{a_2} + \overline{a_4} = 0 \end{cases}\\
&\Leftrightarrow \begin{cases} a_1 = a_2 + a_4 + 2\lambda\\ a_0 = a_2 + a_4 + a_3 + 2\mu \end{cases}
\end{aligned}$$

where λ and μ in $\mathbb{Z}$. We deduce that

$$\begin{aligned}
\alpha \in \mathcal{Q}_2 &\Leftrightarrow \alpha = (a_2 + a_4 + a_3 + 2\mu) + (a_2 + a_4 + 2\lambda)\theta + a_2\theta^2 + a_3\theta^3 + a_4\omega\\
&\Leftrightarrow \alpha = a_2(1 + \theta + \theta^2) + a_3(1 + \theta^3) + a_4(1 + \theta + \omega) + 2\mu + \lambda(2\theta).
\end{aligned}$$

We have

$$\begin{aligned}
1 + \theta + \omega &= \frac{\theta^4 + \theta^3 + \theta^2 + \theta - 1}{3} + \theta + 1\\
&= \frac{\theta^4 + \theta^3 + \theta^2 + 4\theta + 2}{3}\\
&\equiv \theta^4 + \theta^3 + \theta^2 \pmod 2\\
&\equiv 0 \pmod{\mathcal{P}_2}.
\end{aligned}$$

Clearly all the elements $1 + \theta + \theta^2$, $1 + \theta^3$, 2, 2θ and $1 + \theta + \omega$ belong to $\mathcal{P}_2$. Therefore they constitute a set of generators of $\mathcal{P}_2$ as a $\mathbb{Z}$ module. Obviously this set is free over $\mathbb{Z}$. Therefore it is a basis of $\mathcal{P}_2$ over $\mathbb{Z}$. We use the same method for the determination of a $\mathbb{Z}$- basis of $\mathcal{P}_3$. Let $\gamma = a_0 + a_1\theta + a_2\theta^2 + a_3\theta^3 + a_4\omega$ with $a_i \in \mathbb{Z}$ for $i = 0, 1, \ldots, 4$. Then $\gamma \in \mathcal{P}_3$ if and only if

$$\overline{a_0} + \overline{a_1}\tau + \overline{a_2}\tau^2 + \overline{a_3}\tau^3 + \overline{a_4}(\tau^4 + \tau^3 + \tau^2 + \tau + 1) = 0,$$

where τ is a root of $x^3 + x^2 + 1$ in $\mathbb{F}_{2^3}$. We deduce that

$$\begin{aligned}
\gamma \in \mathcal{P}_3 &\Leftrightarrow \overline{a_0} + \overline{a_1}\tau + \overline{a_2}\tau^2 + \overline{a_3}(\tau^2 + 1) + \overline{a_4}(\tau^2 + 1) = 0\\
&\Leftrightarrow \overline{a_0} + \overline{a_3} + \overline{a_4} = 0, \overline{a_1} = 0, \overline{a_2} + \overline{a_3} + \overline{a_4} = 0\\
&\Leftrightarrow a_1 = 2\lambda, a_2 = a_3 + a_4 + 2\mu, a_0 = a_3 + a_4 + 2\nu,
\end{aligned}$$

where λ, μ and $\nu \in \mathbb{Z}$. We conclude (and we omit the proof) that

$$\{2, 2\theta, 2\theta^2, 1+\theta^2+\theta^3, 1+\theta^2+\omega\}$$

is a $\mathbb{Z}$-basis of $\mathcal{P}_3$.

Exercise 10.53.
Let K be a number field, A be its ring of integers,

$$f(x) = a_n x^n + \cdots + a_0,\ g(x) = b_m x^m + \cdots + b_0 \text{ and } h(x) = c_k x^k + \cdots = c_0$$

be polynomials with coefficients in K such that $f(x) = g(x)h(x)$. Consider the fractional ideals of K,

$$\mathcal{A} = (a_0, \ldots, a_n), \quad \mathcal{B} = (b_0, \ldots, b_m) \quad \text{and} \quad \mathcal{C} = (c_0, \ldots, c_k).$$

(1) Show that $\mathcal{A} = \mathcal{B}\mathcal{C}$.
(2) Deduce that if $a_0, \ldots, a_n \in A$, then $b_i c_j \in A$ for any $i \in \{1, \ldots, m\}$ and any $j \in \{1, \ldots, k\}$.
(3) Suppose that all the coefficients a_h belong to A ant that there exists $\alpha \in A$ such that $\alpha \mid a_h$ for $h = 0, \ldots, n$. Show that $b_i c_j \in A$ and $\alpha \mid b_i c_j$ for any (i, j).
(4) Let $F(x)$ be a non constant polynomial with coefficients in A and let $a \in A$ be its leading coefficient. Show that if $F(x)$ is reducible in $K[x]$ then $aF(x)$ is a product of two non constant polynomials in $A[x]$.

Solution 10.53.

(1) Let $\mathcal{P}$ be a prime ideal of A, $e_1 = \nu_{\mathcal{P}}(\mathcal{B})$ and $e_2 = \nu_{\mathcal{P}}(\mathcal{C})$. To get the result, it is sufficient to prove that $\nu_{\mathcal{P}}(\mathcal{A}) = e_1 + e_2$. By definition of e_1 and e_2, for any i and j, $\nu_{\mathcal{P}}(b_i) \geq e_1$, $\nu_{\mathcal{P}}(c_j) \geq e_2$ and there exist i_0 and j_0 such that $\nu_{\mathcal{P}}(b_{i_0}) = e_1$, $\nu_{\mathcal{P}}(c_{j_0}) = e_2$. Moreover we may suppose that i_0 and j_0 are minimal. Since $a_h = \sum_{i+j=h} b_i c_j$, then $\nu_{\mathcal{P}}(a_h) \geq e_1 + e_2$ and

$$\nu_{\mathcal{P}}(a_{i_0+j_0}) = \nu_{\mathcal{P}}(\cdots + b_{i_0-1}c_{j_0+1} + b_{i_0}c_{j_0} + b_{i_0+1}c_{j_0-1} + \cdots) = e_1 + e_2.$$

(2) Since $b_i \in \mathcal{B}$, then $b_i A = \mathcal{B}I$, where I is an ideal of A. Similarly, $c_j A = \mathcal{C}J$, where J is an ideal of A. From **(1)**, we deduce that

$$b_i c_j A = \mathcal{B}\mathcal{C}IJ = \mathcal{A}IJ,$$

hence $b_i c_j \in A$.
(3) Apply the result of **(2)**, for the polynomials $f(x)/\alpha$, $g(x)/\alpha$ and $h(x)$.

(4) Since $F(x)$ is reducible in $K[x]$, then it is so for the monic polynomial $F(x)/a$. Let $G(x)$ and $H(x) \in K[x]$ be monic such that $F(x)/a = G(x)H(x)$. Multiplying by a^2, we get $aF(x) = \big(aG(x)\big)\big(aH(x)\big)$. Set

$$G(x) = b_m x^m + b_{m-1}x^{m-1} + \cdots + b_0 \quad \text{and}$$
$$H(x) = c_k x^k + c_{k-1}x^{k-1} + \cdots + c_0$$

with $b_m = c_k = 1$, then by **(3)** $ab_iac_j \in A$ and $a \mid ab_iac_j$ for any (i,j). In particular $a \mid a^2b_i$ and $a \mid a^2c_j$ for any i,j. We deduce that $ac_i \in A$ and $ac_j \in A$ for any i,j. This implies that $aG(x)$ and $aH(x) \in A[x]$, showing that $aF(x)$ is a product of two non constant polynomials with coefficients in A.

Remarks. (1) generalizes the classical result on the content of a product of polynomials with coefficients in the fraction field of a factorial domain. **(2)** is a generalization of **Exercise 1.10**.

Exercise 10.54.
Let K be a number field and A be its ring of integers. Let $f(x)$ be a polynomial of degree n with integral coefficients, a_n and a_0 be its leading and constant coefficients respectively. Suppose that $f(x)$ is irreducible in $\mathbb{Z}[x]$ and a_n or a_0 is irreducible in A. Show that $f(x)$ is irreducible in $A[x]$.

Solution 10.54.

- **First case: a_n irreducible.** By contradiction, suppose that $f(x)$ is reducible in $A[x]$. If $f(x) = \lambda h(x)$ with $h(x) \in A[x]$ and λ a non zero constant and a non unit. Then λ divides each coefficient a_i of $f(x)$. But since $f(x$ is irreducible over $\mathbb{Z}$, then $f(x)$ is primitive, which implies, the existence of coefficients u_i such that $\sum u_ia_i = 1$. Therefore $\lambda \mid 1$, that is $\lambda = \pm 1$, hence a contradiction. Suppose next that $f(x) = g(x)h(x)$, where $g(x)$ and $h(x)$ are non constant polynomials with coefficients in A. Let b and c be the leading coefficients of $g(x)$ and $h(x)$ respectively, then $a_n = bc$. Since a_n is irreducible in A, then $b = \epsilon$ or $c = \epsilon$ with $\epsilon = \pm 1$. We may suppose that $b = \epsilon$. We have

$$f(x) = \epsilon^{-1}g(x)\epsilon h(x) = g_1(x)h_1(x),$$

where now $g_1(x)$ is monic. Let α be a root of $g_1(x)$, then α is integral over A, hence integral over $\mathbb{Z}$. Since α is a root of $f(x)$ and since this polynomial is irreducible over $\mathbb{Z}$, then $a = \pm 1$ contradicting our assumptions.

- **Second case: a_0 irreducible.** Let $g(x) = x^n f(1/x)$. The leading coefficient of this polynomial is equal to a_0, thus irreducible in A. We apply for $g(x)$, what was proved in the first case and we conclude that $g(x)$ is irreducible in $A[x]$, which in turn implies that $f(x)$ is irreducible in $A[x]$.

Exercise 10.55.
Let K be a number field and A be its ring of integers.

(1) Let h be the class number of K and $\{I_1, \ldots, I_h\}$ be a complete set of representatives of the ideal classes. Let m_0 be a positive integer divisible by $I_1, \ldots, I_h$. For any $g(x) \in A[x]$, let $c(g)$ be the ideal of A generated by the coefficients of $g(x)$. Let $I(g) \in \{I_1, \ldots, I_h\}$ such that $c(g)I(g)$ is principal and let $\alpha(g) \in A$ such that

$$c(g)I(g) = \alpha(g)A. \qquad \text{(Eq 1)}$$

Let $J(g)$ be the ideal of A such that

$$m_0 A = I(g)J(g). \qquad \text{(Eq 2)}$$

(a) Let $\hat{g}(x) = m_0 g(x)/\alpha(g)$. Show that $\hat{g}(x) \in A[x]$ and

$$c(\hat{g}) = J(g). \qquad \text{(Eq 3)}$$

(b) Let $f(x) \in A[x]$ such that $f(x)$ is reducible over K. Show that $m_0^3 f(x)$ may be decomposed in $A[x]$ as a product of two non constant polynomials.

(2) Let $f(x) = 3x^2 + 4x + 3$. Show that f is reducible over $\mathbb{Q}(\sqrt{-5})$, but irreducible over $\mathbb{Z}[\sqrt{-5}]$. Show $5f(x)$ cannot be decomposed as a product of two non constant polynomials in $\mathbb{Z}[\sqrt{-5}][x]$.
(3) Factorize $2f(x)$, $3f(x)$ and $7f(x)$ in $\mathbb{Z}[\sqrt{-5}][x]$.
(4) For any $g(x) \in A[x]$ such that $g(x)$ is reducible over K, let

$$\begin{aligned}\mathcal{D}(g) &= \{0\} \cup \{d \in \mathbb{Z} \setminus \{0\},\\ dg(x) &= g_1(x)g_2(x) \quad \text{with} \quad g_1(x) \quad \text{and} \quad g_2(x) \in A[x] \setminus A\}.\end{aligned}$$

Show that, in general, $\mathcal{D}(g)$ is not an ideal of $\mathbb{Z}$.
(5) Let $g(x)$ be a non constant polynomial with coefficients in A and let a be its leading coefficient. Show that if $g(x)$ is reducible over K, then $a \in \mathcal{D}(g)$.

Solution 10.55.

(1)(a) Set $g(x) = b_k x^k + b_{k-1}x^{k-1} + \cdots + b_0$, where $b_j \in A$ for $j = 0, \ldots, k$. Let $\mathcal{P}$ be a prime ideal of A such that $\mathcal{P}^e \mid\mid \alpha(g)$. From **(Eq 1)**, $\mathcal{P}^e \mid\mid c(g)I(g)$. It follows that $\mathcal{P}^{e_1} \mid\mid c(g)$ and $\mathcal{P}^{e_2} \mid\mid I(g)$, with $e_1 + e_2 = e$. Since $I(g) \mid m_0 A$, then $\mathcal{P}^{e_2} \mid m_0 A$, hence $\mathcal{P}^e \mid m_0 b_j$ for $j = 0, \ldots, k$. Thus $\hat{g}(x) \in A[x]$.
We have $c(\hat{g}) = (m_0 A)c(g)(\alpha(g)A)^{-1}$, hence by **(Eq 1)** and **(Eq 2)**,

$$c(\hat{g}) = I(g)J(g)(\alpha(g)A)(I(g))^{-1}(\alpha(g)A)^{-1} = J(g).$$

(b) Let $I(f)$, $c(f)$, $\alpha(f)$ and $\hat{f}(x)$ as defined in the statement of the exercise. We have $m_0 f(x) = \alpha(f)\hat{f}(x)$. Since $f(x)$ is reducible over K, then so is $\hat{f}(x)$. Let $\hat{f}(x) = g(x)h(x)$ be a non trivial factorization of $\hat{f}(x)$ in $K[x]$, then

$$g(x) = g_1(x)/d \quad \text{and} \quad h(x) = h_1(x)/\delta,$$

where d and δ are positive integers and $g_1(x)$, $h_1(x) \in A[x]$. By (a), we have

$$g_1(x) = \alpha(g_1)\hat{g}_1/m_0 \quad \text{and} \quad h_1(x) = \alpha(h_1)\hat{h}_1/m_0,$$

hence

$$\hat{f} = g_1(x)h_1x)/d\delta = \alpha(g_1)\alpha(h_1)\hat{g}_1(x)\hat{h}_1(x)/m_0^2 d\delta = \lambda\hat{g}_1(x)\hat{h}_1(x),$$

where $\lambda = \alpha(g_1)\alpha(h_1)/m_0^2 d\delta$. Therefore, $c(\hat{f}) = (\lambda A)c(\hat{g}_1)c(\hat{h}_1)$. Since

$$m_0 A = I(g_1)J(g_1) = I(h_1)J(h_1),$$

then $J(g_1)$ and $J(h_1)$ divide $m_0 A$. Since

$$(m_0 A)^2 c(\hat{f})(J(g_1))^{-1}(J(h_1))^{-1} = (m_0 A)^2 \lambda A,$$

then $m_0^2\lambda \in A$. We now have,

$$m_0^3 f(x) = m_0^2 \alpha(f)\hat{f}(x) = \alpha(f)(m_0^2\lambda)\hat{g}_1(x)\hat{h}_1(x),$$

hence $m_0^3 f(x)$ is reducible in $A[x]$.

(2) We have $f(x) = 3(x + \frac{2+\sqrt{-5}}{3})(x + \frac{2-\sqrt{-5}}{3})$ hence $f(x)$ is reducible over $\mathbb{Q}(\sqrt{-5})$. By contradiction, suppose that $f(x)$ is reducible over $\mathbb{Z}[\sqrt{-5}]$. Let $f(x) = (\alpha x + \beta)(\gamma x + \delta)$ be a factorization of $f(x)$ in $A[x]$, then $\alpha\gamma = 3$, $\beta\delta = 3$ and $\alpha\delta + \beta\gamma = 4$. The first two equations imply that

(i) $N_{K/\mathbb{Q}}(\alpha) = 1,\quad N_{K/\mathbb{Q}}(\gamma) = 9$ or
(ii) $N_{K/\mathbb{Q}}(\alpha) = 3,\quad N_{K/\mathbb{Q}}(\gamma) = 3$ or
(iiii) $N_{K/\mathbb{Q}}(\alpha) = 9,\quad N_{K/\mathbb{Q}}(\gamma) = 1$ and
(a) $N_{K/\mathbb{Q}}(\beta) = 1,\quad N_{K/\mathbb{Q}}(\delta) = 9$ or
(b) $N_{K/\mathbb{Q}}(\beta) = 3,\quad N_{K/\mathbb{Q}}(\delta) = 3$ or
(c) $N_{K/\mathbb{Q}}(\beta) = 9,\quad N_{K/\mathbb{Q}}(\delta) = 1.$

Clearly 3 is not the norm of an integer of $\mathbb{Q}(\sqrt{-5})$. Therefore we may exclude (*ii*) and (*b*). By symmetry we must consider only the combinations (*i*)-(*a*) and (*i*)-(*c*).
Suppose that

$$N_{K/\mathbb{Q}}(\alpha) = 1,\quad N_{K/\mathbb{Q}}(\gamma) = 9 \quad \text{and}$$
$$N_{K/\mathbb{Q}}(\beta) = 1,\quad N_{K/\mathbb{Q}}(\delta) = 9.$$

Then $\alpha = \epsilon_1$, $\beta = \epsilon_2$, $\gamma = 3\epsilon_3$ or $\gamma = \epsilon_3(2 + \epsilon_4\sqrt{-5})$ and $\delta = 3\epsilon_5$ or $\delta = \epsilon_5(2 + \epsilon_6\sqrt{-5})$ where $\epsilon_i = \pm 1$. In any case $\alpha x + \beta = \epsilon_1 x + \epsilon_2$, so that ± 1 is a root of $f(x)$, which clearly is not true. Therefore $f(x)$ is irreducible in $\mathbb{Z}(\sqrt{-5})$.
Suppose that $5f(x) = (\alpha x + \beta)(\gamma x + \delta)$ is a factorization of $f(x)$ in $A[x]$ then $\alpha\gamma = 15$, $\beta\delta = 15$ and $\alpha\delta + \beta\gamma = 20$. We omit the proof that the splittings of $5A$ and $3A$ in A are given by $5A = \mathcal{P}^2$ and $3A = \mathcal{Q}\mathcal{Q}'$, where $\mathcal{P}$, $\mathcal{Q}$ and $\mathcal{Q}'$ are prime ideals of A of the first degree. The first two equations may be written in the form $\alpha A\gamma A = \mathcal{P}^2\mathcal{Q}\mathcal{Q}'$ and $\beta\delta = \mathcal{P}^2\mathcal{Q}\mathcal{Q}'$. Suppose that $\alpha = \epsilon$ and $\gamma = 15\epsilon$ with $\epsilon = \pm 1$. The third equation gives $\delta = 20\epsilon - 15\beta$. From the second equation, we obtain the following: $3\beta^2 - 4\epsilon\beta + 3 = 0$. It follows that $\beta = (2\epsilon \pm \sqrt{-5})/3$, which is not an element of A. Thus, this possibility is excluded. Clearly by the same method, we may exclude the possibility $\beta = \epsilon$ (resp. $\gamma = \epsilon$, $\delta = \epsilon$). Since the equation $N_{K/\mathbb{Q}}(a + b\sqrt{-5}) = a^2 + 5b^2 = 3$ has no integral solutions, then αA, βA, γA and δA cannot be equal to $\mathcal{Q}$ nor $\mathcal{Q}'$. It remains to consider the following cases.
(a) $\alpha A = \mathcal{Q}\mathcal{Q}'$ and $\gamma A = \sqrt{-5}A$ or
(b) $\alpha A = \sqrt{-5}A$ and $\gamma A = \mathcal{Q}\mathcal{Q}'$
(c) $\beta A = \mathcal{Q}\mathcal{Q}'$ and $\delta A = \sqrt{-5}A$ or
(d) $\beta A = \sqrt{-5}A$ and $\delta A = \mathcal{Q}\mathcal{Q}'$.
If **(a)** and **(b)** hold, then $\mathcal{Q}$ divides the left hand side of the third equation while this ideal does not divide the right hand side. We thus get a contradiction. Similarly, we may reject the combination **(c)**-**(d)**. If we have the combination **(a)**-**(d)**, then $\sqrt{-5}A$ divides the right hand side of the third equation and also $\beta\gamma$ but not $\alpha\delta$, which is a

contradiction. Similarly, we may reject the combination **(c)-(b)**. Therefore $5f(x)$ is not a product of two non constant polynomials in $A[x]$.

(3) Solving equations, we find the following factorizations of $2f(x)$, $3f(x)$ and $7f(x)$ in $\mathbb{Z}[\sqrt{-5}][x]$:

$$2f(x) = \left((\sqrt{-5}+1)x + (\sqrt{-5}-1)\right)\left((-\sqrt{-5}+1)x - (\sqrt{-5}+1)\right),$$
$$3f(x) = \left(3x+2+\sqrt{-5}\right)\left(3x+2-\sqrt{-5}\right),$$
$$7f(x) = \left((2\sqrt{-5}+1)x + (\sqrt{-5}+4)\right)\left((-2\sqrt{-5}+1)x - (-\sqrt{-5}+4)\right).$$

(4) In **(3)**, we have seen that $2f(x)$ and $3f(x)$ are products of non constant polynomials in $A[x]$, so that 2 and $3 \in \mathcal{D}(f)$. If this set where an ideal, then $f(x)$ would be also an element of this set, contradicting **(2)**. Thus, $\mathcal{D}(f)$ is not an ideal of $\mathbb{Z}$.

(5) See the proof in **Exercise 10.54, (4)**

Questions. **(1)** Let K be a number field such that $h(K) \geq 2$, A be its ring of integers and $n \geq 2$ be an integer. Does there exist a polynomial $g(x) \in A[x]$ of degree n, reducible in $K[x]$ such that $g(x)$ is not a product of two non constant polynomials in $A[x]$?

(2) Let $f(x)$ and $g(x)$ be polynomials with coefficients in A, reducible in $K[x]$. Suppose that they cannot be written as products of non constant polynomials in $A[x]$. What could be said about the relation between $\mathcal{D}(f)$ and $\mathcal{D}(g)$?

(3) Let $g(x) \in A[x]$ be irreducible but reducible in $K[x]$. Let p be a prime number such that any of its prime ideal factors is principal. Is $pf(x)$ a product of non constant polynomials in $A[x]$?

Exercise 10.56.
Let p be an odd prime number and $K = \mathbb{Q}(\sqrt{-p})$. Suppose that $h(K) = 1$.

(1) Show that $p \equiv 3 \pmod 8$.
(2) Let q be a prime number such that $q < (p+1)/4$. Show that $\left(\frac{q}{p}\right) = -1$.
(3) Show that $(p+1)/4$ is prime.
(4) Let $f(X) = X^2 + X + (p+1)/4$. Show that $f(x)$ is a prime number for any $x \in \{0, 1, \ldots, \frac{p-3}{4} - 1\}$.

Solution 10.56.

(1) Let A be the ring of integers of K. Suppose that $p \equiv 1 \pmod 4$, then $-p \equiv 3 \pmod 4$; therefore p is ramified in A. Since $h(K) = 1$, then there exist $a, b \in \mathbb{Z}$ such that $2A = \left((a + b\sqrt{-p})A\right)^2$. It follows that

$2 = a^2 + pb^2$, which is impossible. We conclude that $p \equiv 3 \pmod 4$. Suppose that $p \equiv 7 \pmod 8$, then $-p \equiv 1 \pmod 8$. Therefore $2A = \mathcal{P}\mathcal{P}'$, where $\mathcal{P}$ and $\mathcal{P}'$ are conjugate prime ideals of A having there residual degree equal to 1. Since $h(K) = 1$, then there exist $a, b \in \mathbb{Z}$ such that $\mathcal{P} = \left(a + b\frac{1+\sqrt{-p}}{2}\right)A$. We deduce that

$$2 = \left(a + b\frac{1+\sqrt{-p}}{2}\right)\left(a + b\frac{1-\sqrt{-p}}{2}\right),$$

hence $8 = (2a+b)^2 + pb^2$. Since $p > 7$, this equation is impossible. Therefore $p \equiv 3 \pmod 8$.

(2) Notice that the condition $q < (p+1)/4$ implies that $q \neq p$. Suppose that $(\frac{q}{p}) = 1$. On the one hand, we have $(\frac{-p}{q}) = (\frac{-1}{q})(\frac{p}{q})$. On the other hand, by the quadratic reciprocity law, we have $(\frac{p}{q})(\frac{q}{p}) = (-1)^{\frac{p-1}{2}\frac{q-1}{2}}$. Hence $(\frac{-p}{q}) = (\frac{q}{p}) = 1$. It follows that qA is a product of two conjugate prime ideals of A. Therefore there exist $a, b \in \mathbb{Z}$ such that

$$\mathbb{N}_{K/\mathbb{Q}}\left(a + b\frac{1+\sqrt{-p}}{2}\right) = q.$$

We deduce that $4q = (2a+b)^2 + pb^2 \geq p$. Since $b \neq 0$, this inequality contradicts our assumption. We conclude that $(\frac{q}{p}) = -1$.

(3) Suppose that the integer $(p+1)/4$ is not prime and let q be one of its prime factors. By **(2)**, we have $(\frac{q}{p}) = -1$. Using, as previously the quadratic reciprocity law, we obtain $(\frac{-p}{q}) = -1$. Therefore q is inert in A. Since $q \mid (p+1)/4$ and $(p+1)/4 = \frac{1+\sqrt{-p}}{2}\frac{1-\sqrt{-p}}{2}$, then $q \mid \frac{1+\sqrt{-p}}{2}$ or $q \mid \frac{1-\sqrt{-p}}{2}$. We deduce that there exist $a, b \in \mathbb{Z}$ such that $\frac{1\pm\sqrt{-p}}{2} = q(a + b\frac{1+\sqrt{-p}}{2})$. From this, we obtain $qb = \pm 1$, which is impossible. This proves that $(p+1)/4$ is prime.

(4) Suppose that there exists $x_0 \in \{0, 1, \ldots, \frac{p-3}{4} - 1\}$ such that $f(x_0)$ is not prime and let q be a prime factor of $f(x_0)$. Clearly $f(x_0)$ is odd so q is odd. Let $a \in \mathbb{Z}$, $a \geq 2$ such that $f(x_0) = aq$. We may suppose that

$$q^2 \leq f(x_0) = x_0^2 + x_0 + (p+1)/4.$$

We have

$$\begin{aligned} 4q^2 &\leq 4x_0^2 + 4x_0 + 1 + p \\ &= (2x_0+1)^2 + p < \left(\frac{p-3}{2} + 1\right)^2 + p \\ &= (p^2+1)/4 < \left(\frac{p+1}{2}\right)^2. \end{aligned}$$

Therefore $q < (p+1)/4$ and then, by (2), $(\frac{q}{p}) = -1$. We deduce that

$$\begin{aligned}\left(\frac{-p}{q}\right) &= \left(\frac{-1}{q}\right)\left(\frac{p}{q}\right)\\ &= \left(\frac{-1}{q}\right)\left(\frac{q}{p}\right)(-1)^{\frac{p-1}{2}\frac{q-1}{2}}\\ &= (-1)^{(\frac{p-1}{2}+1)\frac{q-1}{2}}\left(\frac{q}{p}\right) = -1.\end{aligned}$$

On the other hand, we have $4aq = (2x_0+1)^2 + p$, hence $(\frac{-p}{q}) = 1$ which is a contradiction.

Exercise 10.57.
Let p be a prime number such that $p \equiv -1 \pmod 4$, $\theta = e^{2i\pi/p}$ and $K = \mathbb{Q}(\theta)$.

(1) Show that K contains one and only one quadratic subfield, namely $F = \mathbb{Q}(i\sqrt{p})$ and that $\text{Gal}(K, F)$ is formed by the automorphisms σ of K such that $\sigma(\theta) = \theta^r$ with $1 \le r \le p-1$ and $(\frac{r}{p}) = 1$.
(2) Show that -1 is not a square in K.
(3) Suppose that $p \equiv -1 \pmod 8$. Show that -1 is not a sum of two squares in K.

Solution 10.57.

(1) It is known that K is a cyclic extension of $\mathbb{Q}$ of degree $p-1$. The group $G = \text{Gal}(K, \mathbb{Q})$ is formed by the automorphisms σ of K such that $\sigma(\theta) = \theta^r$ for some $r \in \{1, \ldots, p-1\}$. Let

$$H = \left\{\sigma \in G,\, \sigma(\theta) = \theta^r,\, \left(\frac{r}{p}\right) = 1\right\}.$$

It is clear that H is the unique subgroup of G of index 2. Therefore K contains a unique quadratic subfield F, namely the invariant field of H. We show using two different methods that $F = \mathbb{Q}(i\sqrt{p})$.

- **First method.** Let $\phi_p(x)$ be the minimal polynomial of θ over $\mathbb{Q}$, then $\phi_p(x) = x^{p-1} + \cdots + x + 1$. The discriminant of this polynomial is given by

$$\text{Disc}(\phi_p) = \prod_{1 \le i < j \le p-1} (\theta_i - \theta_j)^2 = (-1)^{\frac{p-1}{2}} p^{p-2}.$$

  Since $p \equiv -1 \pmod 4$, then $(p-1)/2$ is odd and we may write the above identity in the form

$$-p(p^{\frac{p-3}{2}})^2 = \prod_{1 \le i < j \le p-1} (\theta_i - \theta_j)^2.$$

We deduce that
$$\pm i\sqrt{p}p^{\frac{p-3}{2}} = \prod_{1\le i<j\le p-1} (\theta_i - \theta_j).$$
This implies that $i\sqrt{p} \in K$ and then $F = \mathbb{Q}(i\sqrt{p})$.

- **Second method.** We use the Gaussian sum
$$S = \sum_{\nu=1,\ldots,p-1} \left(\frac{\nu}{p}\right)\theta^{\nu}$$
[Lang (1965), Chap. 8.3]. It is known that $S^2 = p(\frac{-1}{p})$, hence $S^2 = p(-1)^{\frac{p-1}{2}} = -p$. Therefore $S = \pm i\sqrt{p}$. The proof may be completed by similar lines as in the preceding method.

(2) Suppose that $-1 = \alpha^2$ for some $\alpha \in K$, then $\alpha = \pm i$, hence K contains the quadratic field $\mathbb{Q}(i)$, which contradicts the fact that K contains a unique quadratic subfield, namely $\mathbb{Q}(i\sqrt{p})$. Therefore -1 is not a square in K.

(3) Let R be the set of integers $r \in \{1, \ldots, p-1\}$ such that $(\frac{r}{p}) = 1$. Suppose that $-1 = \alpha^2 + \beta^2$ with α and $\beta \in K$ and $\alpha\beta \neq 0$. We may write α and β in the form $\alpha = f(\theta)$ and $\beta = g(\theta)$, where $f(x)$ and $g(x)$ are polynomials with rational coefficients of degree at most $p-1$. Applying the automorphisms, which belong to H, to both sides of the identity $-1 = (f(\theta))^2 + (g(\theta))^2$, we obtain
$$-1 = (f(\theta^r))^2 + (g(\theta^r))^2, \tag{Eq 1}$$
for any $r \in R$. Set
$$a_r = (f(\theta))^r,\ b_r = (g(\theta))^r \quad \text{and} \quad c_r = b_r/a_r.$$
We have
$$\begin{aligned}\prod_{r\in R}(a_r + b_r i) &= \prod_{r\in R} a_r \prod_{r\in R}(1 + c_r i)\\ &= \left(\prod_{r\in R} a_r\right)\Big(s_{(p-1)/2}i^{(p-1)/2} + s_{(p-1)/2-1}i^{(p-1)/2-1}\\ &\quad + \cdots + s_1 i + 1\Big)\\ &:= A + Bi,\end{aligned}$$
where
$$A = \left(\prod_{r\in R} a_r\right)(1 - s_2 + s_4 - \cdots),$$
$$B = \left(\prod_{r\in R} a_r\right)(s_1 - s_3 + s_5 - \cdots)$$

and s_j is the elementary symmetric function of degree j of the c_k. Notice that for any $j \in \{1, \dots, (p-1)/2\}$, $s_j \in \mathbb{Q}(i\sqrt{p})$. For any $r \in R$, denote by σ_r the element of H such that $\sigma(\theta) = \theta^r$. For any $t \in R$, we have

$$\begin{aligned} \sigma_t(A) &= \sigma_t\left(\prod_{r \in R} f(\theta^r)\right)(1 - \sigma_t(s_2) + \sigma_t(s_4) - \cdots) \\ &= \left(\prod_{r \in R} f(\theta^{tr})\right)(1 - s_2 + s_4 - \cdots) \\ &= A, \end{aligned}$$

therefore $A \in \mathbb{Q}(i\sqrt{p})$. The same conclusion holds for B. Set

$$A = u + vi\sqrt{p} \quad \text{and} \quad B = w + zi\sqrt{p},$$

with $u, v, w, z \in \mathbb{Q}$, then from Equation **(Eq 1)**, we obtain

$$-1 = a_r^2 + b_r^2 = (a_r + ib_r)(a_r - ib_r),$$

for any $r \in R$. We multiply all these equations and we obtain

$$\begin{aligned} (-1)^{(p-1)/2} &= (A + iB)(A - iB) \\ &= A^2 + B^2 \\ &= u^2 + w^2 - p(v^2 + z^2) + 2i\sqrt{p}(uv + wz). \end{aligned}$$

Since $p \equiv 1 \pmod 8$, we deduce that

$$u^2 + w^2 - p(v^2 + z^2) = -1 \quad \text{and} \quad uv + wz = 0.$$

The first of these equations implies that $(v, z) \neq (0, 0)$. Suppose that one of the elements v or z is zero, say $v = 0$, then $w = 0$ and then $pz^2 = u^2 + 1$, which implies that p is a sum of two squares of integers. This claim contradicts the assumption $p \equiv -1 \pmod 4$. Therefore

$vz \neq 0$. Now, using the preceding equations, we have

$$\begin{aligned}
p = \frac{1+u^2+w^2}{v^2+z^2} &= \frac{v^2+z^2+(v^2+z^2)(u^2+w^2)}{(v^2+z^2)^2} \\
&= \left(\frac{v}{v^2+z^2}\right)^2 + \left(\frac{z}{v^2+z^2}\right)^2 + \frac{v^2u^2+v^2w^2+z^2u^2+z^2w^2}{(v^2+z^2)^2} \\
&= \left(\frac{v}{v^2+z^2}\right)^2 + \left(\frac{z}{v^2+z^2}\right)^2 + \frac{w^2z^2+v^2w^2+z^2u^2+z^2w^2}{(v^2+z^2)^2} \\
&= \left(\frac{v}{v^2+z^2}\right)^2 + \left(\frac{z}{v^2+z^2}\right)^2 + \frac{v^2(wz/v)^2+v^2w^2+z^2(wz/v)^2+z^2w^2}{(v^2+z^2)^2} \\
&= \left(\frac{v}{v^2+z^2}\right)^2 + \left(\frac{z}{v^2+z^2}\right)^2 + \frac{w^2(v^4+z^4+2v^2z^2)}{v^2(v^2+z^2)^2} \\
&= \left(\frac{v}{v^2+z^2}\right)^2 + \left(\frac{z}{v^2+z^2}\right)^2 + \left(\frac{w}{v}\right)^2.
\end{aligned}$$

This shows that p is a sum of three squares of rational numbers. Since $p \equiv -1 \pmod 8$, then there exists $(c, d, e) \in \mathbb{Z}^3$ such that

$$-1 \equiv c^2 + d^2 + e^2 \pmod 8.$$

The set of squares in the ring $\mathbb{Z}/8\mathbb{Z}$ is $\{0, 1, 4\}$. Choosing in any way three elements of this set, distinct or not, and adding then, never results -1. Therefore the proof is complete.

Exercise 10.58.
Let K be a number field of degree n and A be its ring of integers.

(1) Let $\mathcal{P}$ be a prime ideal of A and $\mathrm{N}(\mathcal{P}) = |A/\mathcal{P}|$. Show that for any $\alpha \in A$, $\alpha^{\mathrm{N}(\mathcal{P})} \equiv \alpha \pmod{\mathcal{P}}$.

(2) Let I be a non zero ideal of A. Show that the following conditions are equivalent.

  (i) For any $\alpha \in A$, $\alpha^{\mathrm{N}(I)} \equiv \alpha \pmod I$.

  (ii) The ideal I is square free and $\mathrm{N}(\mathcal{P}) - 1 \mid \mathrm{N}(I) - 1$ for any prime ideal $\mathcal{P}$ of A dividing I.

  [A non prime ideal of A satisfying the above equivalent conditions is called a Carmichael ideal.]

(3) Suppose that $K/\mathbb{Q}$ is Galois and let p be prime number, not ramified in K. Show that $\alpha^{N_{K/\mathbb{Q}}(p)} \equiv \alpha \pmod{pA}$. Deduce that pA is prime or Carmichael.

(4) Let n be a composite integer such that $n \geq 3$. Show that there exists a quadratic extension K of $\mathbb{Q}$ such that $n \nmid \mathrm{Disc}(K/\mathbb{Q})$ and $\alpha^{\mathrm{N}_{K/\mathbb{Q}}(n)} \not\equiv \alpha \pmod{nA}$ for some $\alpha \in A$.

Solution 10.58.

(1) Since the congruence is trivial if $\alpha = 0$, we suppose that $\alpha \neq 0$. Since $A/\mathcal{P}$ is a field of cardinality $N(\mathcal{P})$, then $\alpha^{N(\mathcal{P})-1} \equiv 1 \pmod{\mathcal{P}}$, hence the result.

(2) • $(i) \Rightarrow (ii)$. Let $I = \mathcal{P}_1^{h_1} \cdots \mathcal{P}_s^{h_s}$ be the splitting of I as a product of powers of prime ideals of A, for some integer $s \geq 1$. Fix $i \in \{1, \ldots, s\}$ and let α_i such that $\alpha_i + \mathcal{P}_i$ is a generator of the group $(A/\mathcal{P}_i)^\star$. Since $\alpha_i^{\mathrm{N}(I)} \equiv \alpha_i \pmod{I}$, then $\alpha_i^{\mathrm{N}(I)} \equiv \alpha_i \pmod{\mathcal{P}_i}$, hence $\alpha_i^{\mathrm{N}(I)-1} \equiv 1 \pmod{\mathcal{P}_i}$. Therefore, the order of $\alpha_i + \mathcal{P}_i$, which is equal to $\mathrm{N}(\mathcal{P}_i) - 1$ divides $\mathrm{N}(I) - 1$. We now show that I is square free. Suppose that one of the exponents h_i, say h_1, is at least equal to 2. We have

$$|(A/\mathcal{P}_1^2)^\star| = \mathrm{N}(\mathcal{P}_1)\big(\mathrm{N}(\mathcal{P}_1) - 1\big) = p_1^{k_1}(p_1^{k_1} - 1),$$

where p_1 is a prime number and k_1 is a positive integer. Let $\alpha \in A$ such that $\alpha + \mathcal{P}_1^2$ is of order p_1 in the group $(A/\mathcal{P}_1^2)^\star$. Since $\alpha_1^{\mathrm{N}(I)} \equiv \alpha_1 \pmod{I}$, then $\alpha_1^{\mathrm{N}(I)} \equiv \alpha_1 \pmod{\mathcal{P}_1^2}$, hence $\alpha_1^{\mathrm{N}(I)-1} \equiv 1 \pmod{\mathcal{P}_1^2}$. Therefore, the order of $\alpha + \mathcal{P}_1^2$ in $(A/\mathcal{P}_1^2)^\star$, which is equal to p_1 divides $N(I) - 1$. Since

$$\mathrm{N}(I) = \mathrm{N}(\mathcal{P}_1^{h_1}) \cdots \mathrm{N}(\mathcal{P}_s^{h_s}) = p_1^{h_1} \cdots p_s^{h_s},$$

then clearly $p_1 \nmid \mathrm{N}(I) - 1$ and we have reached a contradiction. It follows that $h_1 = \cdots = h_s = 1$ and I is square free.

• $(ii) \Rightarrow (i)$. Since (i) is trivial if $\alpha \in I$, we suppose that $\alpha \notin I$. It follows that, for any prime ideal $\mathcal{P}$ of A dividing I, $\alpha \notin \mathcal{P}$. Hence $\alpha^{\mathrm{N}(\mathcal{P})-1} \equiv 1 \pmod{\mathcal{P}}$. Therefore $\alpha^{N(I)-1} \equiv 1 \pmod{\mathcal{P}}$. Since I is square free, then $\alpha^{\mathrm{N}(I)-1} \equiv 1 \pmod{I}$. We conclude that $\alpha^{\mathrm{N}(I)} \equiv \alpha \pmod{I}$.

(3) Suppose that pA is not prime and let $pA = \mathcal{P}_1 \cdots \mathcal{P}_r$ be its splitting as a product of distinct and conjugate prime ideals of A having a common residual degree over $p\mathbb{Z}$ equal to f. Then $\mathrm{N}(\mathcal{P}_1) = \ldots = \mathrm{N}(\mathcal{P}_r) = p^f$ and for any $i \in \{1, \ldots, r\}$, $p^f - 1 = N(\mathcal{P}_i) - 1$ divides $p^{rf} - 1 = N(pA) - 1$. Therefore pA is Carmichael by **(2)**.

(4) If n is not square free, then in any number field K whose ring of integers is A the ideal nA is not square free. Therefore, the conclusion follows in any quadratic field K satisfying $n \nmid \mathrm{Disc}(K/\mathbb{Q})$. Suppose that n

is square free and let p be a prime factor (necessarily odd) of n. Let $K = \mathbb{Q}(\sqrt{(-1)^{(p-1)/2}p})$, then $\mathrm{Disc}(K/\mathbb{Q}) = p$ which proves that $n \nmid \mathrm{Disc}(K/\mathbb{Q})$. Since p is ramified in K, then nA is not square free in A an the conclusion follows from **(2)**.

Exercise 10.59.

Let l and p be odd prime numbers, r an integer such that $\gcd(p, r) = 1$. Suppose that the integer $d := -(4p^l - r^2)$ is negative and square free. Let $K = \mathbb{Q}(\sqrt{d})$, A be its ring of integers and $h(K)$ be the number of ideal classes in K.

(1) Suppose that one of the prime ideal factors of pA is not principal. Show that $l \mid h(K)$.
(2) Show that the hypothesis in **(1)**, on the prime ideal factors of pA, holds if we suppose that $4p$ is not representable in the form $4p = a^2 - db^2$.
(3) Suppose that $r = 1$. Show that $4p$ is not representable in the form $4p = a^2 - db^2$.

Solution 10.59.

(1) We have $(\frac{d}{p}) = (\frac{r^2}{p}) = 1$, hence the splitting of pA is given by $pA = \mathcal{P}_1\mathcal{P}_2$, where $\mathcal{P}_1$ and $\mathcal{P}_2$ are distinct conjugate prime ideals of A. We may suppose that $\mathcal{P}_1$ is not principal. From the assumptions we see that r is odd and $d \equiv 1 \pmod 4$. An integral basis of K is given by $\{1, (1 + \sqrt{d})/2\}$ and we have $(r \pm \sqrt{d})/2 = \frac{r-1}{2} + \frac{1 \pm \sqrt{d}}{2}$. Since $4p^l = r^2 - d$, then $p^l = \frac{r+\sqrt{d}}{2}\frac{r-\sqrt{d}}{2}$, hence

$$p^l A = \frac{r + \sqrt{d}}{2} A \frac{r - \sqrt{d}}{2} A = \mathcal{P}_1^l \mathcal{P}_2^l.$$

Suppose that some ideal I divides both $\frac{r+\sqrt{d}}{2}$ and $\frac{r-\sqrt{d}}{2}$, then

$$I \mid \left(\frac{r + \sqrt{d}}{2} + \frac{r - \sqrt{d}}{2}\right) = r \quad \text{and}$$

$$I \mid \left(\frac{r + \sqrt{d}}{2} - \frac{r - \sqrt{d}}{2}\right)^2 = d.$$

But since $\gcd(p, r) = 1$, then $\gcd(d, r) = 1$. Therefore $I = A$. We conclude that $\mathcal{P}_1^l = \frac{r+\sqrt{d}}{2}A$ or $\mathcal{P}_1^l = \frac{r-\sqrt{d}}{2}A$. This shows that $\mathcal{P}_1^l$ is principal. Let e be the order of $\mathcal{P}_1$ in the group of the ideal classes, then $e \geq 2$ by assumption and $e \mid l$ since $\mathcal{P}_1^l$ is trivial in this group. Since l is prime then $e = l$. On the other hand $e \mid h(K)$, hence $l \mid h(K)$.

(2) Suppose that $4p$ is not representable in the form $4p = a^2 - db^2$. In the proof of **(1)**, we have seen that $pA = \mathcal{P}_1\mathcal{P}_2$, where $\mathcal{P}_1$ and $\mathcal{P}_2$ are distinct conjugate prime ideals of A. We must show that $\mathcal{P}_1$ or $\mathcal{P}_2$ is not principal. This statement is equivalent to $\mathcal{P}_1$ is not principal. By contradiction, suppose that $\mathcal{P}_1 = \frac{a+b\sqrt{d}}{2}A$, with $a, b \in \mathbb{Z}$ and $a \equiv b \pmod 2$, then since d is negative, we have

$$\mathrm{N}(\mathcal{P}_1) = p = |(a^2 - db^2)/4| = (a^2 - db^2)/4.$$

Hence $4p = a^2 - db^2$, which is a contradiction.

(3) Suppose that $4p = a^2 - db^2$ with $a, b \in \mathbb{Z}$, then

$$4p = a^2 - (1 - 4p^l)b^2 > 4p^l b^2 - b^2,$$

hence $4p + b^2 > 4p^l b^2$. Dividing the two sides of this last inequality by $4pb^2$ leads to $\frac{1}{b^2} + \frac{1}{4p} > p^{l-1}$, which is impossible.

Exercise 10.60.

Let $a \in \mathbb{Z}$ such that $a \geq 2$, $d = (27a^4 + 1)/4$ and

$$f(x) = x^3 - x^2 - \frac{9a^4 - 1}{4}x + a^4.$$

(1) Show that $f(x)$ is irreducible over $\mathbb{Q}$.
Hint. One may use the identity:

$$f(x) = (x - 1/3)^3 - \frac{d}{3}(x - 4/9).$$

(2) Show that $\mathrm{Disc}(f) = a^4 d^2$. Deduce that $\mathrm{Gal}(f, \mathbb{Q}) \simeq \mathbb{Z}/3\mathbb{Z}$.
[Recall that the discriminant of the polynomial $g(x) = a_0x^3 + a_1x^2 + a_2x + a_3$ is given by:

$$\mathrm{Disc}(g) = a_1^2a_2^2 - 4a_0a_2^3 - 4a_1^3a_3 - 27a_0^2a_3^2 + 18a_0a_1a_2a_3].$$

(3) Let α be a root of f and $K = \mathbb{Q}(\alpha)$. Show that $f(u(x)) \equiv 0 \pmod{f(x)\mathbb{Q}[x]}$, where $u(x) = (x^2 - \frac{a^2+1}{2}x - \frac{3a^4-a^2}{2})/a^2$. Deduce that the roots of f are α, $\beta = u(\alpha)$ and $\gamma = 1 - \alpha - \beta$. Let $g(x) = f(u(x))/f(x)$. Show that this polynomial is irreducible over $\mathbb{Q}$ and compute its roots.

(4) Show that $\{1, \alpha, \beta\}$ is a basis of K over $\mathbb{Q}$ and that $\mathrm{Disc}(1, \alpha, \beta) = d^2$.

(5) Suppose that d is square free. Show that no prime divisor of d is a factor of the index of α. Deduce that $\{1, \alpha, \beta\}$ is an integral basis of K and that $\mathrm{Disc}(K) = d^2$.

Solution 10.60.

(1) Suppose that the polynomial $f(x)$ is reducible over $\mathbb{Q}$, then it has a root $x_0 \in \mathbb{Z}$ an this root satisfies $x_0 = 0$ or $x_0 \mid a^4$. We have $f(0) = a^4 \neq 0$, hence $x_0 \neq 0$ and $x_0 \mid a^4$. We use the identity $(3x_0 - 1)^3 = d(9x_0 - 4)$. It is clear that $9x_0 - 4 \notin \{0, -1, 1\}$. Let p be a prime divisor of $9x_0 - 4$, then $p \neq 3$ and $p \mid 3x_0 - 1$. It follows that $x_0 \equiv 4/9 \pmod{p}$ and $x_0 \equiv 3/9 \pmod{p}$. Therefore $4 \equiv 3 \pmod{p}$ which is a contradiction. We conclude that $f(x)$ is irreducible over $\mathbb{Q}$.

(2) Straightforward computations.

(3) We omit the verification that $f(x) \mid f(u(x))$ in $\mathbb{Q}[x]$. Let $g(x) \in \mathbb{Q}[x]$ such that $f(u(x)) = f(x)g(x)$, then $f(u(\alpha)) = f(\alpha)g(\alpha) = 0$, hence $\beta = u(\alpha)$ is a root of $f(x)$. Since α is not a root of a quadratic equation, then $\beta \neq \alpha$. Let γ be the third root of $f(x)$. Since $\alpha + \beta + \gamma = 1$, then $\gamma = 1 - \alpha - \beta$. We have

$$f(u(x)) = (u(x) - \alpha)(u(x) - \beta)(u(x) - \gamma) = f(x)g(x).$$

Suppose that $g(x)$ is reducible over $\mathbb{Q}$, then it has a rational root, say r. This implies that $u(r) = \alpha$ or $u(r) = \beta$ or $u(r) = \gamma$, which contradicts the irreducibility of $f(x)$. Thus $g(x)$ is irreducible over $\mathbb{Q}$. Since $u(\alpha) = \beta$, $u(\beta) = \gamma$ and $u(\gamma) = \alpha$, then the roots of $g(x)$ are given by

$$(a^2 + 1)/2 - \alpha, \quad (a^2 + 1)/2 - \beta \quad \text{and} \quad (a^2 + 1)/2 - \gamma.$$

(4) Since $\beta = \left(\alpha^2 - \frac{a^2+1}{2}\alpha - \frac{3a^4-a^2}{2}\right)/a^2$, then it is clear that we may express $1, \alpha, \alpha^2$ in terms of $1, \alpha, \beta$, hence $\{1, \alpha, \beta\}$ is a basis of K over $\mathbb{Q}$. We have [Marcus (1977), Th. 6, Chap. 2]

$$\operatorname{Disc}(1, \alpha, \beta) = \begin{vmatrix} \operatorname{Tr}(1) & \operatorname{Tr}(\alpha) & \operatorname{Tr}(\beta) \\ \operatorname{Tr}(\alpha) & \operatorname{Tr}(\alpha^2) & \operatorname{Tr}(\beta\alpha) \\ \operatorname{Tr}(\beta) & \operatorname{Tr}(\alpha\beta) & \operatorname{Tr}(\beta^2) \end{vmatrix}.$$

Here, Tr denotes the trace of elements of K over $\mathbb{Q}$. From the minimal polynomial of α, it is seen that $\operatorname{Tr}(\alpha) = \operatorname{Tr}(\beta) = 1$. Let $\mu = \alpha^2$. Since

$$\alpha^3 - \alpha^2 - \alpha(9a^4 - 1)/4 + a^4 = 0,$$

then

$$\alpha(\mu - (9a^4 - 1)/4) = \mu - a^4,$$

hence $\mu(\mu - (9a^4 - 1)/4)^2 = (\mu - a^4)^2$. Therefore,

$$\mu\left(\mu - (9a^4 - 1)/4\right)^2 - (\mu - a^4)^2 = 0.$$

From this equation it is seen that

$$\mathrm{Tr}(\beta^2) = \mathrm{Tr}(\alpha^2) = \mathrm{Tr}(\mu) = (1 + 9a^4)/2.$$

We have

$$\alpha\beta = \left(\frac{1 - a^2}{2}\alpha^2 + \frac{3a^4 + 2a^2 - 1}{4}\alpha - a^4\right)/a^2,$$

hence

$$\mathrm{Tr}(\alpha\beta) = \frac{1}{a^2}\left(\frac{1 - a^2}{2}\,\mathrm{Tr}(\alpha^2) + \frac{3a^4 + 2a^2 - 1}{4}\,\mathrm{Tr}(\alpha) - a^4\right) = (1 - 9a^4)/4.$$

It follows that

$$\mathrm{Disc}(1, \alpha, \beta) = \begin{vmatrix} 3 & 1 & 1 \\ 1 & (1 + 9a^4)/2 & (1 - 9a^4)/4 \\ 1 & (1 - 9a^4)/4 & (1 + 9a^4)/2 \end{vmatrix} = \left(\frac{1 + 27a^4}{4}\right)^2 = d^2.$$

(5) Let p be a prime factor of d. Obviously, $p \neq 3$. From the identity given in **(1)**, it is seen that $f(x) \equiv (x-1/3)^3 \pmod p$ and $x - 1/3 \nmid \frac{d}{3p}(x-4/9)$ in $\mathbb{F}_p[x]$, thus $p \nmid I(\alpha)$ by **Exercise 10.30**. We have $\mathbb{Z}[\alpha] \subset \mathbb{Z}[\alpha, \beta] \subset A$. Using the values of the discriminants computed above, we obtain

$$\mathrm{Disc}(1, \alpha, \alpha^2) = \mathrm{Disc}(f) = a^4 d^2 = I(\alpha)^2\,\mathrm{Disc}(K) \quad \text{and}$$

$$\mathrm{Disc}(1, \alpha, \beta) = d^2 = (A : \mathbb{Z}[\alpha, \beta])^2\,\mathrm{Disc}(K) = \left(\frac{I(\alpha)}{(A : \mathbb{Z}[\alpha, \beta])}\right)^2 \mathrm{Disc}(K).$$

Since $\gcd(d, I(\alpha)) = 1$, then $\gcd(d, \frac{I(\alpha)}{(A:\mathbb{Z}[\alpha,\beta])}) = 1$, hence $\mathrm{Disc}(K) = d^2$ and $\{1, \alpha, \beta\}$ is a basis of A over $\mathbb{Z}$.

Exercise 10.61.

Let d be a cube free integer such that $d \geq 2$. Write d in the form $d = ab^2$, where $a, b \in \mathbb{N}$, $\gcd(a, b) = 1$ and ab square free. Let $\alpha = \sqrt[3]{d}$, $\beta = \sqrt[3]{a^2 b}$, $K = \mathbb{Q}(\alpha)$, and A be its ring of integers.

(1) Show that $\{1, \alpha, \beta\}$ is a basis of K over $\mathbb{Q}$ and that $\mathbb{Z}[\alpha, \beta] = \mathbb{Z} + \mathbb{Z}\alpha + \mathbb{Z}\beta$.
(2) Show that $\mathrm{Disc}(1, \alpha, \beta) = -27a^2b^2$ and $(A : \mathbb{Z}[\alpha, \beta]) \mid 3ab$.
(3) Let p be a prime divisor of ab. Show that p is totally ramified in A.
(4) Show that

$$3A = \begin{cases} \mathcal{P}^3 & \text{if } a^2 \not\equiv b^2 \pmod 9 \\ \mathcal{P}_1^2\mathcal{P}_2 & \text{if } a^2 \equiv b^2 \pmod 9 \end{cases},$$

where $\mathcal{P}$, $\mathcal{P}_1$ and $\mathcal{P}_2$ are prime ideals of A having their residual degree equal to 1.

(5) Let p be a prime divisor of ab. Show that $p \nmid (A : \mathbb{Z}[\alpha, \beta])$.
(6) If $a^2 \not\equiv b^2 \pmod 9$, show that $A = \mathbb{Z} + \mathbb{Z}\alpha + \mathbb{Z}\beta$ and $\operatorname{Disc}(K) = -27a^2b^2$. If $a^2 \equiv b^2 \pmod 9$, show that $A = \mathbb{Z}\alpha + \mathbb{Z}\beta + \mathbb{Z}\gamma$ and $\operatorname{Disc}(K) = -3a^2b^2$, where $\gamma = (1 + a\alpha + b\beta)/3$.

Solution 10.61.

(1) We have $\beta = \alpha^2/b$. Since $\{1, \alpha, \alpha^2\}$ is a basis of K over $\mathbb{Q}$, then so is $\{1, \alpha, \beta\}$. It is clear that $\mathbb{Z} + \mathbb{Z}\alpha + \mathbb{Z}\beta \subset Z[\alpha, \beta]$. To get the other inclusion it is sufficient to show that α^2, β^2 and $\alpha\beta \in \mathbb{Z} + \mathbb{Z}\alpha + \mathbb{Z}\beta$. This claim is verified since $\alpha^2 = b\beta$, $\beta^2 = a\alpha$ and $\alpha\beta = ab$.
(2) Let j be primitive cube root of unity in $\mathbb{C}$. The embeddings $\sigma_1, \sigma_2, \sigma_3$ of K into $\mathbb{C}$ are given by $\sigma_1 = Id_K$, $\sigma_2(\alpha) = j\alpha$ and $\sigma_3(\alpha) = j^2\alpha$. Since $\beta = \alpha^2/b$, then $\sigma_2(\beta) = j^2\alpha^2/b = j^2\beta$ and $\sigma_3(\beta) = j\alpha^2/b = j\beta$. It follows that [Marcus (1977), Chap. 2]

$$\operatorname{Disc}(1, \alpha, \beta) = \begin{vmatrix} 1 & \alpha & \beta \\ 1 & j\alpha & j^2\beta \\ 1 & j^2\alpha & j\beta \end{vmatrix}^2 = \alpha^2\beta^2 \begin{vmatrix} 1 & 1 & 1 \\ 1 & j & j^2 \\ 1 & j^2 & j \end{vmatrix}^2$$
$$= a^2b^2\Big((j-1)(j^2-1)(j^2-j)\Big)^2 = -27a^2b^2.$$

We use the formula relating the absolute discriminant of a number field, the index of an order of this field and the discriminant of a $\mathbb{Z}$-basis of this order. We obtain

$$-27a^2b^2 = \operatorname{Disc}(1, \alpha, \beta) = (A : \mathbb{Z}[\alpha, \beta])^2 \operatorname{Disc}(K),$$

hence

$$3^4a^2b^2 = (A : \mathbb{Z}[\alpha, \beta])^2(-3\operatorname{Disc}(K)).$$

It follows that $-3\operatorname{Disc}(K)$ is a square and $(A : \mathbb{Z}[\alpha, \beta])^2 \mid 3^2a^2b^2$. Therefore $(A : \mathbb{Z}[\alpha, \beta]) \mid 3ab$.
(3) Suppose that $p \mid a$ and let $\mathcal{P}$ be a prime ideal of A lying over $p\mathbb{Z}$. Since $\alpha^3 = ab^2$, then $\mathcal{P} \mid \alpha$, hence $\mathcal{P}^3 \mid ab^2$. Since $\gcd(a, b) = 1$, then $ab^2 = pc$, where $c \in \mathbb{Z}$ and $p \nmid c$. We have $\mathcal{P}^3 \mid pc$, hence $\mathcal{P}^3 \mid p$. We conclude that $pA = \mathcal{P}^3$. If $p \mid b$, we use a similar proof on replacing α by β.
(4) Since $-3\operatorname{Disc}(K)$ is a squares, then $3 \mid \operatorname{Disc}(K)$, therefore 3 is ramified in A and the splitting of $3A$ in A must have one of the following forms $3A = \mathcal{P}^3$ or $3A = \mathcal{P}_1^2\mathcal{P}_2$, where the ideals $\mathcal{P}_1$, $\mathcal{P}_2$ and $\mathcal{P}$ are of the first degree and $\mathcal{P}_1 \neq \mathcal{P}_2$. If $3 \mid ab$, then since ab is square free, we have

$a^2 \not\equiv b^2 \pmod 9$. Therefore by **(3)**, 3 is totally ramified in A and the proof is complete in this case. Suppose now that $3 \nmid ab$, then $a^2 \equiv b^2 \pmod 3$. Let $\theta = \alpha - a$, then θ satisfies the following equation

$$\theta^3 + 3a\theta\alpha + a(a^2 - b^2) = 0. \qquad \text{(Eq 1)}$$

Suppose first that $3A = \mathcal{P}^3$ and let $s = \nu_{\mathcal{P}}(\theta)$. We show that $1 \leq s \leq 2$. From this equation it is seen that $3 \mid \theta^3$, hence $\mathcal{P} \mid \theta$. Therefore $s \geq 1$. Suppose that $3 \mid \theta$, then $\theta/3$ is an algebraic integer whose minimal polynomial over $\mathbb{Q}$ is given by

$$g(x) = x^3 + \frac{a}{3}x^2 + \frac{3a^2}{9}x + \frac{a(a^2 - b^2)}{27} = 0.$$

It follows that $3 \mid a$, which is a contradiction. We conclude that $s \leq 2$. Suppose that $s = 2$. From the identity $\alpha^3 = ab^2$, we deduce that $\mathcal{P} \nmid \alpha$. We have $\nu_{\mathcal{P}}(\theta^3) = 6$, $\nu_{\mathcal{P}}(3a\theta\alpha) = 5$, which implies, by using **(Eq 1)**, $\nu_{\mathcal{P}}(a^2 - b^2) \geq 5$. Therefore $3^2 \mid a^2 - b^2$. **(Eq 1)** shows the incompatibility of of these $\mathcal{P}$-adic valuations. We conclude that $s = 1$ and again, by **(Eq 1)**, $a^2 \not\equiv b^2 \pmod 9$. Suppose now that $3A = \mathcal{P}_1^2\mathcal{P}_2$, where the ideals $\mathcal{P}_1$, $\mathcal{P}_2$ are of the first degree and $\mathcal{P}_1 \neq \mathcal{P}_2$. From Equation (4) we conclude that $\mathcal{P}_2 \mid \theta$, hence $\nu_{\mathcal{P}_2}(\theta^3 + 3a\theta\alpha) \geq 2$. Using **(Eq 1) (4)**, we obtain $\nu_{\mathcal{P}_2}(a^2 - b^2) \geq 2$. We conclude that $a^2 \equiv b^2 \pmod 9$.

(5) By contradiction, suppose that $p \mid (A : \mathbb{Z}[\alpha, \beta])$. Then, by Lagrange's Theorem, there exists $\omega \in A \setminus \mathbb{Z}[\alpha, \beta]$ such that $p\omega \in \mathbb{Z}[\alpha, \beta]$. Set

$$p\omega = x + y\alpha + z\beta \qquad \text{(Eq 2)}$$

with $x, y, z \in \mathbb{Z}$. Since $p \mid ab$, then by **(3)**, p is totally ramified in A. Let $\mathcal{P}$ be the unique prime ideal of A lying over $p\mathbb{Z}$. Suppose that $p \mid a$. Since $\alpha^3 = ab^2$, then $\nu_{\mathcal{P}}(\alpha) = 1$. Similarly, we have $\nu_{\mathcal{P}}(\beta) = 2$. **(Eq 2)** implies $p \mid x$ and then $p \mid y$ and then $p \mid z$. It follows that

$$\omega = (x/p) + (y/p)\alpha + (z/p)\beta \in \mathbb{Z}[\alpha, \beta],$$

which is a contradiction. Using similar arguments, one reach a contradiction in the case where $p \mid b$. We conclude that $p \nmid (A : \mathbb{Z}[\alpha, \beta])$.

(6) Recall that

$$-27a^2b^2 = \operatorname{Disc}(1, \alpha, \beta) = (A : \mathbb{Z}[\alpha, \beta])^2 \operatorname{Disc}(K).$$

Using **(5)**, we conclude that $a^2b^2 \mid \operatorname{Disc}(K)$. **(4)** implies that $3 \mid \operatorname{Disc}(K)$. We deduce that $(A : \mathbb{Z}[\alpha, \beta]) = 1$, $\operatorname{Disc}(K) = -27a^2b^2$ and then $A = \mathbb{Z} + \mathbb{Z}\alpha + \mathbb{Z}\beta$ or $(A : \mathbb{Z}[\alpha, \beta]) = 3$, $\operatorname{Disc}(K) = -3a^2b^2$ and A strictly contains $\mathbb{Z} + \mathbb{Z}\alpha + \mathbb{Z}\beta$. Suppose that $a^2 \not\equiv b^2 \pmod 9$, then by **(4)**, 3 is totally ramified in A. If moreover $3 \mid ab$, then by **(5)**, $3 \nmid (A : \mathbb{Z}[\alpha, \beta])$. Therefore $A = \mathbb{Z}[\alpha, \beta]$ and the proof is complete in this case. Now, if moreover $3 \nmid ab$, then $a^2 \equiv b^2 \pmod 3$. Equation **(Eq 1)** shows that $\mathcal{P} \mid \theta$ but $\mathcal{P}^2 \nmid \theta$. Let $\omega \in A$. We want to show that $\omega \in \mathbb{Z} + \mathbb{Z}\alpha + \mathbb{Z}\beta$. Since $3\omega \in \mathbb{Z} + \mathbb{Z}\alpha + \mathbb{Z}\beta$, set $3\omega = x + y\alpha + z\beta$ with $x, y, z \in \mathbb{Z}$. We have

$$\begin{aligned} 3b\omega &= bx + by(\theta + a) + z(\theta + a)^2 \\ &= bx + bya + za^2 + (by + 2za)\theta + z\theta^2 \\ &:= A + B\theta + C\theta^2, \end{aligned}$$

hence $3 \mid A$ and then $3 \mid B$. We deduce that $3 \mid C$. It follows that $x \equiv y \equiv z \equiv 0 \pmod 3$. Therefore $\omega \in \mathbb{Z}[\alpha, \beta]$ and then $A = \mathbb{Z}[\alpha, \beta]$ and $\operatorname{Disc}(K) = -27a^2b^2$.
Suppose now that $a^2 \equiv b^2 \pmod 9$ (which implies $3 \nmid ab$). By **(4)**, the splitting of $3A$ is given by $3A = \mathcal{P}_1^2\mathcal{P}_2$, where the prime ideals $\mathcal{P}_1$ and $\mathcal{P}_2$ are distinct. **(Eq 1)** shows that $\mathcal{P}_1\mathcal{P}_2 \mid \theta$ and then $\mathcal{P}_1^2\mathcal{P}_2^2 \mid \theta^2$, hence $3 \mid \theta^2$. Since

$$\begin{aligned} \theta^2 &= (\alpha - a)^2 \\ &= \alpha^2 - 2a\alpha + a^2 \\ &= b\beta - 2a\alpha + a^2 \\ &= (1 + a\alpha + b\beta) + (a^2 - 1 - 3a\alpha) \end{aligned}$$

and since $3 \mid a^2 - 1$, then $1 + a\alpha + b\beta \equiv 0 \pmod 3$. It follows that the number $\gamma = (1 + a\alpha + b\beta)/3$ is an algebraic integer and $\gamma \notin \mathbb{Z} + \mathbb{Z}\alpha + \mathbb{Z}\beta$. We have $\mathbb{Z}[\alpha, \beta] \subsetneq \mathbb{Z}[\alpha, \beta, \gamma] \subset A$ and since $(A : \mathbb{Z}[\alpha, \beta]) \leq 3$, it follows that this index is equal to 3, $A = \mathbb{Z}[\alpha, \beta, \gamma]$ and $\operatorname{Disc}(K) = -3a^2b^2$. It remains to show that $\mathbb{Z}[\alpha, \beta, \gamma] = \mathbb{Z}\alpha + \mathbb{Z}\beta + \mathbb{Z}\gamma$. Clearly the set appearing on the right side of this equality is contained in the other. To get the reverse inclusion it is sufficient to prove that the elements $1, \alpha^2, \beta^2, \gamma^2, \alpha\beta, \alpha\gamma, \beta\gamma$ belong to the set on the right side. It is easy to show the following relations:

$$\begin{aligned}
1 &= 3\gamma - a\alpha - b\beta \\
\alpha^2 &= b\beta \\
\beta^2 &= a\alpha \\
\alpha\beta &= ab(3\gamma - a\alpha - b\beta) \\
\alpha\gamma &= \frac{1-a^2b^2}{3}\alpha + \frac{ab(1-b^2}{3}\beta + ab^2\gamma \\
\beta\gamma &= \frac{1-a^2b^2}{3}\beta + \frac{ab(1-a^2}{3}\beta + a^2b\gamma \\
\gamma^2 &= \frac{a(1+b^2-2a^2b^2)}{9}\alpha + \frac{b(1+a^2-2a^2b^2)}{9}\beta + \frac{1+2a^2b^2}{3}\gamma.
\end{aligned}$$

Clearly $1 - a^2b^2 \equiv 0 \pmod 3$. It remains to prove that

$$1 + b^2 - 2a^2b^2 \equiv 0 \pmod 9 \quad \text{and}$$
$$1 + a^2 - 2a^2b^2 \equiv 0 \pmod 9.$$

By symmetry it is sufficient to prove the first congruence. This follows easily from the following: $a^2 \equiv b^2 \pmod 9 \equiv 1$ or 4 or $-2 \pmod 9$.

Exercise 10.62.

(1) Let K be a number field of degree n, A be its ring of integers, $\{\omega_1, \ldots \omega_n\}$ be an integral basis and let p be a prime number which is not ramified in A. Show that the following assertions are equivalent.

 (i) For any $i \in \{1, \ldots, n\}$, $\omega_i^p \equiv \omega \pmod{pA}$.
 (ii) For any $\theta \in A$, $\theta^p \equiv \theta \pmod{pA}$.
 (iii) The prime p completely (totally) splits in A.

(2) Suppose that the preceding equivalent conditions hold and that $p < n$. Show that, for any $\theta \in A$, $p \mid I(\theta)$.

(3) Let $f(x) = x^3 - x^2 - 2x - 8$, $\theta \in \mathbb{C}$ be a root of f and $E = \mathbb{Q}(\theta)$. Show that

 (a) $f(x)$ is irreducible over $\mathbb{Q}$.
 (b) $\mathrm{Disc}(f) = -2^2 \cdot 503$.
 (c) $2 \mid I(\theta)$ and $\mu := \theta(\theta - 1)/2$ is an algebraic integer.
 Hint. One may use **Exercise 10.30**.
 (d) $\{1, \theta, \mu\}$ is an integral basis.
 (e) The prime 2 completely splits in E and $2 \mid I(\theta)$ for any algebraic integer $\theta \in E$.

(f) Let $\mathcal{P}_1, \mathcal{P}_2, \mathcal{P}_3$ be the prime ideals of E lying over $2\mathbb{Z}$. Show that we may suppose that they are numbered in order to satisfy the following conditions:

$$\begin{cases} \theta & \equiv 0 \pmod{\mathcal{P}_1} \\ \theta & \equiv 0 \pmod{\mathcal{P}_2} \\ \theta & \equiv 1 \pmod{\mathcal{P}_3} \end{cases} \quad \text{and} \quad \begin{cases} \mu & \equiv 0 \pmod{\mathcal{P}_1} \\ \mu & \equiv 1 \pmod{\mathcal{P}_2} \\ \mu & \equiv 1 \pmod{\mathcal{P}_3} \end{cases}.$$

For each of these three prime ideals of E, determine a basis over $\mathbb{Z}$.

(g) Let $I = \theta A + \mu A$. Show that $I = \mathcal{P}_1$.

Solution 10.62.

(1) • $(i) \Rightarrow (ii)$. Let $\theta = \sum_{i=1}^{n} a_i\omega_i$ be an element of A, then

$$\theta^p \equiv \sum_{i=1}^{n} a_i\omega_i^p \equiv \theta \pmod{p}.$$

• $(ii) \Rightarrow (iii)$. Since the prime p is not ramified in A, its splitting in A has the form $pA = \mathcal{P}_1 \cdots \mathcal{P}_r$, where r is a positive integer and $\mathcal{P}_1, \dots \mathcal{P}_r$ are distinct prime ideals of A. Let $j \in \{1, \dots, r\}$. For any $\theta \in A$, $\theta^p \equiv \theta \pmod{\mathcal{P}_j}$. This may be written in the form: for any $\bar{\theta} \in A/\mathcal{P}_j$, $\overline{\theta^p} = \bar{\theta}$. This implies that the field $A/\mathcal{P}_j$ has cardinality equal to p. Therefore $A/\mathcal{P}_j \simeq \mathbb{F}_p$ and the residual degree of $\mathcal{P}_j$ is equal to 1.

• $(iii) \Rightarrow (i)$. Let $pA = \mathcal{P}_1 \cdots \mathcal{P}_r$ be the splitting of p in A, where r is a positive integer and $\mathcal{P}_1, \dots \mathcal{P}_r$ are distinct prime ideals of A all of residual degree equal to 1. Fix $i \in \{1, \dots, n\}$. For any $j \in \{1, \dots, r\}$, we have $\omega_i^p \equiv \omega \pmod{\mathcal{P})_|}$, hence $\omega_i^p \equiv \omega \pmod{pA}$.

(2) By (ii), we have for any $\theta \in A$, $\theta^p \equiv \theta \pmod{pA}$, hence modulo p, θ satisfies an algebraic equation of degree less than n. Therefore $p \mid I(\theta)$ by **Exercise 10.29**.

(3)(a) If $f(x)$ where reducible over $\mathbb{Q}$, then it will have a root $n = 0$ or $n \in \mathbb{Z}$ with $n \mid -8$. The possible values of n are $n = \pm1, \pm2, \pm4, \pm8$. It is straightforward to verify that $f(x)$ does not vanish for any of these values of n nor for 0. Therefore $f(x)$ is irreducible over $\mathbb{Q}$.

(b) Let $\theta, \theta^{'}, \theta^{''}$ be the roots of $f(x)$. We have

$$\begin{aligned} \mathrm{Disc}(f) &= -\mathrm{N}_{E/\mathbb{Q}}(f^{'}(\theta)) \\ &= -\mathrm{N}_{E/\mathbb{Q}}(3\theta^2 - 2\theta - 2) \\ &= -(3\theta^2 - 2\theta - 2)(3\theta'^2 - 2\theta' - 2)(3\theta''^2 - 2\theta'' - 2). \end{aligned}$$

It is possible to expand this product and to express it as a polynomial in the elementary symmetric functions of the roots of $f(x)$ and then find the value of $\mathrm{Disc}(f)$. We will not go further in this way. Instead of that, set $\alpha = 3\theta^2 - 2\theta - 2$ and multiply this equation successively by θ and θ^2. Using the equation satisfied by θ, we get the following equations:

$$\begin{aligned} 3\theta^2 - 2\theta + (-2-\alpha) &= 0 \\ \theta^2 + (4-\alpha)\theta + 24 &= 0 \\ (5-\alpha)\theta^2 + 26\theta + 8 &= 0. \end{aligned}$$

This implies that $x_1 = 1$, $x_2 = \theta$, $x_3 = \theta^3$ is a non zero solution of the homogeneous system of equations

$$\begin{aligned} 3x_3 - 2x_2 + (-2-\alpha)x_1 &= 0 \\ x_3 + (4-\alpha)x_2 + 24x_1 &= 0 \\ (5-\alpha)x_3 + 26x_2 + 8x_1 &= 0. \end{aligned}$$

It follows that the determinant of this system is 0, that is

$$\begin{vmatrix} 3 & -2 & (-2-\alpha) \\ 1 & (4-\alpha) & 24 \\ (5-\alpha) & 26 & 8 \end{vmatrix} = 0.$$

We get the equation $\alpha^3 - 7\alpha^2 - 2^2 \cdot 503 = 0$. Therefore

$$\mathrm{N}_{E/\mathbb{Q}}(f^{'}(\theta)) = 2^2 \cdot 503$$

and $\mathrm{Disc}(f) = -2^2 \cdot 503$.

(c) We may write $f(x)$ in the form

$$f(x) = x^2(x-1) - 2(x+4)$$

and we see that $x \mid x+4$ in $\mathbb{F}_2[x]$. Therefore, by **Exercise 10.30**, $2 \mid I(\theta)$. Using the same method as in **(2),(b)**, we compute the minimal polynomial $g(x)$ over $\mathbb{Q}$, of μ and we find $g(x) = x^3 - 2x^2 + 3x - 10$. This polynomial is monic and has integral coefficients, hence μ is an algebraic integer.

(d) Consider the inclusions $\mathbb{Z}[\theta] \subset \mathbb{Z}[\theta, \mu] \subset A$. By **(b)** and **(c)**, the index of $\mathbb{Z}[\theta]$ in A is equal to 2. The index of $\mathbb{Z}[\theta]$ in $\mathbb{Z}[\theta, \alpha]$ is equal to the determinant of the matrix expressing $1, \theta, \theta^2$ in terms of $1, \theta, \alpha$. This determinant is equal to 2, hence $A = \mathbb{Z}[\theta, \alpha]$. Therefore $\{1, \theta, \alpha\}$ is an integral basis.

(e) Since the rational prime 2 is not ramified, we may use **(1)**. We have $\theta^2 = \theta + 2\mu$, hence $\theta^2 \equiv \theta \pmod 2$. Straightforward computations, lead to $\mu^2 = -2 + 2\theta + \mu$, hence $\mu^2 \equiv \mu \pmod 2$. According to **(1)**, the prime 2 completely splits in E.

(f) From **(e)**, we have $2A = \mathcal{P}_1\mathcal{P}_2\mathcal{P}_3$, where the prime ideals $\mathcal{P}_i$, $i = 1, 2, 3$ are distinct and have their residual degree equal to 1. The minimal polynomial $f(x)$ of θ over $\mathbb{Q}$ may be written in the form: $f(x) = x^2(x-1) - 2(x-4)$. According to **Exercise 10.30**, $2 \mid I(\theta)$ and we may suppose that

$$\begin{cases} \theta & \equiv 0 \pmod{\mathcal{P}_1} \\ \theta & \equiv 0 \pmod{\mathcal{P}_2} \\ \theta & \equiv 1 \pmod{\mathcal{P}_3} \end{cases}. \qquad \text{(Eq 1)}$$

Similarly, the minimal polynomial $g(x) = x^3 - 2x^2 + 3x - 10$ of μ over $\mathbb{Q}$, may be written in the form

$$g(x) = x(x+1)^2 + 2(-2x^2 + x - 5),$$

hence $\begin{cases} \mu & \equiv 1 \pmod{\mathcal{P}_{i_1}} \\ \mu & \equiv 1 \pmod{\mathcal{P}_{i_2}} \\ \mu & \equiv 0 \pmod{\mathcal{P}_{i_3}} \end{cases}$, where the indices i_1, i_2, i_3 are distinct and belong to $\{1, 2, 3\}$. Since

$$\theta\mu = \theta^2(\theta - 1)/2 = (\theta^3 - \theta^2)/2 = \theta + 4$$

and since $\theta \not\equiv 0 \pmod{\mathcal{P}_3}$, then $\theta\mu \not\equiv 0 \pmod{\mathcal{P}_3}$. We deduce that $\mu \not\equiv 0 \pmod{\mathcal{P}_3}$, thus $\mu \equiv 1 \pmod{\mathcal{P}_3}$. Therefore, we may suppose that

$$\begin{cases} \mu \equiv 0 \pmod{\mathcal{P}_1} \\ \mu \equiv 1 \pmod{\mathcal{P}_2} \\ \mu \equiv 1 \pmod{\mathcal{P}_3} \end{cases}. \qquad \text{(Eq 2)}$$

We begin with the computation of a $\mathbb{Z}$-basis of $\mathcal{P}_1$. Since $A/\mathcal{P}_1$ is isomorphic to $\mathbb{Z}/2\mathbb{Z}$, then there exists a morphism of rings $\phi : A \to \mathbb{Z}/2\mathbb{Z}$ whose kernel is equal to $\mathcal{P}_1$. Since $A = \mathbb{Z} + \mathbb{Z}\theta + \mathbb{Z}\mu$, then in order to define ϕ, it is sufficient to define ϕ on $\mathbb{Z}$ and to find explicitly the respective values of $\phi(\theta)$ and $\phi(\mu)$. Obviously, for any $a \in \mathbb{Z}$, we must have $\phi(a) = \bar{a}$, where $\bar{a} = 0$ or $\bar{a} = 1$, according to a is even or a is odd. From the congruences satisfied by θ and μ described above, we must have $\phi(\theta) = \phi(\mu) = 0$. Let $\gamma = a + b\theta + c\mu$ be an element of A, then

$$\begin{aligned}\gamma \in \mathcal{P}_1 &\Leftrightarrow \phi(\gamma = 0)\\ &\Leftrightarrow \bar{a} = 0\\ &\Leftrightarrow a \equiv 0 \pmod 2\\ &\Leftrightarrow a = 2\lambda \quad \text{with} \quad \lambda \in \mathbb{Z}\\ &\Leftrightarrow \gamma = 2\lambda + b\theta + c\mu \quad \text{with} \quad \lambda \in \mathbb{Z}.\end{aligned}$$

Thus $\mathcal{P}_1$ is generated over $\mathbb{Z}$ by $\{2, \theta, \mu\}$. Obviously this set is free over $\mathbb{Z}$, hence $\{2, \theta, \mu\}$ is a basis of $\mathcal{P}_1$.
We use a similar reasoning for $\mathcal{P}_2$ and keep the same notations as above. We have

$$\begin{aligned}\gamma \in \mathcal{P}_2 &\Leftrightarrow \phi(\gamma = 0)\\ &\Leftrightarrow \bar{a} + \bar{c} = 0\\ &\Leftrightarrow a + c \equiv 0 \pmod 2\\ &\Leftrightarrow a = c + 2\lambda \quad \text{with} \quad \lambda \in \mathbb{Z}\\ &\Leftrightarrow \gamma = 2\lambda + b\theta + c(1 + \mu) \quad \text{with} \quad \lambda \in \mathbb{Z}.\end{aligned}$$

Hence $\mathcal{P}_2$ is generated over $\mathbb{Z}$ by $\{2, \theta, \mu + 1\}$. Obviously this set is free over $\mathbb{Z}$, hence $\{2, \theta, \mu + 1\}$ is a basis of $\mathcal{P}_2$. We omit the computations for $\mathcal{P}_3$. It is seen that $\{2, \theta + 1, \mu + 1\}$ is a basis of $\mathcal{P}_3$.

(g) From **(Eq 1)**, and since $\mathrm{N}_{E/\mathbb{Q}}(\theta) = 8$, we have $\theta A = {\mathcal{P}_1}^{e_1}{\mathcal{P}_2}^{e_2}$ with $e_1, e_2 \geq 1$ and $e_1 + e_2 = 3$. From **(Eq 2)**, and since $\mathrm{N}_{E/\mathbb{Q}}(\mu) = 10$, we have $\mu A = \mathcal{P}_1\mathcal{Q}$, where $\mathcal{Q}$ is a prime ideal lying over $5\mathbb{Z}$. We use the following identity:

$$\theta\mu = \theta + 4. \tag{Eq 3}$$

It is easy to show that $N_{E/\mathbb{Q}}(\theta + 4) = 5.2^4$. Suppose that $e_1 = 1$. Then $\nu_{\mathcal{P}_1}(\theta\mu) = 2$ and $\nu_{\mathcal{P}_1}(\theta + 4) = 1$, which is a contradiction to **(Eq 3)**. It follows that $e_1 = 2$, $e_2 = 1$, $\theta A = {\mathcal{P}_1}^2\mathcal{P}_2$ and $\mu A = \mathcal{P}_1\mathcal{Q}$, thus $I = \mathcal{P}_1$.

Remark. We may prove the same result as follows. Let

$$\gamma = a + b\theta + c\mu \in A,$$

then $\gamma \equiv a \pmod I$. We have

$$2 = \mu(10/\mu) - \theta(8/\theta) = \mu(\mu^2 - 2\mu + 3) - \theta(\theta^2 - \theta - 2),$$

hence $2 \in I$. We deduce that $2\mathbb{Z} \subset I$. Let $n = 2k + 1$ be an odd integer, then $n \equiv 1 \pmod I$. The question now reduces to does 1 belong to I? Or do we have the equality $I = A$? Since $\theta A \subset \mathcal{P}_1$ and $\mu A \subset \mathcal{P}_1$ (see **(Eq 1)**, **(Eq 2)**), then the answer is negative and then $A/I \simeq \mathbb{Z}/2\mathbb{Z}$. It follows that I is a prime ideal lying over $2\mathbb{Z}$. By **(Eq 1)**, **(Eq 2)** again we conclude that $I = \mathcal{P}_1$.

Exercise 10.63.

Let K be a number field of degree n, A be its ring of integers, p be a prime number and $\mathcal{P}$ be a prime ideal of A lying over $p\mathbb{Z}$. For any rational integer $m \geq 1$ and any $\alpha \in K$, let

$$\binom{\alpha}{m} = \frac{\alpha(\alpha-1)\cdots(\alpha-(m-1))}{m!}.$$

Let

$$K_{\mathcal{P}} = \{\alpha \in K, \nu_{\mathcal{P}}(\alpha) \geq 0\}.$$

(1) Show that the following propositions are equivalent.
  (i) For all integers $m \geq 1$ and all $\alpha \in K_{\mathcal{P}}$, $\binom{\alpha}{m} \in K_{\mathcal{P}}$.
  (ii) The residual degree of $\mathcal{P}$ is equal to 1 and $\mathcal{P}^2 \nmid p$.

(2) For any $\alpha \in K$, denote by $Den(\alpha)$ the denominator of α, that is the smallest positive integer d such that $d\alpha$ is integral. Let
$$K_p = \{\alpha \in K, p \nmid Den(\alpha)\}.$$
Show that $K_p = \cap_{\mathcal{P}|p} K_{\mathcal{P}}$. Deduce that the following assertions are equivalent.
  (iii) For all integers $m \geq 1$ and all $\alpha \in K_p$, $\binom{\alpha}{m} \in K_p$.
  (iv) The splitting of pA as a product of prime ideals of A is given by $pA = \mathcal{P}_1\mathcal{P}_2\cdots\mathcal{P}_n$, where the ideals $\mathcal{P}_i$, $i = 1, \ldots, r$ are distinct and of the first degree over $p\mathbb{Z}$.

Solution 10.63.

(1)
- *(i)* $\Rightarrow$ *(ii)*. Suppose that $\mathcal{P}$ is of degree $f \geq 2$, then there exists $\alpha \in A \subset K_{\mathcal{P}}$ such that $\alpha \not\equiv k \pmod{\mathcal{P}}$ for $k = 0, 1\ldots, p-1$. Therefore $\binom{\alpha}{p} \notin K_{\mathcal{P}}$. Suppose that $\mathcal{P}$ is of the first degree and $\mathcal{P}^2 \mid p$. Let $\alpha \in A$ such that $\alpha A + pA = \mathcal{P}$. Consider $\binom{\alpha}{p}$. The numerator of this fraction is divisible by $\mathcal{P}$ but not by $\mathcal{P}^2$, while the denominator is divisible by $\mathcal{P}^2$. Therefore $\binom{\alpha}{p} \notin K_{\mathcal{P}}$.
- *(ii)* $\Rightarrow$ *(i)*. We first show that for any positive integer $s \geq 1$, the numbers $0, 1, \ldots, p^s - 1$ form a complete residue system modulo $\mathcal{P}^s$. We have

$$|A/\mathcal{P}^s| = N_{K/\mathbb{Q}}(\mathcal{P}^s) = p^s,$$

hence it is sufficient to prove that the elements in the preceding list are distinct modulo $\mathcal{P}^s$. If not, then $p^s \mid m$ for some m, $1 \leq m < p^s$, which is a contradiction. Let $\alpha \in K_{\mathcal{P}}$ and let $m \geq 1$ be an integer. It follows from the preceding that for any $s \geq 1$, the numbers $\alpha, \alpha - 1, \ldots, \alpha - (p^s - 1)$ form a complete system of residues modulo $\mathcal{P}^s$. Consider the fraction $\binom{\alpha}{m}$. In the sequence $\alpha, \alpha - 1, \ldots, \alpha - (m-1)$, there are $\lfloor m/p \rfloor$ multiples of $\mathcal{P}$, $\lfloor m/p^2 \rfloor$ multiples of $\mathcal{P}^2$ and so on. Therefore the numerator of the fraction is divisible by $\mathcal{P}^k$, where

$$k = \lfloor m/p \rfloor + \lfloor m/p^2 \rfloor + \cdots.$$

It is known that $m!$ is divisible by exactly p^k [Ribenboim (2001), Lemma 3, Chap. 17.2]. It follows that $\binom{\alpha}{m} \in K_{\mathcal{P}}$.

(2) Let $\alpha \in K_p$, then $\alpha = \theta/d$, where θ is an integer of K, d is the denominator of α and $p \nmid d$. It follows that for any prime ideal $\mathcal{P}$ lying over $p\mathbb{Z}$, we have $\mathcal{P} \nmid d$, hence $\alpha \in \cap_{\mathcal{P}|p} K_{\mathcal{P}}$. We prove the reverse inclusion. Let α be an element of this intersection, then $\nu_{\mathcal{P}}(\alpha) \geq 0$ for any $\mathcal{P} \mid p$, hence $p \nmid Den(\alpha)$, thus $\alpha \in K_p$. Now the equivalence of **(iii)** and **(iv)** follows from **(1)**.

Exercise 10.64.

Let K be a field, $f(x) = x^n + a_{n-1}x^{n-1} + \cdots + +a_0$ be a polynomial with coefficients in K. The matrix

$$M = \begin{pmatrix} 0 & 0 & \ldots & 0 & -a_0 \\ 1 & 0 & \ldots & 0 & -a_1 \\ 0 & 1 & \ldots & 0 & -a_2 \\ \ldots & \ldots & \ldots & \ldots & \ldots \\ 0 & 0 & \ldots & 1 & -a_{n-1} \end{pmatrix}$$

is called the companion matrix of $f(x)$. For any matrix $N \in M_n(K)$, we denote by $\chi(N)$ the characteristic polynomial of N.

(1) Show that $\chi(M) = f(x)$.
(2) Suppose that $f(x)$ is irreducible over K and let α be a root of $f(x)$ in an algebraic closure of K. Let $N \in M_n(K)$ such that $f(N) = 0$. Show that $K[\alpha] \simeq K[N]$.
(3) Show that the finite field $\mathbb{F}_8$ may be represented as a field of matrices belonging to $M_3(\mathbb{F}_2)$. Write explicitly these matrices.
(4) Suppose that $K = \mathbb{Q}$ and let $f(x) = x^n + a_{n-1}x^{n-1} + \cdots + a_0 \in \mathbb{Z}[x]$, α be a root of $f(x)$ in $\mathbb{C}$ and $N \in M_n(\mathbb{Z})$ such that $f(N) = 0$. Suppose

that $f(x)$ is irreducible over $\mathbb{Q}$. Let $u(\alpha) \in \mathbb{Q}[\alpha]$, where $u(x) \in \mathbb{Q}[x]$ and $\deg u \leq n-1$. Show that $u(\alpha)$ is an algebraic integer if and only if $\chi(u(N)) \in \mathbb{Z}[x]$.

(5) Let $g(x) \in \mathbb{Z}[x]$ be monic, irreducible of degree n and θ be a root of $g(x)$. Let $F = \mathbb{Q}(\theta)$ and A be the ring of integers of F. Show that the following conditions are equivalent.

  (i) $A \neq \mathbb{Z}[\theta]$.

  (ii) There exist a prime number p, a positive integer k, matrices $M, N \in M_n(\mathbb{Z})$ and $b_0, \dots, b_k \in \mathbb{Z}$ such that $b_k \neq 0$, $g(M) = 0$ and
$$b_0 I + b_1 M + \cdots + b_k M^k = pN.$$

Solution 10.64.

(1) • **First method.** We have
$$\chi(M) = \mathrm{Det}(xI - M) = \mathrm{Det}\begin{pmatrix} x & 0 & \dots & 0 & a_0 \\ -1 & x & \dots & 0 & a_1 \\ 0 & -1 & \dots & 0 & a_2 \\ \dots & \dots & \dots & \dots & \dots \\ 0 & 0 & \dots & -1 & x + a_{n-1} \end{pmatrix}.$$
Expanding this determinant through its first row, we obtain
$$\begin{aligned}\chi(M) &= \mathrm{Det}(xI - M) \\ &= x\,\mathrm{Det}\begin{pmatrix} x & 0 & \dots & 0 & a_1 \\ -1 & x & \dots & 0 & a_2 \\ 0 & -1 & \dots & 0 & a_3 \\ \dots & \dots & \dots & \dots & \dots \\ 0 & 0 & \dots & -1 & x + a_{n-1} \end{pmatrix} \\ &\quad + (-1)^{n+1} a_0\,\mathrm{Det}\begin{pmatrix} -1 & x & \dots & 0 & 0 \\ 0 & -1 & \dots & 0 & 0 \\ 0 & 0 & \dots & 0 & 0 \\ \dots & \dots & \dots & \dots & \dots \\ 0 & 0 & \dots & 0 & -1 \end{pmatrix} \\ &= a_0 + x\,\mathrm{Det}\begin{pmatrix} x & 0 & \dots & 0 & a_1 \\ -1 & x & \dots & 0 & a_2 \\ 0 & -1 & \dots & 0 & a_3 \\ \dots & \dots & \dots & \dots & \dots \\ 0 & 0 & \dots & -1 & x + a_{n-1} \end{pmatrix}\end{aligned}$$
and the proof may be completed by induction.

- **Second method.** Let $\tilde{a}_0, \dots, \tilde{a}_{n-1}$ be indeterminates algebraically independent over K, $E = K(\tilde{a}_0, \dots, \tilde{a}_{n-1})$,

$$\tilde{f}(x) = x^n + \tilde{a}_{n-1}x^{n-1} + \cdots + \tilde{a}_0$$

and $\tilde{M}$ be the matrix obtained from M by replacing a_i by $\tilde{a}_i$ for $i = 0, \dots, n-1$. Let $\mathcal{B} = \{e_1, \dots, e_n\}$ be the canonical basis of E^n over E and let $T : E^n \to E^n$ be the unique E-linear map such that $T(e_i) = e_{i+1}$ for $i = 1, \dots, n-1$ and $T(e_n) = -\tilde{a}_0 e_1 - \cdots - \tilde{a}_{n-1}e_n$, then clearly the matrix of T relative to the basis $\mathcal{B}$ is equal to $\tilde{M}$. We first show that $\tilde{f}(T) = 0$. We proceed by induction. We have

$$\tilde{f}(T)(e_1) = (-\tilde{a}_0 Id_{E^n} - \tilde{a}_1 T - \cdots - \tilde{a}_{n-1}T^{n-1} - T^n)(e_1) = 0.$$

Suppose that $\tilde{f}(T)(e_j) = 0$ for $j = 1, \dots, k$. Then

$$\tilde{f}(T)(e_{k+1}) = \tilde{f}(T)(T(e_k)) = T(\tilde{f}(T)(e_k)) = T(0) = 0.$$

It follows that $\tilde{f}(T) = 0$, hence $\tilde{f}(\tilde{M}) = 0$. Let $\tilde{P}(x)$ be the minimal polynomial of $\tilde{M}$ over E. On the one hand $\tilde{P}(x) \mid \tilde{f}(x)$ in $E[x]$. Since $\tilde{f}(x)$ is irreducible in $K[\tilde{a}_0, \dots, \tilde{a}_{n-1}, x]$, then it is irreducible in $E[x]$. It follows that $\tilde{f}(x) = \tilde{P}(x)$. On the other hand, $\tilde{P}(x) \mid \mathrm{Det}(xI - \tilde{M})$ in $E[x]$. Since these polynomials are both monic of the same degree n, then

$$\tilde{f}(x) = \tilde{P}(x) = \mathrm{Det}(xI - \tilde{M}).$$

Consider the unique morphism of rings $\phi : K[\tilde{a}_0, \dots, \tilde{a}_{n-1}, x] \to K$ such that $\phi(\lambda) = \lambda$ for any $\lambda \in K$, $\phi(x) = x$ and $\phi(\tilde{a}_i) = a_i$ for $i = 0, \dots, n-1$. Applying ϕ to the preceding identity, we obtain

$$f(x) = \mathrm{Det}(xI - M) = \chi(M).$$

(2) Consider the maps $\sigma : K[x] \to \bar{K}$ and $\tau : K[x] \to M_n(K)$, defined by $\sigma(u(x)) = u(\alpha)$ and $\tau(u(x)) = u(M)$, then clearly σ and τ are morphisms of rings,

$$\mathrm{Ker}\,\sigma = \mathrm{Ker}\,\tau = f(x)K[x], \quad \mathrm{Im}\,\sigma = K[\alpha] \quad \text{and} \quad \mathrm{Im}\,\tau = K[M].$$

The first isomorphism theorem implies that

$$K[M] \simeq K[x]/f(x)K[x] \simeq K[\theta].$$

(3) The polynomial $f(x) = x^3 - x + 1$ is irreducible over $\mathbb{F}_2$. Let θ be a root of $f(x)$ in an algebraic closure of $\mathbb{F}_2$, then

$$\mathbb{F}_8 \simeq \mathbb{F}_2[\theta] \simeq \mathbb{F}_2[M],$$

where M is any matrix belonging to $M_3(\mathbb{F}_2)$ such that $f(M) = 0$. We may take M to be the companion matrix of $f(x)$, that is $M = \begin{pmatrix} 0 & 0 & -1 \\ 1 & 0 & 1 \\ 0 & 1 & 0 \end{pmatrix}$. Then

$$\mathbb{F}_8 \simeq \{0, I, M, I+M, M^2, I+M^2, M+M^2, I+M+M^2\}.$$

(4) We use the isomorphism of fields $\psi : \mathbb{Q}[\theta] \to \mathbb{Q}[M]$ such that $\psi(\theta) = M$. Suppose that $\beta := u(\theta)$ is integral over $\mathbb{Z}$, then $g(\beta) = 0$, where $g(x) \in \mathbb{Z}[x]$ is the minimal polynomial of β over $\mathbb{Q}$. Applying ψ, we obtain $g(u(N)) = 0$. Therefore $g(x)$ is the minimal polynomial of $u(N)$. Since $g(x)$ is irreducible, then $\chi(u(N))$ is a power of $g(x)$, hence $\chi(u(N)) \in \mathbb{Z}[x]$. We prove the converse. Suppose that $\chi(u(N)) \in \mathbb{Z}[x]$, then $\chi(u(N))(N) = 0$ and applying ψ^{-1}, we obtain $\chi(u(N))(\beta) = 0$, which implies that β is integral over $\mathbb{Z}$.

(5) • *(i)* $\Rightarrow$ *(ii)*. Let $\mathcal{B} = \{\omega_1, \ldots, \omega_n\}$ be an integral basis of F. Since $A \neq \mathbb{Z}[\theta]$, there exists $\beta \in A \setminus \mathbb{Z}[\theta]$ of the form

$$\beta = (b_0 + b_1\theta + b_k\theta^k)/p,$$

where p is a prime number and $b_0, \ldots, b_k$ are integers such that $b_k \not\equiv 0 \pmod{p}$. Let M be the matrix in base $\mathcal{B}$ of the $\mathbb{Q}$-endomorphism m_θ of F such that $m_\theta(x) = \theta x$. **Exercise 2.5** shows that

$$N := (b_0 + b_1 M + \cdots + b_k M^k)/p$$

is the matrix representing the endomorphism multiplication by β in the basis $\mathcal{B}$. So it has integral entries and

$$b_0 + b_1 M + \cdots + b_k M^k = pN.$$

We show that $g(M) = 0$. We have

$$\chi(M)(x) = \mathrm{Det}(xI - M) = \mathrm{Det}(xId_F - m_\theta),$$

hence

$$\chi(M)(\theta) = \mathrm{Det}(\theta Id_F - m_\theta).$$

Since $(\theta Id_F - m_\theta)(1) = 0$, then $\chi(M)(\theta) = 0$. Therefore $g(x) \mid \chi(M)(x)$. Since these polynomials are both monic of degree n, then $g(x) = \chi(M)(x)$, which implies that $g(M) = 0$.

- $(ii) \Rightarrow (i)$. Let a prime number p, matrices $M, N \in M_n(\mathbb{Z})$ and $b_0, \ldots, b_k \in \mathbb{Z}$ such that $b_k \not\equiv 0 \pmod{p}$, $g(M) = 0$ and

$$b_0 I + b_1 M + \cdots + b_k M^k = pN.$$

Let

$$u(x) = (b_0 + b_1 x + \cdots + b_k x^k)/p,$$

then

$$\chi(u(M)(x) = \chi(N)(x) \in \mathbb{Z}[x],$$

hence $u(\theta)$ is an algebraic integer by **(4)**. Moreover $u(\theta) \notin \mathbb{Z}[\theta]$, hence $A \neq \mathbb{Z}[\theta]$.

Exercise 10.65.

(1) Let A be a Dedekind domain, $\mathcal{B}_1, \ldots, \mathcal{B}_k$, be ideals of A. Show that

$$\cap_{i=1}^{k-1}(\mathcal{B}_i + \mathcal{B}_k) = \cap_{i=1}^{k-1}\mathcal{B}_i + \mathcal{B}_k.$$

(2) Let $a_1, \ldots, a_k \in A$. Show that the following system of congruence equations $x \equiv a_i \pmod{\mathcal{B}_i}$ for $i = 1, \ldots, k$ is solvable in A if and only if for any i, j, $i \neq j$, $a_i \equiv a_j \pmod{\gcd(\mathcal{B}_i, \mathcal{B}_j)}$.

(3) Let (a_i, b_i) for $i = 1, \ldots, k$ be distinct couples of integers such that $0 \leq a_i < b_i$. We say that this set of couples is a disjoint covering system if for any non negative integer n, there exists a unique $i \in \{1, \ldots, k\}$ such that $n \equiv a_i \pmod{b_i}$. Show that (a_i, b_i), for $i = 1, \ldots, k$, is a disjoint covering system if and only if the following conditions hold.

   (i) $\sum 1/b_i = 1$.
   (ii) For any i, $j \in \{1, \ldots, k\}$, $i \neq j$, we have $a_i \equiv a_j \pmod{\gcd(b_i, b_j)}$.

(4) Let (a_i, b_i) for $i = 1, \ldots, k$ be distinct couples of integers such that $0 \leq a_i < b_i$. For $i = 1, \ldots, m$, let $a_i' = b_i - a_i - 1$. Show that (a_i, b_i) is a disjoint covering system if and only if (a_i', b_i) satisfies the same property.

(5) Let (a_i, b_i) for $i = 1, \ldots, k$ be distinct couples of integers such that $0 \leq a_i < b_i$ and let $N = \operatorname{lcm}(b_1, \ldots, b_k)$.

   (a) Show that (a_i, b_i), for $i = 1, \ldots, k$, is a disjoint covering system if and only if for any integer n, $0 \leq n \leq N - 1$, there exists a unique $i \in \{1, \ldots, k\}$ such that $n \equiv a_i \pmod{b_i}$.

(b) Let $\alpha \in \mathbb{C}$ be transcendental or algebraic over $\mathbb{Q}$ of degree greater than N. Prove the following identities.

$$\sum_{t=0}^{N-1} \alpha^t = (\alpha^N - 1)/(\alpha - 1),$$

$$\sum_{j=1}^{k} \left(\sum_{s=0}^{N/b_j - 1} \alpha^{a_j + sb_j} \right) = \sum_{j=1}^{k} \alpha^{a_j} (\alpha^N - 1)/(\alpha^{b_j} - 1).$$

Deduce that (a_i, b_i), for $i = 1, \dots, k$, is a disjoint covering system if and only if

$$\alpha^{a_1}/(\alpha^{b_1} - 1) + \cdots + \alpha^{a_k}/(\alpha^{b_k} - 1) = 1/(\alpha - 1).$$

Solution 10.65.

(1) For any $i = 1, \dots, k-1$, $\mathcal{B}_i \subset \mathcal{B}_i + \mathcal{B}_k$, hence

$$\cap_{i=1}^{k-1} \mathcal{B}_i \subset \cap_{i=1}^{k-1} (\mathcal{B}_i + \mathcal{B}_k).$$

Moreover since $\mathcal{B}_k \subset \mathcal{B}_i + \mathcal{B}_k$ for $i = 1, \dots, k-1$, then

$$\mathcal{B}_k \subset \cap_{i=1}^{k-1} (\mathcal{B}_i + \mathcal{B}_k).$$

We deduce that

$$\cap_{i=1}^{k-1} \mathcal{B}_i + \mathcal{B}_k \subset \cap_{i=1}^{k-1} (\mathcal{B}_i + \mathcal{B}_k).$$

We prove the reverse inclusion. Let $\mathcal{P}$ be a prime ideal dividing $\cap_{i=1}^{k-1} (\mathcal{B}_i + \mathcal{B}_k)$, with a positive exponent equal to e. Then there exists $i_0 \in \{1, \dots, k-1\}$ such that $\mathcal{P}^e \mid \mathcal{B}_{i_0} + \mathcal{B}_k$. It follows that $\mathcal{P}^e \mid \mathcal{B}_{i_0}$ and $\mathcal{P}^e \mid \mathcal{B}_k$. We deduce that $\mathcal{P}^e \mid \cap_{i=1}^{k-1} \mathcal{B}_i$ and $\mathcal{P}^e \mid \mathcal{B}_k$ and then the desired inclusion.

(2) • **Necessity of the conditions.** Let $x \in A$ be a solution of the system of congruence equations and let i and $j \in \{1, \dots, k\}$ such that $i \neq j$, then $a_i \equiv x \equiv a_j \pmod{\gcd(\mathcal{B}_i, \mathcal{B}_j)}$.

• **Sufficiency of the conditions.** Notice that $\gcd(\mathcal{B}_i, \mathcal{B}_j) = \mathcal{B}_i + \mathcal{B}_j$. We prove the result by induction on k. Suppose first that $k = 2$. Since $a_1 \equiv a_2 \pmod{\mathcal{B}_i + \mathcal{B}_j}$, then $a_1 - a_2 = u + v$, where $u \in \mathcal{B}_1$ and $v \in \mathcal{B}_2$. Let $x = a_1 - u = a_2 + v$, then x satisfies the congruence equations. Suppose that the result holds for any number of ideals of A smaller than k. Then, there exists $c \in A$ such that $c \equiv a_i \pmod{\mathcal{B}_i}$ for $i = 1, \dots, k-1$. Consider the following system consisting of two congruence equations

$$x \equiv y \pmod{\cap_{i=1}^{k-1} \mathcal{B}_i}$$
$$x \equiv a_n \pmod{\mathcal{B}_k}.$$

To use the result proved for two ideals, we must show that

$$y \equiv a_n \pmod{\cap_{i=1}^{k-1} \mathcal{B}_i + \mathcal{B}_k}.$$

For any $i = 1, \ldots, k-1$, we have

$$y - a_k = (y - a_i) + (a_i - a_k) \in \mathcal{B}_i + (\mathcal{B}_i + \mathcal{B}_k) = \mathcal{B}_i + \mathcal{B}_k.$$

Hence by **(1)**, we have

$$y - a_k \in \cap_{i=1}^{k-1}(\mathcal{B}_i + \mathcal{B}_k) = \cap_{i=1}^{k-1}\mathcal{B}_i + \mathcal{B}_k$$

and this proves the verification to be done.

(3) We begin with a remark. Let $b = \operatorname{lcm}(b_1, \ldots, b_k)$, $P_i = a_i + b_i\mathbb{Z}$ for $i = 1, \ldots, k$ and $A = \{0, 1, \ldots, b-1\}$, then

$$|P_i \cap A|/|A| = |\{a_i, a_i + b, \ldots, a_i + (q_i - 1)b\}|/b,$$

where q_i is defined by the relation $b = b_i q_i$. Hence $|P_i \cap A|/|A| = 1/b_i$.

- **Necessity of the conditions.** The conditions (ii) follows from **(1)**. Since the covering system is disjoint, then $A = \cup_{i=1}^{k}(P_i \cap A)$ and this union is a disjoint one. It follows that

$$1 = |\cup_{i=1}^{k} (P_i \cap A)|/|A| = \left(\sum_{i=1}^{k} |(P_i \cap A)|\right)/|A| \sum_{i=1}^{k} 1/b_i,$$

hence (i).

- **Sufficiency of the conditions.** (ii) shows that (a_i, b_i) for $i = 1, \ldots, k$ is a covering system of $\mathbb{Z}$. (i) implies that the covering system is disjoint on A, hence disjoint on $\mathbb{Z}$.

(4) We use the result in **(3)**. It is clear that the condition (i) is satisfied by both systems or none. The second condition (ii), follows from the identities $a_i' - a_j' = b_i - b_j - (a_i - a_j)$ for any $i \neq j$.

(5)(a) The necessity of the condition is obvious. We prove the sufficiency. Let $n \in \mathbb{Z}$, $0 \leq r < N$ such that $n \equiv r \pmod{N}$ and let i be the unique integer in $\{1, \ldots, k\}$ such that $r \equiv a_i \pmod{b_i}$. Then

$$\begin{aligned} n &\equiv r \pmod{N} \\ &\equiv r \pmod{b_i} \\ &\equiv a_i \pmod{b_i}. \end{aligned}$$

Suppose that $n \equiv a_i \pmod{b_i}$ and $n \equiv a_j \pmod{b_j}$, then $r \equiv a_i \pmod{b_i}$ and $r \equiv a_j \pmod{b_j}$, hence $i = j$.

(b) The first identity is obvious, so we omit its proof. We have

$$\sum_{j=1}^{k}\left(\sum_{s=0}^{N/b_j-1}\alpha^{a_j+sb_j}\right)=\sum_{j=1}^{k}\alpha^{a_j}\left(\sum_{s=0}^{N/b_j-1}(\alpha^{b_j})^s\right)$$
$$=\sum_{j=1}^{k}\alpha^{a_j}(\alpha^N-1)/(\alpha^{b_j}-1),$$

which proves the second identity.

Suppose that (a_i, b_i), for $i = 1, \ldots, k$, is a disjoint covering system, then

$$\sum_{t=0}^{N-1}\alpha^t=\sum_{j=1}^{k}\left(\sum_{s=0}^{N/b_j-1}\alpha^{a_j+sb_j}\right),$$

hence using the two identities, we get the following relation.

$$\alpha^{a_1}/(\alpha^{b_1}-1)+\cdots+\alpha^{a_k}/(\alpha^{b_k}-1)=1/(\alpha-1).$$

Conversely suppose that this relation holds, then according to the two identities, we obtain

$$\sum_{t=0}^{N-1}\alpha^t=\sum_{j=1}^{k}\left(\sum_{s=0}^{N/b_j-1}\alpha^{a_j+sb_j}\right).$$

This shows that α is a root of the following polynomial with integral coefficients

$$f(x)=\sum_{t=0}^{N-1}x^t-\sum_{j=1}^{k}\left(\sum_{s=0}^{N/b_j-1}x^{a_j+sb_j}\right).$$

This polynomial is the zero polynomial or its degree is less than N. The assumptions made on α implies that $f(x) = 0$, thus

$$\sum_{t=0}^{N-1}x^t=\sum_{j=1}^{k}\left(\sum_{s=0}^{N/b_j-1}x^{a_j+sb_j}\right).$$

The polynomial on the left hand side of this identity is of degree $N-1$ and contains N non zero terms. The same must be true for the polynomial appearing on the right hand side. It follows that for any $t \in \{0, \ldots, N-1\}$, there exists a unique $j \in \{1, \ldots, k\}$ such that $t = a_j + sb_j$.

Exercise 10.66.
Let K be a number field, A be its ring of integers, $\mathcal{F}$, $\mathcal{P}$ be the sets of fractional ideals of K and principal fractional ideals respectively. Let $\mathcal{H} = \mathcal{F}/\mathcal{P}$. For any ideal $I \in \mathcal{F}$, denote by $[I]$ be its class modulo $\mathcal{P}$.

(1) Let $\mathcal{A}$ be a fractional ideal of K and let d be the order of $[\mathcal{A}]$ in $\mathcal{H}$. Let $\omega \in K$ such that $\mathcal{A}^d = \omega A$, $\rho = \omega^{1/d}$, $E = K(\rho)$ and B be the ring of integers of E. Show that

(a) $\mathcal{A}B = \rho B$.

(b) If for some number field F containing K, with ring of integers C, we have $\mathcal{A}C = \alpha C$, with $\alpha \in F$, then $\alpha = \mu\rho$, where μ is an algebraic unit.

(2) Let h be the class number of K and let $\mathcal{A}_1, \ldots, \mathcal{A}_h$ be a complete set of representatives of the classes of ideals modulo $\mathcal{P}$. Let $d_1, \ldots, d_h$ be the respective orders of $[\mathcal{A}_1], \ldots, [\mathcal{A}_h]$. For $i = 1, \ldots, h$ let $\omega_i \in K$ such that $\mathcal{A}_i^{d_i} = \omega_i A$ and let $\rho_i = \omega_i^{1/d_i}$. Let $E = K(\rho_1, \ldots, \rho_h)$ and B be its ring of integers. Show that for any fractional ideal I of K, IB is principal.

(3) Let γ and η be algebraic numbers. Show that there exists an algebraic number ϕ such that the following relation holds $a\gamma + b\eta = \phi$, where a and b are algebraic integers. Show that ϕ is unique up to multiplication by an algebraic unit. The number ϕ will be called the gcd of γ and η.

Solution 10.66.

(1)(a) We have

$$(\mathcal{A}B)^d = \mathcal{A}^d B = \omega B = (\rho B)^d,$$

hence by the uniqueness of the factorization in E into a product of prime ideals, we obtain $\mathcal{A}B = \rho B$.

(b) It is known that any fractional ideal of K may be generated by two elements. So let a_1, a_2 be elements of K such that $\mathcal{A} = a_1 A + a_2 A$, then $\alpha C = a_1 C + a_2 C$, thus $\alpha = a_1 c_1 + a_2 c_2$ with $c_1, c_2 \in C$. We have $a_1 = \rho b_1$ and $a_2 = \rho b_2$, where $b_1, b_2 \in B$, hence $\alpha = \rho(b_1 c_1 + b_2 c_2)$. We obtain a similar relation by permuting ρ and α. Thus $b_1 c_1 + b_2 c_2$ is an algebraic unit.

(2) For any $i \in \{1, \ldots, h\}$, let $E_i = K(\rho_i)$ and B_i be its ring of integers. According to **(1)**, $\mathcal{A}_i B_i = \rho_i B_i$, hence $\mathcal{A}_i B = \rho_i B$, that is $\mathcal{A}_i B$ is principal. Since I is equivalent to some $\mathcal{A}_i$, say $\mathcal{A}_{i_0}$, then $I = \mathcal{A}_{i_0}(\alpha A)$,

where $\alpha \in K$. We deduce that

$$IB = \mathcal{A}_{i_0}B(\alpha B) = \rho_{i_0}B\alpha B = \rho_{i_0}\alpha B,$$

thus IB is principal.

Exercise 10.67.

Let d be a non zero square free integer, $K = \mathbb{Q}(\sqrt{d})$ and A be the ring of integers of K. Recall that $A = \mathbb{Z}[\omega]$, where

$$\omega = \begin{cases} \sqrt{d} & \text{if } d \equiv 2,3 \pmod 4 \\ (1+\sqrt{d})/2 & \text{if } d \equiv 1 \pmod 4 \end{cases}.$$

Let I be a non zero ideal of A.

(1) Let $\delta = \gcd\{b, \text{ there exists } a \in \mathbb{Z} \text{ such that } a + b\omega \in I\}$. Show that there exists $a_0 \in \mathbb{Z}$ such that $a_0 + \delta\omega \in I$.
(2) For any $\alpha = a + b\omega \in I$, let $c(\alpha) = a + b\omega - (b/\delta)(a_0 + \delta\omega)$. Show that $c(\alpha) \in \mathbb{Z} \cap I$ and I is generated over $\mathbb{Z}$ by $a_0 + \delta\omega$ and $\{c(\alpha), \alpha \in I\}$.
(3) Let $c = \gcd\{c(\alpha), \alpha \in I\}$. Show that I is generated over $\mathbb{Z}$ by c and $a_0 + \delta\omega$. Show that c is the smallest positive integer which belongs to I.
(4) Let q and $r \in \mathbb{Z}$ such that $a_0 = cq + r$ with $0 \leq r < c$. Show that I is generated by c and $r + \delta\omega$. Show that $\delta \mid c$ and $\delta \mid r$.
(5) Let $\mathcal{P}$ be a prime ideal of A lying over some rational prime p such that $A/\mathcal{P} \simeq \mathbb{F}_p$. Show that $\mathcal{P} = (p, r + \omega)$ with $0 \leq r < p$.

Solution 10.67.

(1) Let $(a_i, b_i) \in \mathbb{Z}^2$, $i = 1, \ldots, n$ such that $a_i + b_i\omega \in \mathbb{Z}$ and $\delta = \sum_{i=1}^n x_i b_i$ with $x_i \in \mathbb{Z}$. Since $\sum_{i=1}^n x_i(a_i + b_i\omega) \in I$, then $a_0 + \delta\omega \in I$ with $a_0 = \sum_{i=1}^n x_i a_i$.
(2) We have $c(\alpha) = a - (b/\delta)a_0$, hence $c(\alpha) \in \mathbb{Z}$. Obviously $c(\alpha) \in I$, thus $c(\alpha) \in I \cap \mathbb{Z}$. Any $\alpha = a + b\omega \in I$ may be written in the form
$$\alpha = c(\alpha) + (b/\delta)(a_0 + \delta\omega),$$
hence $a_0 + \delta\omega$ with the set $\{c(\alpha), \alpha \in I\}$ generate I.
(3) The first part is obvious. Let $d \in I$ be a positive integer, then $d = c(d)$, hence $c \mid d$, that is c is the smallest positive integer which belongs to I.
(4) We have
$$r + \delta\omega = (a_0 - cq) + \delta\omega = a_0 + \delta\omega - cq \in I \quad \text{and}$$
$$a_0 + \delta\omega = r + \delta\omega + qc,$$
hence c and $r + \delta\omega$ generate I over $\mathbb{Z}$. Since $c \in I$ and $r + \delta\omega \in I$, then $\delta \mid c$ and $\delta \mid r$.

(5) Since $\mathrm{N}_{K/\mathbb{Q}}(\mathcal{P}) = p$, then $p \in \mathcal{P}$. Since $1 \notin \mathcal{P}$, then the element c defined in **(2)** is equal to p. Hence I has the form $I = (p, r + \delta\omega)$, with $0 \leq r < p$. Since $\delta \mid r$ and $\delta \mid p$, then $\delta = 1$. Thus $I = (p, r + \omega)$.

Chapter 11

Derivations

Exercise 11.1.
Let p be a prime number, K be a perfect field of characteristic p and E be an extension of K. Let $\alpha_1, \ldots, \alpha_n$ be elements of E such that $\alpha_i = \sum_{j=1}^{n} c_{ij}\alpha_j^p$ for $i = 1, \ldots, n$, where $c_{ij} \in K$ for any $(i, j) \in \{1, \ldots, n\}^2$. Show that $\alpha_1, \ldots, \alpha_n$ are algebraic over K.

Solution 11.1.
Suppose that there exists $i \in \{1, \ldots, n\}$ such that α_i is transcendental over K. Let $r \geq 1$ be the transcendence degree of $K(\alpha_1, \ldots, \alpha_n)$ over K. We may suppose that $\{\alpha_1, \ldots, \alpha_r\}$ is a transcendence basis and $\alpha_{r+1}, \ldots, \alpha_n$ are algebraic over $K(\alpha_1, \ldots, \alpha_r)$. Let $d : K(\alpha_1, \ldots, \alpha_r) \to K(\alpha_1, \ldots, \alpha_n)$ be the K-derivation such that $d(f(\alpha_1, \ldots, \alpha_r)) = \frac{\partial f(\alpha_1, \ldots, \alpha_r)}{\partial \alpha_1}$. Since K is perfect and $K(\alpha_1, \ldots, \alpha_n)$ is algebraic and separable over $K(\alpha_1, \ldots, \alpha_r)$, then d has one and only one extension $\hat{d} : K(\alpha_1, \ldots, \alpha_n) \to K(\alpha_1, \ldots, \alpha_n)$ [Lang (1965), Th. 7, Chap. 10.7]. Since $\alpha_1 = \sum_{j=1}^{n} c_{1j}\alpha_j^p$, then

$$1 = \hat{d}(\alpha_1) = \sum_{j=1}^{n} c_{ij} p \alpha_j^{p-1} \hat{d}(\alpha_j) = 0,$$

which is a contradiction. Therefore $\alpha_1, \ldots, \alpha_n$ are algebraic over K.

Exercise 11.2.
Let K be a field of characteristic 0, $f(x, y)$ be an irreducible polynomial with coefficients in K. Let α be an element of an algebraic closure of $K(x)$, $E = K(x, \alpha)$ and K_0 be the algebraic closure of K in E. Let d be the derivation of $K(x)$ such that $d(u(x)) = u'(x)$ and let $\tilde{d}$ be the unique extension of d to E. Show that $\operatorname{Ker} \tilde{d} = K_0$.

Solution 11.2.
Let $\beta \in K_0$ and let $g(x)$ be its minimal polynomial over K. We have

$$0 = \tilde{d}(g(\beta)) = g'(\beta)\tilde{d}(\beta).$$

If $g'(\beta) = 0$, then $g'(x) = 0$, hence $g(x)$ is a constant, which is excluded. It follows that $\tilde{d}(\beta) = 0$ which means $\beta \in \operatorname{Ker} \tilde{d}$. Let $\beta \in \operatorname{Ker} \tilde{d}$ and let $h(x, y) \in K[x, y]$ be the unique (up to a multiplication by a non zero constant in K) irreducible polynomial satisfying $h(x, \beta) = 0$. We have

$$0 = \tilde{d}(h(x, \beta)) = \frac{\partial h(x, \beta)}{\partial x} + \tilde{d}(\beta)\frac{\partial h(x, \beta)}{\partial y}.$$

We deduce that $\frac{\partial h(x,\beta)}{\partial x} = 0$, hence $h(x, y) \mid \frac{\partial(h(x,y))}{\partial x}$ in $K(x)[y]$. Taking into account of the respective degrees in y of h and $\frac{\partial h}{\partial x}$, we conclude that there exists $c(x) \in K(x)$ such that $\frac{\partial(h(x,y))}{\partial x} = c(x)h(x, y)$. Set

$$h(x, y) = h_m(x)y^m + \cdots + h_0(x)$$

and $c(x) = a(x)/b(x)$, where $h_0(x), \ldots, h_m(x), a(x)$ and $b(x) \in K[x]$, $h_m(x) \neq 0$, $b(x) \neq 0$ and $\gcd(a(x), b(x)) = 1$. Using the above relation between $h(x, y)$ and $\frac{\partial h(x,y)}{\partial x}$, we get the equations: $b(x)h_j'(x) = a(x)h_j(x)$ for $j = 0, \ldots, m$. This implies that $b(x)$ divides all the polynomials $h_j(x)$ hence $b(x)$ is a constant in K. It follows that $a(x) = 0$ and $h_j'(x) = 0$ for $j = 0, \ldots, m$. Since the characteristic of K is 0, we conclude that $h_j(x)$ is a constant in K for $j = 0, \ldots, m$ which means $\beta \in K_0$.

Exercise 11.3.
Let E be a field of characteristic $\neq 2$. Suppose that there exists a map $\sigma : E \Rightarrow E$ satisfying the following conditions.

(1) For any $x, y \in E$, $\sigma(x + y) = \sigma(x) + \sigma(y)$.
(2) For any $x \in E$, $\sigma(x^2) = 2x(\sigma(x))$.

Show that σ is a derivation of E.

Solution 11.3.
The proof is similar to the one given in **Exercise 1.13**.

Exercise 11.4.
Let d be the K-derivation of $K[x, y]$ defined by $d(f) = x\frac{\partial f}{\partial x} + y\frac{\partial f}{\partial y}$. Show that $\operatorname{Ker} d = K$.

Solution 11.4.
Let $f(x,y) = a_n(y)x^n + \cdots + a_0(y)$ be an element of $\operatorname{Ker} d$ with $a_n(y) \neq 0$ and let $m = \deg a_n(y)$, then the coefficient of x^n of $d(f)$ is equal to $na_n(y) + ya_n'(y)$, hence $na_n(y) + ya_n'(y) = 0$. Set $a_n(y) = b_m y^m + \cdots + b_0$, where $b_m \neq 0$, then the coefficient of y^m in the equation relating a and a' is equal to $(n+m)b_m$, hence $n + m = 0$ which implies $n = m = 0$ and $f \in K$.

Exercise 11.5.
Let K be a field of characteristic 0, $\vec{x} = (x_1, \ldots, x_n)$ and $f(\vec{x}) \in K[\vec{x}]$ irreducible. Let δ be a K-derivation of $K[\vec{x}]$ such that $\operatorname{Ker} \delta$ contains $n-1$ algebraically independent polynomials over K. Show that there exists $T(\vec{x}) \in K[\vec{x}]$ such that $\delta(f) = Tf$ if and only if for any $i \in \{1, \ldots, n\}$, $f \mid \delta(x_i)$ or there exists $P(\vec{x}) \in \operatorname{Ker} \delta$ such that $f \mid P$.

Solution 11.5.

- **Necessity of the conditions.**
Suppose that there exists some $i \in \{1, \ldots, n\}$, say $i = n$, such that $f \nmid \delta(x_i)$ and let $P_1, \ldots, P_{n-1}$ be algebraically independent elements over K of $\operatorname{Ker} d$. Let $A = K[\vec{x}]/(f)$ and E the fraction field of A. Since the transcendence degree of E over K is equal to $n-1$, then the elements of A: $\bar{x}_n, \overline{P_1}, \ldots, \overline{P_{n-1}}$ are algebraically dependent over K, hence there exists $R(y_1, \ldots, y_n) \in K[y_1, \ldots, y_n]$ such that $R(\overline{P_1}, \ldots, \overline{P_{n-1}}, \bar{x}_n) = \bar{0}$. We deduce that there exists $q(\vec{x}) \in K[\vec{x}]$ such that $R(P_1, \ldots, P_{n-1}, x_n) = fq$. We may write this identity in the form
$$a_k(P_1, \ldots, P_{n-1})x_n^k + \cdots + a_1(P_1, \ldots, P_{n-1})x_n + a_0(P_1, \ldots, P_{n-1}) = fq,$$
where k is supposed to be minimal. Suppose that $k \geq 1$, then applying δ to both sides of this identity, yields
$$\begin{aligned}\delta(x_n)\left(ka_k(P_1, \ldots, P_{n-1})x_n^{k-1} + \cdots + a_1(P_1, \ldots, P_{n-1})\right) &= f\delta(q) + q\delta(f)\\ &= f(\delta(q) + Tq).\end{aligned}$$
Since $f \nmid \delta(x_n)$, then there exists $g(\vec{x}) \in K[\vec{x}]$ such that
$$ka_k(P_1, \ldots, P_{n-1})x_n^{k-1} + \cdots + a_1(P_1, \ldots, P_{n-1}) = fg,$$
which contradicts the minimality of k. Therefore $k = 0$ and $a_0(P_1, \ldots, P_{n-1}) \equiv 0 \pmod{f}$ and the proof is complete.
- **Sufficiency of the conditions.**
Suppose that $f \mid \delta(x_i)$ for $i = 1, \ldots, n$. We have $\delta(f) = \sum \frac{\partial f}{\partial x_i}\delta(x_i)$, hence $f \mid \delta(f)$ in $K[\vec{x}]$, thus $\delta(f) = Tf$ for some $T \in K[\vec{x}]$. If

$f(\vec{x}) \mid P(x)$ for some $P(x) \in K[\vec{x}]$, let $g(\vec{x}) \in K[\vec{x}]$ such that $P(\vec{x}) = f(\vec{x})^e g(\vec{x})$, where $g(\vec{x}) \in K[\vec{x}]$, e is a positive integer and $\gcd(f(\vec{x}), g(\vec{(x)}))$. Then

$$0 = \delta(P) = ef(\vec{x})^{e-1}g(\vec{x})\delta(f) + f(\vec{x})\delta(g).$$

We deduce that

$$eg(\vec{x})\delta(f) + f(\vec{x})\delta(g) = 0,$$

which shows that $f(\vec{x}) \mid \delta(f)$ in $K[\vec{x}]$. Therefore, there exists $T(\vec{x}) \in K[\vec{x}]$ such that $\delta(f) = Tf$.

Exercise 11.6.

(1) Let A be an integrally closed domain, $d : A \to A$ a derivation and $\tilde{d}$ its extension to $\mathrm{Frac}(A)$. Show that $\mathrm{Ker}\,\tilde{d} = \mathrm{Frac}(\mathrm{Ker}\,d)$ if and only if $\mathrm{Ker}\,\tilde{d}$ is algebraic over $\mathrm{Frac}(\mathrm{Ker}\,d)$.
(2) Let K be a field of characteristic 0, $A = K[x_1, \ldots, x_n]$, d a K-derivation of A and $\tilde{d}$ its extension to $\mathrm{Frac}(A) = K(x_1, \ldots, x_n)$. If $\mathrm{Ker}\,d$ contains $n-1$ algebraically independent polynomials over K, show that $\mathrm{Ker}\,\tilde{d} = \mathrm{Frac}(\mathrm{Ker}\,d)$.
(3) Let K be a field of characteristic 0, $d : K[x_1, x_2] \to K[x_1, x_2]$ the derivation such that $d = x_1\frac{\partial}{\partial x_1} + x_2\frac{\partial}{\partial x_2}$. Show that $\mathrm{Ker}\,d = K$ and $x_2/x_1 \in \mathrm{Ker}\,\tilde{d}$. Conclude that the assumption on the transcendence degree in **(2)** cannot be omitted.

Solution 11.6.

(1) • **Necessity of the condition.**
Let $\alpha \in \mathrm{Ker}\,\tilde{d}$, then $\alpha = u/v$, where $u, v \in \mathrm{Ker}\,d$ and $v \neq 0$, hence $\alpha - u/v = 0$ which shows that α is algebraic over $\mathrm{Frac}(\mathrm{Ker}\,d)$.
• **Sufficiency of the condition.**
The inclusion $\mathrm{Frac}(\mathrm{Ker}\,d) \subset \mathrm{Ker}\,\tilde{d}$ is obvious. We prove the reverse inclusion. Let $\alpha \in \mathrm{Ker}\,\tilde{d}$, then there exists a positive integer n and $u_0/v_0, u_1/v_1, \ldots, u_{n-1}/v_{n-1} \in \mathrm{Frac}(\mathrm{Ker}\,d)$ such that

$$\alpha^n + (u_{n-1}/v_{n-1})\alpha^{n-1} + \cdots + (u_0/v_0) = 0.$$

Let $v = v_0 \cdots v_{n-1}$, then $v \in \mathrm{Ker}\,d$ and $v\alpha$ is integral over A, which is integrally closed. Hence $v\alpha \in A$. Set $v\alpha = a \in A$, then $d(a) = \tilde{d}(a) = 0$. Therefore $\alpha = a/v \in \mathrm{Frac}(\mathrm{Ker}\,d)$.

(2) We apply **(1)** for $A = K[x_1, \ldots, x_n]$. Here A is a unique factorization domain, hence integrally closed. From the assumptions, we conclude that the transcendence degree of $\operatorname{Ker} \tilde{d}$ over K is equal to n or $n-1$. In the first case, $K(x_1, \ldots, x_n)$ is algebraic separable over $\operatorname{Ker} \tilde{d}$, hence $\tilde{d} = 0$. Therefore $d = 0$, $\operatorname{Ker} d = K[x_1, \ldots, x_n]$, $\operatorname{Ker} \tilde{d} = K(x_1, \ldots, x_n)$ and the result is proved in this case. In the second case, let t be the transcendence degree of $\operatorname{Frac}(\operatorname{Ker} d)$ over K. Since $\operatorname{Frac}(\operatorname{Ker} d) \subset \operatorname{Ker} \tilde{d}$, then $n-1 \le t \le n-1$, hence $t = n-1$. It follows that $\operatorname{Ker} \tilde{d}$ is algebraic over $\operatorname{Frac}(\operatorname{Ker} d)$. By **(1)**, we conclude that $\operatorname{Ker} \tilde{d} = \operatorname{Frac}(\operatorname{Ker} d)$.

(3) Let $f(x_1, x_2) \in K[x_1, x_2]$,

$$f(x_1, x_2) = \sum a_{i_1 i_2} x_1^{i_1} x_2^{i_2},$$

then

$$d(f) = \sum (i_1 + i_2) a_{ij} x_1^{i_1} x_2^{i_2},$$

hence $d(f) = 0$ if and only if $f \in K$. Therefore $\operatorname{Ker} d = K$. It is easy to check that $\tilde{d}(x_2/x_1) = 0$. This shows that $\operatorname{Ker} \tilde{d} \neq \operatorname{Frac}(\operatorname{Ker} d)$. It is obvious that, in this example, the assumptions of **(2)** are not satisfied.

Exercise 11.7.

Let K be a field of characteristic $p \ge 0$, $\vec{X} = (x_1, \ldots x_n)$ be an n-tuple of variables, y another variable such that $x_1, \ldots, x_n, y$ are algebraically independent over K. Let $f(\vec{X}, y) \in K[\vec{X}, y]$ be irreducible and such that $\deg_y f \ge 1$ and $\deg_{x_i} f \ge 1$ for $i = 1, \ldots, n$. Let $\alpha \in \overline{K[\vec{X}]}$ such that $f(\vec{X}, \alpha) = 0$ and let $E = K(\vec{X}, \alpha)$. Let K_0 be the algebraic closure of K in E. If $p > 0$, we assume that $f \notin K[x_1^p, \ldots, x_n^p, y^p]$.

(1) Show that $[K_0 : K] \mid \deg_y f$.
(2) If f is absolutely irreducible, show that $K_0 = K$.
(3) Let $f_1(\vec{X}, y) \in \bar{K}[\vec{X}, y]$ be an irreducible factor of f such that $f_1(\vec{X}, \alpha) = 0$. Suppose that the leading coefficient of f_1 for the lexicographic order is equal to 1. Let K_1 be the extension of K generated by the coefficients of f_1.

   (a) If $p > 0$, let K_1^s be the separable closure of K in K_1 and let $p^e = [K_1 : K]_i = [K_1 : K_1^s]$. Let k be the smallest non negative integer such that $k \le e$ and $f_1^{p^k}(\vec{X}, y) \in K_1^s[\vec{X}, y]$. Let $g_1(\vec{X}, y) = f_1^{p^k}(\vec{X}, y)$.

   - Show that g_1 is irreducible over K_1^s.

•• Let $d_1 = [K_1^s : K]$ and let $\sigma_1, \ldots, \sigma_{d_1} : K_1^s \to \bar{K}$ be the distinct embeddings. Let $g(\vec{X}, y) = \prod_{i=1}^{d_1} g_1^{\sigma_i}(\vec{X}, y)$. Show that $f(\vec{X}, y) = cg(\vec{X}, y)$ for some $c \in K^\star$.

••• Deduce that $k = 0$, K_1/K is separable and $g_1 = f_1$.

(b) In any case $p = 0$ or $p > 0$, show that $K_1 = K_0$.

(c) Deduce that f is absolutely irreducible if and only if $K_0 = K$.

(4) Let F be the separable closure of $K(\vec{X})$ in E. Show that $K_0 \subset F$.

(5) From now on, we suppose that $n = 1$ and $p = 0$. We set $x_1 = x$ and we suppose that $f(x, y) \notin K[x^p, y]$ and $f(x, y) \notin K[x, y^p]$. Let $d : K(x) \to K(x)$ be the derivation such that for any $u(x) \in K(x)$, we have $d(u(x)) = u'(x)$.

• Show that d extends, in a unique way, as a derivation $\tilde{d}$ of E and compute $\tilde{d}(\alpha)$.

•• Show that $\operatorname{Ker} \tilde{d} = K_0$.

(6) Let $m = \deg_y f$. For any $\beta \in E$ we denote by β' the element $\tilde{d}(\beta)$ and generally $\beta^{(k)}$ the k-th derivative of β. Show that there exist $a_0(x), a_1(x), \ldots, a_m(x) \in K[x]$ not all 0 such that

$$a_0(x)\alpha + a_1(x)\alpha' + \cdots + a_m(x)\alpha^{(m)} = 0.$$

(7) • Let β be a conjugate of α over $K(x)$ and let $\sigma : E \to \overline{K(x)}$ be the $K(x)$-embedding such that $\sigma(\alpha) = \beta$. Show that $\sigma(\alpha^{(k)}) = \beta^{(k)}$.

•• Let $k \leq m$ be the smallest positive integer such that

$$b_0(x)\alpha + b_1(x)\alpha' + \cdots + b_k(x)\alpha^{(k)} = 0,$$

where $b_i(x) \in K[x]$ for $i = 0, \ldots, k$ and $b_k(x) \neq 0$. Show that all the conjugates of α over $K(x)$ satisfy this differential equation.

Solution 11.7.

(1) Consider the diagram.

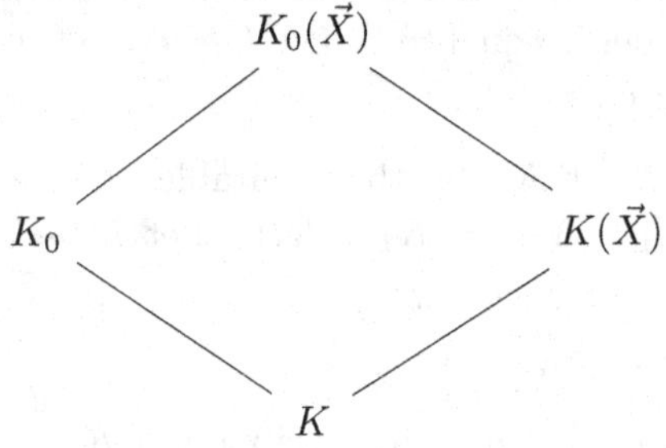

K_0/K is algebraic while $K(\vec{X})/K$ is purely transcendental, hence K_0 and $K(\vec{X})$ are linearly disjoint over K [Lang (1965), Pro. 3, Chap. 10.5]. Therefore $[K_0(\vec{X}) : K(\vec{X})] = [K_0 : K]$. It follows that

$$\begin{aligned}\deg_y f &= [E : K(\vec{X})] = [E : K_0(\vec{X})][K_0(\vec{X}) : K(\vec{X})] \\ &= [E : K_0(\vec{X})][K_0 : K],\end{aligned}$$

which proves that $[K_0 : K] \mid \deg_y f$.

(2) Suppose that $f(\vec{X}, y)$ is absolutely irreducible. We have

$$[K(\vec{X}, \alpha) : K_0(\vec{X})] = \deg_y f = [K(\vec{X}, \alpha) : K(\vec{X})].$$

Therefore $K_0(\vec{X}) = K(\vec{X})$ and then $K_0 = K$.

(3)(a) • Suppose that g_1 is reducible over K_1^s, then there exists $h < k$ such that $f^{p^h} \in K_1^s[x, y]$, which contradicts the minimality of k. Hence g_1 is irreducible over K_1^s.

• • We have $f(\vec{X}, \alpha) = g_1(\vec{X}, \alpha) = 0$, the two polynomials f and g_1 have their coefficients in K_1^s and the second is irreducible over this field. It follows that $g_1(\vec{X}, \alpha) \mid f(\vec{X}, \alpha)$ in $K_1^s[x, y]$. Therefore, for any K-embedding $\sigma : K_1^s \to \bar{K}$, $g_1^\sigma(\vec{X}, \alpha) \mid f(\vec{X}, \alpha)$ in $K_1^s[x, y]$. Since $g_1^\sigma \neq g_1^\tau$ for $\sigma \neq \tau$, then $\prod_{i=1}^{d_1} g_1^{\sigma_i}(\vec{X}, y) \mid f(\vec{X}, y)$ in $K_1^s[\vec{X}, y]$. The first polynomial has its coefficients in K and the second is irreducible over K, hence there exists $c \in K^\star$ such that $f(\vec{X}, y) = cg(\vec{X}, y)$.

• • • We have

$$f(\vec{X}, y) = c\prod_{i=1}^{d_1}(f_1^{p^k}(\vec{X}, y))^{\sigma_i} = c\prod_{i=1}^{d_1}(\tilde{f_{1,\sigma_i}}(\vec{X}, y))^{p^k},$$

where $\tilde{f_{1,\sigma_i}}(\vec{X}, y)$ is the polynomial obtained from $f_1^{\sigma_i}(\vec{X}, y)$ by replacing any of its coefficient a by a^{1/p^k}. If $k \geq 1$ then $f \in K[x_1^p, \ldots, x_n^p, y^p]$, which contradicts the assumptions. Hence $k = 0$, $K_1^s = K_1$, K_1/K is separable and $g_1 = f_1$.

(b) We have shown that if $p > 0$, then K_1/K is separable. This claim is also true if $p = 0$. We have in any case $f(\vec{X}, y) = c\prod_{i=1}^{d_1}(f_1(\vec{X}, y))^{\sigma_i}$, where the product runs over the distinct K-embeddings $\sigma_i : K_1^s \to \bar{K}$. We deduce that $\deg_y f = \deg_y f_1[K_1 : K]$. Consider the diagram.

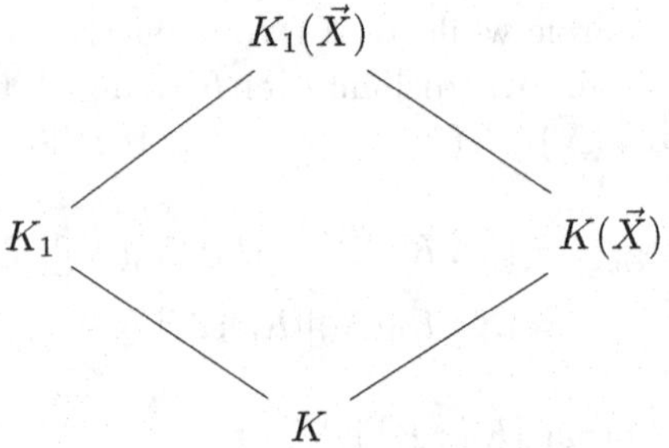

On the one hand, we have

$$\begin{aligned}[K_1(\vec{X},\alpha):K(\vec{X})] &= [K_1(\vec{X},\alpha):K_1(\vec{X})][K_1(\vec{X}):K(\vec{X})]\\ &= [K_1(\vec{X},\alpha):K_1(\vec{X})][K_1:K]\\ &= \deg_y f_1[K_1:K]\\ &= \deg_y f.\end{aligned}$$

On the other hand, we have

$$\begin{aligned}[K_1(\vec{X},\alpha):K(\vec{X})] &= [K_1(\vec{X},\alpha):K(\vec{X},\alpha)][K(\vec{X},\alpha):K(\vec{X})]\\ &= [K_1(\vec{X},\alpha):K(\vec{X},\alpha)]\deg_y f.\end{aligned}$$

Hence $K_1(\vec{X},\alpha) = K(\vec{X},\alpha)$, which implies $K_1 \subset K(\vec{X},\alpha)$ and then $K_1 \subset K_0$. Since f_1 is absolutely irreducible, then by **(2)**, we have $K_1^0 = K_1$. Since $K_1 \subset K_0$, then $K^0 \subset K_1^0 = K_1 \subset K^0$, hence $K_1 = K_0$.

(c) We have proved, in **(2)**, that if f is absolutely irreducible, then $K^0 = K$. We prove the converse. Suppose that f is not absolutely irreducible and let f_1 be an irreducible factor in $\bar{K}[x,y]$ of f. Let K_1 be the field generated by the coefficients of f_1 over K, then $K \subsetneq K_1 = K_0$ by **(3) (b)**, which is a contradiction.

(4) Let $a \in K_0$ and $g(x) = \mathrm{Irr}(a,K)$, then g is separable over K because K_0/K is separable. Since $\mathrm{Irr}(a,K(\vec{X})) = \mathrm{Irr}(a,K)$, then a is separable over $K(\vec{X})$. Therefore $a \in F$.

(5) • Suppose that an extension $\tilde{d}$ exists, then since $f(x,\alpha) = 0$, we have

$$\frac{\partial f}{\partial x}(x,\alpha) + \frac{\partial f}{\partial y}(x,\alpha)\tilde{d}(\alpha) = 0.$$

Since $\frac{\partial f}{\partial y}(x,\alpha) \neq 0$, then $\tilde{d}(\alpha) = -\frac{\frac{\partial f}{\partial x}(x,\alpha)}{\frac{\partial f}{\partial y}(x,\alpha)}$. Since $\tilde{d}$ is completely determined by $\tilde{d}(\alpha)$, then $\tilde{d}$ is unique. Moreover, this formula for $\tilde{d}(\alpha)$ allows one to define $\tilde{d}(\beta)$ for any $\beta \in E$.

•• It is clear that $K_0 \subset \operatorname{Ker} \tilde{d}$. We prove the reverse inclusion. Let $\beta \in \operatorname{Ker} \tilde{d}$ and

$$g(x,y) = y^q + b_{q-1}(x)y^{q-1} + \cdots + b_0(x)$$

be its minimal polynomial over $K(x)$. Since $g(x,\beta) = 0$, then

$$\frac{\partial g}{\partial x}(x,\beta) + \frac{\partial g}{\partial y}(x,\beta)\tilde{d}(\beta) = 0.$$

Hence $\frac{\partial g}{\partial x}(x,\beta) = 0$. Therefore, $g((x,y) \mid \frac{\partial g}{\partial x}(x,y)$ in $K(x)[y]$. It follows that $\frac{\partial g}{\partial x}(x,y) = 0$ and then $b'(x) = 0$ for $i = 0,\ldots,q-1$. We deduce that $b(x) \in K$ for $i = 0,\ldots,q-1$, which implies that β is algebraic over K. Therefore $\beta \in K_0$.

(6) Since E is a vector space over $K(x)$ of dimension m, then any family of $m+1$ elements is linearly dependent. Therefore, there exist $a_0(x), a_1(x), \ldots, a_m(x) \in K[x]$ not all 0 such that

$$a_0(x)\alpha + a_1(x)\alpha' + \cdots + a_m(x)\alpha^{(m)} = 0.$$

(7) • It is sufficient to prove that if $\sigma(\alpha) = \beta$, then $\sigma(\alpha') = \beta'$. From **(5)**, we have

$$\alpha' = \tilde{d}(\alpha) = -\frac{\frac{\partial f}{\partial x}(x,\alpha)}{\frac{\partial f}{\partial y}(x,\alpha)},$$

hence

$$\sigma(\alpha') = -\frac{\frac{\partial f}{\partial x}(x,\sigma(\alpha))}{\frac{\partial f}{\partial y}(x,\sigma(\alpha))} = \beta'.$$

•• It is clear that the integer k satisfying the given properties exists. Applying σ to the differential equation satisfied by α:

$$b_0(x)\alpha + b_1(x)\alpha' + \cdots + b_k(x)\alpha^{(k)} = 0$$

and using **(7) •**, we conclude that β satisfies the same equation.

Exercise 11.8.
Let K be a field and A be a K-algebra without zero divisors. Let Δ be a set of K-derivations of A and

$$A^{\Delta} = \{a \in A, d(a) = 0, \text{ for any } d \in \Delta\}.$$

Show that $\operatorname{Frac}(A^{\Delta}) \cap A = A^{\Delta}$.

Solution 11.8.
Let

$$\bar{\Delta} = \{\delta, \text{ derivation of Frac}(A) \text{ extending some derivation of } A\}$$

and

$$\text{Frac}(A)^{\bar{\Delta}} = \{x \in \text{Frac}(A), \delta(a) = 0, \text{ for any } \delta \in \bar{\Delta}\}.$$

We have $\text{Frac}(A^{\Delta}) \subseteq \text{Frac}(A)^{\bar{\Delta}}$ and $\text{Frac}(A)^{\bar{\Delta}} \cap A = A^{\Delta}$, hence

$$A^{\Delta} \subseteq \text{Frac}(A)^{\Delta} \cap A \subseteq \text{Frac}(A)^{\bar{\Delta}} \cap A = A^{\Delta}.$$

Therefore $\text{Frac}(A^{\Delta}) \cap A = A^{\Delta}$.

Exercise 11.9.
Let R and S be integral domains such that $S \subset R$ and let F and E be their respective fraction fields. Suppose that E is a Galois extension of F, R is integrally closed and that $R \cap F = S$. Let $D : E \to E$ be a derivation such that $D(F) \subset F$ and $D(R) \subset R$. Let $\bar{S}$ be the integral closure of S in E.

(1) Show that for any $\sigma \in \text{Gal}(E, F)$, $\sigma D \sigma^{-1}$ is a derivation of E and the restrictions to F of $\sigma D \sigma^{-1}$ and D are equal.
(2) Deduce that for any $\sigma \in \text{Gal}(E, F)$, $\sigma D \sigma^{-1} = D$.
(3) Show that for any $\sigma \in \text{Gal}(E, F), \sigma(\bar{S}) = \bar{S}$.
(4) Show that $D(\bar{S}) \subset \bar{S}$.

Solution 11.9.

(1) Obvious.
(2) D_F is a derivation of F and $D, \sigma D \sigma^{-1}$ are extensions of D_F to E, where D_F is the restriction of D to F. Since E/F is algebraic, this extension is unique, hence $\sigma D \sigma^{-1} = D$.
(3) It is sufficient to prove that $\sigma(\bar{S}) \subset \bar{S}$ for any $\sigma \in \text{Gal}(E, F)$. Since $\sigma(\bar{S}) \subset E$ and $\sigma(\bar{S})$ is integral over S, then $\sigma(\bar{S}) \subset \bar{S}$.
(4) Let $\bar{s} \in \bar{S}$, then its conjugates over F, $\sigma(\bar{s})$, $\sigma \in G$, belong to $\bar{S}$ by **(3)**, hence any elementary symmetric function of these elements belongs to $\bar{S}$. Since $\bar{S} \subset \bar{R}$, where $\bar{R}$ denotes the integral closure of R in E and since R is integrally closed, then $\bar{R} = R$, hence all these elementary symmetric functions belong to R. Since they also belong to F, they are elements of $R \cap F$, which is equal to S. It follows that $D(\bar{s})$ is integral over S, that is $D(\bar{s}) \in \bar{S}$, thus $D(\bar{S}) \subset \bar{S}$.

Exercise 11.10.

Let K be a perfect field, E be an algebraic extension of K, $d : K \to K$ be a derivation and $\hat{d} : E \to E$ be its unique extension. Suppose that $\operatorname{Ker} d$ is algebraically closed in E. Show that $\operatorname{Ker} \hat{d} = \operatorname{Ker} d$.

Solution 11.10.

Obviously, we have $\operatorname{Ker} d \subset \operatorname{Ker} \hat{d}$. For the reverse inclusion, let $\alpha \in \operatorname{Ker} \hat{d}$ and let

$$f(x) = \operatorname{Irr}(\alpha, K) = x^n + a_{n-1}x^{n-1} + \cdots + a_0.$$

Since $f(\alpha) = 0$, then

$$\sum_{i=0}^{n-1} \hat{d}(a_i)\alpha^i + \hat{d}(\alpha)f'(\alpha) = 0,$$

hence

$$\sum_{i=0}^{n-1} \hat{d}(a_i)\alpha^i = \sum_{i=0}^{n-1} d(a_i)\alpha^i = 0.$$

It follows that $a_i \in \operatorname{Ker} d$ for any $i \in \{0, \ldots, n-1\}$ which implies that α is algebraic over $\operatorname{Ker} d$. Since $\operatorname{Ker} d$ is algebraically closed in E, then $\alpha \in \operatorname{Ker} d$.

Exercise 11.11.

(1) Let K be a field, A be a K-algebra, Δ be a set of K-derivations of A and

$$A^{\Delta} = \{a \in A, d(a) = 0 \text{ for any } d \in \Delta\}.$$

Show that $\operatorname{Frac}(A^{\Delta}) \subset (\operatorname{Frac} A)^{\bar{\Delta}}$, where

$$\bar{\Delta} = \{\delta \text{ derivation of } \operatorname{Frac} A, \delta \text{ is a prolongation of some } d \in \Delta\}.$$

(2) Show that the reverse inclusion does not hold in the following case: K is a field of characteristic 0, $A = K[x, y]$, $\Delta = \{d\}$, where d is the unique K-derivation of A such that $d(x) = x$ and $d(y) = y$.

Solution 11.11.

(1) Since $(\operatorname{Frac} A)^{\bar{\Delta}}$ is a field, it is sufficient to prove that $A^{\Delta} \subset (\operatorname{Frac} A)^{\bar{\Delta}}$. Let $a \in A^{\Delta}$ and $\delta \in \bar{\Delta}$, then $a = \frac{a}{1} \in \operatorname{Frac} A$ and

$$\delta(a) = \frac{\delta(a) \cdot 1 - a\delta(1)}{1^2} = \frac{d(a) - ad(1)}{1^2} = 0,$$

hence $a \in (\operatorname{Frac}(A))^{\bar{\Delta}}$.

(2) We show that $\frac{x}{y} \in (\operatorname{Frac} A)^{\bar{\Delta}}$ but $\frac{x}{y} \notin \operatorname{Frac}(A^{\Delta})$. It is easy to see that for any $f(x,y) \in K[x,y]$, $d(f) = \frac{\delta f}{\delta x}x + \frac{\delta f}{\delta y}y$. Let $\hat{d}$ be the unique extension of d to $K(x,y) = \operatorname{Frac} A$, then

$$\hat{d}(\frac{x}{y}) = \frac{y\hat{d}(x) - x\hat{d}(y)}{y^2} = \frac{yd(x) - xd(y)}{y^2} = 0,$$

thus $\frac{x}{y} \in (\operatorname{Frac} A)^{\bar{\Delta}}$. Let $f(x,y) \in A^{\Delta}$, then

$$x\frac{\delta f}{\delta y} + y\frac{\delta f}{\delta x} = 0, \qquad \text{(Eq 1)}$$

hence $x|\frac{\delta f}{\delta y}$. Set $\frac{\delta f}{\delta y} = xg(x,y)$, where $g(xy) \in K[x,y]$, then $\frac{\delta f}{\delta x} = -yg(x,y)$. It follows that $\frac{\delta f}{\delta y}/x = -\frac{\delta f}{\delta x}/y$. Set

$$f(x,y) = a_n(x)y^n + \cdots + a_1(x)y + a_0(x).$$

Since $y \mid \frac{\delta f}{\delta x}$, then $a_0(x) = 0$.
We deduce that

$$n\frac{a_n(x)}{x}y^{n-1} + \cdots + \frac{a_1(x)}{x} = -[a_n'(x)y^{n-1} - \cdots - a_1'(x)]$$

and then $n\frac{a_n(x)}{x} = -a_n'(x)$. Set

$$a_n(x) = b_m x^m + b_{m-1}x^{m-1} + \cdots + b_1 x$$

with $b_m \neq 0$, then

$$\begin{aligned} &nb_m x^{m-1} + nb_{m-1}x^{m-2} + \cdots + nb_1 \\ &= -mb_m x^{m-1} - (m-1)b_{m-1}x^{m-2} - \cdots - b_1. \end{aligned}$$

It follows that $(n+m)b_m = 0$, which implies $n = m = 0$. Therefore, $f(x,y) \in K[x]$ and then from **(1)**, $\frac{\delta f}{\delta x} = 0$ which means $f(x,y) \in K$. We have proved that $A^{\Delta} = K$, hence $\operatorname{Frac}(A^{\Delta}) = K$ and $\frac{x}{y} \in K$. It follows that $\operatorname{Frac}(A^{\Delta}) \subsetneq (\operatorname{Frac} A)^{\Delta}$.

Exercise 11.12.
Let K be a field of characteristic 0 and Δ be a K-algebra without zero divisors. Let Δ be a set of derivations of A and

$$A^{\Delta} = \{a \in A, d(a) = 0 \text{ for any } d \in \Delta\}.$$

Show that A^{Δ} is integrally closed in A.

Solution 11.12.
Let $B = A^\Delta$ and let $a \in A$ be integral over B. Then there exist $b_0, b_1, \ldots, b_n - 1 \in B$ such that $a^n + b_{n-1}a^{n-1} + \cdots + b_0 = 0$ and we may suppose that n is minimal. We deduce that for any $d \in \Delta$,

$$d(a)\left(na^{n-1} + (n-1)b_{n-1}a^{n-2} + \cdots + b_1\right) = 0,$$

hence $d(a) = 0$, that is $a \in A^\Delta$. Therefore, A^Δ is integrally closed in A.

Exercise 11.13.
Let K be a field, n be an integer at least equal to 2, $A = K[x_1, \ldots, x_n]$ and B the integral closure of $K[x_1, x_1x_2]$ in A. Show that for any non empty set Δ of derivations of A, we have $B \neq A^\Delta$, where A^Δ is the set of elements $a \in A$ such that $d(a) = 0$ for any $d \in \Delta$. Show that $B \subset K[x_1, x_2]$.

Solution 11.13.
Suppose that there exists a non empty set Δ of derivations of A such that $B = A^\Delta$. For any $d \in \Delta$, we have

$$d(x_1) = 0 = d(x_1x_2) = x_1d(x_2),$$

hence $d(x_2) = 0$, which implies $x_2 \in B$. We show that this leads to a contradiction. There exist $a_0, a_1, \ldots, a_{m-1} \in K[x_1, x_1x_2]$ such that $x_2^m + a_{m-1}x_2^{m-1} + \cdots + a_0 = 0$. Write the left side of this identity as a sum of monomials $bx_1^ix_2^j$. Then it is seen that the coefficient of x_2^m is equal to 1. This means that the polynomial on the left side of the identity is non zero, hence x_1, x_2 are algebraically dependent over K, which is a contradiction. We have $K[x_1, x_1x_2] \subset K[x_1, x_2] \subset A$, hence $B \subset \overline{K[x_1, x_2]}$, where $\overline{K[x_1, x_2]}$ is the integral closure of $K(x_1, x_2)$ in A. It is easy to show that $K[x_1, x_2]$ is integrally closed in A, hence $B \subset K[x_1, x_2]$.

Exercise 11.14.
Let K be a field of characteristic 0, d be a K-derivation of $K[x_1, \ldots, x_n]$. Suppose that $d \neq 0$.

(1) Show that the number of polynomials algebraically independent over K and belonging to $\operatorname{Ker} d$ is at most equal to $n - 1$.
(2) Show that the condition on the characteristic of K cannot be omitted.

Solution 11.14.

(1) Suppose that $\operatorname{Ker} d$ contains n polynomials

$$f_1(x_1, \ldots, x_n), \ldots, f_n(x_1, \ldots, x_n)$$

algebraically independent over K and let $g(x_1,\ldots,x_n)\in K[x_1,\ldots,x_n]$. Then clearly $f_1,\ldots,f_n,g$ are algebraically dependent over K. Therefore, there exists $\phi(y_1,\ldots,y_n,z)\in K[y_1,\ldots,y_n,z]$ such that ϕ is irreducible over K and $\phi(f_1,\ldots,f_n,g)=0$. Set

$$\phi(y_1,\ldots,y_n,z)=\sum_{i=0}^{m}\phi_i(y_1,\ldots,y_n)z^i,$$

where m is a positive integer, $\phi_i\in k[y_1,\ldots,y_n]$ for $i=0,\ldots,m$ and $\phi_m\neq 0$.

Since $\sum_{i=0}^{m}\phi_i(f_1,\ldots,f_n)g^i=0$, then

$$\sum_{i=0}^{m}d(\phi_i(f_1,\ldots,f_n))g^i+\left(\sum_{i=1}^{m}i\phi_i(f_1,\ldots,f_n)g^{i-1}\right)d(g)=0,$$

hence

$$\left(\sum_{i=1}^{m}i\phi_i(f_1,\ldots,f_n)g^{i-1}\right)d(g)=0.$$

Since K is of characteristic 0, $f_1,\ldots,f_n$ are algebraically independent over K and $\phi_m(y_1,\ldots,y_n)\neq 0$, then $m\phi_m(f_1,\ldots,f_n)\neq 0$, hence $\sum_{i=1}^{m}i\phi_i(y_1,\ldots,y_n)y^{i-1}\neq 0$. Therefore $\sum_{i=1}^{m}i\phi_i(f_1,\ldots,f_n)g^{i-1}\neq 0$ and then $d(g)=0$.

Since g was arbitrary, then $d=0$, which is a contradiction to our assumptions. We conclude that the number of elements of $\operatorname{Ker} d$, algebraically independent over K is at most equal to $n-1$.

(2) Let K be a field of characteristic p and let $d: K[x]\to K[x]$ such that $d(f(x))=f'(x)$. Then clearly $d(x^p)=px^{p-1}=0$. Therefore $\operatorname{Ker} d$ contains a transcendental element over K.

Exercise 11.15.

Let K be a field of characteristic 0 and $P(x,y)$ be a non constant polynomial with coefficients in K. Let $D: K(x,y)\to K(x,y)$ be the map such that for any $f(x,y)\in K(x,y)$,

$$D(f)=\begin{vmatrix}\frac{\partial p}{\partial x} & \frac{\partial p}{\partial y}\\ \frac{\partial f}{\partial x} & \frac{\partial f}{\partial y}\end{vmatrix}$$

(1) Show that D is a K-derivation.
(2) Let $f(x,y)\in K(x,y)$. Show that the following propositions are equivalent.

   (i) $f\in\operatorname{Ker} D$.
   (ii) f and P are algebraically dependent over K.

Solution 11.15.

(1) Let $f, g \in K(x,y)$ and $\lambda, \mu \in K$.
Then

$$\begin{aligned} D(\lambda f+\mu g) &= \begin{vmatrix} \frac{\partial p}{\partial x} & \frac{\partial p}{\partial y} \\ \lambda\frac{\partial f}{\partial x}+\mu\frac{\partial g}{\partial x} & \lambda\frac{\partial f}{\partial y}+\mu\frac{\partial g}{\partial y} \end{vmatrix} \\ &= \frac{\partial p}{\partial x}\left(\lambda\frac{\partial f}{\partial y}+\mu\frac{\partial g}{\partial y}\right)-\frac{\partial p}{\partial y}\left(\lambda\frac{\partial f}{\partial x}+\mu\frac{\partial g}{\partial x}\right) \\ &= \lambda\left(\frac{\partial p}{\partial x}\frac{\partial f}{\partial y}-\frac{\partial p}{\partial y}\frac{\partial f}{\partial x}\right)+\mu\left(\frac{\partial p}{\partial x}\frac{\partial g}{\partial y}-\frac{\partial p}{\partial y}\frac{\partial g}{\partial x}\right) \\ &= \lambda D(f)+\mu D(g), \end{aligned}$$

hence D is K-linear.
We have

$$\begin{aligned} D(fg) &= \begin{vmatrix} \frac{\partial p}{\partial x} & \frac{\partial p}{\partial y} \\ g\frac{\partial f}{\partial x}+f\frac{\partial g}{\partial x} & g\frac{\partial f}{\partial y}+f\frac{\partial g}{\partial y} \end{vmatrix} \\ &= \frac{\partial p}{\partial x}\left(g\frac{\partial f}{\partial y}+f\frac{\partial g}{\partial y}\right)-\frac{\partial p}{\partial y}\left(g\frac{\partial f}{\partial x}+f\frac{\partial g}{\partial x}\right) \\ &= f\left(\frac{\partial p}{\partial x}\frac{\partial g}{\partial y}-\frac{\partial p}{\partial y}\frac{\partial g}{\partial x}\right)+g\left(\frac{\partial p}{\partial x}\frac{\partial f}{\partial y}-\frac{\partial p}{\partial y}\frac{\partial f}{\partial x}\right) \\ &= fD(g)+gD(f). \end{aligned}$$

Therefore D is a derivation. Obviously $D(a)=0$ for any $a \in K$, hence D is K-derivation.
Remark. We may write the map D in the form $D = \frac{\partial p}{\partial x}\frac{\partial}{\partial y} - \frac{\partial p}{\partial y}\frac{\partial}{\partial x}$, therefore D is a K-derivation.

(2) • $(i) \Rightarrow (ii)$. Suppose that f and P are algebraically independent over K. Let $Q(x,y) \in K(x,y)\backslash K$, then f, P and Q are algebraically dependent over K. Let $R(x,y,z) \in K[x,y,z]\backslash\{0\}$ such that $R(P,f,Q)=0$. We may write this identity in the form: $\sum_{i=0}^{n} R_i(P,f)Q^i = 0$, where $n \geq 1$, $R_1(x,y) \in K[x,y]$ and $R_n(x,y) \neq 0$. Moreover, we may suppose that n is minimal. We have

$$0 = D(R(P,f,Q)) = \sum_{i=0}^{n} D(R_i(P,f))Q^i + \left(\sum_{i=1}^{n} R_i(P,f)iQ^{i-1}\right)D(Q).$$

Since $D(P)=D(f)=0$, then

$$D(Q)\sum_{i=1}^{n} R_i(P,f)iQ^{i-1} = 0.$$

If $n \geq 2$, then since n is minimal, we conclude that $D(Q) = 0$. If $n = 1$, then $R_1(P,Q)D(Q) = 0$ and since $R_1(P,Q) \neq 0$, we conclude that $D(Q) = 0$. In any case we obtain $D(Q) = 0$ for any $Q \in K[x,y]\backslash K$. Therefore $D = 0$. We deduce that

$$0 = D(x) = -\frac{\partial P}{\partial y} \quad \text{and} \quad 0 = D(y) = \frac{\partial P}{\partial x}.$$

It follows that $P(x,y) \in K$, a contradiction.

- $(ii) \Rightarrow (i)$. Obviously if $f \in K$, then $f \in \operatorname{Ker} D$, hence we may suppose that $f \notin K$. Since $P(x,y)$ and $f(x,y)$ are algebraically dependent over K, there exist a positive integer n and $a_0(x), a_1(x), \ldots, a_n(x) \in k[x]$ such that $a_n(x) \neq 0$, $\sum_{i=0}^n a_i(f(x,y))P(x,y)^i = 0$ and the polynomial $\sum_{i=0}^n a_i(x)y^i$ is irreducible over K. We deduce that

$$\sum_{i=0}^n D(a_i(f))P^i + D(P)\sum_{i=1}^n ia_i(f)P^{i-1} = 0.$$

Since $D(P) = 0$, then

$$D(f)\sum_{i=0}^n a_i'(f)P^i = 0.$$

Suppose that $\sum_{i=0}^n a_i'(f)P^i = 0$, then $\sum_{i=0}^n a_i(x)y^i$ divides $\sum_{i=0}^n a_i'(x))y^i$. It follows that $a_i'(x) = 0$ for $i = 0, \ldots, n$, hence $a_i(x) \in K$ for $i = 0, \ldots, n$. Therefore $P(x,y) \in K$ which is a contradiction. We conclude that $D(f) = 0$, that is $f \in \operatorname{Ker} D$.

Exercise 11.16.
Let R and S be integral domains such that $S \subset R$ and let E and F be their fraction fields respectively. Suppose that R is integrally closed, E is a finite Galois extension of F and $R \cap F = S$. Let R_0 be the integral closure of S in E. Let d be a derivation of E such that $d(F) \subset F$ and $d(R) \subset R$.

(1) Show that $R_0 \subset R$.
(2) Show that $d(R_0) \subset R_0$.

Solution 11.16.

(1) The following diagram may be helpful.

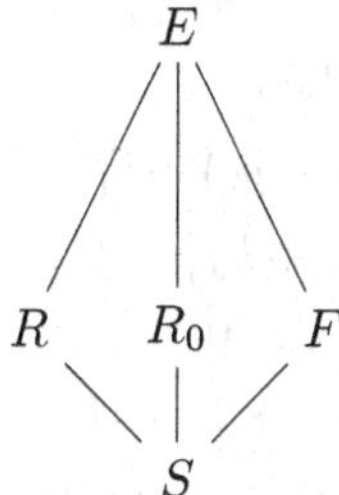

Let R_1 be the integral closure of R in E. Since R is integrally closed, then $R_1 = R$. On the other hand, since $S \subset R$, then $R_0 \subset R_1$, hence $R_0 \subset R$.

(2) Let $G = \{\sigma_1, \ldots, \sigma_n\}$ be the Galois group of E over F and let $\alpha \in R_0$. We must show that $d(\alpha) \in R_0$. This is equivalent to show that the elements $s_k := \sum \sigma_{i_1}(d(\alpha)) \cdots \sigma_{i_k}(d(\alpha))$, $(k = 1, \ldots, n)$, which represent, up to multiplication by -1, the coefficients of the characteristic polynomial of $d(\alpha)$ over F, except the leading one, belong to $S = R \cap F$. Obviously, they belong to F. We show that they belong to R. For any $\sigma \in G, \sigma d \sigma^{-1}$ is a derivation of E and $\sigma d \sigma^{-1}_{IF} = d_{IF}$. Since E/F is algebraic and separable, then any derivation of F has one and only one extension to E. It follows that $\sigma d \sigma^{-1} = d$, hence $\sigma d = d\sigma$. Since $\alpha \in R_0$, then $\sigma(\alpha) \in R_0$ and $\sigma(\alpha) \in R$ by **(1)**. We conclude that $s_k \in R$.

Exercise 11.17.
Let A be an integral domain containing $\mathbb{Q}$ and $d : A[x] \to A[x]$ be the derivation defined by $d(f(x)) = f'(x)$ for any $f(x) \in A[x]$.

(1) Show that d is surjective.
(2) Let I and J be ideals of $A[x]$ such that $I \subset J$. Show that $J + d(I)$ is an ideal of $A[x]$. Show that $d(I)$ is not always an ideal of $A[x]$.
(3) Let I be an ideal of $A[x]$ and let $c(I)$ be the ideal of A generated by all the coefficients of the polynomials $f(x) \in I$. Show that $d^{-1}(I)$ is a subring of $A[x]$ and $A \subset d^{-1}(I) \subset A + c(I)A[x]$.

Solution 11.17.

(1) Let $g(x) = \sum_{i=0}^{n} a_i x^i$ be an element of $A[x]$, then

$$g(x) = d\left(\sum_{i=0}^{n} \frac{1}{i+1} a_i x^{i+1}\right),$$

hence d is surjective.

(2) Let $g_1 + f_1'$, $g_2 + f_2'$ be elements of $J + d(I)$, where f_1, $f_2 \in I$ and g_1, $g_2 \in J$ and let $h \in R[x]$, then

$$g_1 + f_1' + g_2 + f_2' = (g_1 + g_2) + (f_1 + f_2)' \in J + d(I) \quad \text{and}$$
$$h(g_1 + f_1') = hg_1 + hf_1' = (hg_1 - h'f_1) + (fh)' \in J + d(I),$$

hence $J + d(I)$ is an ideal of $A[x]$.
Let $A = \mathbb{Q}$ and $I = (x^2 + x + 1)A[x]$, then

$$d(I) = \{[(x^2 + x + 1)f(x)]', f(x) \in A[x]\},$$

hence $[x(x^2 + x + 1)]' \in d(I)$, that is $3x^2 + 2x + 1 \in d(I)$. We show that $x(3x^2 + 2x + 1) \notin d(I)$ which will prove that $d(I)$ is not an ideal of $\mathbb{Q}[x]$. Suppose that $x(3x^2 + 2x + 1) \in d(I)$, then

$$3x^3 + 2x^2 + x = [(x^2 + x + 1)f(x)]'$$

for some $f(x) \in \mathbb{Q}[x]$. Since

$$3x^3 + 2x^2 + x = \left[\frac{3x^4}{4} + \frac{2x^2}{3} + \frac{x^2}{2}\right]',$$

then

$$(x^2 + x + 1)f(x) = \frac{3}{4}x^4 + \frac{2}{3}x^3 + \frac{1}{2}x^2 + a,$$

where $a \in \mathbb{Q}$. We deduce that $x^2 + x + 1$ divides $\frac{3}{4}x^4 + \frac{2}{3}x^3 + \frac{1}{2}x^2 + a$ in $\mathbb{Q}[x]$. Performing an Euclidean division, we obtain

$$\frac{3}{4}x^4 + \frac{2}{3}x^3 + \frac{1}{2}x^2 + a = (x^2 + x + 1)\left(\frac{3}{4}x^2 - \frac{1}{12}x^2 - \frac{1}{12}x\right) + \frac{1}{4}x + \frac{1}{6} + a.$$

Since for any $a \in \mathbb{Q}$, $x^2 + x + 1 \nmid \frac{3}{4}x^4 + \frac{2}{3}x^3 + \frac{1}{2}x^2 + a$, then we get a contradiction. We conclude that $x(3x^2 + 2x + 1) \notin d(I)$.

(3) Let $f_1, f_2 \in d^{-1}(I)$, then

$$d(f_1 - f_2) = (f_1 - f_2)' = f_1' - f_2' \in I \quad \text{and}$$
$$d(f_1 \cdot f_2) = f_1 f_2' + f_2 f_1' \in I,$$

hence $f_1 - f_2$ and $f_1 f_2 \in d^{-1}(I)$. Therefore $d^{-1}(I)$ is a subring of $A[x]$. Let $a \in A$, then $d(a) = 0 \in I$, hence $a \in d^{-1}(I)$. We deduce that $A \subset d^{-1}(I)$. Let $f(x) \in d^{-1}(I)$ and let $a = f(0)$, then $d(f) = f' \in I$. Set $f'(x) = \sum_{i=0}^{n} a_i x^i$, then $a_i \in c(I)$ and

$$f(x) - f(0) = f(x) - a = \sum_{i=0}^{n} \frac{a_i}{i+1} x^{i+1} \in c(I)A[x].$$

Therefore

$$f(x) = a + (f(x) - a) \in A + c(I)A[x].$$

Exercise 11.18.

Let A be an integral domain containing $\mathbb{Q}$, K be its fraction field, R be a subring of some extension E of K satisfying the condition $R \cap K = A$. Suppose that there exist $t, u \in R$ and an A-derivation d of R such that $R \subset K[u]$ and $d(t) = 1$.

(1) Show that u is transcendental over K.
(2) Show that for any $x, y \in R, xy \in A \Rightarrow x \in A$ and $y \in A$.
(3) Show that $R \subset K[t]$.
(4) Deduce that $R = A[t]$.

Solution 11.18.

(1) The following diagram may be useful.

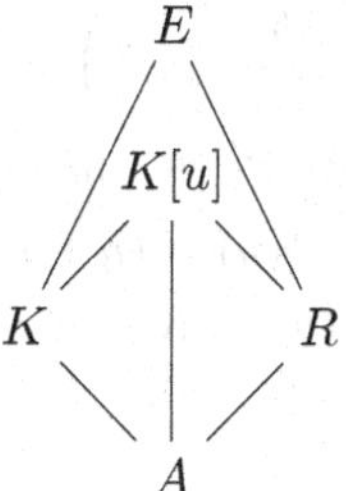

Suppose that u is algebraic over K and let

$$f(x) = x^n + a_{n-1}x^{n-1} + \cdots + a_0$$

be its minimal polynomial over K. Since $f(u) = 0$, then $f'(u)d(u) = 0$. If $f'(u) = 0$, then $f(x)|f'(x)$ in $K[x]$ and then $f'(x) = 0$ which contradicts the fact that the characteristic of K is equal to 0. Therefore $d(u) = 0$. We deduce that for any $g(u) \in R \subset K[u]$, we have $d(g(u)) = g'(u)d(u) = 0$, which contradicts the existence of $t \in R$, such $d(t) = 1$. We conclude that u is transcendental over K.

(2) Let $x, y \in R$ such that $xy \in A$. We may write x and y in the form $x = f_1(u)/c_1$ and $y = f_2(u)/c_2$, where $c_1, c_2 \in A$ and $f_1(u), f_2(u) \in A[u]$. We have $f_1(u)f_2(u) = c_1c_2xy \in A$. Since u is transcendental over K, then $f_1(u)$ and $f_2(u) \in A$. We deduce that x and $y \in R \cap K = A$.

(3) Since $R \subset K[u]$, it is sufficient to show that $u \in K[t]$. Since $t \in R$, then $t \in K[u]$, hence there exist $a \in A\backslash\{0\}$ and $F(u) \in A[u]$ such that $at = F(u)$. Since d is an A-derivation, we deduce that

$$d(at) = ad(t) = a = F'(u)d(u) \in A.$$

By **(2)**, we conclude that $F'(u)$ and $d(u) \in A$. Let $b \in A$ such that $F'(u) = b$, then $F(u) = bu + c$ with $c \in A$. We deduce that $at = bu + c$ and this shows that $b \neq 0$. We now have $a = F'(u)d(u) = bd(u)$, hence $ab^{-1} = d(u) \in R \cap K$, thus $ab^{-1} \in A$. We deduce that $cb^{-1} = ab^{-1}t - u \in R \cap K$. Since $R \cap K = A$, then $u = ab^{-1}t - cb^{-1} \in A[t]$. We conclude that $R \subset K[u] \subset K[t]$.

(4) The inclusion $A[t] \subset R$ is obvious. We prove the reverse inclusion. Let $x \in R \backslash \{0\}$. Since $R \subset K[t]$, then $x \in K[t]$, hence there exists $a \in A$ such that $ax \in A[t]$. Set $ax = b_0 t^n + b_1 t^{n-1} + \cdots + b_n$, where n is a positive integer, $b_0, b_1, \ldots, b_n \in A$ and $b_0 \neq 0$. Let d^n be the n-th iterate of d, then $d^n(ax) = ad^n(x) = n! b_0$. Since $\mathbb{Q} \subset A$, then $n!$ is a unit of A, hence $a | b_0$ in A. We complete the proof that $a | b_i$ for $i = 0, \ldots, n$ by induction. Suppose that $a | b_i$ for $i = 0, \ldots, m$ for $m \geq 1$, then
$$a\left(x - \frac{b_0}{a}t^n - \cdots - \frac{b_m}{a}t^{n-m}\right) = b_{m+1}t^{n-m-1} + \cdots + b_n.$$
Let $y = x - \frac{b_0}{a}t^n - \cdots - \frac{b_m}{a}t^{n-m}$, then $y \in R$ and we have
$$ad^{n-m-1}(y) = (n-m-1)! b_{m+1}.$$
This shows that $d^{n-m-1}(y) \in K \cap R = A$, hence $a \mid b_{m+1}$ in A and then $x \in A[t]$ as desired.

Exercise 11.19.
Let K be a field, d be a K-derivation of $K(x, y)$ such that $d(x)$ and $d(y) \in K[x, y]$.

(1) Show that the map $\tau : K(x, y) \setminus \{0\} \to K(x, y)$ defined by $\tau(f) = d(f)/f$ for any $f(x, y) \in K(x, y)^* \setminus \{0\}$ is a homomorphism of the multiplicative group into the additive group.
(2) Let $f_1, f_2 \in K[x, y] \setminus \{0\}$ such that $\gcd(f_1, f_2) = 1$. Suppose that $\tau(f_1 f_2) \in K[x, y]$ or $\tau(f_1/f_2) \in K[x, y]$. Show that $\tau(f_1)$ and $\tau(f_2) \in K[x, y]$.

Solution 11.19.

(1) Let $f_1, f_2 \in K(x, y) \setminus \{0\}$, then
$$\tau(f_1 f_2) = d(f_1 f_2)/f_1 f_2 = \frac{f_1 d(f_2) + f_2 d(f_1)}{f_1 f_2} = \tau(f_1) + \tau(f_2)$$
and
$$\tau(f_1^{-1}) = \frac{d(f_1^{-1})}{f_1^{-1}} = f_1 d(1/f_1) = f_1(-d(f_1)/f_1^2) = -d(f_1)/f_1 = -\tau(f_1),$$
hence τ is a homomorphism of groups.

(2) If $\tau(f_1 f_2) \in K[x,y]$, then from the identity

$$\tau(f_1 f_2) \cdot f_1 f_2 = f_1 d(f_2) + f_2 d(f_1),$$

we see that $f_1 \mid d(f_1)$ and $f_2 \mid d(f_2)$. Hence $\tau(f_1) \in K[x,y]$. If $\tau(f_1/f_2) \in K[x,y]$, we use the identity

$$f_1 f_2 d(f_1/f_2) = d(f_1) f_2 - d(f_2) f_1$$

and we obtain the same conclusions as above: $f_1 \mid d(f_1)$ and $f_2 \mid d(f_2)$. Therefore $\tau(f_1) \in K[x,y]$ and $\tau(f_2) \in K[x,y]$.

Exercise 11.20.

Let K be a field of characteristic 0, d be a derivation of K and K_0 be a subfield of K containing $\operatorname{Ker} d$ and satisfying the condition $d(K_0) \subset K_0$.

(1) Let $\{f_1, f_2, f_3, \ldots\}$ be a subset of K such that $d(f_i) \in K_0$ for any i and let

$$T_0 = \left\{ \sum_{i=1}^{n} c_i f_i, n \geq 1, (c_1, \ldots, c_n) \in (\operatorname{Ker} d)^n \backslash \{0\} \right\}.$$

If $T_0 \cap K_0 = \phi$, show that $1, f_1, f_2, \ldots$ are linearly independent over K_0.

(2) Let F be a subfield of K such that $K_0 \subset F$ and $d(F) \subset F$. Let $L \in T_0$ such that $1, L$ are linearly independent over F.

(3) Deduce that if $K_0 \cap T_0 = \phi$, then for any $L \in T_0$ and any integer $n \geq 1, 1, L, \ldots, L^n$ are linearly independent over K_0.

(4) Let F be as in **(2)** and let $L \in T_0$. If $\lambda + \mu L = 0$, where $\lambda, \mu \in F$, show that $d(\lambda)\mu - \lambda(\mu) + \mu^2 d(L) = 0$.

(5) Let $A_0 = K_0$ and $A_m = K_0[f_1, \ldots, f_m]$ for any integer $m \geq 1$. Let K_m be the fractions field of A_m.

  (a) Show that $d(A_m) \subset A_m$ and $d(K_m) \subset K_m$ for any $m \geq 0$.

  (b) For any $m \geq 1$ let

$$T_m = \left\{ \sum_{i=m+1}^{n} c_i f_i, n \geq m+1, (c_{m+1}, \ldots, c_n) \in (\operatorname{Ker} d)^{n-m} \backslash \{0\} \right\}.$$

  Suppose that $T_0 \cap K_0 \neq \phi$ and let $L \in T_m$ for some $m \geq 0$. Show that $1, L, \ldots, L^n$ are linearly independent over K_m for any $n \geq 1$.

(6) Let $f_1, f_2, \ldots,$ and T_0 as in **(1)**. Show that $f_1, f_2, f_3, \ldots$ are algebraically independent over K_0 if and only if $T_0 \cap K_0 \neq \phi$.

(7) Let F be a field of characteristic 0, $K = F((x)), K_0 = F(x)$ and d be the derivation of K such that $d(u(x)) = \frac{d}{dx}(u(x))$ for any $u(x) \in K$.

(a) Let $f(x) \in K\backslash K_0$ such that $d(f) \in K_0$. Show that $f(x)$ is transcendental over K_0.

(b) Show that $f(x) = \sum_{n=1}^{\infty} \frac{x^n}{n}$ satisfies the conditions of **(a)** and then it is transcendental over $F(x)$.

(8) Let F, d, K and K_0 as in **(7)**.

(a) For $g(x) \in K_0$ and $\alpha \in \bar{F}$ write $g(x)$ in the form $g(x) = \sum_{i=-m}^{\infty} a_i(x-\alpha)^i$, where m is a non negative integer and $a_i \in \bar{F}$ for any. The coefficient a_{-1} is called the residue of $g(x)$ at α. Show that the residue of $g'(x)$ at α is equal to 0.

(b) Let $f_1(x), f_2(x), \ldots$ be elements of K such that $d(f(x)) \in K_0$. Suppose that there exist $\alpha_1, \alpha_2, \ldots \in \bar{F}$ such that for any integer $n \geq 1$, $\text{Det}(r_{ij})$, where r_{ij} is the residue of $d(f_i(x))$ at α_j. Show that $f_1(x), f_2(x), \ldots$ are algebraically independent over K_0.

(9) Let F be a field of characteristic $0, g_1(x), g_2(x), \ldots$ be pairwise relatively prime and non constant polynomials with coefficients in F. Suppose that $g_i(0) = 1$ for any $i \geq 1$. Let

$$f_i(x) = Log(g_i(x)) := \sum_{n=1}^{\infty} \frac{(1-g_i(x))^n}{n}.$$

Show that the $g_i (i \geq 1)$ are algebraically independent over $F(x)$.

Solution 11.20.

(1) We first show that $1, f_1$ are linearly independent over K_0. Suppose the contrary. Then there exist $(\lambda, \mu) \in K_0^2\backslash\{0\}$ such that $\lambda + \mu f_1 = 0$, hence $\lambda = -\mu f_1 \in K_0 \cap T_0$, which is a contradiction. Therefore $1, f_1$ are linearly independent over K_0. Let $n \geq 2$, we show by induction on n that $1, f_1, \ldots, f_n$ are linearly independent over K_0. Suppose that

$$\lambda_0 + \sum_{i=1}^{n} \lambda_i f_i = 0, \tag{Eq 1}$$

where $\lambda_j \in K_0$ for $j = 0, \ldots, n$. We assume that $\lambda_n \neq 0$ and we will get a contradiction. Differentiating the above identity we obtain:

$$d(\lambda_0) + \sum_{i=1}^{n} d(\lambda_i) f_i + \sum_{i=1}^{n} \lambda_i d(f_i) = 0. \tag{Eq 2}$$

Eliminating f_n from the relations (1) and (2), yields to the equation:

$$\mu_0 + \sum_{i=1}^{n-1} (\lambda_n d(\lambda_i) - d(\lambda_n)\lambda_i) f_i = 0, \tag{Eq 3}$$

where

$$\mu_0 = \lambda_n d(\lambda_0) - d(\lambda_n)\lambda_0 + \lambda_n \sum_{i=1}^{n} \lambda_i d(f_i).$$

The assumptions on the λ_j and on the f_i show that μ_0 and $\lambda_n d(\lambda_i) - d(\lambda_n)\lambda_i = 0$ for $i = 1, \ldots, n-1$.

These equations may be written in the form:
$\mu_0 = 0$ and $d(\frac{\lambda_i}{\lambda_n}) = 0$ for $i = 1, \ldots, n-1$. We deduce that for any $i = 1, \ldots, n-1$, there exists $c_i \in \operatorname{Ker} d$ such that $\lambda_i = c_i\lambda_n$. Notice that we also have $\lambda_n = c_n\lambda_n$ with $c_n = 1$. Replacing these values of the λ_i in the equation $\mu = 0$ and dividing by λ_n^2, we obtain:

$$\frac{\lambda_n d(\lambda_0) - d(\lambda_n)\lambda_0}{\lambda_n^2} + \sum_{i=1}^{n} c_i d(f_i) = 0,$$

hence $d(\frac{\lambda_0}{\lambda_n} + \sum_{i=1}^{n} c_i f_i) = 0$. This implies that there exists $c \in \operatorname{Ker} d$ such that $\frac{\lambda_0}{\lambda_n} + \sum_{i=1}^{n} c_i f_i = c$.

It follows that $c - \frac{\lambda_0}{\lambda_n} = \sum_{i=1}^{n} c_i f_i \in K_0 \cap T_0$, which is a contradiction.

(2) Let $n \geq 2$ be an integer. Suppose that

$$\sum_{i=0}^{n} \lambda_i L^i = 0, \qquad \text{(Eq 4)}$$

where $\lambda_i \in F$ for $i = 0, \ldots, n$ and $\lambda_n \neq 0$. We will get a contradiction. Differentiating the above relation, we obtain:

$$\sum_{i=0}^{n} d(\lambda_i)L^i + \left(\sum_{i=0}^{n} i\lambda_i L^{i-1}\right) d(L) = 0.$$

We may write this identity in the form:

$$d(\lambda_n)L^n + \sum_{i=0}^{n-1} (d(\lambda_i) + (i+1)d(L)\lambda_{i+1})L^i = 0. \qquad \text{(Eq 5)}$$

Eliminating L^n from **(4)** and **(5)**, we get the following identity:

$$\sum_{i=0}^{n-1} \left(\lambda_n[d(\lambda_i) + (i+1)d(L)\lambda_{i+1}] - d(\lambda_n)\lambda_i\right) L^i = 0.$$

Notice that $d(L) \in F$ since $d(f_i) \in K_0 \subset F$ for any i so that the coefficients of L^i belong to F. Using the inductive hypothesis, we conclude that, in particular,

$$\lambda_n d(\lambda_{n-1}) - d(\lambda_n)\lambda_{n-1} + nd(L)\lambda_n^2 = 0.$$

Dividing by λ_n^2, we get:

$$\frac{\lambda_n d(\lambda_{n-1}) - d(\lambda_n)\lambda_{n-1}}{\lambda_n^2} + nd(L) = 0,$$

that is $d(\frac{\lambda_{n-1}}{\lambda_n} + nL) = 0$.

We deduce that there exists $c \in \operatorname{Ker} d$ such that $\frac{\lambda_{n-1}}{\lambda_n} + nL = c$. Therefore $\frac{\lambda_{n-1}}{\lambda_n} - c + nL = 0$, which shows that $1, L$ are linearly independent over F, a contradiction.

(3) We apply **(2)** for the particular case $F = K_0$. For this aim, we must show that $1, L$ are linearly independent over K_0. Set $L = \sum_{i=1}^n c_i f_i$, where $c_i \in \operatorname{Ker} d$ for $i = 1, \dots, n$ and $(c_1, \dots, c_n) \neq 0$ and suppose that there exist $\lambda, \mu_n \in K_0$ such that $\lambda \cdot 1 + \mu L = 0$. Then $\lambda + \sum_{i=1}^n \mu c_i f_i = 0$, hence by **(1)**, $\lambda = 0$ and $\mu c_1 = 0$ for $i = 1, \dots, n$. Since one of the c_i is non zero, then $\lambda = \mu = 0$. Now the condition in **(2)** is fulfilled and we conclude that $1, L, \dots, L^n$ are linearly independent over K_0.

(4) Differentiating the identity $\lambda + \mu L = 0$, we obtain

$$d(\lambda) + d(\mu)L + \mu d(L) = 0.$$

We eliminate L from these two relations and we get the result.

(5)(a) Since $d(f_i) \in K_0$ for any $i \geq 1$, then

$$d(A_m) = d(K_0)[d(f_1), \dots, d(f_m)] \subset K_0 \subset A_m \quad \text{and}$$
$$d(K_m) = d(K_0)[d(f_1), \dots, d(f_m)] \subset K_0 \subset K_m.$$

(b) We proceed by induction on m. If $m = 0$, the result follows from **(3)**. Let $m \geq 1$ and suppose that the result is true for $m - 1$. According to **(2)** with $F = K_m$ it is sufficient to prove that $1, L$ are linearly independent over K_m or equivalently over A_m. Suppose that there exist $\lambda, \mu \in A_m \backslash \{0\}$ such that $\lambda + \mu L = 0$. If none of λ and μ depends of f_m, then we may apply the inductive hypothesis since $T_m \subset T_{m-1}$, and we get a contradiction. In the other case we set $\lambda = \sum_{i=0}^n a_i f_m^i$ and $\mu = \sum_{i=0}^n b_i f_m^i$, where n is an integer $\geq 1, a_i, b_i \in A_{m-1}$ for $i = 0, \dots, n-1, a_n \neq 0$ or $b_n \neq 0$. By **(4)**, we have $d(\lambda)\mu - \lambda d(\mu) + \mu^2 d(L) = 0$, hence

$$\sum_{i=0}^n b_i f_m^i \cdot \sum_{i=0}^n d(a_i) f_m^i - \sum_{i=0}^n a_i f_m^i \sum_{i=0}^n d(b_i) f_m^i$$
$$+ d(f_m) \left[\sum_{i=0}^n b_i f_m^i \cdot \sum_{i=1}^n i a_i f_m^{i-1} - \sum_{i=0}^n a_i f_m^i \cdot \sum_{i=1}^n b_i f_m^{i-1} \right]$$
$$+ \left(\sum_{i=0}^n b_i f_m^i \right)^2 d(L) = 0.$$

We consider this relation as an equation of the form $\sum_{i=0}^{2n} c_i f_m^i$, where $c_i \in K_{m-1}$ for $i = 0, 1, \ldots, 2n$. Since $f_m \in T_{m-1}$, then we apply the inductive hypothesis and conclude that $c_i = 0$ for $i = 0, \ldots, n$ and this will lead to a contradiction. We have

$$c_{2n} = b_n d(a_n) - a_n d(b_n) + d(L) \cdot b_n^2 = 0.$$

Suppose that $b_n \neq 0$, then

$$c_{2n} = \frac{b_n d(a_n) - a_n d(b_n)}{b_n^2} + d(L) = 0,$$

hence $d(\frac{a_n}{b_n} + L) = 0$. Therefore, there exists $c \in \operatorname{Ker} d$ such that $\frac{a_n}{b_n} + L = c$.
Since $L \in T_m \subset T_{m-1}$ and $\frac{a_n}{b_n} - c \in K_{m-1}$, this equality contradicts our inductive hypothesis. Thus $b_n = 0$ and then $a_n \neq 0$. We have:

$$c_{2n+1} = b_{n-1} d(a_n) - a_n d(b_{n-1}) = 0,$$

hence $d(\frac{b_{n-1}}{a_n}) = 0$. Therefore $\frac{b_{n-1}}{a_n} \in \operatorname{Ker} d$. Thus $b_{n-1} = ea_n$ with $e \in \operatorname{Ker} d$. Using the relations $b_n = 0$ and $c_{2n-2} = 0$, we obtain the following equation:

$$\begin{aligned} &b_{n-2} d(a_n) - a_n d(b_{n-2}) \\ &\quad + b_{n-1} d(a_{n-1}) - a_{n-1} d(b_{n-1}) \\ &\quad + d(f_m) b_{n-1} a_n + d(L) b_{n-1}^2 = 0. \end{aligned}$$

Substituting ea_n for b_{n-1}, we get:

$$\begin{aligned} &b_{n-2} d(a_n) - a_n d(b_{n-2}) \\ &\quad + ea_n d(a_{n-1}) - ea_{n-1} d(a_n) \\ &\quad + ed(f_m) a_n^2 + e^2 a_n^2 d(L) = 0. \end{aligned}$$

Dividing by a_n^2, we obtain:

$$\frac{b_{n-2} d(a_n) - a_n d(b_{n-2})}{a_n^2} + e \frac{a_n d(a_{n-1}) - a_{n-1} d(a_n)}{a_n^2} + ed(f_m) + e^2 d(L) = 0,$$

hence

$$d\left(-\frac{b_{n-2}}{a_n} + e\frac{a_{n-1}}{a_n} + ef_m + e^2 L\right) = 0.$$

We conclude that there exists $h \in \operatorname{Ker} d$ such that

$$\frac{-b_{n-2} + ea_{n-1}}{a_n} + e(f_m + eL) = h.$$

Thus

$$-b_{n-2} + ea_{n-1} - ha_n + ea_n(f_m + eL) = 0.$$

This equation has the form: $\lambda + \mu(f_m + eL) = 0$, where λ and $\mu \in K_{m-1}$. Since $f_m + eL \in T_{m-1}$, the inductive hypothesis implies $\lambda = \mu = 0$. Therefore $ea_n = 0$. Since $b_{n-1} = ea_n$, then $b_{n-1} = 0$. Suppose that $b_n = b_{n-1} = \ldots = b_{n-(r-1)} = 0$, for $r > 1$, we show that $b_{n-r} = 0$. We have

$$c_{2n-1} = b_{n-r}d(a_n) - d(b_{n-r})a_n = 0,$$

hence $d(\frac{b_{n-r}}{a_n}) = 0$. Thus $b_{n-r} = ka_n$ for some $k \in \operatorname{Ker} d$. We have

$$\begin{aligned} c_{2n-(r+1)} = b_{n-(r+1)}&d(a_n) \\ &- d(b_{n-(r+1)})a_n + b_{b-r}d(a_{n-1}) \\ &- d(b_{n-r})a_{n-1} + rd(f_m)b_{n-r}a_n = 0. \end{aligned}$$

We substitute ka_n for b_{n-r} and we obtain:

$$\begin{aligned} &b_{n-(r+1)}d(a_n) - d(b_{n-(r+1)})a_n \\ &\quad + ka_nd(a_{n-1}) - ka_{n-1}d(a_n) \\ &\quad + rka_n^2d(f_m) = 0. \end{aligned}$$

Dividing by a_n^2, we get

$$\frac{b_{n-(r+1)}d(a_n) - d(b_{n-(r+1)})a_n}{a_n^2} + k\frac{a_nd(a_{n-1}) - a_{n-1}d(a_n)}{a_n^2} + rkd(f_m) = 0,$$

hence

$$d\left(\frac{b_{n-(r+1)}}{a_n} + k\frac{a_{n-1}}{a_n} + rkf_m\right) = 0.$$

It follows that there exists $l \in \operatorname{Ker} d$ such that

$$b_{n-(r+1)} - la_n + ka_{n-1} + rka_nf_m = 0.$$

This identity has the form $\lambda + \mu f_m = 0$, where $\lambda, \mu \in K_{m-1}$ and $f_m \in \mathbb{F}_{m-1}$. The inductive hypothesis implies, in particular that $rka_n = 0$. Since $b_{b-r} = ka_n$ and since the characteristic of K is 0, then $b_{n-r} = 0$.

(6) Suppose that $T_0 \cap K_0 \neq \phi$ and let g be an element of this intersection, then $g \in K_0$, and there exist $n \geq 1$ and $(c_1, \ldots, c_n) \in \operatorname{Ker} d^n$ such that $\sum_{i=1}^n c_if_i - g = 0$. Therefore $f_1, \ldots, f_n$ are algebraically dependent over K_0. Suppose that $T_0 \cap K_0 \neq \phi$. To conclude for the algebraic independence over K_0 of the f_i, it is sufficient to prove that for any $m \geq 1$ f_m is transcendental over K_{m-1}, that is $1, f_m, \ldots, f_m^n$ are linearly independent over K_{m-1} for any $n \geq 1$. By **(5)** this claim is true, hence the f_i are algebraically independent over K_0.

(7)(a) Here $\operatorname{Ker} d = F$ and $T_0 = \{\lambda f(x), \lambda \in F^*\}$. We use **(6)**. It is clear that for any $\lambda \in F^*$ that $\lambda f(x) \notin K_0$ since $f(x) \in K \backslash K_0$. Therefore $T_0 \cap K_0 = \phi$ and the conclusion follows.

(b) We must show that $f(x) \notin K_0$ and $d(f(x)) \in K_0$. We have

$$d(f(x)) = \sum_{n=0}^{\infty} x^n = \frac{1}{1-x},$$

hence $d(f(x)) \in F(x)$. Suppose that $f(x) \in F(x)$ and let $u(x), v(x) \in F[x]$ such that $f(x) = \frac{u(x)}{v(x)}$ and $\gcd(u(x), v(x)) = 1$, then

$$d(f) = \frac{1}{1-x} = \frac{d(u(x))v(x) - d(v(x))u(x)}{v^2(x)}.$$

We deduce that

$$v^2(x) = (1-x)(d(u(x))v(x) - d(v(x))u(x)).$$

Suppose that there exists $P(x) \in f[x]$ such that $P(x)|v(x)$, $P(x) \neq 1-x$ and $P(x)$ is irreducible over F, then from the above identity, we conclude that $P(x)|d(v(x))$. Denote by ν_p, the valuation at p and let $e = \nu_p(v(x))$, then

$$\nu_p[d(u(x))v(x) - d(v(x))u(x)] = e - 1.$$

But $\nu_p(v^2(x)) = 2e$ and we have reached a contradiction. From the identity, we see that $1 - x|v(x)$. Therefore $v(x) = \lambda(1-x)^e$ with $\lambda \in F$ and $e \geq 1$. Using valuations at $1-x$ as previously we obtain a contradiction, hence $f(x) \notin F(x)$.

(8)(a) We may suppose that $\alpha = 0$. Then $g(x)$ has the form:

$$g(x) = ax^k \sum_{j=0}^{\infty} b_j x^j,$$

where $b_j \in F$ and $b_0 = 1$ and k an integer. If $k \geq 0$, then obviously the residue of $g'(x)$ at 0 is 0. If $k < 0$, then

$$g'(x) = a\left(b_0 x^{k-1} + \sum_{j=1}^{\infty}(k+j)b_j x^{k+j-1}\right).$$

If $k + j - 1 =$ then $j = -k$ and $a_{-1} = (k+j)b_j = 0$.

(b) We apply **(6)**. Suppose that for some positive integer n, there exist $c_1, c_2, \dots, c_n \in \operatorname{Ker} d = F$ and $g(x) \in K_0 = F$ such that $\sum_{i=1}^n c_i f_i(x) = g(x)$, then $\sum_{i=1}^n c_i d(f_i(x)) = d(g(x))$.
By **(8)** **(a)** For any $j = 1, \dots, n$ the residue of $d(g(x))$ at α_j is zero, hence $\sum_{i=1}^n c_i r_{ij} = 0$ for $j = 1, \dots, n$. This is a homogeneous linearly system of equations in $c_1, \dots, c_n$, whose determinant is by assumption, non zero. Hence $c_1 = \dots = c_n = 0$. We deduce that $T_0 \cap K_0 = \phi$ and then by **(6)**, $f_1, f_2, \dots$ are algebraically independent over $F(x)$.

(9) We show that the assumptions of **(8)** **(b)** are fulfilled. We have

$$\frac{d}{dx}(g_i(x))\frac{1}{1-g_i(x)} = \frac{d}{dx}f_i(x),$$

hence $\frac{d}{dx}(f_i(x)) \in F(x)$. For any $i \geq 1$ select a root α_i of $g_i(x)$ in $\bar{F}$, then the preceding computations show that α_i is a simple pole of $\frac{d}{dx}(f_i(x))$ and α_i is not a pole of $\frac{d}{dx}(f_j(x))$ for any $j \neq i$. Therefore the residue of $\frac{d}{dx}(f_j(x))$ at α_i is 0 if $i \neq j$. We show that the residue of $\frac{d}{dx}(f_i(x))$ at α_i is equal to the multiplicity of α_i as a root of $g_i(x)$. Set $g_1(x) = (x-\alpha_i)^{m_i}Q(x)$, where m_i is a positive integer, $Q(x) \in \bar{F}[x]$ and $Q(\alpha_i) \neq 0$. Then

$$\frac{d}{dx}g_1(x) = m_i(x-\alpha_i)^{m_i-1}Q(x) + (x-\alpha_i)^{m_i}\frac{d}{dx}Q(x),$$

hence

$$\frac{d}{dx}(f_1(x)) = \frac{d}{dx}(g_i(x))/g_i(x) = \frac{m_i Q(x) + (x-\alpha)\frac{d}{dx}Q(x)}{(x-\alpha)Q(x)}.$$

Therefore the value at $x = \alpha$ of $(x-\alpha)\frac{d}{dx}f_i(x)$ is equal to m_i and then the residue of $\frac{d}{dx}(f_i(x))$ at α_i is equal to m_i. It follows that $\operatorname{Det}(r_{ij})_{\substack{1\leq i\leq n \\ 1\leq j\leq n}} = \prod_{i=1}^n m_i \neq 0$, and the hypotheses of **(8)** are satisfied.
Therefore the $f_i(x)(i \geq 1)$ are algebraically independent over $F(x)$.

Notations

$a \mid b$	a divides b
$a \nmid b$	a does not divide b
$p^e \mid\mid a$	p^e is the exact power of p dividing a
$\nu_p(a), \quad \nu_{\mathcal{P}}(I)$	p-adic valuation of the integer a and $\mathcal{P}$-adic valuation of the ideal I respectively
$\gcd(a_1, \ldots, a_n)$	greatest common divisor of the integers $a_1, \ldots, a_n$
$\mathrm{lcm}(a_1, \ldots, a_n)$	least common multiple of the integers $a_1, \ldots, a_n$
$a \equiv b \pmod q$	a is congruent to b modulo q
$\bar{a}, \quad \overline{f(x)}$	class modulo an ideal of the integer a and class of the polynomial $f(x)$ respectively
$\binom{n}{m}$	binomial coefficient
$(\frac{q}{p})$	quadratic character of q modulo p
$A \setminus B$	complementary set of B in A
$\max(A)$	supremum of the elements of A
$\lvert A \rvert$	cardinal of the set A
$\mathbb{N}$	set of non negative rational integers
$\mathbb{Z}$	set of rational integer
$\mathbb{Q}$	set of rational numbers
$\mathbb{R}$	set of real numbers
$\mathbb{C}$	set of complex numbers
$\overline{\mathbb{Q}}$	set of algebraic numbers
$h(K)$	class number of the number field K
$K[x]$	ring of polynomials with coefficients in K
$K(x)$	field of rational functions over K

$K[[x]]$	ring of formal power series with coefficients in K
$K((x))$	field of formal series with coefficients in K
$\mathcal{S}_n$	symmetric group operating on a set of n elements
$\mathcal{A}_n$	alternating group operating on a set of n elements
$\emptyset$	empty set
E^n	Cartesian product of the set E with itself n times or set of x^n when x runs in E
$\operatorname{Ker}\phi$, $\operatorname{Im}\phi$, $\operatorname{Rank}\phi$	kernel, image and rank of ϕ respectively
$E \simeq F$	E is isomorphic to F
$\operatorname{Det}(A)$	determinant of A
A^t	transpose of the matrix A
$M_n(K)$	set of all $n \times n$ matrices with coefficients in K
$GL_n(K)$	linear group of K^n
$\operatorname{Dim}_K E$	dimension over K of E
$\operatorname{Dim} R$	Krull dimension of R
$\operatorname{Frac}(R)$	fraction field of the integral domain R
$[E:K]$	degree of the extension E/K
$[E:K]_s$	degree of separability of the extension E/K
$[E:K]_i$	degree of inseparability of the extension E/K
$\operatorname{Trdeg} E/K$	transcendence degree of E/K
$(G:H)$	index in G of the subgroup H
$\operatorname{Tr}_{E/K}(\alpha)$	trace of α in E/K
$\operatorname{N}_{E/K}(\alpha)$	norm of α in E/K
$\operatorname{Gal}(E,K)$, $\operatorname{Gal}(E/K)$	Galois group of the extension E/K
$\operatorname{Inv}(H)$	field or ring of invariants of the group H
$K(\alpha)$	field obtained by adjoining α to K
$K(A)$	field obtained by adjoining the elements of A to K
$E.F$	composite field of E and F
$\deg f$	degree of the polynomial f
$\operatorname{Disc}(f)$	discriminant of the polynomial f
$\deg_y f$	degree of the multivariate polynomial f relatively to y
$\operatorname{cont}(f)$	content of the polynomial f

$f'(x)$	derivative of f
$f^{(k)}(x)$	derivative of order k of f
$\frac{\partial \phi}{\partial x}$	partial derivative relatively to x of ϕ
$\text{Res}_x(f(x), g(x))$	resultant of the polynomials $f(x)$ and $g(x)$
$\text{Irr}(\alpha, K, x)$	minimal polynomial of α over K
$\text{Char}(\alpha, K, x)$	characteristic polynomial of α over K
$\text{I}(\alpha)$	index of α
$\text{Disc}_{L/K}(\alpha_1, \dots, \alpha_n)$	discriminant of the elements of L, $\alpha_1, \dots, \alpha_n$
$E \bigotimes_K F$	tensor product of E and F over K

Bibliography

Alaca, S. and Williams, K. S. (2003). *Introductory Algebraic Number Theory*, (Cambridge University Press).

Ayad, M. (1997). *Théorie de Galois 122 Exercices corrigés, Niveau I*, (Ellipses, Paris).

Ayad, M. (1997). *Théorie de Galois 115 Exercices corrigés, Niveau II*, (Ellipses, Paris).

Baker, A. (2012). *A Comprehensive Course in Number Theory*, (Cambridge University Press).

Bewersdorf, J. and Kramer, D. (2006). *Galois Theory for Beginners: A Historical Perspective*, (American Mathematical Society, Providence).

Borevich, Z. I. and Shafarevich I. R. (1966). *Number Theory*, (Academic Press).

Bourbaki, N. (1950). *Algèbre Chap. V (Corps commutatifs)*, (Hermann, Paris).

Clark, P. L. (1993). `ftp://math.uga.edu/~pete/FieldTheory.pdf`.

Cox, D. A. (2012). *Galois Theory*, 2nd edn. (Wiley).

Escofier, J. P. and Schneps, L. (2012). *Galois Theory*, (Graduate Texts in Math.) (Springer).

Flath, D. F. (1989). *Introduction to Number Theory*, (Wiley, New York).

Hancock, H. (1932). *Foundations of the Theory of Algebraic Numbers, Vol. 2*, (Dover Pub. Inc., New York).

Hasse, H. (1980). *Number Theory*, (Springer-Verlag).

Howie, J. M. (2010). *Fields and Galois Theory*, (Springer).

Hua, L. K. (1982). *Introduction to Number Theory*, (Springer-Verlag, New York).

Ireland, K. and Rosen I. M. (1990). *Classical Introduction to Modern Number Theory*, (Springer-Verlag).

Janusz, G. J. (1973). *Algebraic Number Fields*, (Academic Press, New York).

Jarvis, F. (2014). *Algebraic Number Theory*, (Springer).

Lang, S. (1965). *Algebra*, (Addison-Wesley).

Lang, S. (1994). *Algebraic Number Theory*, (Springer-Verlag, New York).

Lidl, R. and Niederreiter H. (2008). *Finite fields* (Encyclopedia of Mathematics and its Applications), (Cambridge University Press).

Marcus, D. A. (1977). *Number fields*, (Springer-Verlag).

Mollin, R. A. (2011). *Algebraic Number Theory*, 2nd edn. (Chapman and Hall CRC).
Murty, M. R. and Esmonde J. (2006). *Problems in Algebraic Number Theory*, 2nd edn. (Springer).
Narkievicz, K. (1990). *Elementary and Analytic Theory of Algebraic Numbers*, (Springer-Verlag).
Postnikov, M. M. (2004). *Foundations of Galois Theory*, (Dover Publications).
Ribenboim, P. (2001). *Classical Theory of Algebraic Numbers*, (Springer).
Roitman, M. (1997). *On Zsigmondy primes*, (Proc. A. M. S., 125, 1913–1919).
Rotman, J. (2013). *Galois Theory*, 3rd edn. (Springer).
Samuel, P. (1970). *Algebrai Theory of Numbers*, (Houghton-Mifflin).
Schinzel, A. (2000). *Polynomials with special regard to reducibility*, (Cambridge University Press).
Small, C. (1991). *Arithmetic of finite fields*, (Marcel Dekker).
Stewart, I. N. (2015). *Galois Theory*, (CRS Press).
Stewart, I. and Tall D. (2015). *Algebraic Number Theory and Fermat's Last Theorem*, 4th edn. (Chapman and Hall CRC).
Swallow, J. (2004). *Exploratory Galois Theory*, (Cambridge University Press).
Weil, A. (1967). *Basic Number Theory*, (Springer-Verlag).
Weintraub, S. (2007). *Galois Theory*, (Springer).